COMPUTATIONAL AND MATHEMATICAL POPULATION DYNAMICS

COMPUTATIONAL AND MATHEMATICAL POPULATION DYNAMICS

Editors

Necibe Tuncer
Florida Atlantic University, USA

Maia Martcheva
University of Florida, USA

Olivia Prosper
University of Tennessee, Knoxville, USA

Lauren Childs
Virginia Polytechnic Institute and State University, USA

World Scientific

NEW JERSEY • LONDON • SINGAPORE • BEIJING • SHANGHAI • HONG KONG • TAIPEI • CHENNAI • TOKYO

Published by

World Scientific Publishing Co. Pte. Ltd.
5 Toh Tuck Link, Singapore 596224
USA office: 27 Warren Street, Suite 401-402, Hackensack, NJ 07601
UK office: 57 Shelton Street, Covent Garden, London WC2H 9HE

Library of Congress Cataloging-in-Publication Data
Names: Tuncer, Necibe, editor. | Martcheva, M. (Maia), editor. |
Prosper, Olivia, editor. | Childs, Lauren, editor.
Title: Computational and mathematical population dynamics / editors Necibe Tuncer
(Florida Atlantic University, USA), Maia Martcheva (University of Florida, USA),
Olivia Prosper (University of Tennessee, Knoxville, USA),
Lauren Childs (Virginia Polytechnic Institute and State University, USA).
Description: New Jersey : World Scientific, [2023] | Includes bibliographical references and index.
Identifiers: LCCN 2022051286 | ISBN 9789811263026 (hardcover) |
ISBN 9789811263033 (ebook) | ISBN 9789811263040 (ebook other)
Subjects: LCSH: Epidemiology--Mathematical models. | Communicable diseases--
Mathematical models. | Medicine--Mathematical models.
Classification: LCC RA652.2.M3 C66 2023 | DDC 610.1/5118--dc23/eng/20221128
LC record available at https://lccn.loc.gov/2022051286

British Library Cataloguing-in-Publication Data
A catalogue record for this book is available from the British Library.

For any available supplementary material, please visit
https://www.worldscientific.com/worldscibooks/10.1142/13045#t=suppl

Desk Editors: Logeshwaran Arumugam/Nijia Liu

Typeset by Stallion Press
Email: enquiries@stallionpress.com

https://doi.org/10.1142/9789811263033_fmatter

Preface

This contributed volume is an outgrowth and extension of papers presented at the Fifth Computational and Mathematical Population Dynamics Conference (CMPD5), which was held from May 19 to 24, 2019, in Fort Lauderdale, Florida, USA. CMPD5 brought together a distinguished set of researchers to exchange ideas on advancements in computational and mathematical modeling with applications to biology, ecology, environmental science, epidemiology, immunology, medicine, and more. Some of the advances presented at CMPD5 and extensions upon them are included in this book, which comprises 10 chapters.

Classical applications of population dynamics appear in ecology, immunology, and epidemiology. This book includes work across these areas as well as spans a range of mathematical modeling tools and formulations, such as stochastic models and differential equation systems, both ordinary and partial. A spectrum of mathematical tools are brought to bear on the various models introduced. Differential equation stability analysis and bifurcation analysis are common devices employed, but other techniques such as optimal control theory and identifiability analysis also arise.

A common theme throughout the book is infectious disease dynamics with chapters discussing a range of diseases, specifically those of Zika, HIV, opioid abuse, and COVID-19. Korb and Martcheva assess the ability to reduce Zika transmission through the use of a naturally-occurring bacterium, known to have suppressive effects on the mosquito host. Using a system of differential equations,

Timsina and Tuncer examine the interplay of HIV and opioids, determining effective controls for the US population. They consider multiple controls such as education and drug treatment, and minimize a range of disease metrics including number of infected or addicted individuals at different stages, using data from the Center for Disease Control. Nadima, Sahab, and Chattopadhyaya use a stochastic epidemic model to determine the effect of lockdowns. In particular, they capture the lack of uniformity between entering and exiting quarantine, depending on the period of lockdown. Through the introduction of a novel maximum stability index, they propose an effective lockdown strategy.

While many of the works focus on epidemiological patterns, Meadows and Schwartz look within the human body and analyze a model of virus infection and components of the immune response, including antibodies and cytotoxic T cells. Their analysis indicates that for a stable biologically relevant equilibrium with both antibodies and cytotoxic T cells, the cytotoxic cells must be boosted more so than the antibodies. This has important implications for selection of interventions, with emphasis on stimulation over cytotoxic responses over antibodies. Nemeth, Tuncer, and Martcheva also build on within-host dynamics with their assessment of a multi-scale model of virus infection and population spread. Using a combination of viremia and incidence data, they determine the structural and practical identifiability in the context of multi-scale data, and find enhanced practical identifiability with raw incidence data over cumulative data. They additionally assess the importance of the order of numerical method used, and show for the PDE system the choice is not significantly important for identifiability of parameters.

Beyond epidemiology, two chapters focus on ecological dynamics, particularly related to disease-transmitting *Aedes* mosquito populations. Walker, Robert, and Childs use a PDE framework to track the age- and mass-structured development of mosquitoes. As mosquito size has significant impacts on life history traits and behavior, accurately tracking their size has important implications on their potential to spread disease. This chapter examines how the resources available to young mosquitoes alter the distribution in age and mass of adult mosquitoes. They develop a numerical scheme and simulate mosquito population dynamics under oscillatory and decaying resources. Korb and Martcheva also examine *Aedes* mosquitoes, but with particular emphasis on the control of the spread of Zika through a bacterium, *Wolbachia*. Using their ODE system with infected and uninfected *Wolbachia* populations in competition, they determine stability of equilibria and the basic reproductive number. Through numerical simulation, they also assess the role of *Wolbachia* persistence and transmission on Zika prevalence.

Other chapters take a more abstracted approach to studying infectious disease. Martcheva, Yakubu, and Tuncer introduce a novel hybrid discrete-continuous model to examine epidemiology. They develop techniques to study the model and compute the basic reproductive number as well as introduce numerical methods that allow for simulation of the system. Such a hybrid system, previously used with applications to energy and transportation, has the potential to revolutionize biological modeling. Childs and Prosper extend the analysis of PDE models involving multiple traits, such as time, mass, and age. They show conditions under which analytical solutions are possible and introduce a numerical scheme that simulates these higher order PDEs.

Analytical methods are also introduced and used to study other areas of mathematical biology. Atkins, Agahee, Martcheva, and Hager introduce a nonlinear polyhedral constrained optimization solver known as Polyhedral Active Set Algorithm (PASA). With this method, a general optimal control program can be discretized. Furthermore, they show a way to regularize control problems where there are wild fluctuations in the control through the addition of a term to the cost functional that involves a tuning parameter. This work also sheds light on how to appropriately choose the tuning parameter. Using Lyapunov Exponents, Berringer, Vasudevan, Braverman, Cavers, and Federico analyze phase synchronization in the brain networks of epileptic patients. These dynamical systems, originating from intracranial electroencephalogram data from seizures, exhibit deterministic chaos. From this analysis, they hypothesize on the origins of such seizure behavior and the relationship to phase synchronization.

This Preface provides a brief introduction to all the works found in this book, but additional details and features of the works are found in the chapter abstracts. Each chapter brings advancements in computational and mathematical modeling of population dynamics, with much of this work originating from the CMPD5 Conference. The works demonstrate a range of computational and mathematical advancements to the study of population dynamics with applications across mathematical biology.

https://doi.org/10.1142/9789811263033_fmatter

About the Editors

Necibe Tuncer is an Associate Professor in the Department of Mathematical Sciences at Florida Atlantic University. She received her PhD from Auburn University in 2007. Her postdoctoral position was as John Thompson Assistant Professor at the University of Florida from 2008 to 2011. Before joining Florida Atlantic University in 2014, she worked as an Assistant Professor of Mathematics at the University of Tulsa. She is a member of the editorial boards of *Mathematical Biosciences* and *Journal of Biological Systems*. Since 2012, the National Science Foundation has been funding her research in mathematical epidemiology.

Maia Martcheva is a Professor of Mathematics at the University of Florida. She obtained her PhD from Purdue University in 1998. After that she was a postdoctoral associate at IMA, University of Minnesota, Arizona State University and an NSF Advance Fellow at Cornell University in 2002-2003. Since 2003, she has been an Assistant, Associate, and Full Professor at the Department of Mathematics, University of Florida. Maia Martcheva has published over 130 papers. She has also published three books: *Gender Structured Population Modeling* (2005, SIAM), *An Introduction to Mathematical Epidemiology* (2015, Springer), and *Age Structured Population Modeling* (2020, Springer). Her research has been supported by the National Science Foundation. In 2016–2018, Maia Martcheva was a Managing Editor of the *Journal of Biological Systems*. Currently, she serves on the editorial boards of the *Journal of Biological Systems*, the *Journal of Biological Dynamics*, and the *Journal of Difference Equations with Applications*.

Olivia Prosper is an Associate Professor of Mathematics at the University of Tennessee. She earned her PhD from the University of Florida in 2012, held the postdoctoral position of Instructor of Applied and Computational Mathematics at Dartmouth College from 2012 to 2015, and spent four years as an Assistant Professor of Mathematics at the University of Kentucky before joining the Department of Mathematics at the University of Tennessee in 2019. Since 2018, her research has been supported by NSF grants, including the NSF CAREER Award, awarded in 2021. In 2021, she also received the Intercollegiate Biomathematics Alliance Excellence in Research Award.

Lauren M. Childs is an Associate Professor in the Department of Mathematics at Virginia Tech. She received her BS in Mathematics and Chemistry from Duke University followed by her MA and PhD in Applied Mathematics from Cornell University. Lauren Childs was a Postdoctoral Fellow at Georgia Tech joint between the School of Biology and the School of Mathematics. Following her work at Georgia Tech, she was a MIDAS Postdoctoral Fellow and Research Scientist in the Center for Communicable Disease Dynamics at the Harvard T.H. Chan School of Public Health. In 2016, she joined the faculty at Virginia Tech as an Assistant Professor and was promoted to Associate Professor in 2022. Her research program involves the development, analysis, and simulation of mathematical and computational models to examine biologically motivated questions. Her work includes modeling with diverse techniques for a wide variety of applications in epidemiology, immunology, and ecology. Lauren Childs has published over 35 peer-reviewed articles and book chapters. Her research has been supported by the National Science Foundation, the National Institutes of Health, the Simons Foundation, and the Jeffress Trust including an NSF CAREER Award in 2022.

https://doi.org/10.1142/9789811263033_fmatter

Contents

https://doi.org/10.1142/9789811263033_0001

Chapter 1

Wolbachia Invasion and Establishment in *Aedes aegypti* Populations to Suppress Zika Transmission

Cristina Korb[*,‡] **and Maia Martcheva**[†,§]

[*]*Department of Mathematics, Northwestern University, Evanston, Illinois, USA*

[†]*Department of Mathematics, University of Florida, Gainesville, Florida, USA*

[‡]*cristina.korb@northwestern.edu*

[§]*maia@ufl.edu*

Arboviral diseases such as dengue and Zika are diseases that pose a threat to health globally. *Wolbachia*-based control is an eco-friendly strategy that is carried out by infecting wild mosquitoes with a *Wolbachia* strain and then strategically releasing the *Wolbachia*-infected mosquitoes with the goal of reducing disease transmission. In this study, we develop and analyze an ordinary differential equation model to quantify the effectiveness of different release strategies of *Wolbachia*-infected mosquitoes in order to create a sustained infection of *Wolbachia* in the mosquito population and reduce Zika transmission. The model accounts for mating between mosquitoes, assumes complete cytoplasmic incompatibility, and allows for different parameters related to vector-borne transmission. We compute all the reproduction numbers and derive analytic forms of equilibria, where possible. Then local stability analysis is performed for these equilibria. Using numerical simulations we investigate different release strategies of *Wolbachia*-infected mosquitoes and observe that there are multiple ways to reach persistence of *Wolbachia* mosquitoes. We perform sensitivity analysis on the reproduction numbers to determine parameters' relative importance to *Wolbachia* transmission and Zika prevalence. Lastly, we study the effects of seasonal variations on the spread of Zika and *Wolbachia* infection invasion and establishment.

Keywords: Wolbachia, Zika, cytoplasmic incompatibility, seasonality, male-female mosquito mating, aquatic stage mosquitoes, reproduction numbers

1.1 Introduction

1.1.1 *Zika background*

Zika virus (ZIKV) is a mosquito-borne viral disease that is mainly transmitted by *Aedes aegypti* mosquitoes. Isolation of the virus from different *Aedes* species has been demonstrated in laboratory, but the focus of attention on ZIKV vectors in the Americas has been on *A. aegypti*, which is also the vector of dengue virus (DENV), chikungunya (CHIKV), and urban Yellow Fever. A potential secondary vector and even more invasive species is *Aedes albopictus*, a mosquito present in temperate regions of Europe and North America, currently inhabiting 28 countries beyond its native tropical range in Southeast Asia.[1] *Aedes albopictus* may become a significant vector of ZIKV in the future if the virus were to adapt to them through genome microevolution as occurred with CHIKV in La Réunion 2005–2006 during the Indian Ocean outbreak.[2] A number of Zika vaccines have shown significant promise in phase 1 and phase 2 human clinical trials.[3] However, the widespread distribution of *Aedes* mosquitoes, along with the fact that as of April 2021 no Zika vaccines have been brought up to licensure, make control of the mosquito populations the most effective tool in combating Zika and other arboviral diseases. Traditional control measures are the use of insecticides and reduction of breeding sites. Some novel technologies to control mosquito populations have been developed more recently and include the release of genetically modified mosquitoes and the use of *Wolbachia*, a maternally inherited bacterium that once established within the mosquito population can help suppress arboviral diseases.

ZIKV was first discovered in rhesus monkeys in 1947 while researchers were studying yellow fever in Zika forest, Uganda, and then in humans in Nigeria in 1952.[4] Its potential effect on public health was not recognized until the outbreaks on Yap Island in 2007 and then in French Polynesia in 2013. During these early outbreaks, symptoms were mild including fever, rash, arthritis/arthralgia, and conjunctivitis.[5] The virus then migrated to Latin America with the first autochthonous transmission detected in 2015 in Brazil.[6] The incidences of ZIKV infections in the Americas peaked in early 2016 with the cumulative number of documented and suspected cases exceeding 1 million.[3] The number of incidences in the Americas and the world has waned significantly after 2017 with only 43 reported cases in US states and territories as of December 3, 2020.[7] However, outbreaks and infection clusters continue to occur in some regions, such as India and Southeast Asia.[8]

As the virus moved from Africa to America, some unexpected symptoms/complications related to ZIKV have emerged. The ZIKV epidemic in Brazil was linked to microcephaly in newborns, especially when the mother acquired the infection in the first trimester of pregnancy.[9] Also, ZIKV infections were found to be associated with Guillain–Barré syndrome (GBS) in adults, an auto-immune disease of the peripheral nerves that can result in muscle weakness and paralysis.[10] This dramatic increase in microcephaly cases and GBS led to a declaration of public health emergency of international concern by WHO in February 2016.

In addition to vector transmission, Zika has also other modes of transmission. What makes Zika unique among arboviral diseases is that it can be transmitted through sexual contact. Although female-to-male and male-to-male transmission is possible, the most common sexual transmission is from male-to-female.[11] Male-to-female sexual transmission can occur regardless if the male shows symptoms. Zika can be detected in the semen up to 370 days after infection, while shedding of the infected virions is most likely to happen up to 30 days from infection.[12] Vertical transmission can occur at any time during pregnancy in symptomatic or asymptomatic mothers.[13] Vertical transmission has been estimated to occur in 26% of fetuses of Zika-infected mothers in French Guiana. This percentage is similar to transmission percentages that have been observed for other congenital infections.[14] Zika can also be transmitted through blood transfusion,[15] and although infective Zika particles have been detected in breast milk, milkborne transmission has not been confirmed.[16]

1.1.2 *Wolbachia background*

Wolbachia is a bacterium that occurs naturally in many different insect species, but it is not present in the *A. aegypti* mosquitoes, the main vector of Zika. It was first identified in 1920s but did not capture attention until 1971, when UCLA researchers discovered that *Culex pipiens* mosquito eggs were killed when the sperm of *Wolbachia*-infected males fertilized *Wolbachia*-free eggs.[17] *Wolbachia*-infected females will transmit the bacterium to their offspring. Complete maternal transmission means all the offspring of *Wolbachia*-infected females are *Wolbachia*-infected. Imperfect maternal transmission means that some of the offspring of *Wolbachia*-infected females do not have *Wolbachia*. When deliberately introduced into *A. aegypti*, *Wolbachia* disrupts the reproductive cycle of hosts through a cytoplasmic incompatibility (CI) between the sperm of *Wolbachia*-infected males and eggs of *Wolbachia*-free females. The cross between *Wolbachia*-infected males and wild females produces embryos that die before hatching because of complete CI as depicted in Table 1.1. Therefore, CI produces a reproductive advantage for *Wolbachia*-infected females, leading the *Wolbachia* infection to establish itself within mosquito populations.[18] Bidirectional CI can also alter the reproductive cycle of mosquitoes when a male is infected with a different strain of *Wolbachia* to that of the female.[19] Other features of *Wolbachia* infection include imperfect maternal transmission, loss of *Wolbachia* infection,[20] and coinfection of two strains of *Wolbachia*.

In general, the *Wolbachia* strain is named based on the source insect. For example, the *Wolbachia* variant *wMel* was originally found in natural *D. melanogaster* populations. Depending on the *Wolbachia* strain (*wMel, wMelPop, wAlB, wStri*, and more recently *wAu*), infection with *Wolbachia* can enhance viral blockage in *A. aegypti* mosquitoes while also imposing additional fitness costs such as

Table 1.1. Impact of *Wolbachia* infection on different male–female mosquito couplings.

	♀ Female mosquito	
♂ Male mosquito	*Wolb*-infected	*Wolb*-free
Wolb-infected	*Wolb*-infected offspring	No offspring due to CI
Wolb-free	*Wolb*-infected offspring	*Wolb*-free offspring

Table 1.2. *Wolbachia* strains characteristics in *Aedes* mosquitoes as defined in Ref. 23.

Strain	CI	Viral blockage	Maternal transmission	Loss of infection	Fitness cost
wAu	None	High	High	Low	Medium
wMel	High	Medium	High	High	Medium
wMelPop	High	High	High	High	High
wAlbA	High	Medium	High	High	High
wAlbB	High	Medium	High	Medium	Medium

Note: Effect size is denoted as: High (>90%), medium (20–90%), low (<20%), and none (no detectable effects).

reduced life span, reduced fecundity, increased egg and larval development time, and reduced survival of desiccated eggs. It has been shown that the *wMel* strain reduces the capacity to transmit ZIKV and CHIKV in *A. aegypti* mosquitoes[21] and this made the *wMel* strain the most commonly used strain for field releases. The novel strain, *wAu*, provides strong blocking of Dengue and Zika virus transmission while offering greater stability at higher temperatures when compared to *wMel*, but it does not induce CI.[22] The difference in the most commonly used *Wolbachia* strains are listed in Table 1.2.

Traditional control methods against mosquito populations that transmit arboviral diseases have involved the use of larvicides and removal of breeding sites and preventive measures such as bed nets and indoor or personal sprays. These methods show short-term and small scale efficacy and should still be continued, but they should be complemented with longer lasting and larger scale methods. One issue with these control measures is that they require major human intervention and approval, and the mosquitoes are becoming resistant to insecticides. Therefore, there is a need for more powerful tools to fight Zika such as the use of (RIDL) release of insects with dominant lethality and the use of *Wolbachia*. These novel methods, especially RIDL, are met by a resistance from policymakers and the community.[24]

1.1.3 *Zika modeling background*

Models developed for Zika transmission vary in complexity and methods used. They include compartmental, spatial, metapopulation, network, and agent-based models.[25] Here we will introduce the methods and results from a few deterministic models developed so far. For an overview of models that include more sophisticated methods and integrate real-world data, the reader can refer to Wiratsudakul *et al.*[25] Some of the first Zika transmission models appeared in 2016 (Champagne *et al.*,[27] Funk *et al.*,[26] and Kucharski *et al.*[28]). These models were compartmental models that included the vector transmission and focused on a particular outbreak. For example, Funk *et al.* compared three outbreaks of dengue and Zika virus in two different island settings in Micronesia, the Yap Main Islands and Fais, making full use of commonalities in disease and setting between the outbreaks. They found that the estimated reproduction numbers for Zika and dengue were similar when considered in the same setting, but that, conversely, reproduction number for the same disease can vary considerably by setting.[26]

Gao *et al.*[29] were the first to include sexual transmission alongside with vector transmission. They compute the basic reproduction number to be $R_0 = 2.055$ in which the percentage contribution of sexual transmission is 3.044%. Sensitivity analysis indicates that R_0 is most sensitive to the mosquito parameters while sexual transmission increases the risk of infection and epidemic size, and prolongs the outbreak. They conclude that prevention and control efforts against Zika should include both control of mosquitoes and reduction of sexual transmission.

In addition to sexual transmission, Maxian *et al.* (2017), develop an age- and sex-structured mathematical model that describes the transmission dynamics of Zika. Since Zika was found to persist in semen long after it was undetectable in blood, in this model sexually active males have an extended period of sexual transmission. Instead of moving directly to the recovered class, they enter two additional infectious classes (asymptomatic infectious or symptomatic infectious) during which they are infectious to humans but not mosquitoes. The authors conclude that the sexual contribution to the reproduction number is 4.8%, which is too minor to independently sustain an outbreak and therefore vector transmission is the main driver of the then ongoing epidemic.[30]

Since Zika was linked to microcephaly in newborns, including pregnant females in models became necessary in order to project

Zika virus infections in childbearing women in the Americas. Tuncer *et al.* 2018 introduce six models of Zika, starting from the very generic vector–host model and incorporating distinct features of Zika one by one, such as asymptomatic infections, sexual transmission, and separate class for pregnant women. The models were fit to time-series data of cumulative incidences and pregnant infections from the Florida Department of Health Daily Zika Update Reports. The structural and practical identifiability of the models was tested in order to find whether unknown model parameters can uniquely be determined. Some of their conclusions are that direct transmission rates are not practically identifiable and that the reproduction numbers are most sensitive to mosquito parameters and therefore control measures should be targeted toward controlling the mosquito population.[31]

In more recent papers,[32,33] optimal control strategies are investigated. Azahrani *et al.* formulate a mathematical model on Zika virus with mutations and present their analysis in the presence of three controls: preventions through bednets for humans and pregnant women, the possible treatments for the infected compartments and the use of insecticide spraying on mosquitoes. They also use Colombia real data of Zika virus for the year 2016 and estimate and fit the model parameters as well as present a mathematical control problem for the elimination of the Zika virus infection in the community. They conclude that while every strategy has some limitations, the combined control was able to reduce infections in humans.[32] Gonzales *et al.* also use data from Colombia and consider control strategies such as awareness and spraying campaigns. They found that the educational campaign (the use of insect repellent, bednets, and appropriate clothing) reduced the number of infected people, but not as well as the insecticide campaign.[33]

1.1.4 *Wolbachia modeling background*

Beginning in 1950s, mathematical models that model the spread of *Wolbachia* infection within a wild-type mosquito population have been proposed and studied in the literature. These models take into account the trade-off between the fitness benefits (CI) and costs (reduced life span and decreased fecundity of *Wolbachia*-infected mosquitoes) and they can be categorized into those that take into account the population dynamics and those that don't. Some examples of models that neglect changes in the population size are Caspari and Watson 1959,[34] Turelli and Hoffman 1991,[35] and Schofield 2002.[36] The first model of *Wolbachia* infection appeared in 1959 and used a discrete generation population genetic model. The authors concluded that the trade-off between the benefits and costs of *Wolbachia* infection results in a bistable dynamics, where two stable equilibria exist: one where infection frequency is zero, and one where there is a high proportion of infected mosquitoes.[34] In order to reach the non-zero equilibrium, infection frequency must exceed a critical threshold value, determined by the trade-off between the reduction in fecundity of *Wolbachia*-infected females and the intensity of CI.

Local establishment of *Wolbachia* does not necessarily guarantee spatial spread. *Wolbachia* spread beyond the local environment depends on the initial infection frequency, the critical threshold frequency, the dispersal behavior of the mosquito populations, and the environment. Initial analysis by Turelli and Hoffmann showed that following a local establishment, a critical frequency threshold of less than 0.5 is necessary for spatial spread to occur.[35] Later they show that as the critical threshold approaches 0.5, wave speed slows dramatically, suggesting that a critical threshold value of 0.35 or less is most likely necessary for spatial spread.[37]

1.1.5 *Modeling work combining Zika and Wolbachia (and modeling work with Dengue)*

Many mathematical models have explored the impact of *Wolbachia* on dengue transmission. For example, Hughes and Britton[38] investigated the potential impact of a *Wolbachia* strain with perfect maternal transmission and CI on the transmission of a single-strain dengue virus. They concluded that *Wolbachia* has excellent potential for dengue control in areas where the reproduction number of dengue is not too large. Another study by Ndii *et al.* formulated a mathematical model that considered the competition for persistence between non-*Wolbachia* and *Wolbachia*-infected mosquitoes. To do this, the authors derived the steady state solutions of the model and showed that vertical transmission of *Wolbachia*, death, maturation, and reproductive rates determine the dominance of *Wolbachia*-infected mosquitoes.[39] In Ref. 40 the authors build a model of *Wolbachia* infection into an *A. aegypti* population of mosquitoes and then couple it with a classical dengue model. Their results show that if a sufficiently large number of *Wolbachia*-infected mosquitoes are released, then dengue will disappear and they use real data from field releases in Australia to calibrate their model. For a more extensive review of the mathematical models that give insight into the dynamics of the spread of *Wolbachia* and the potential impact of *Wolbachia* on dengue transmission and other arboviral diseases, the reader can refer to Refs. 41 and 42.

To date, few mathematical models have been developed to investigate the impact of introducing *Wolbachia*-infected mosquitoes into wild mosquito population in order to suppress Zika transmission. Wang *et al.*[43] formulate a differential equations model for Zika transmission without any control measures and two control models to study the impact of releasing *Wolbachia* mosquitoes on the transmission of Zika in Brazil. The first control model considers the strategy of releasing *Wolbachia* harboring female and male mosquitoes while the second strategy considers releasing only *Wolbachia* harboring male mosquitoes. They use an SEIR model for humans and an SEI model for the mosquitoes. Furthermore, the infected humans are divided into three classes: suspected cases, confirmed cases, and asymptomatic cases. They combine the egg, larval, and pupal stages as one aquatic stage. Their analysis suggests that releasing both *Wolbachia* harboring female and male mosquitoes will replace the wild mosquito population, while releasing only *Wolbachia*-infected male mosquitoes will suppress or even eradicate the wild mosquitoes.[43]

To understand the transmission dynamics of Zika, Xue *et al.*[44] develop a deterministic and a stochastic model that takes into account direct transmission and the release of *Wolbachia*-infected male mosquitoes. Their deterministic model uses an SEIR model for humans, and mosquitoes are grouped into six compartments: aquatic stage, susceptible males, susceptible females, exposed, infected, and *Wolbachia*-infected male mosquitoes. Through numerical simulations they have found that the basic reproduction number is the most sensitive to the death rate of adult mosquitoes and thus reducing the lifespan of mosquitoes can reduce the basic reproduction number dramatically. Also, they conclude that the role of sexual transmission is not negligible and mitigation strategies for the transmission of Zika virus should not ignore sexual transmission. The release of

Wolbachia-infected male mosquitoes is cost-effective if the ratio of the release rate of *Wolbachia*-infected male mosquitoes over the number of wild mosquitoes at the initial state is between 0.1308 and 0.3750.[44]

1.1.6 *Seasonality background*

Since the transmission of Zika virus is affected by periodic seasonality, it is worth further exploring how temperature fluctuations affect the dynamics of Zika transmission periodically. The impact of climate variability on vectorborne diseases can be explained by the fact that the arthropod vectors of these diseases are cold-blooded. This means that fluctuating temperatures and rainfall can impact the development, reproduction (including availability of breeding sites), behavior, and population dynamics of mosquitoes.[45] However, this impact cannot be easily predicted. We can thus conclude that higher temperatures and increased rainfall leading to increased transmission of arboviral diseases must be accompanied by a more careful and thoughtful analysis of the interplay between climate and human behavior.[46] For example, extremely high temperatures can increase mosquito mortality[47] and heavy rainfall can wash out mosquito breeding sites.[48] Also, during hot weather humans may seek refuge in air-conditioned buildings and thus avoid mosquito bites.[49] During the dry season, domestic water storage containers can provide more breeding sites for *A. aegypti* mosquitoes and thus cause the incidence of the diseases they transmit to rise.[50]

A. aegypti mosquitoes originated from Africa, but now they are present in tropical, subtropical, and temperate zones throughout the world. Large portions of the Americas, including the southernmost part of the eastern United States, the Caribbean, Central America, and lower elevation areas in Mexico and South America, have warm and humid climates well suited for proliferation of *A. aegypti*.[51] The lower temperature limit for *A. aegypti* is around 10°C, a temperature below which mosquitoes become inactive and unable to move. However, the lower temperature limit at which female *A. aegypti* has been found to cease biting is 15°C, both in the field and experimentally in the lab.[52] *A. aegypti* are most active at 28°C[53] and females fed faster between 26°C and 35°C.[52] The upper temperature limit for blood-feeding is above 36°C, and the mosquitoes die at 40°C.[54] CDC has updated the estimated range map for *A. aegypti* by using county-level records along with historical records.[55] According to CDC, the US regions where these mosquitoes can be found are expanding to include regions with more temperate climate rather than just subtropical or tropical climate.[55] Thus, it is important to investigate the effects of seasonality when the releasing of *Wolbachia*-infected mosquitoes spans multiple seasons.

1.1.7 *Goals of the present work*

The main goal of this model is to investigate how *Wolbachia*-infected mosquitoes can be released in wild mosquito populations in order for *Wolbachia* to establish itself in the population and suppress Zika transmission. We are particularly interested in comparing what control measures should be taken when *Wolbachia* is established within the wild mosquito population vs when it is not established. Previous models investigating the effect of *Wolbachia* on Zika have concentrated mainly on exploring the release of *Wolbachia*-infected males or a combination of *Wolbachia*-infected females and males. In addition to exploring the release strategies of *Wolbachia*-infected males or simultaneous *Wolbachia*-infected females and males, this model explores additional scenarios that include releasing *Wolbachia*-infected aquatic stage mosquitoes. We also investigate the effects of seasonal variations on the spread of Zika and *Wolbachia infection* by running numerical simulations for a nonautonomous model with seasonal variation into the birth rates, transitioning rates, and death rates of mosquitoes.

1.2 Model Derivation

Here we consider a deterministic model of Zika transmission. As a vector-borne disease, Zika is spread primarily by mosquitoes from the *A. aegypti* genus.[56] Reports have shown that in humans Zika can be transmitted both sexually by males[57] and from mothers to newborns. Zika infection during pregnancy has been linked to severe birth defects.[58] Along with vector transmission, our model includes direct transmission. The human population is modeled using a susceptible-exposed-infectious-removed (SEIR) model. To investigate the effect of *Wolbachia* on the dynamics of Zika spread, we consider two sub-populations of mosquitoes: (i) wild mosquitoes or *Wolbachia*-free mosquitoes, and (ii) *Wolbachia*-infected mosquitoes. Thus, the total mosquito population $N_v(t)$ consists of *Wolbachia*-free mosquitoes plus the *Wolbachia*-infected mosquitoes.

The *Wolbachia*-free mosquito population is divided into an aquatic stage $A_{wf}(t)$, susceptible *Wolbachia*-free females $S_{wf}(t)$, Zika-infected *Wolbachia*-free females $I_{wf}(t)$, and *Wolbachia*-free males, $M_{wf}(t)$. Similarly, the *Wolbachia*-infected mosquito population is divided into an aquatic stage, $A_{wi}(t)$, susceptible *Wolbachia*-infected females $S_{wi}(t)$, Zika-infected *Wolbachia*-infected females $I_{wi}(t)$ and *Wolbachia*-infected males $M_{wi}(t)$.

The total human population $N_h(t)$ is subdivided into four categories: (i) susceptible humans $S_h(t)$, (ii) exposed $E_h(t)$, (iii) infected $I_h(t)$, and (iv) recovered $R_h(t)$. Susceptible humans can become infected via sexual contact with an infected human, or through a bite from an infected female mosquito (*Wolbachia*-free or *Wolbachia*-infected). Since we don't have separation of sexes for the human population, our direct transmission term is represented by a single value which is an average value between male-to-male and male-to-female transmission. We included this term because human-to-human transmission of Zika makes the model more realistic and because we want to compare the importance of direct transmission in the presence and in the absence of *Wolbachia*-infected mosquitoes.

Wolbachia-free parents produce *Wolbachia*-free aquatic stage mosquitoes which become susceptible *Wolbachia*-free females and *Wolbachia*-free males. We assume that the probability of an encounter of a *Wolbachia*-free female with a *Wolbachia*-free male is given by M_{wf}/N_v. Then the per capita rate at which *Wolbachia*-free females are fertilized by *Wolbachia*-free males is $\eta M_{wf}/N_v$, where η is the egg-laying rate of *Wolbachia*-free mosquitoes. A susceptible *Wolbachia*-free female can become infected with Zika by biting an infected human.

Wolbachia-infected susceptible females come from mating involving a *Wolbachia*-infected mother (the father may be either *Wolbachia*-infected or *Wolbachia*-free). It is known that *Wolbachia*-harboring mosquitoes are highly resistant to infection with two currently circulating Zika virus isolates from the recent Brazilian epidemic.[59] Thus, we consider that Zika-infected and *Wolbachia*-infected female mosquitoes transmit the virus at a lower rate. *Wolbachia* also imposes various fitness costs, so we assume a different death rate for *Wolbachia*-infected mosquitoes.[60] The model assumes complete CI, so there is no mosquito offspring from *Wolbachia*-infected males with *Wolbachia*-free females. The maternal transmission is assumed to be perfect. All offsprings from *Wolbachia*-infected female mosquitoes have inherited *Wolbachia*, regardless of the status of the male.

The equations that govern the dynamics are as follows.

Human population (SEIR):

$$\dot{S}_{\mathrm{h}} = \Lambda - (\beta_{\mathrm{vh}} I_{\mathrm{wf}} + \beta_{\mathrm{vh}}^{\mathrm{w}} I_{\mathrm{wi}} + \beta_{\mathrm{hh}} I_{\mathrm{h}}) \frac{S_{\mathrm{h}}}{N_{\mathrm{h}}} - \mu_{\mathrm{h}} S_{\mathrm{h}} \tag{1.1}$$

$$\dot{E}_{\mathrm{h}} = (\beta_{\mathrm{vh}} I_{\mathrm{wf}} + \beta_{\mathrm{vh}}^{\mathrm{w}} I_{\mathrm{wi}} + \beta_{\mathrm{hh}} I_{\mathrm{h}}) \frac{S_{\mathrm{h}}}{N_{\mathrm{h}}} - (\nu_{\mathrm{h}} + \mu_{\mathrm{h}}) E_{\mathrm{h}} \tag{1.2}$$

$$\dot{I}_{\mathrm{h}} = \nu_{\mathrm{h}} E_{\mathrm{h}} - (\gamma_{\mathrm{h}} + \mu_{\mathrm{h}}) I_{\mathrm{h}} \tag{1.3}$$

$$\dot{R}_{\mathrm{h}} = \gamma_{\mathrm{h}} I_{\mathrm{h}} - \mu_{\mathrm{h}} R_{\mathrm{h}} \tag{1.4}$$

Mosquito population *Wolbachia*-free — SI for female mosquitoes

$$\dot{A}_{\mathrm{wf}} = \eta \frac{S_{\mathrm{wf}} M_{\mathrm{wf}}}{N_{\mathrm{v}}} \left(1 - \frac{A_{\mathrm{wf}} + A_{\mathrm{wi}}}{K}\right) - (\gamma_{\mathrm{wf}} + \mu_{\mathrm{A}}) A_{\mathrm{wf}} \tag{1.5}$$

$$\dot{S}_{\mathrm{wf}} = \alpha \gamma_{\mathrm{wf}} A_{\mathrm{wf}} - \beta_{\mathrm{hv}} \frac{I_{\mathrm{h}}}{N_{\mathrm{h}}} S_{\mathrm{wf}} - \mu_{\mathrm{v}} S_{\mathrm{wf}} \tag{1.6}$$

$$\dot{I}_{\mathrm{wf}} = \beta_{\mathrm{hv}} \frac{I_{\mathrm{h}}}{N_{\mathrm{h}}} S_{\mathrm{wf}} - \mu_{\mathrm{v}} I_{\mathrm{wf}} \tag{1.7}$$

$$\dot{M}_{\mathrm{wf}} = (1 - \alpha) \gamma_{\mathrm{wf}} A_{\mathrm{wf}} - \mu_{\mathrm{v}} M_{\mathrm{wf}} \tag{1.8}$$

Mosquito population *Wolbachia*-infected — SI for female mosquitoes

$$\dot{A}_{\mathrm{wi}} = S_{\mathrm{wi}} \frac{(q_1 M_{\mathrm{wi}} + q_2 M_{\mathrm{wf}})}{N_{\mathrm{v}}} \left(1 - \frac{A_{\mathrm{wf}} + A_{\mathrm{wi}}}{K}\right) - (\gamma_{\mathrm{wi}} + \mu_{\mathrm{Ai}}) A_{\mathrm{wi}} \tag{1.9}$$

$$\dot{S}_{\mathrm{wi}} = \alpha \gamma_{\mathrm{wi}} A_{\mathrm{wi}} - \beta_{\mathrm{hv}}^{\mathrm{w}} \frac{I_{\mathrm{h}}}{N_{\mathrm{h}}} S_{\mathrm{wi}} - \mu_{\mathrm{vi}} S_{\mathrm{wi}} \tag{1.10}$$

$$\dot{I}_{\mathrm{wi}} = \beta_{\mathrm{hv}}^{\mathrm{w}} \frac{I_{\mathrm{h}}}{N_{\mathrm{h}}} S_{\mathrm{wi}} - \mu_{\mathrm{vi}} I_{\mathrm{wi}} \tag{1.11}$$

$$\dot{M}_{\mathrm{wi}} = (1 - \alpha) \gamma_{\mathrm{wi}} A_{\mathrm{wi}} - \mu_{\mathrm{vi}} M_{\mathrm{wi}} \tag{1.12}$$

where $N_h = S_h + E_h + I_h + R_h$ and $N_v = S_{wi} + S_{wf} + M_{wi} + M_{wf} + I_{wi} + I_{wf}$.

The parameters pertaining to humans are defined in Table 1.3 along with their estimated values, and the parameters pertaining to mosquitoes are given in Table 1.4. The diagram of the model is given in Figure 1.1.

Table 1.3. Definition of human parameters used in the model framework.

Parameters	Description	Value	Units	Range	Ref.
Λ	Recruitment as susceptible per unit time	696	People per day		
μ_h	Per capita natural death rate humans	$\frac{1}{78.8\times 365}$	Day^{-1}	$\left[\frac{1}{78.5\times 365}, \frac{1}{79\times 365}\right]$	61
ν_h	Average incubation rate for humans	$\frac{1}{10}$	Day^{-1}	$\left[\frac{1}{14}, \frac{1}{3}\right]$	61
γ_h	Per capita recovery rate humans	$\frac{1}{5}$	Day^{-1}	$\left[\frac{1}{7}, \frac{1}{2}\right]$	61
b	Mosquito biting rate	0.5	Bites per mosq per day	[0.3, 1.5]	29
p_{vh}	Prob of transmission from wild vector to susceptible human per bite	0.4	Unitless	[0.1, 0.75]	29
β_{vh}	Transmission rate from wild vector to susceptible human per bite	bp_{vh}	Day^{-1}		
β_{vh}^{w}	Transmission rate from *Wolbachia*-infected vector to susceptible human per bite	$0.042\beta_{vh}$	Day^{-1}		62
β_{hh}	Transmission rate from infected human to susceptible human	0.05	Day^{-1}	[0.001, 0.1]	29
p_{hv}	Prob of transmission from infected human to susceptible mosquito	0.5	Unitless	[0.3, 0.75]	29
β_{hv}	Transmission rate from infected human to wild mosquito per bite	bp_{hv}	Day^{-1}		
β_{hv}^{w}	Transmission rate from infected human to *Wolbachia*-infected mosquito per bite	$0.042\beta_{hv}$	Day^{-1}		

Table 1.4. Definition of mosquito parameters in the model framework.

Parameters	Description	Value	Units	Range	Ref.
K	Carrying capacity of aquatic stage	10^6	Num of aquatic stage mosq	$[10^4, 10^9]$	Assumed
μ_{v}	Death rate wild mosq	0.061	Day^{-1}	[0.02–0.09]	60
μ_{vi}	Death rate *Wolb*-infected mosq	0.068	Day^{-1}	[0.03–0.14]	62, 63
γ_{wf}	Transitioning rate of aquatic stage wild mosq	0.11	Day^{-1}	[0.1, 0.12]	64
γ_{wi}	Transitioning rate of aquatic stage *Wolb*-infected mosq	0.11	Day^{-1}	[0.1, 0.12]	64
μ_{A}	Death rate of aquatic stage wild mosq	0.02	Day^{-1}		65
μ_{Ai}	Death rate of aquatic stage *Wolb*-infected mosq	0.2	Day^{-1}		65
η	Egg laying rate of *Wolb*-free females (per day)	13	Number of eggs per mosq per day	[12-18]	60, 64
α	Fraction of births that are female mosq	0.5	Unitless	[0.34, 0.6]	66
q_1	Egg laying rate of *Wolb*-infected females mating with *Wolb*-infected males	11	Number of eggs per mosq per day	[8, 12]	60, 64
q_2	Egg laying rate of *Wolb*-infected females mating with *Wolb*-free males	10	Number of eggs per mosq per day	[8, 12]	60, 64

1.3 Model Analysis

Since the model simulates the dynamics of humans and mosquito populations, all the state variables and parameters must be non-negative. To show that the model is well posed, we need to show that when starting with non-negative initial values we remain with non-negative values for the variables for all future times.

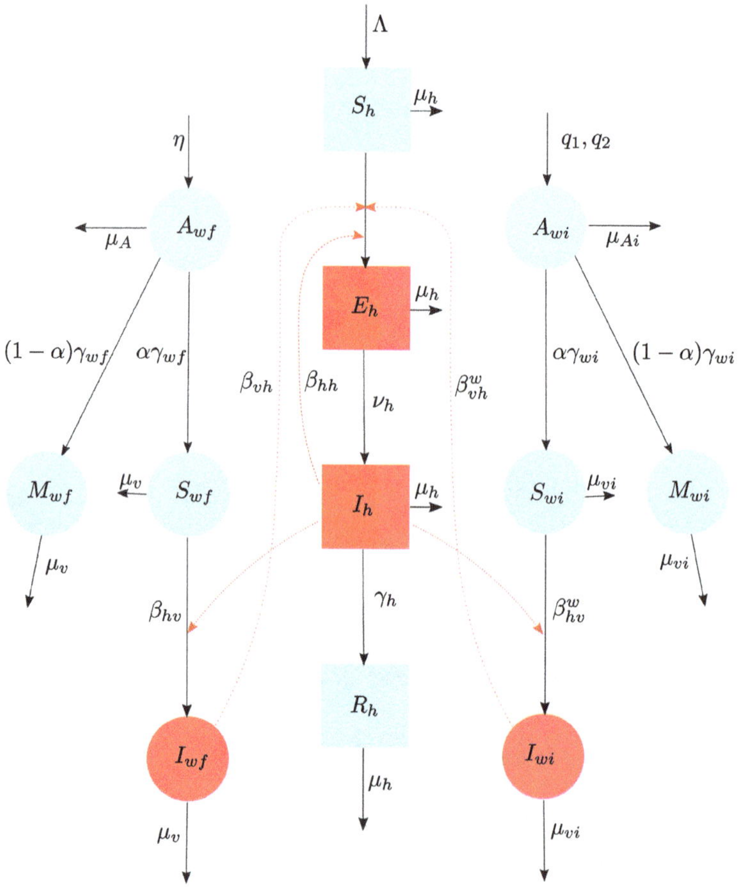

Fig. 1.1. Diagram of the model. Squares represent human compartments and circles represent vector compartments. Infected compartments are colored red and infection pathway is represented by the red dotted lines.

1.3.1 *Well-posedness of the model*

Theorem 1.1. *Let* $F : D \to \mathbb{R}^{12}$ *be* $F(t,x) = (F_1(t,x), \ldots, F_n(t,x))$ *where*

$$D = \{(t, S_{\rm h}, E_{\rm h}, I_{\rm h}, R_{\rm h}, A_{\rm wf}, S_{\rm wf}, I_{\rm wf}, M_{\rm wf}, A_{\rm wi}, S_{\rm wi}, I_{\rm wi}, M_{\rm wi}) \in \mathbb{R}^{13}_{+} | \\ S_{\rm h} \geq \epsilon_0, \epsilon_0 \leq N_{\rm h} \leq \Lambda/\mu_{\rm h}, A_{\rm wf}, A_{\rm wi} \leq K, N_{\rm v}^{\rm wf} \\ \leq \gamma_{\rm wf} K/\mu_{\rm v}, N_{\rm v}^{\rm wi} \leq \gamma_{\rm wi} K/\mu_{\rm vi}\}$$

for some $0 < \epsilon_0 < \dfrac{\Lambda}{\mu_{\mathrm{h}}}$ *and where* $N_{\mathrm{h}} = S_{\mathrm{h}} + E_{\mathrm{h}} + I_{\mathrm{h}} + R_{\mathrm{h}}$, $N_{\mathrm{v}}^{\mathrm{wf}} = S_{\mathrm{wf}} + I_{\mathrm{wf}} + M_{\mathrm{wf}}$ *and* $N_{\mathrm{v}}^{\mathrm{wi}} = S_{\mathrm{wi}} + I_{\mathrm{wi}} + M_{\mathrm{wi}}$. *The system* (1.1)–(1.12) *is epidemiologically and mathematically well-posed in the valid domain* D.

Proof. Since F is continuously differentiable in D, we have that F is locally Lipschitz in D. Then for any (t_0, x_0) in D, there exists a unique solution passing through (t_0, x_0).

Now, assume that we start from positive values and at some point in time t_1 we have that $x_j = 0$. Then as seen in what follows, we have that $F_j(t_1, x) \geq 0$, which means that x_j is nondecreasing and therefore returns to the positive quadrant or remains 0:

$$\Lambda \geq 0, (\beta_{\mathrm{vh}} I_{\mathrm{wf}} + \beta_{\mathrm{vh}}^{\mathrm{w}} I_{\mathrm{wi}} + \beta_{\mathrm{hh}} I_{\mathrm{h}}) \frac{S_{\mathrm{h}}}{N_{\mathrm{h}}} \geq 0,$$

$$\nu_{\mathrm{h}} E_{\mathrm{h}} \geq 0, \gamma_{\mathrm{h}} I_{\mathrm{h}} \geq 0, \eta \frac{S_{\mathrm{wf}} M_{\mathrm{wf}}}{N_{\mathrm{v}}} \left(1 - \frac{A_{\mathrm{wi}}}{K}\right) \geq 0,$$

$$A_{\mathrm{wi}} \leq K, \alpha \gamma_{\mathrm{wf}} A_{\mathrm{wf}} \geq 0,$$

$$\beta_{\mathrm{hv}} \frac{I_{\mathrm{h}}}{N_{\mathrm{h}}} S_{\mathrm{wf}} \geq 0, (1 - \alpha) \gamma_{\mathrm{wf}} A_{\mathrm{wf}} \geq 0,$$

$$S_{\mathrm{wi}} \frac{(q_1 M_{\mathrm{wi}} + q_2 M_{\mathrm{wf}})}{N_{\mathrm{v}}} \left(1 - \frac{A_{\mathrm{wf}}}{K}\right) \geq 0,$$

$$A_{\mathrm{wf}} \leq K, \alpha \gamma_{\mathrm{wi}} A_{\mathrm{wi}} \geq 0, \beta_{\mathrm{hv}}^{\mathrm{w}} \frac{I_{\mathrm{h}}}{N_{\mathrm{h}}} S_{\mathrm{wi}} \geq 0, (1 - \alpha) \gamma_{\mathrm{wi}} A_{\mathrm{wi}} \geq 0.$$

Now we have that $0 \leq S_{\mathrm{h}}, E_{\mathrm{h}}, I_{\mathrm{h}}, R_{\mathrm{h}} \leq N_{\mathrm{h}}(t)$. Adding the first four equations, we get $\dot{N}_{\mathrm{h}}(t) = \Lambda - \mu_{\mathrm{h}} N_{\mathrm{h}}(t)$. Integrating and taking lim inf and lim sup for $t \to \infty$, we have

$$N_{\mathrm{h}}(0) \leq \liminf N_{\mathrm{h}}(t) = \limsup N_{\mathrm{h}}(t) = \frac{\Lambda}{\mu_{\mathrm{h}}}$$

which implies that $\lim N_{\mathrm{h}}(t) = \frac{\Lambda}{\mu_{\mathrm{h}}}$.

Now we will show the boundness of A_{wf}. We claim that $A_{\mathrm{wf}}(t) \leq K$ for all t. Suppose there exists a time t_1 such that $A_{\mathrm{wf}}(t_1) > K$. Then $\dot{A}_{\mathrm{wf}}(t_1) < 0$, which means that A_{wf} is decreasing near t_1. Then we must have that $A_{\mathrm{wf}}(t_1 + \epsilon) < K$ for some $\epsilon > 0$ and thus $A_{\mathrm{wf}}(t_1) \leq K$, which is a contradiction. Therefore, $A_{\mathrm{wf}}(t) \leq K$ for all t. Similarly, we have that $A_{\mathrm{wi}}(t) \leq K$ for all t.

Adding Eqs. (1.6)–(1.8) we have the following $\dot{N}_{\mathrm{V}}^{\mathrm{wf}}(t) = \gamma_{\mathrm{wf}} A_{\mathrm{wf}}(t) - \mu_{\mathrm{v}} N_{\mathrm{V}}^{\mathrm{wf}}(t)$. Since $A_{\mathrm{wf}}(t) \leq K$, we get that $\dot{N}_{\mathrm{V}}^{\mathrm{wf}}(t) \leq \gamma_{\mathrm{wf}} K - \mu_{\mathrm{v}} N_{\mathrm{V}}^{\mathrm{wf}}(t)$. Separating variables and solving for $N_{\mathrm{V}}^{\mathrm{wf}}(t)$, we have

$$N_{\mathrm{v}}(t) \leq \frac{\gamma_{\mathrm{wf}} K}{\mu_{\mathrm{v}}} + e^{-\mu_{\mathrm{v}} t}\left(N_{\mathrm{V}}^{\mathrm{wf}}(0) - \frac{\gamma_{\mathrm{wf}} K}{\mu_{\mathrm{v}}}\right).$$

If $N_{\mathrm{V}}^{\mathrm{wf}}(0) \leq \frac{\gamma_{\mathrm{wf}} K}{\mu_{\mathrm{v}}}$, then, by previous inequality we have $N_{\mathrm{v}}(t) \leq \frac{\gamma_{\mathrm{wf}} K}{\mu_{\mathrm{v}}}$ and thus D is a positively invariant set. If $N_{\mathrm{V}}^{\mathrm{wf}}(0) > \frac{\gamma_{\mathrm{wf}} K}{\mu_{\mathrm{v}}}$, then we have that $\dot{N}_{\mathrm{V}}^{\mathrm{wf}} \leq 0$ and the wild mosquito population is decreasing. Also, as t goes to infinity, we have that $N_{\mathrm{V}}^{\mathrm{wf}}(t)$ approaches $\frac{\gamma_{\mathrm{wf}} K}{\mu_{\mathrm{v}}}$. A similar argument holds for $N_{\mathrm{V}}^{\mathrm{wi}}(t)$. Therefore, the solutions either enter D in finite time or $N_{\mathrm{V}}^{\mathrm{wf}}(t)$ approaches $\frac{\gamma_{\mathrm{wf}} K}{\mu_{\mathrm{v}}}$ and $N_{\mathrm{V}}^{\mathrm{wi}}(t)$ approaches $\frac{\gamma_{\mathrm{wi}} K}{\mu_{\mathrm{vi}}}$, thus D is an attracting set. □

1.3.2 *Disease-free and only wild mosquitoes present*

When there is no Zika present in the vectors or human population and no mosquitoes are infected with *Wolbachia*, we obtain the following disease-free equilibrium point $E^{(1)}$:

$$\left(\frac{\Lambda}{\mu_{\mathrm{h}}}, 0, 0, 0, K\left(1 - \frac{1}{R}\right), \frac{\alpha\gamma_{\mathrm{wf}}}{\mu_{\mathrm{v}}} K\left(1 - \frac{1}{R}\right), 0, \frac{(1-\alpha)\gamma_{\mathrm{wf}}}{\mu_{\mathrm{v}}} K \right.$$
$$\left. \times \left(1 - \frac{1}{R}\right), 0, 0, 0, 0\right)$$

where $R = \frac{\eta\alpha(1-\alpha)\gamma_{\mathrm{wf}}}{\mu_{\mathrm{v}}(\gamma_{\mathrm{wf}} + \mu_{\mathrm{A}})}$ denotes the offspring reproduction number of wild mosquitoes.

Theorem 1.2. *If* $R > 1$, *then the equilibrium* $E^{(1)}$ *exists. If* $R_{\mathrm{W}} = \frac{q_2 \alpha(1-\alpha)\gamma_{\mathrm{wi}}}{\mu_{\mathrm{vi}}(\gamma_{\mathrm{wi}} + \mu_{\mathrm{Ai}})R} < 1$ *and* $R_Z = \frac{\beta_{\mathrm{vh}}\beta_{\mathrm{hv}}\nu_{\mathrm{h}}\alpha\mu_{\mathrm{h}}\gamma_{\mathrm{wf}}}{\Lambda\mu_{\mathrm{v}}^2(\mu_{\mathrm{h}}+\nu_{\mathrm{h}})(\gamma_{\mathrm{h}}+\mu_{\mathrm{h}})} K\left(1 - \frac{1}{R}\right) + \frac{\nu_{\mathrm{h}}\beta_{\mathrm{hh}}}{(\mu_{\mathrm{h}}+\nu_{\mathrm{h}})(\gamma_{\mathrm{h}}+\mu_{\mathrm{h}})} < 1$, *then* $E^{(1)}$ *is locally asymptotically stable. If any of the last two inequalities is reversed, then* $E^{(1)}$ *is unstable.*

Proof. The Jacobian for the system evaluated at the equilibrium $E^{(1)}$ is a block matrix $J^{(1)} = \left(\begin{array}{c|c} B & \star \\ \hline 0 & C \end{array}\right)$ where

$$B = \begin{pmatrix} -\mu_h & 0 & -\beta_{hh} & 0 & 0 & 0 & -\beta_{vh} & 0 \\ 0 & -e & \beta_{hh} & 0 & 0 & 0 & \beta_{vh} & 0 \\ 0 & \nu_h & -f & 0 & 0 & 0 & 0 & 0 \\ 0 & 0 & \gamma_h & -\mu_h & 0 & 0 & 0 & 0 \\ 0 & 0 & 0 & 0 & -b-d & \frac{(1-\alpha)^2\eta}{R} & c & \frac{\alpha^2\eta}{R} \\ 0 & 0 & -a & 0 & \alpha\gamma_{wf} & -\mu_v & 0 & 0 \\ 0 & 0 & a & 0 & 0 & 0 & -\mu_v & 0 \\ 0 & 0 & 0 & 0 & (1-\alpha)\gamma_{wf} & 0 & 0 & -\mu_v \end{pmatrix} \quad \text{and}$$

$$C = \begin{pmatrix} -\mu_{Ai} - \gamma_{wi} & \frac{(1-\alpha)q_2}{R} & 0 & 0 \\ \alpha\gamma_{wi} & -\mu_{vi} & 0 & 0 \\ 0 & 0 & -\mu_{vi} & 0 \\ (1-\alpha)\gamma_{wi} & 0 & 0 & -\mu_{vi} \end{pmatrix}.$$

Here the following notations were made first: $a = \frac{\alpha K(R-1)\mu_h\beta_{hv}\gamma_{wf}}{\Lambda R\mu_v}$, $b = \frac{(1-\alpha)\alpha\eta(R-1)\gamma_{wf}}{R\mu_v}$, $c = -\frac{(1-\alpha)\alpha\eta}{R}$, $d = \mu_A + \gamma_{wf}$, $e = \mu_h + \nu_h$, $f = \gamma_h + \mu_h$, $g = \frac{(1-\alpha)q_2}{R}$ and $h = -\mu_{Ai} - \gamma_{wi}$.

The eigenvalues of the Jacobian will be the eigenvalues of matrix B and matrix C. In order for the eigenvalues of C to have negative real part, the determinant must be positive. This means that $\frac{\eta\mu_{vi}\gamma_{wf}(\gamma_{wi}+\mu_{Ai})}{q_2\gamma_{wi}\mu_v(\gamma_{wf}+\mu_A)} > 1$. This can be written as $\frac{R_w}{R} < 1$, where $R_w = \frac{q_2\alpha(1-\alpha)\gamma_{wi}}{\mu_{vi}(\gamma_{wi}+\mu_{Ai})}$ and can be interpreted as the invasion number of *Wolbachia*-infected mosquitoes in absence of disease.

The characteristic polynomial corresponding to B is $P(\lambda) = (\mu_v+\lambda)P_1(\lambda)P_2(\lambda)$ where $P_1(\lambda) = -(b+d+\lambda)(\mu_v+\lambda)+\mu_v(\gamma_{wf}+\mu_A)$ and $P_2(\lambda) = \beta_{vh}\nu_h a - (\mu_v+\lambda)\left[(e+\lambda)(f+\lambda) - \nu_h\beta_{hh}\right]$. After substitutions, we have that all roots of $P_1(\lambda) = 0$ have negative real part.

Rewriting $P_2(\lambda) = 0$, we get that $(\mu_v+\lambda)(e+\lambda)(f+\lambda) = \beta_{vh}\nu_h\frac{\alpha K(R-1)\mu_h\beta_{hv}\gamma_{wf}}{\Lambda R\mu_v} + (\mu_v+\lambda)\nu_h\beta_{hh}$. Then we will have

$$\frac{\beta_{\text{vh}}\nu_{\text{h}}\alpha K(R-1)\mu_{\text{h}}\beta_{\text{hv}}\gamma_{\text{wf}}}{\Lambda R\mu_{\text{v}}(\mu_{\text{v}}+\lambda)(e+\lambda)(f+\lambda)}+\frac{(\mu_{\text{v}}+\lambda)\nu_{\text{h}}\beta_{\text{hh}}}{(\mu_{\text{v}}+\lambda)(e+\lambda)(f+\lambda)}=1.$$

Define

$$\begin{aligned}G(\lambda)&=\frac{\beta_{\text{vh}}\nu_{\text{h}}\alpha K(R-1)\mu_{\text{h}}\beta_{\text{hv}}\gamma_{\text{wf}}}{\Lambda R\mu_{\text{v}}(\mu_{\text{v}}+\lambda)(e+\lambda)(f+\lambda)}+\frac{\nu_{\text{h}}\beta_{\text{hh}}}{(e+\lambda)(f+\lambda)}\\&=\frac{\beta_{\text{vh}}\beta_{\text{hv}}\alpha\mu_{\text{h}}\gamma_{\text{wf}}}{\Lambda\mu_{\text{v}}(\mu_{\text{v}}+\lambda)(e+\lambda)(f+\lambda)}K\left(1-\frac{1}{R}\right)+\frac{\nu_{\text{h}}\beta_{\text{hh}}}{(e+\lambda)(f+\lambda)}.\end{aligned}$$

We define the reproduction number of Zika in absence of *Wolbachia*-infected mosquitoes $R_Z=G(0)$, that is,

$$R_Z=\frac{\beta_{\text{vh}}\beta_{\text{hv}}\nu_{\text{h}}\alpha\mu_{\text{h}}\gamma_{\text{wf}}}{\Lambda\mu_{\text{v}}^2(\mu_{\text{h}}+\nu_{\text{h}})(\gamma_{\text{h}}+\mu_{\text{h}})}K\left(1-\frac{1}{R}\right)+\frac{\nu_{\text{h}}\beta_{\text{hh}}}{(\mu_{\text{h}}+\nu_{\text{h}})(\gamma_{\text{h}}+\mu_{\text{h}})}.$$

Furthermore, let $\frac{\nu_{\text{h}}\beta_{\text{vh}}\beta_{\text{hv}}\alpha\mu_{\text{h}}\gamma_{\text{wf}}}{\Lambda\mu_{\text{v}}^2(\mu_{\text{h}}+\nu_{\text{h}})(\gamma_{\text{h}}+\mu_{\text{h}})}K\left(1-\frac{1}{R}\right)=R_Z^{\text{wf}}$ and $\frac{\nu_{\text{h}}\beta_{\text{hh}}}{(\mu_{\text{h}}+\nu_{\text{h}})(\gamma_{\text{h}}+\mu_{\text{h}})}=R_d$. Then we have that $R_Z=G(0)=R_Z^{\text{wf}}+R_d$. If $R_Z>1$, then the characteristic equation has a real positive root because $G(0)=R_Z>1$ and $G(\lambda)$ is a decreasing function of λ with $\lim_{\lambda\to\infty}G(\lambda)=0$ where λ is assumed to be real. However, if $R_Z<1$, then all roots have negative real parts. To see this, we assume that there is a λ with $\text{Re}(\lambda)\geq 0$. Then we have the following:

$$\begin{aligned}|G(\lambda)|\leq{}&\frac{\beta_{\text{vh}}\beta_{\text{hv}}\alpha\mu_{\text{h}}\gamma_{\text{wf}}}{\Lambda\mu_{\text{v}}(|\mu_{\text{v}}+\lambda|)(|e+\lambda|)(|f+\lambda|)}K\left(1-\frac{1}{R}\right)\\&+\frac{\nu_{\text{h}}\beta_{\text{hh}}}{(|e+\lambda|)(|f+\lambda|)}\\&\leq G(\text{Re}(\lambda))\leq G(0)=R_Z<1\end{aligned}$$

Therefore, $|G(\lambda)|<1$ and such a λ with nonnegative real part cannot be a solution to the characteristic equation. □

1.3.3 *Disease-free and only Wolbachia-infected mosquitoes present*

When there is no Zika present in the vectors or human population and all mosquitoes are infected with *Wolbachia*, we have the following disease-free equilibrium point $E^{(2)}$:

$$\left(\frac{\Lambda}{\mu_{\rm h}}, 0, 0, 0, K\left(1-\frac{1}{M}\right), \frac{\alpha\gamma_{\rm wi}}{\mu_{\rm vi}}K\left(1-\frac{1}{M}\right), 0, \right.$$
$$\left. \times\ \frac{(1-\alpha)\gamma_{\rm wi}}{\mu_{\rm vi}}K\left(1-\frac{1}{M}\right), 0, 0, 0, 0\right)$$

where $M = \frac{q_1\alpha(1-\alpha)\gamma_{\rm wi}}{\mu_{\rm vi}(\gamma_{\rm wi}+\mu_{\rm Ai})}$ is the offspring reproduction number of *Wolbachia*-infected mosquitoes. Note we rearranged the equations such that the equations for *Wolbachia*-infected mosquitoes appear before the *Wolbachia*-free mosquitoes.

Theorem 1.3. *If $M > 1$, then $E^{(2)}$ exists. If*

$$R_Z^i = \frac{\beta_{\rm vh}^{\rm w}\beta_{\rm hv}^{\rm w}\nu_{\rm h}\alpha\mu_{\rm h}\gamma_{\rm wi}}{\Lambda\mu_{\rm vi}^2(\mu_{\rm h}+\nu_{\rm h})(\gamma_{\rm h}+\mu_{\rm h})}K\left(1-\frac{1}{M}\right)+\frac{\nu_{\rm h}\beta_{\rm hh}}{(\mu_{\rm h}+\nu_g)(\gamma_{\rm h}+\mu_{\rm h})} < 1,$$

then $E^{(2)}$ is locally asymptotically stable. If $R_Z^i > 1$, then $E^{(2)}$ is unstable.

Proof. The Jacobian of the system evaluated at $E^{(2)}$ is $J^{(2)} = \left(\begin{array}{c|c} D & \star \\ \hline 0 & E\end{array}\right)$, where

$$D = \begin{pmatrix} -\mu_{\rm h} & 0 & -\beta_{\rm hh} & 0 & 0 & 0 & -\beta_{\rm vh}^{\rm w} & 0 \\ 0 & -e & \beta_{\rm hh} & 0 & 0 & 0 & \beta_{\rm vh}^{\rm w} & 0 \\ 0 & \nu_{\rm h} & -f & 0 & 0 & 0 & 0 & 0 \\ 0 & 0 & \gamma_{\rm h} & -\mu_{\rm h} & 0 & 0 & 0 & 0 \\ 0 & 0 & 0 & 0 & -(n+h) & \dfrac{(1-\alpha)^2 q_1}{M} & -r & \frac{\alpha^2 q_1}{M} \\ 0 & 0 & -g & 0 & \alpha\gamma_{\rm wi} & -\mu_{\rm vi} & 0 & 0 \\ 0 & 0 & g & 0 & 0 & 0 & -\mu_{\rm vi} & 0 \\ 0 & 0 & 0 & 0 & (1-\alpha)\gamma_{\rm wi} & 0 & 0 & -\mu_{\rm vi} \end{pmatrix} \quad \text{and}$$

$$E = \begin{pmatrix} -d & 0 & 0 & 0 \\ \alpha\gamma_{\rm wf} & -\mu_{\rm v} & 0 & 0 \\ 0 & 0 & -\mu_{\rm v} & 0 \\ (1-\alpha)\gamma_{\rm wf} & 0 & 0 & -\mu_{\rm v} \end{pmatrix}.$$

The matrix E has four obvious negative eigenvalues equal to the diagonal entries $-d$ and $-\mu_{\rm v}$. The matrix D has two negative eigenvalues equal to $-\mu_{\rm h}$. Let D^* be the matrix D after removing the column and row corresponding to the two eigenvalues equal to $-\mu_{\rm h}$.

The characteristic polynomial of the matrix D^* is $P(\lambda) = (\mu_{\text{vi}} + \lambda)P_1(\lambda)P_2(\lambda)$, where

$$P_1(\lambda) = \frac{q_1\alpha(1-\alpha)\gamma_{\text{wi}}}{M} - (n + h + \lambda)(\mu_{\text{vi}} + \lambda)$$

and

$$P_2(\lambda) = \beta^{\text{w}}_{\text{vh}}\nu_{\text{h}}g - (\mu_{\text{vi}} + \lambda)[(e + \lambda)(f + \lambda) - \nu_{\text{h}}\beta_{\text{hh}}].$$

It can be easily shown that the roots of $P_1(\lambda) = 0$ are negative when $M > 1$.

The characteristic equation corresponding to $P_2(\lambda)$ is

$$P_2(\lambda) = \beta^{\text{w}}_{\text{vh}}\nu_{\text{h}}g - (\mu_{\text{vi}} + \lambda)[(e + \lambda)(f + \lambda) - \nu_{\text{h}}\beta_{\text{hh}}] = 0$$

and it can be written as

$$\frac{\beta^{\text{w}}_{\text{vh}}\nu_{\text{h}}g}{(\mu_{\text{vi}} + \lambda)(e + \lambda)(f + \lambda)} + \frac{\nu_{\text{h}}\beta_{\text{hh}}}{(e + \lambda)(f + \lambda)} = 1.$$

Define $G(\lambda) = \frac{\beta^{\text{w}}_{\text{vh}}\nu_{\text{h}}g}{(\mu_{\text{vi}}+\lambda)(e+\lambda)(f+\lambda)} + \frac{\nu_{\text{h}}\beta_{\text{hh}}}{(e+\lambda)(f+\lambda)}$. We define the reproduction number of Zika in absence of wild mosquitoes as $R^i_Z = G(0)$. Substituting the values of e, f, and $g = \frac{\alpha K(M-1)\mu_{\text{h}}\gamma_{\text{wi}}\beta^{\text{w}}_{\text{hv}}}{\Lambda M\mu_{\text{vi}}}$, we get

$$R^i_Z = \frac{\beta^{\text{w}}_{\text{vh}}\beta^{\text{w}}_{\text{hv}}\nu_{\text{h}}\alpha\mu_{\text{h}}\gamma_{\text{wi}}}{\Lambda\mu^2_{\text{vi}}(\mu_{\text{h}} + \nu_{\text{h}})(\gamma_{\text{h}} + \mu_{\text{h}})}K\left(1 - \frac{1}{M}\right) + \frac{\nu_{\text{h}}\beta_{\text{hh}}}{(\mu_{\text{h}} + \nu_g)(\gamma_{\text{h}} + \mu_{\text{h}})}.$$

Furthermore, let $\frac{\beta^{\text{w}}_{\text{vh}}\beta^{\text{w}}_{\text{hv}}\nu_{\text{h}}\alpha\mu_{\text{h}}\gamma_{\text{wi}}}{\Lambda\mu^2_{\text{vi}}(\mu_{\text{h}}+\nu_{\text{h}})(\gamma_{\text{h}}+\mu_{\text{h}})}K\left(1 - \frac{1}{M}\right) = R^{\text{wi}}_Z$. Then we have that $R^i_Z = R^{\text{wi}}_Z + R_d$. If $R^i_Z > 1$, then the characteristic equation has a real positive root because $G(0) = R^i_Z > 1$ and $G(\lambda)$ is a decreasing function of λ with $\lim_{\lambda\to\infty} G(\lambda) = 0$ where λ is assumed to be real. However, if $R^i_Z < 1$, then all roots have negative real parts. To see this we assume there is a λ with $Re(\lambda) \geq 0$. Then we have the following:

$$\begin{aligned}|G(\lambda)| &\leq \frac{\beta^{\text{w}}_{\text{vh}}\nu_{\text{h}}g}{(|\mu_{\text{vi}} + \lambda|)(|e + \lambda|)(|f + \lambda|)} + \frac{\nu_{\text{h}}\beta_{\text{hh}}}{(|e + \lambda|)(|f + \lambda|)} \\ &\leq G(Re(\lambda)) \leq G(0) = R^i_Z < 1.\end{aligned}$$

Therefore, $G(\lambda) < 1$ and such a λ with a nonnegative real part cannot be a solution to the characteristic equation. □

1.3.4 *Disease-free and both types of mosquitoes present*

When there is no Zika present in the vectors or human population but some mosquitoes are infected with *Wolbachia*, we have the following disease-free equilibrium point $E^{(3)}$:

$$\left(\frac{\Lambda}{\mu_{\mathrm{h}}}, 0, 0, 0, A_{\mathrm{wf}}^{(3)}, \frac{\alpha\gamma_{\mathrm{wf}}A_{\mathrm{wf}}^{(3)}}{\mu_{\mathrm{v}}}, 0, \frac{(1-\alpha)\gamma_{\mathrm{wf}}A_{\mathrm{wf}}^{(3)}}{\mu_{\mathrm{v}}}, CA_{\mathrm{wf}}^{(3)}, \frac{\alpha\gamma_{\mathrm{wi}}CA_{\mathrm{wf}}^{(3)}}{\mu_{\mathrm{vi}}}, 0, \right.$$
$$\left. \times\ \frac{(1-\alpha)\gamma_{\mathrm{wi}}CA_{\mathrm{wf}}^{(3)}}{\mu_{\mathrm{vi}}}\right),$$

where $C = \frac{\gamma_{\mathrm{wf}}\mu_{\mathrm{vi}}R}{\gamma_{\mathrm{wi}}\mu_{\mathrm{v}}M}\left(1 - \frac{R_{\mathrm{w}}}{R}\right)$, $A_{\mathrm{wf}}^{(3)} = \frac{K}{1+C}\left(1 - \frac{1}{R} - \frac{1}{M}\left(1 - \frac{R_{\mathrm{w}}}{R}\right)\right)$, and for $A_{\mathrm{wf}}^{(3)}$ to exist we must have that $\frac{1}{R} + \frac{1}{M}\left(1 - \frac{R_{\mathrm{w}}}{R}\right) < 1$.

The stability of this disease-free equilibrium can be established using the next generation matrix. The infected compartments are E_{h}, I_{h}, I_{wf}, and I_{wi} ordered $(E_{\mathrm{h}}, I_{\mathrm{h}}, I_{\mathrm{wf}}, I_{\mathrm{wi}})$. The nonlinear terms with new infections $\mathcal{F}$ and the outflow term $\mathcal{V}$ are given by

$$\mathcal{F} = \begin{pmatrix} (\beta_{\mathrm{vh}}I_{\mathrm{wf}} + \beta_{\mathrm{vh}}^{\mathrm{w}}I_{\mathrm{wi}} + \beta_{\mathrm{hh}}I_{\mathrm{h}})\dfrac{S_{\mathrm{h}}}{N_{\mathrm{h}}} \\ 0 \\ \beta_{\mathrm{hv}}\dfrac{I_{\mathrm{h}}}{N_{\mathrm{h}}}S_{\mathrm{wf}} \\ \beta_{\mathrm{hv}}^{\mathrm{w}}\dfrac{I_{\mathrm{h}}}{N_{\mathrm{v}}}S_{\mathrm{wi}} \end{pmatrix} \quad \text{and}$$

$$\mathcal{V} = \begin{pmatrix} (\nu_{\mathrm{h}} + \mu_{\mathrm{h}})E_{\mathrm{h}} \\ -\nu_{\mathrm{h}}E_{\mathrm{h}} + (\gamma_{\mathrm{h}} + \mu_{\mathrm{h}})I_{\mathrm{h}} \\ \mu_{\mathrm{v}}I_{\mathrm{wf}} \\ \mu_{\mathrm{vi}}I_{\mathrm{wi}} \end{pmatrix}.$$

The next generation matrix is

$$\mathcal{K} = FV^{-1} = \begin{pmatrix} \dfrac{\beta_{\rm hh}\nu_{\rm h}}{(\gamma_{\rm h}+\mu_{\rm h})(\nu_{\rm h}+\mu_{\rm h})} & \dfrac{\beta_{\rm hh}}{\gamma_{\rm h}+\mu_{\rm h}} & \dfrac{\beta_{\rm vh}}{\mu_{\rm v}} & \dfrac{\beta_{\rm vh}^{\rm w}}{\mu_{\rm vi}} \\ 0 & 0 & 0 & 0 \\ \dfrac{\nu_{\rm h}\beta_{\rm hv}\mu_{\rm h}S_{\rm wf}^{(3)}}{\Lambda(\gamma_{\rm h}+\mu_{\rm h})(\nu_{\rm h}+\mu_{\rm h})} & \dfrac{\beta_{\rm hv}\mu_{\rm h}S_{\rm wf}^{(3)}}{\Lambda(\gamma_{\rm h}+\mu_{\rm h})} & 0 & 0 \\ \dfrac{\nu_{\rm h}\beta_{\rm hv}^{\rm w}\mu_{\rm h}S_{\rm wi}^{(3)}}{\Lambda(\gamma_{\rm h}+\mu_{\rm h})(\nu_{\rm h}+\mu_{\rm h})} & \dfrac{\beta_{\rm hv}^{\rm w}\mu_{\rm h}S_{\rm wi}^{(3)}}{\Lambda(\gamma_{\rm h}+\mu_{\rm h})} & 0 & 0 \end{pmatrix}.$$

The reproduction of Zika in presence of both types of mosquitoes denoted by $R_{0Z}^{\rm NG}$ is given by the spectral radius of $\mathcal{K}$. So $R_{0Z}^{\rm NG} = \rho(\mathcal{K})$.

The characteristic polynomial corresponding to $\mathcal{K}$ is $t^2 - R_d t - q = 0$, where

$$q = \frac{1}{1+C}\left(1 - \frac{1}{R} - \frac{1}{M}\left(1 - \frac{R_{\rm w}}{R}\right)\right)\left(\frac{R_Z^{\rm wf}}{1-1/R} + C\frac{R_Z^{\rm wi}}{1-1/M}\right)$$

and it is clearly positive when $R > 1$, $M > 1$ and $\frac{1}{R} + \frac{1}{M}\left(1 - \frac{R_{\rm w}}{R}\right) < 1$. The solutions for this quadratic are given by $t = \frac{R_d \pm \sqrt{R_d^2 + 4q}}{2}$. Therefore, the reproduction number of Zika in presence of both types of mosquitoes defined using the next generation method, $R_{0Z}^{\rm NG}$, is the spectral radius of $\mathcal{K}$, which equals the largest positive solution of the characteristic equation. Thus, we have that

$$R_{0Z}^{\rm NG} = \frac{R_d + \sqrt{R_d^2 + \frac{4}{1+C}\left(1 - \frac{1}{R} - \frac{1}{M}\left(1 - \frac{R_{\rm w}}{R}\right)\right)\left(\frac{R_Z^{\rm wf}}{1-1/R} + C\frac{R_Z^{\rm wi}}{1-1/M}\right)}}{2}.$$

This reproduction number of Zika in presence of both types of mosquitoes derived by the next-generation approach serves as a threshold condition for the stability of the disease-free equilibrium, but it is not an easy task to interpret it epidemiologically. From Ref. 67 we have

Theorem 1.4. *If $R_{0Z}^{\rm NG} < 1$, the disease-free equilibrium $E^{(3)}$ is locally asymptotically stable; otherwise, it is unstable.*

1.3.5 *Zika present and no Wolbachia-infected mosquitoes present*

We express all the state variables in terms of I_{h} and then arrive at an equation in terms of R_Z and I_{h}, as follows:

$$AI_{\mathrm{h}}^2 + BI_{\mathrm{h}} + Rb\Lambda^3\mu_{\mathrm{v}}^3(R_Z - 1) = 0,$$

where

$$A = b^2R^2\mu_{\mathrm{v}}^2\Lambda\mu_{\mathrm{h}}\beta_{\mathrm{hv}}\left(\frac{R_Z/R - R_d}{\nu_{\mathrm{h}}(R-1)}\right) \quad \text{and}$$

$$B = -\frac{\Lambda^2\beta_{\mathrm{hv}}\mu_{\mathrm{h}}R\mu_{\mathrm{v}}^2(bR_Z - \nu_{\mathrm{h}}\beta_{\mathrm{hh}})}{(R-1)} - \frac{b\Lambda^2\mu_{\mathrm{v}}^3R(bR_Z - \nu_{\mathrm{h}}\beta_{\mathrm{hh}})}{\nu_{\mathrm{h}}} + \beta_{\mathrm{hh}}R\mu_{\mathrm{v}}^2\Lambda^2\nu_{\mathrm{h}}\beta_{\mathrm{hv}}\mu_{\mathrm{h}} - b\beta_{\mathrm{hh}}\mu_{\mathrm{v}}^3\Lambda^2R - bR\mu_{\mathrm{v}}^2\Lambda^2\beta_{\mathrm{hv}}\mu_{\mathrm{h}}.$$

Using the Sign of the Derivative Method,[68] we differentiate the quadratic equation implicitly with respect to R_Z and evaluate at $I_{\mathrm{h}} = 0$ and $R_Z = 1$, to obtain the following $\frac{\partial I_{\mathrm{h}}}{\partial R_Z}B_* + Rb\Lambda^3\mu_{\mathrm{v}}^3 = 0$, where B_* is the coefficient of I_{h} evaluated at $R_Z = 1$.

Furthermore, calculations give that $B_* = Rb\Lambda^2\mu_{\mathrm{v}}^2\left[\frac{\beta_{\mathrm{hv}}\mu_{\mathrm{h}}R(R_d-1)}{R-1} - \frac{\mu_{\mathrm{v}}(\gamma_{\mathrm{h}}+\mu_{\mathrm{h}})(\nu_{\mathrm{h}}+\mu_{\mathrm{h}})}{\nu_{\mathrm{h}}}\right]$.

Solving for the derivative we get $\frac{\partial I_{\mathrm{h}}}{\partial R_Z} = \frac{-\Lambda\mu_{\mathrm{v}}\nu_{\mathrm{h}}(R-1)}{\beta_{\mathrm{hv}}\mu_{\mathrm{h}}\nu_{\mathrm{h}}R(R_d-1) - \mu_{\mathrm{v}}(\gamma_{\mathrm{h}}+\mu_{\mathrm{h}})(\nu_{\mathrm{h}}+\mu_{\mathrm{h}})(R-1)}$ and observe that the derivative is always positive since $R > 1$ and $R_d < 1$.

Theorem 1.5. *In presence of wild mosquitoes only the system admits a unique endemic equilibrium,* $E^{(4)}$*, when* $R_Z > 1$.

1.3.6 *Zika present and no wild mosquitoes present*

When there is Zika present in the human population and the mosquito population is composed of only *Wolbachia*-infected mosquitoes, we have the following equilibrium point:

$$E^{(5)} = (S_{\mathrm{h}}^{(5)}, E_{\mathrm{h}}^{(5)}, I_{\mathrm{h}}^{(5)}, R_{\mathrm{h}}^{(5)}, A_{\mathrm{wi}}^{(5)}, S_{\mathrm{wi}}^{(5)}, I_{\mathrm{wi}}^{(5)}, M_{\mathrm{wi}}^{(5)}, 0, 0, 0, 0).$$

Note that we rearranged the system again as it was done in a previous section. We express all the state variables in terms of I_{h} and then arrive at an equation in terms of R_Z^i and I_{h}, as follows:

$$I_{\mathrm{h}}^2\left(b^2M^2\mu_{\mathrm{vi}}^2\Lambda\mu_{\mathrm{h}}\beta_{\mathrm{hv}}^{\mathrm{w}}\left(\frac{R_Z^i/M-R_d}{\nu_{\mathrm{h}}(M-1)}\right)\right)$$
$$+\ I_{\mathrm{h}}(-\Lambda\beta_{\mathrm{hv}}^{\mathrm{w}}\mu_{\mathrm{h}}M\Lambda\mu_{\mathrm{vi}}^2\frac{(bR_Z^i-\nu_{\mathrm{h}}\beta_{\mathrm{hh}})}{(M-1)}-b\Lambda M\Lambda\mu_{\mathrm{vi}}^3\frac{(bR_Z^i-\nu_{\mathrm{h}}\beta_{\mathrm{hh}})}{\nu_{\mathrm{h}}}$$
$$+\ \beta_{\mathrm{hh}}\mu_{\mathrm{vi}}^2\Lambda^2\nu_{\mathrm{h}}M\mu_{\mathrm{h}}\beta_{\mathrm{hv}}^{\mathrm{w}}-b\beta_{\mathrm{hh}}\Lambda^2\mu_{\mathrm{vi}}^3M-b\Lambda^2\mu_{\mathrm{vi}}^2M\mu_{\mathrm{h}}\beta_{\mathrm{hv}}^{\mathrm{w}})$$
$$+\ Mb\Lambda^3\mu_{\mathrm{vi}}^3(R_Z^i-1)=0$$

Theorem 1.6. *In presence of only Wolbachia infected mosquitoes the system admits a unique endemic equilibrium, $E^{(5)}$, when $R_Z^i>1$.*

1.3.7 *Zika present and both types of mosquitoes present*

When there is Zika present in the human population and the mosquito population is composed of both *Wolbachia*-free and *Wolbachia*-infected mosquitoes, we have the following equilibrium point

$$E^{(6)}=(S_{\mathrm{h}}^{(6)},E_{\mathrm{h}}^{(6)},I_{\mathrm{h}}^{(6)},R_{\mathrm{h}}^{(6)},A_{\mathrm{wf}}^{(6)},S_{\mathrm{wf}}^{(6)},I_{\mathrm{wf}}^{(6)},M_{\mathrm{wf}}^{(6)},$$
$$A_{\mathrm{wi}}^{(6)},S_{\mathrm{wi}}^{(6)},I_{\mathrm{wi}}^{(6)},M_{\mathrm{wi}}^{(6)}).$$

We will explore this equilibrium using numerical simulations since this equilibrium is too complex to deal with analytically.

1.3.8 *Overview of equilibria*

Here is an overview of the equilibria with conditions for existence and stability listed in Table 1.5.

$$E^{(1)}=\left(\frac{\Lambda}{\mu_{\mathrm{h}}},0,0,0,K\left(1-\frac{1}{R}\right),\frac{\alpha\gamma_{\mathrm{wf}}}{\mu_{\mathrm{v}}}K\left(1-\frac{1}{R}\right),0,\right.$$
$$\left.\times\ \frac{(1-\alpha)\gamma_{\mathrm{wf}}}{\mu_{\mathrm{v}}}K\left(1-\frac{1}{R}\right),0,0,0,0\right).$$

$$E^{(2)} = \left(\frac{\Lambda}{\mu_{\text{h}}}, 0, 0, 0, K\left(1 - \frac{1}{M}\right), \frac{\alpha\gamma_{\text{wi}}}{\mu_{\text{vi}}}K\left(1 - \frac{1}{M}\right), 0,\right.$$
$$\left.\times\ \frac{(1-\alpha)\gamma_{\text{wi}}}{\mu_{\text{vi}}}K\left(1 - \frac{1}{M}\right), 0, 0, 0, 0\right).$$

$$E^{(3)} = \left(\frac{\Lambda}{\mu_{\text{h}}}, 0, 0, 0, A_{\text{wf}}^{(3)}, \frac{\alpha\gamma_{\text{wf}}A_{\text{wf}}^{(3)}}{\mu_{\text{v}}}, 0, \frac{(1-\alpha)\gamma_{\text{wf}}A_{\text{wf}}^{(3)}}{\mu_{\text{v}}}, CA_{\text{wf}}^{(3)},\right.$$
$$\left.\times\ \frac{\alpha\gamma_{\text{wi}}CA_{\text{wf}}^{(3)}}{\mu_{\text{vi}}}, 0, \frac{(1-\alpha)\gamma_{\text{wi}}CA_{\text{wf}}^{(3)}}{\mu_{\text{vi}}}\right).$$

$$E^{(4)} = (S_{\text{h}}^{(4)}, E_{\text{h}}^{(4)}, I_{\text{h}}^{(4)}, R_{\text{h}}^{(4)}, A_{\text{wf}}^{(4)}, S_{\text{wf}}^{(4)}, I_{\text{wf}}^{(4)}, M_{\text{wf}}^{(4)}, 0, 0, 0, 0).$$
$$E^{(5)} = (S_{\text{h}}^{(5)}, E_{\text{h}}^{(5)}, I_{\text{h}}^{(5)}, R_{\text{h}}^{(5)}, A_{\text{wi}}^{(5)}, S_{\text{wi}}^{(5)}, I_{\text{wi}}^{(5)}, M_{\text{wi}}^{(5)}, 0, 0, 0, 0).$$
$$E^{(6)} = (S_{\text{h}}^{(6)}, E_{\text{h}}^{(6)}, I_{\text{h}}^{(6)}, R_{\text{h}}^{(6)}, A_{\text{wf}}^{(6)}, S_{\text{wf}}^{(6)}, I_{\text{wf}}^{(6)}, M_{\text{wf}}^{(6)},$$
$$A_{\text{wi}}^{(6)}, S_{\text{wi}}^{(6)}, I_{\text{wi}}^{(6)}, M_{\text{wi}}^{(6)}).$$

Table 1.5 summarizes the results of the analysis. Besides simulations, no analysis was done for equilibria $E^{(4)}$, $E^{(5)}$, and $E^{(6)}$, hence the n/a values listed in the table.

1.4 Numerical Simulations

The model is simulated using the parameter values and ranges from Tables 1.3 and 1.4. We are assuming that there are already some *Wolbachia*-infected mosquitoes present and we are trying to determine what ratio of *Wolbachia*-infected to *Wolbachia*-free mosquitoes should be attained at the time of release of additional *Wolbachia*-infected mosquitoes in order to get a desired outcome.

1.4.1 *Numerical simulation for disease-free equilibria*

First, we begin with the disease-free equilibria. Our goal is to determine under what circumstances *Wolbachia* establishes itself in the wild mosquito population, and if it does, how does that affect Zika dynamics within the human and mosquito populations.

Table 1.5. Overview of equilibria and their conditions for stability.

Equilibrium	Description	Existence conditions	Stability conditions	Instability conditions
$E^{(1)}$	DF, WF	$R > 1$	$R_w/R < 1$, $R_Z < 1$	$R_w/R > 1$ or $R_Z > 1$
$E^{(2)}$	DF, WI	$M > 1$	$R_Z^i < 1$	$R_Z^i > 1$
$E^{(3)}$	DF, WF, WI	$R_w/R < 1$, $R > 1$, $M > 1$	$R_{0Z} < 1$	$R_{0Z} > 1$
$E^{(4)}$	DP, WF	$R > 1$, $R_Z > 1$	n/a	n/a
$E^{(5)}$	DP, WI	$M > 1$, $R_Z^i > 1$	n/a	n/a
$E^{(6)}$	DP, WF, WI	$R > 1$, $M > 1$	n/a	n/a

Notes: DF, Disease-free;
WF, *Wolbachia*-free mosquitoes present;
DP, Disease present;
WI, *Wolbachia*-infected mosquitoes present.

Table 1.6. Initial conditions for disease-free simulations.

Variable	Initial value	Variable	Initial value
$S_h(0)$	10,000,000	$I_{wf}(0)$	50,000
$E_h(0)$	3,000,000	$M_{wf}(0)$	250,000
$I_h(0)$	2,500,000	$A_{wi}(0)$	500,000
$R_h(0)$	10,000	$S_{wi}(0)$	250,000
$A_{wf}(0)$	500,000	$I_{wi}(0)$	50,000
$S_{wf}(0)$	250,000	$M_{wi}(0)$	250,000

1.4.1.1 *Wolbachia fails to establish when starting with same amount of wild and Wolbachia-infected mosquitoes*

Running simulations using the baseline values for parameters from Tables 1.3 and 1.4 and the initial conditions listed in Table 1.6 yields to failure of *Wolbachia* infection to establish itself in the wild mosquito population. Note that we start with the same amount of wild mosquitoes and *Wolbachia*-infected mosquitoes. Also, we chose to start with 2.5 mil Zika-infected humans to emphasize convergence to the disease-free equilibrium even for high level of initial infection.

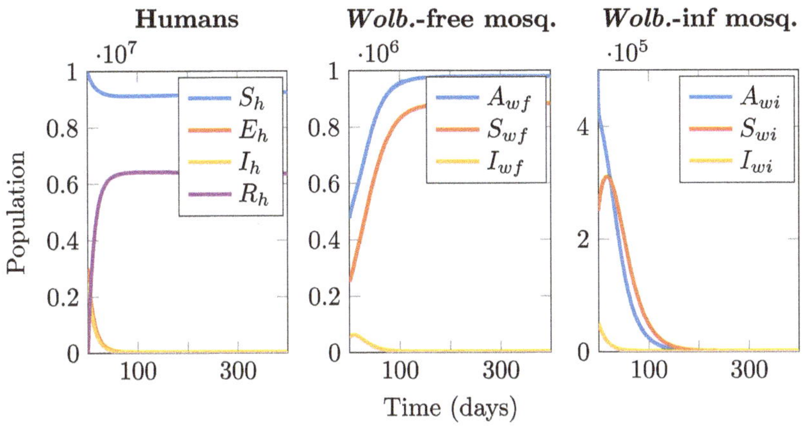

Fig. 1.2. Dominance of *Wolbachia*-free mosquitoes. Figure obtained using baseline values for parameters from Tables 1.3 and 1.4 and the initial conditions listed in Table 1.6.

Note in Figure 1.2 that *Wolbachia*-free mosquitoes persist and the *Wolbachia*-infected mosquitoes are eliminated in approximately 200 days. The disease is eradicated in approximately 250 days in both humans and wild mosquitoes. This corresponds to steady state $E^{(1)}$ where wild mosquitoes persist and *Wolbachia* infection fails to establish.

1.4.1.2 *Wolbachia is established when more Wolbachia-infected susceptible females are released*

Increasing the amount of *Wolbachia*-infected susceptible female mosquitoes can lead to the second steady state, $E^{(2)}$, where *Wolbachia*-infected mosquitoes persist and wild mosquitoes are eliminated. More specifically, increasing the initial amount of *Wolbachia*-infected females from $S_{\text{wi}}(0) = 250{,}000$ to 850,000 causes the *Wolbachia*-infected mosquitoes to persist and the *Wolbachia*-free mosquito population to die out in about 330 days. The disease again is eradicated in both humans and wild mosquitoes in approximately 250 days and in 180 days in *Wolbachia*-infected mosquitoes as seen in Figure 1.3.

Simulations suggest that one would need to start with approximately 3.4 times as many *Wolbachia*-infected females, compared with *Wolbachia*-free females in order to allow for the *Wolbachia*-infected

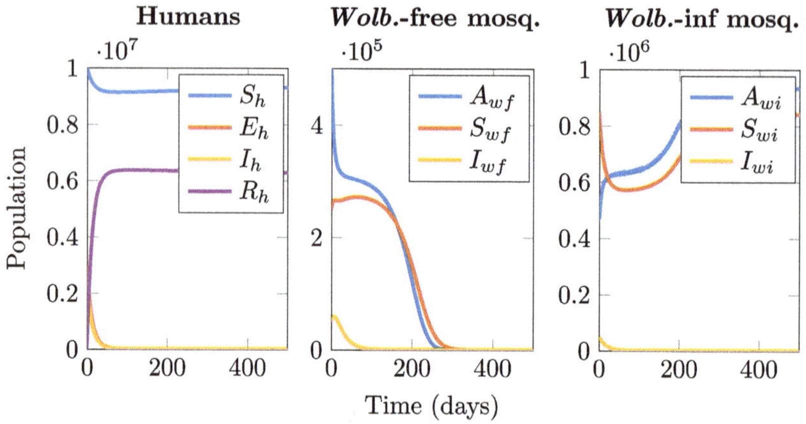

Fig. 1.3. Dominance of *Wolbachia*-infected mosquitoes when we start with more *Wolbachia*-infected susceptible females. Figure obtained using baseline values for parameters from Tables 1.3 and 1.4 and the initial conditions listed in Table 1.6 with $S_{\mathrm{wi}}(0) = 850{,}000$.

mosquitoes to establish in the population in roughly around 1 year. In this case where *Wolbachia*-infected mosquitoes persist when we start with more *Wolbachia*-infected susceptible female mosquitoes, the sexual transmission component becomes more important than in any other case. If the sexual transmission parameter, β_{hh}, is set to 0, the *Wolbachia*-infected mosquito population fails to establish itself as seen in Figure 1.4. All other simulations did not change when setting $\beta_{\mathrm{hh}} = 0$.

These two semitrivial equilibria $E^{(1)}$ and $E^{(2)}$ correspond to the dominance of each type of mosquito. The winner is determined by the initial conditions. So depending on the initial conditions, we have that either *Wolbachia*-free mosquitoes persist or *Wolbachia*-infected mosquitoes persist as time goes to infinity.

1.4.1.3 *Wolbachia is established when more Wolbachia-infected aquatic stage mosquitoes are released*

The steady state where some mosquitoes carry *Wolbachia* in the long run is the desirable one, thus, it is important to develop mathematical models that suggest the optimal release strategy of *Wolbachia*-infected mosquitoes (adults or aquatic stage) in order to drive the mosquito population close to the steady state where only susceptible

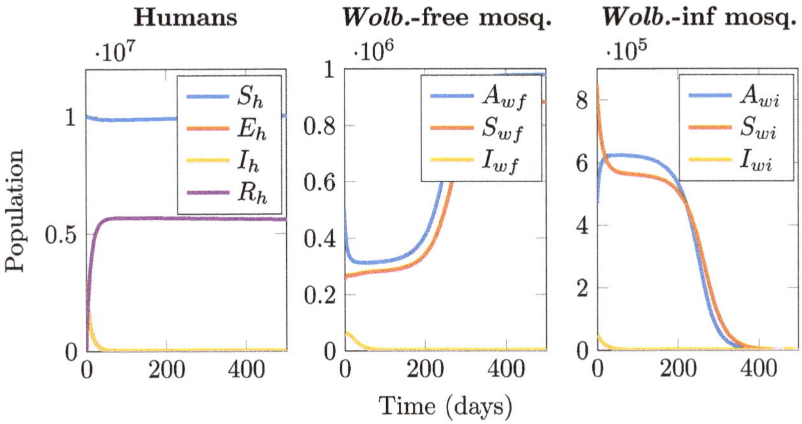

Fig. 1.4. *Wolbachia*-infected mosquito population dies when we start with more *Wolbachia*-infected susceptible females but sexual transmission is ignored. Figure obtained using baseline values for parameters from Tables 1.3 and 1.4 and the initial conditions listed in Table 1.6 with $S_{\text{wi}}(0) = 850{,}000$ and $\beta_{\text{hh}} = 0$.

humans and *Wolbachia*-infected mosquitoes exist. As we will see from simulations, there are multiple ways to reach this outcome. We are interested in the release strategy that also allows for the fastest establishment of *Wolbachia*-infected mosquitoes.

Field studies have used both approaches of releasing *Wolbachia*-infected adult mosquitoes and *Wolbachia*-infected aquatic stage mosquitoes with successful outcomes. *Wolbachia* releases are now ongoing or planned in 12 countries.[69] Recently, in Indonesia *Wolbachia* carrying mosquitoes were released as eggs using mosquito release containers, which were 2-l plastic buckets each containing one oviposition strip with 100–150 eggs, food, and 1 l of water. These containers were covered and placed outside houses, protected from direct sun and rain with holes drilled near the top of the bucket walls to allow *Wolbachia*-infected adult mosquitoes to escape once they emerged.[70]

We are interested to see whether or not *Wolbachia* infection can be established by releasing *Wolbachia*-infected mosquitoes in aquatic stage. Increasing the initial amount of *Wolbachia*-infected aquatic stage from $A_{\text{wi}}(0) = 500{,}000$ to $A_{\text{wi}}(0) = 1{,}850{,}000$ allows the *Wolbachia*-infected mosquito to persist as well as seen in Figure 1.5.

Our simulations suggest that one would need to release 3.7 times as many *Wolbachia*-infected mosquitoes at the aquatic stage

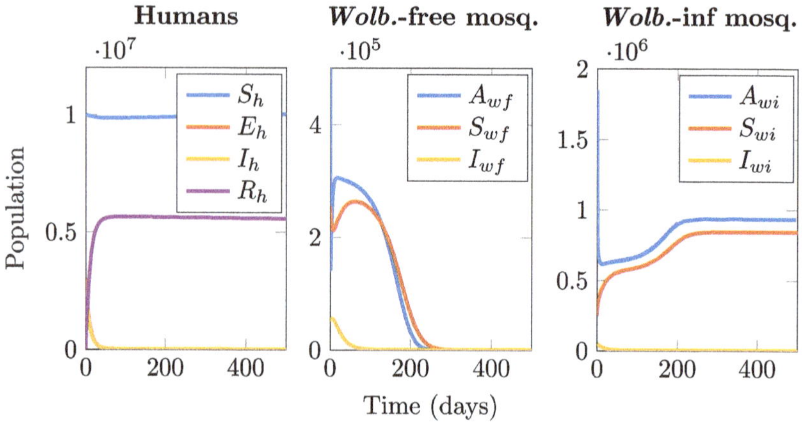

Fig. 1.5. Dominance of *Wolbachia*-infected mosquitoes when we start with more *Wolbachia*-infected aquatic stage. Figure obtained using baseline values for parameters from Tables 1.3 and 1.4 and the initial conditions listed in Table 1.6 with $A_{\mathrm{wi}}(0) = 1{,}850{,}000$.

compared to *Wolbachia*-free aquatic stage. The wild mosquitoes are eliminated and the *Wolbachia*-infected mosquitoes will establish in a little over 250 days, faster than in the case when more *Wolbachia*-infected females are released as seen in Section 1.4.1.2. The disease is completely eradicated in humans and wild mosquitoes in around 230 days, a little earlier than in the case where more *Wolbachia* females were released, and in 180 days in *Wolbachia*-infected mosquitoes.

1.4.1.4 *Wolbachia is established when more Wolbachia-infected males are released*

Persistence of *Wolbachia*-infected mosquitoes by releasing more *Wolbachia*-infected males can be achieved as well, but it requires 5.8 as many *Wolbachia*-infected males compared with *Wolbachia*-free males which is much higher than the required number of females as seen in Section 1.4.1.2. In Figure 1.6, we observe that increasing the initial amount of *Wolbachia*-infected males from $M_{\mathrm{wi}}(0) = 250{,}000$ to $M_{\mathrm{wi}}(0) = 1{,}450{,}000$ allows the *Wolbachia*-infected mosquito population to establish itself in the population of mosquitoes in roughly 250 days, the same amount of time it takes using more aquatic stage mosquitoes as seen in Section 1.4.1.3. Disease is again eradicated in approximately 230 days.

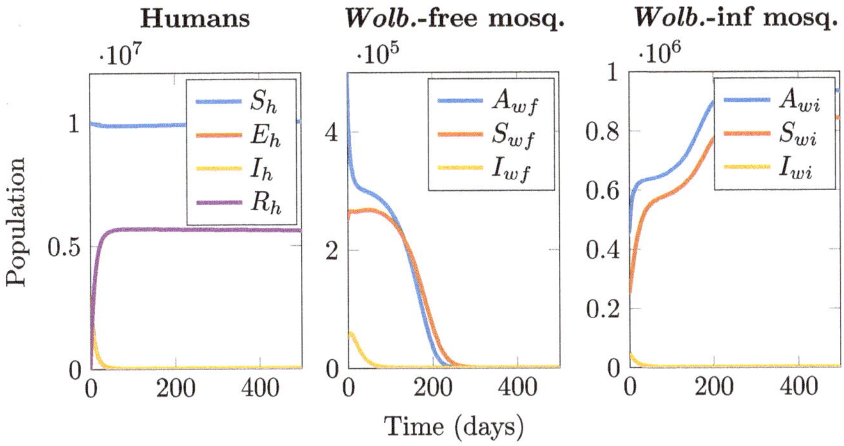

Fig. 1.6. Dominance of *Wolbachia*-infected mosquitoes when starting with more *Wolbachia*-infected males. Figure obtained using baseline values for parameters from Tables 1.3 and 1.4 and the initial conditions listed in Table 1.6 with $M_{\mathrm{wi}}(0) =$ 1,450,000.

1.4.1.5 *Wolbachia is established when more Wolbachia-infected aquatic stage and females are released simultaneously*

Not surprisingly, a combination of more *Wolbachia*-infected aquatic stage and more *Wolbachia*-infected females also allows the *Wolbachia*-infected mosquito population to establish itself. In Figure 1.7, we note that if we release twice as many *Wolbachia*-infected aquatic stage and three times as many *Wolbachia*-infected females compared to their *Wolbachia*-free counterparts, then the wild mosquitoes population is eliminated and the *Wolbachia*-infected mosquitoes persist after roughly 250 days. Disease is again eradicated in approximately 230 days.

1.4.1.6 *Wolbachia is established when more Wolbachia-infected aquatic stage mosquitoes and males are released simultaneously*

Similarly, a combination of aquatic stage mosquitoes and male mosquitoes works as well. If twice as many *Wolbachia*-infected aquatic stage are released and 4.4 as many times *Wolbachia*-infected males, the same outcome is achieved. The *Wolbachia*-infected mosquito

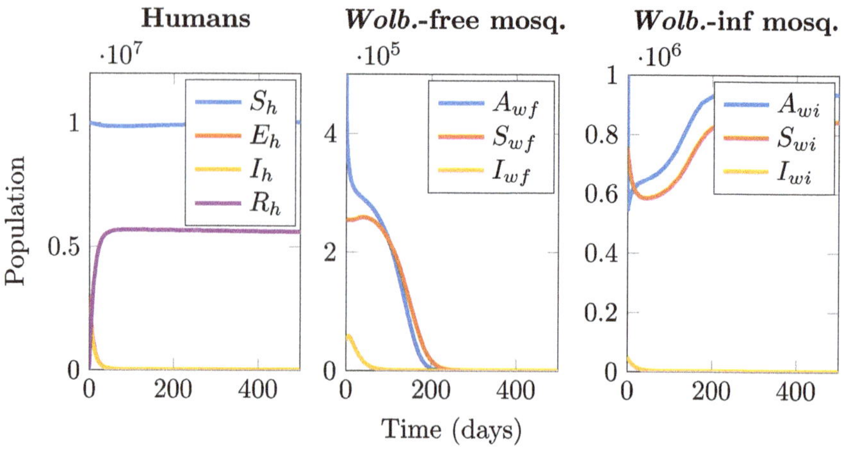

Fig. 1.7. Dominance of *Wolbachia*-infected mosquitoes when we start with more *Wolbachia*-infected aquatic stage mosquitoes and more females. Figure obtained using baseline values for parameters from Tables 1.3 and 1.4 and the initial conditions listed in Table 1.6 with $A_{\mathrm{wi}}(0) = 1{,}000{,}000$ and $S_{\mathrm{wi}}(0) = 750{,}000$.

population establishes itself in 250 days in this case as seen in Figure 1.8. Disease is again eradicated in approximately 230 days.

This strategy of releasing a combination of both types of mosquito stages (aquatic and adults) might be a better choice since one can survey the land to estimate the number of wild aquatic stage mosquitoes. Also, since the hatching rate of the aquatic stage depends on environmental parameters such as humidity, precipitation, and temperature, introducing adults can offset the unhatched aquatic stage mosquitoes.

As seen in Sections 1.4.1.2–1.4.1.6, *Wolbachia* establishment can be achieved by different release strategies. The establishment takes roughly around one year in all scenarios, but the disease takes longer to be eradicated when *Wolbachia*-infected susceptible females are released when compared to all other strategies.

For realistic parameter values, we have not been able to find a situation in which both types of mosquitoes coexist. This suggests that for realistic parameter values the two mosquito populations are in a competitive-exclusion regime where only one of the species will persist. These results are consistent with other models that have considered complete cytoplasmic incompatibility and

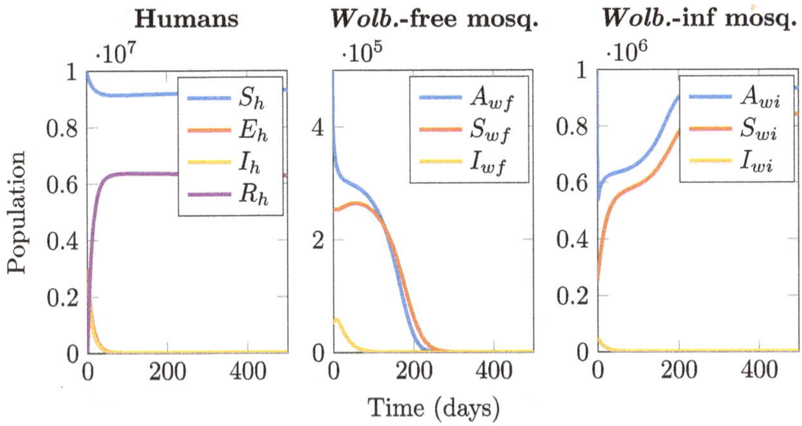

Fig. 1.8. Dominance of *Wolbachia*-infected mosquitoes when we start with more *Wolbachia*-infected aquatic stage mosquitoes and more males. Figure obtained using baseline values for parameters from Tables 1.3 and 1.4 and the initial conditions listed in Table 1.6 with $A_{\rm wi}(0) = 1{,}000{,}000$ and $S_{\rm wi}(0) = 1{,}100{,}000$.

perfect maternal transmission such as in Refs. 38, 40 and 71. Incomplete cytoplasmic incompatbility refers to when a fraction of the offspring resulting from *Wolbachia*-infected males with *Wolbachia*-free females survives. Incomplete CI and imperfect maternal transmission are two mechanisms by which *Wolbachia*-free offspring are produced. Therefore, models that include incomplete CI and/or imperfect maternal transmission of *Wolbachia* can observe a coexistence equilibrium. For example, in Ref. 71 the authors consider both cases when the maternal transmission of *Wolbachia* is perfect and imperfect. In the case where they consider imperfect maternal transmission, there is no complete *Wolbachia*-infected equilibrium achieved but two endemic equlibria: a high-infection stable endemic equilibrium and a low-infection unstable endemic equilibrium. Also in Ref. 72 the results indicate that when incomplete cytoplasmic incompatibility and imperfect maternal transmission are taken into account, four steady states that are biologically feasible are observed: all mosquitoes dying out, only *Wolbachia*-free mosquitoes surviving, and two steady states where *Wolbachia*-free and *Wolbachia*-infected mosquitoes coexist. The stability of the coexistence steady states is analyzed numerically with only one of them being a physically realistic stable steady state.

Table 1.7. Initial conditions for disease present simulations.

Variable	Initial value	Variable	Initial value
$S_h(0)$	8,000,000	$I_{wf}(0)$	50,000
$E_h(0)$	1,000,000	$M_{wf}(0)$	250,000
$I_h(0)$	900,000	$A_{wi}(0)$	500,000
$R_h(0)$	100,000	$S_{wi}(0)$	250,000
$A_{wf}(0)$	2,500,000	$I_{wi}(0)$	0
$S_{wf}(0)$	250,000	$M_{wi}(0)$	250,000

1.4.2 *Numerical simulations for disease present equilibria*

In this section, we perform numerical simulations for the model when the disease is endemic. The initial conditions used for the simulations are given in Table 1.7. Note that we assume that we start with no Zika-infected *Wolbachia* females. We again use the baseline values from Tables 1.3 and 1.4 with two changes. First, the biting rate of the mosquitoes is set to the higher end of the range $b = 1.25$, and second the carrying capacity of the aquatic stage is set to $K = 10^9$.

1.4.2.1 *Zika is present but Wolbachia infection is not established*

When the biting rate of the mosquitoes and the carrying capacity of the aquatic environment is set to the higher end of their range, we get that the disease is endemic as we can see in Figure 1.9. Zika persists both in humans and wild mosquitoes, and the *Wolbachia*-infected mosquitoes are eliminated in 400 days. The peak of humans infected with Zika is reached in 25 days with approximately 1.8 million infected humans. Zika stays endemic in humans with 3,500 humans infected and in the *Wolbachia*-free mosquitoes with approximately 2.4 mil Zika-infected wild mosquitoes.

1.4.2.2 *Zika is eradicated when Wolbachia infection is established*

If we allow for the *Wolbachia*-infected mosquitoes to establish in the wild mosquito population, we get that the disease is eradicated.

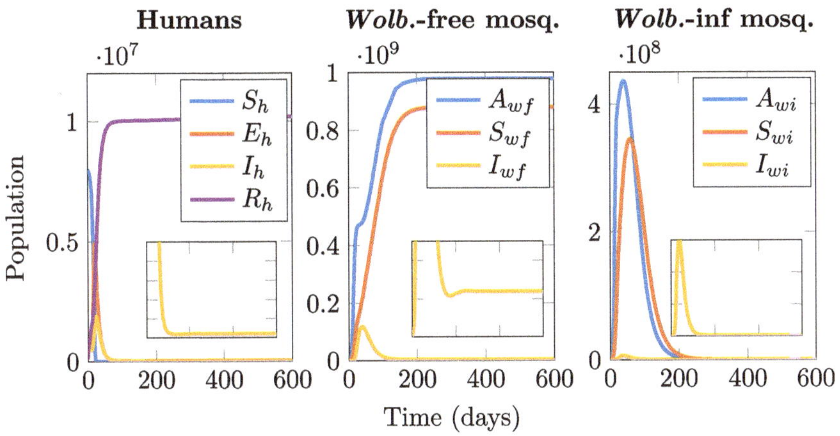

Fig. 1.9. Endemic Zika in humans and *Wolbachia*-free mosquitoes. Figure is obtained using initial conditions given in Table 1.7 and baseline values from Tables 1.3 and 1.4 with two changes ($b = 1.25$, $K = 10^9$).

We keep the same parameters as in Section 1.4.2.1 and change only the initial condition from $A_{\text{wf}}(0) = 2{,}500{,}000$ to $A_{\text{wf}}(0) = 500{,}000$ and $A_{\text{wi}}(0) = 500{,}000$ to $A_{\text{wi}}(0) = 1{,}500{,}000$. We can see in Figure 1.10 that the *Wolbachia*-infected mosquitoes persist and the disease is eradicated in humans and wild mosquitoes in around 300 days. Zika is eradicated in *Wolbachia*-infected mosquitoes in around 380 days. The peak of 1 million Zika-infected humans is reached in 40 days. In the field this would correspond to a strategy of decreasing the number of breeding sites of wild mosquitoes before releasing *Wolbachia*-infected aquatic stage mosquitoes.

1.4.2.3 *Zika is present and Wolbachia infection is established when Wolbachia-infected vectors are more competent*

In this section, we are interested to see if Zika can stay endemic even when *Wolbachia* infection is established. In order to investigate this we keep the same initial conditions as in Section 1.4.2.2, same parameter values as in Section 1.4.2.1 with the change that we increase the competence of *Wolbachia*-infected mosquitoes $\beta_{\text{hv}}^{\text{w}} = 0.042\beta_{\text{hv}}$ to $\beta_{\text{hv}}^{\text{w}} = 0.042\beta_{\text{hv}}$. We note in Figure 1.11 that in this case the *Wolbachia*-infected mosquito population persists and the disease persists in both humans and *Wolbachia*-infected mosquitoes. The wild

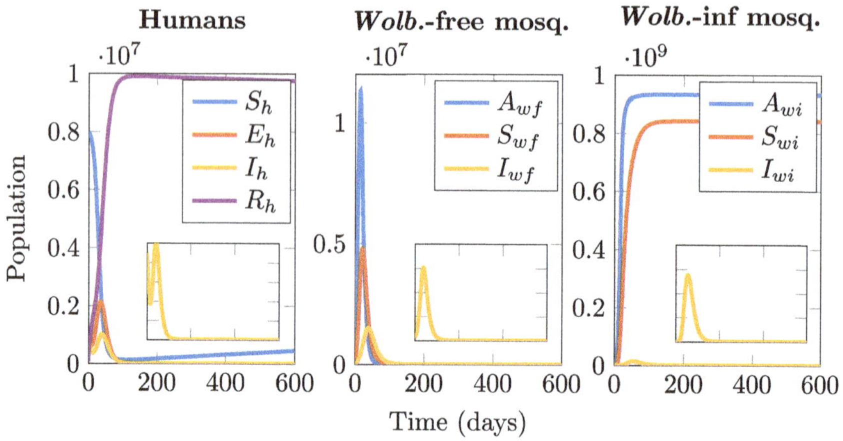

Fig. 1.10. Disease dies out when *Wolbachia* mosquitoes persist. Figure is obtained using initial conditions given in Table 1.7 with two changes ($A_{\rm wf}(0) =$ 500,000, $A_{\rm wi}(0)$ = 1,500,000) and baseline values from Tables 1.3 and 1.4 with two changes ($b = 1.25$ and $K = 10^9$).

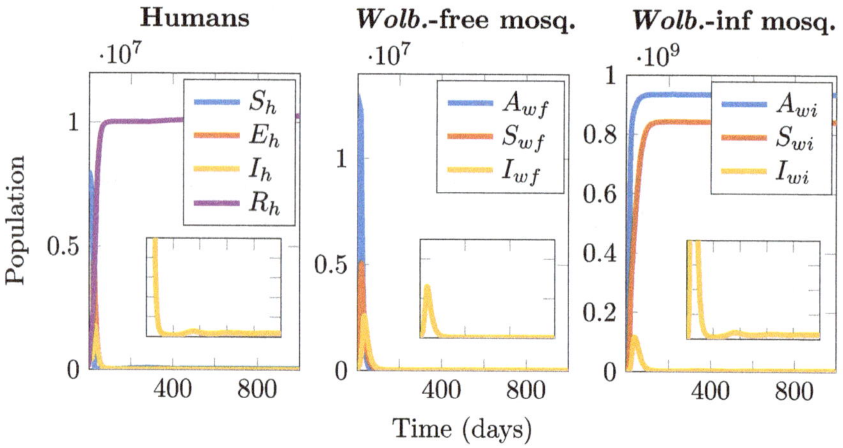

Fig. 1.11. Disease persists when *Wolbachia* mosquitoes are more competent. Figure is obtained using initial conditions given in Table 1.7 with two changes ($A_{\rm wf}(0)$ = 500,000, $A_{\rm wi}(0)$ = 1,500,000) and baseline values from Tables 1.3 and 1.4 with three changes ($b = 1.25$, $K = 10^9$, $\beta^{\rm w}_{\rm hv} = 0.042\beta_{\rm hv}$).

mosquito population is eliminated in around 330 days. The number of Zika-infected humans reach a peak of 1.8 million in around 35 days and then settle to around 3,500. As we can see in Figure 1.11

even though the decrease in the number of Zika-infected *Wolbachia*-infected mosquitoes is very substantial (from 2.4 millions to around 850,000), the impact in humans seems very minimal.

Thus, if the carrying capacity of the environment is large enough and the mosquito biting rate is high, Zika can be endemic in the human population and in wild mosquitoes. If *Wolbachia*-infected mosquitoes persist, Zika is eradicated. However, if *Wolbachia*-infected mosquitoes lose some of their ability to block Zika virus (due to loss of *Wolbachia* or higher temperatures), Zika can become endemic even when *Wolbachia*-infected mosquitoes dominate.

1.5 Elasticity of the Reproduction Numbers

In this section, we investigate the model dynamics over a wide range of feasible parameters to help better understand the model response under different assumptions. More specifically, we explore the elasticity of the reproduction numbers. The following is a list of all the reproduction numbers. The elasticity of a quantity Q with respect to a parameter p is given by

$$\epsilon_p^Q = \frac{\partial Q}{\partial p}\frac{p}{Q}.$$

The elasticities give the percentage change in the quantity Q due to a 1% increase in the parameter p.[73] We will investigate how the reproduction numbers from the model change in response to 1% increase in various parameters.

Here is a summary of all the reproduction numbers:

- $R = \frac{\eta\alpha(1-\alpha)\gamma_{\mathrm{wf}}}{\mu_{\mathrm{v}}(\gamma_{\mathrm{wf}}+\mu_{\mathrm{A}})}$ is the offspring reproduction number of wild mosquitoes.
- $R_{\mathrm{w}} = \frac{q_2\alpha(1-\alpha)\gamma_{\mathrm{wi}}}{\mu_{\mathrm{vi}}(\gamma_{\mathrm{wi}}+\mu_{\mathrm{Ai}})}$ is the invasion number of *Wolbachia*-infected mosquitoes in absence of disease.
- $R_Z = R_Z^{\mathrm{wf}} + R_d$ is the reproduction number of Zika in absence of *Wolbachia*-infected mosquitoes where $R_Z^{\mathrm{wf}} = \frac{\nu_{\mathrm{h}}\beta_{\mathrm{vh}}\beta_{\mathrm{hv}}\alpha\mu_{\mathrm{h}}\gamma_{\mathrm{wf}}}{\Lambda\mu_{\mathrm{v}}^2(\mu_{\mathrm{h}}+\nu_{\mathrm{h}})(\gamma_{\mathrm{h}}+\mu_{\mathrm{h}})}K\left(1-\frac{1}{R}\right)$ and $R_d = \frac{\nu_{\mathrm{h}}\beta_{\mathrm{hh}}}{(\mu_{\mathrm{h}}+\nu_{\mathrm{h}})(\gamma_{\mathrm{h}}+\mu_{\mathrm{h}})}$.
- $M = \frac{q_1\alpha(1-\alpha)\gamma_{\mathrm{wi}}}{\mu_{\mathrm{vi}}(\gamma_{\mathrm{wi}}+\mu_{\mathrm{Ai}})}$ is the offspring reproduction number of *Wolbachia*-infected mosquitoes.

- $R^i_Z = R^{\text{wi}}_Z + R_d$ is the reproduction number of Zika in absence of *Wolbachia*-free mosquitoes where $R^{\text{wi}}_Z = \frac{\beta^{\text{w}}_{\text{vh}}\beta^{\text{w}}_{\text{hv}}\nu_{\text{h}}\alpha\mu_{\text{h}}\gamma_{\text{wi}}}{\Lambda\mu^2_{\text{vi}}(\mu_{\text{h}}+\nu_{\text{h}})(\gamma_{\text{h}}+\mu_{\text{h}})}K\left(1-\frac{1}{M}\right)$.
- $R^{\text{NG}}_{0Z} = \frac{R_d+\sqrt{R_d^2+\frac{4}{1+C}\left(1-\frac{1}{R}-\frac{1}{M}\left(1-\frac{R_{\text{w}}}{R}\right)\right)\left(\frac{R^{\text{wf}}_Z}{1-1/R}+C\frac{R^{\text{wi}}_Z}{1-1/M}\right)}}{2}$ is the reproduction number of Zika in presence of both types of mosquitoes where

$$q = \frac{1}{1+C}\left(1-\frac{1}{R}-\frac{1}{M}\left(1-\frac{R_{\text{w}}}{R}\right)\right)\left(\frac{R^{\text{wf}}_Z}{1-1/R}+C\frac{R^{\text{wi}}_Z}{1-1/M}\right)$$

and

$$C = \frac{\gamma_{\text{wf}}\mu_{\text{vi}}R}{\eta\gamma_{\text{wi}}\mu_{\text{v}}M}\left(1-\frac{R_{\text{w}}}{R}\right).$$

We begin with the elasticity of the reproduction offspring numbers of the mosquitoes R, M, and R_{w}. Looking through Figure 1.12, we can observe some common trends. All of the mosquitoes offspring numbers, R, M, R_{w} are most sensitive to the mosquito death rates, μ_{v} and μ_{vi}, and to the corresponding mosquito egg laying rates η, q_1, and q_2.

The elasticity of the offspring numbers to the death rate of the mosquitoes are approximately 1%, meaning that 1% increase in the parameter results in 1% decrease in the offspring numbers. The elasticity of the offspring numbers to the egg laying rates are also approximately 1%, meaning that 1% increase in the parameter results in 1% increase in the offspring numbers. This suggests that measures to control the population of mosquitoes should be targeted toward decreasing the lifespan of the mosquitoes and decreasing the egg laying rates. The offspring numbers are not sensitive at all to the proportion of adult females arising from the aquatic stage, α. Also, the transition rate from aquatic stage to adult mosquito, γ_{wi}, and the death rate of the aquatic stage, μ_{Ai}, are more influential for the *Wolbachia*-infected mosquitoes. As we can observe in Figures 1.12(b) and 1.12(c) the reproduction offspring numbers M and R_{w} are more sensitive to γ_{wi} and μ_{Ai} compared to the sensitivity of the reproduction offspring number R with respect to γ_{wf} and μ_A.

Next, we investigate the elasticity of the reproduction numbers of Zika in presence of only one type of mosquitoes. In Figure 1.13(a), we

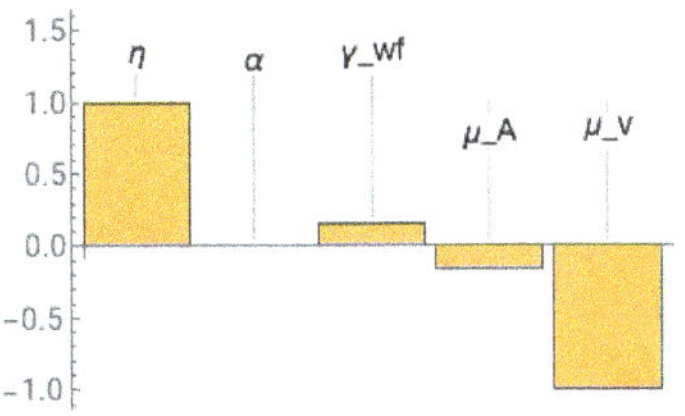

(a) Elasticity of R

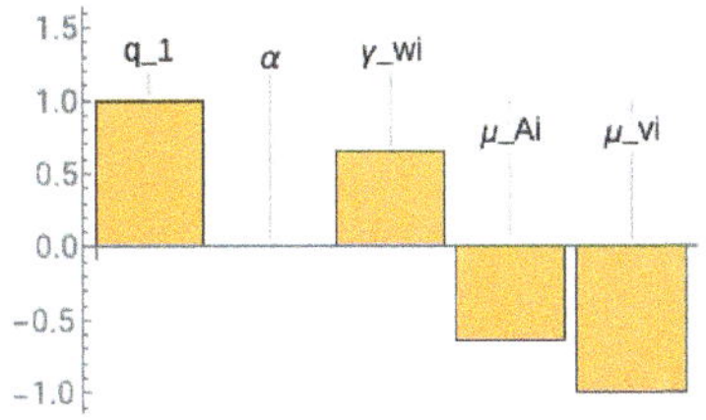

(b) Elasticity of M

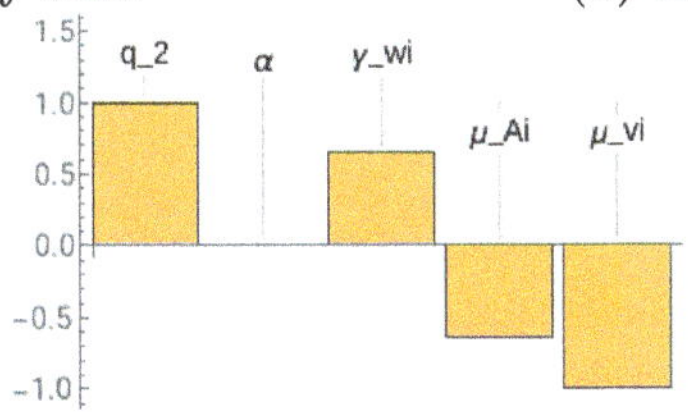

(c) Elasticity of R_{w}

Fig. 1.12. Elasticities of the reproduction offspring numbers.

note that the reproduction number of Zika in absence of *Wolbachia*-infected mosquitoes, R_Z, is most sensitive to the mosquito death rate, μ_{v}, and somewhat sensitive to γ_{h}, β_{vh}, β_{hv}, K, and γ_{wf}. On the other hand, the reproduction number R_Z is not sensitive at all to ν_{h} and depends very little on β_{hh}, the direct transmission parameter. Figure 1.13(b) shows the elasticities of R_{Zi}, the reproduction number of Zika in absence of *Wolbachia*-free mosquitoes. Note that treatment of humans and controlling the sexual transmission are most influential because of the high elasticity of γ_{h} and β_{hh} parameters. That suggests that the presence of *Wolbachia*-infected mosquitoes changes the way Zika must be controlled, making mosquito control less important. R_{Zi} shows very little sensitivity to the death rate of mosquitoes, meaning that killing the *Wolbachia*-infected mosquitoes is not a good control strategy. Also, in the presence of *Wolbachia*-infected mosquitoes we see more sensitivity to the direct transmission parameter.

The elasticity of the reproduction number of Zika in presence of both types of mosquitoes with respect to other offspring/reproduction numbers is investigated next. As we note in Figure 1.13(c), the reproduction number in presence of both types

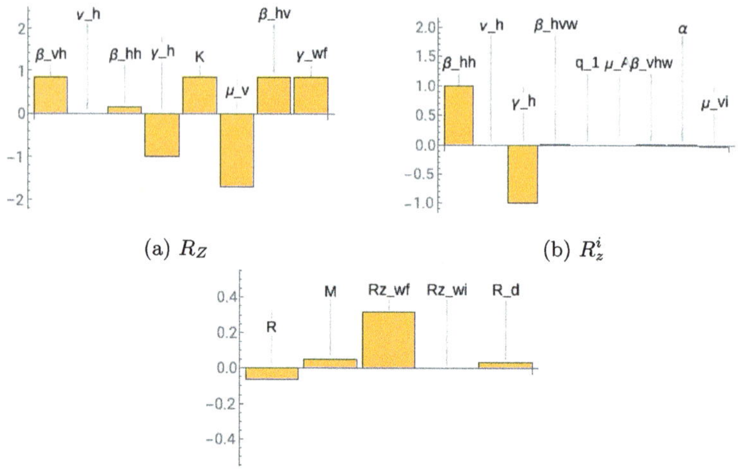

(a) R_Z (b) R_z^i

(c) R_{0Z}^{NG}

Fig. 1.13. Elasticities of reproduction numbers of Zika in presence of one type of mosquito and in presence of both types of mosquitoes with respect to other reproduction numbers.

of mosquitoes, R_{0Z}^{NG}, is most sensitive to R_z^{wf}, the reproduction number in absence of *Wolbachia*-infected mosquitoes. A 1% increase in R_z^{wf} results in a 0.32% increase in R_{0Z}^{NG}. Lastly, we investigate the elasticity of the reproduction number of Zika in presence of both types of mosquitoes with respect to parameters in Figure 1.14. In terms of parameters, R_{0Z}^{NG} is most sensitive again to the death rate of the *Wolbachia*-free mosquitoes and the direct transmission. A 1% increase in β_{hh} results in a 3.358% increase in R_{0Z}^{NG}. We note again that in the presence of *Wolbachia*-infected mosquitoes, the direct transmission becomes somewhat more influential. This might be happening due to the fact that *Wolbachia*-infected mosquitoes have a lower transmission rate. Parameters related to the vector-borne transmission in *Wolbachia*-free mosquitoes are also influential.

The presence of *Wolbachia*-infected mosquitoes switches the control strategies. When no *Wolbachia*-infected mosquitoes are present, the control should be targeted toward decreasing the lifespan of the mosquitoes and when *Wolbachia*-infected mosquitoes are present, the efforts should be concentrated on treatment of humans to increase the recovery rate and sexual transmission control to decrease the direct transmission rate.

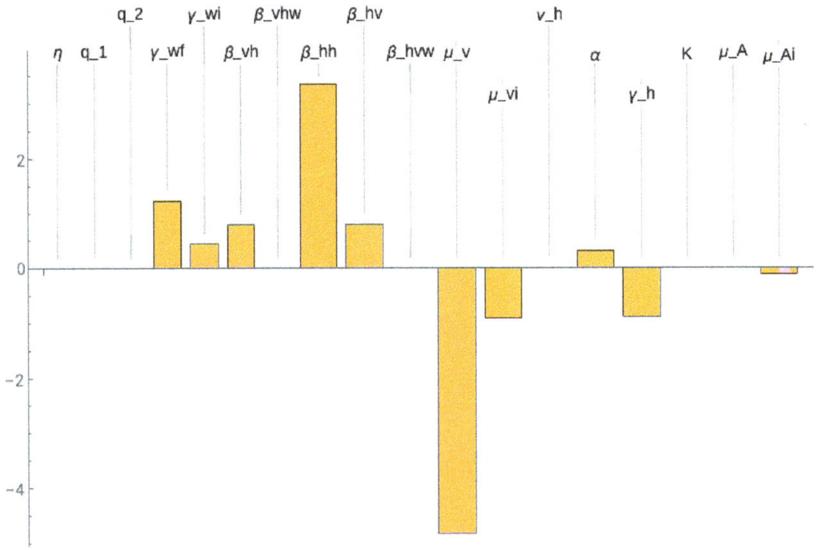

Fig. 1.14. Elasticity of R_{0Z}^{NG}, Zika reproduction number in presence of both types of mosquitoes with respect to parameters.

1.6 Non-Autonomous Model Simulations

In this section, we study the effects of seasonal variations on the spread of Zika and *Wolbachia* infection. Since the mosquito population varies periodically, we introduce a seasonal variation into the birth rates, transitioning rates, and death rates of mosquitoes. The egg laying rates are not constant any more. We assume $\eta(t)$, $q_1(t)$, $q_2(t)$ to be periodically forced. These periodic functions will take larger values during the wet season and smaller values during the dry season. In particular, we take the egg laying rate of *Wolbachia* free females to have the form $\eta(t) = 11\sin(2\pi(t-91)/365) + 13$. This function assumes a 365-day period and has an amplitude of 11, vertical shift of 13, and phase shift of 91 days. Note that the choice of this simple sinusoidal function assures that $\eta(t) \geq 0$ for all time. Similarly, the egg laying rates of *Wolbachia*-infected females when mating with wild males, $q_1(t)$, and when mating with *Wolbachia*-infected males, $q_2(t)$, have the following form $q_1(t) = q_2(t) = 9\sin(2\pi(t-91)/365) + 11$. The lower amplitude and vertical shift ensure that *Wolbachia*-infected females lay fewer eggs.

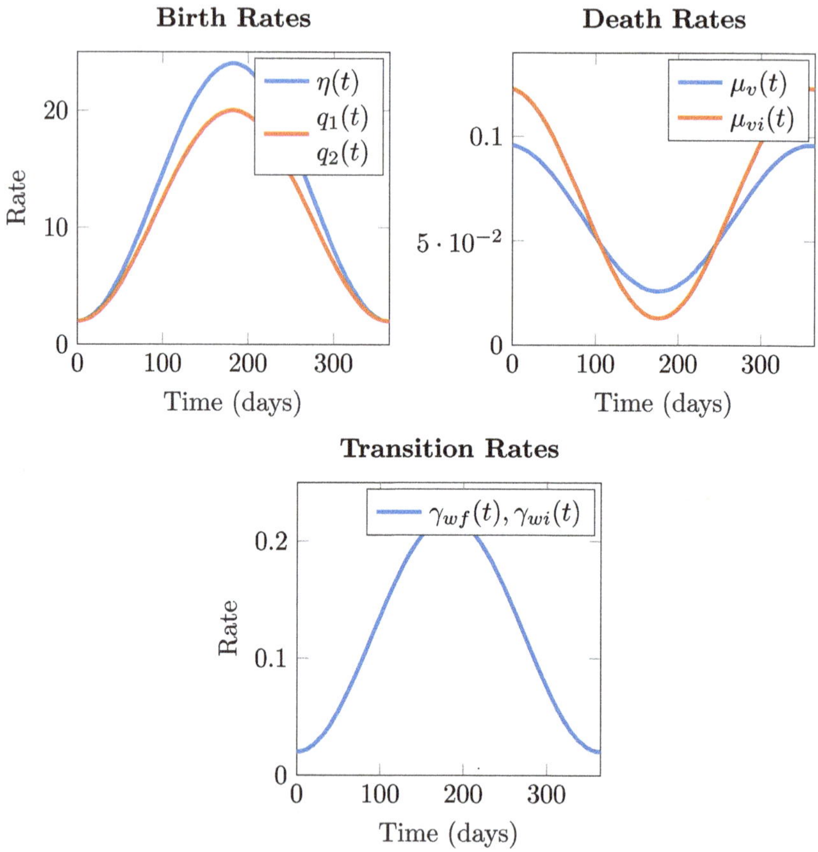

Fig. 1.15. Seasonality of rates.

The periodic death rate of wild mosquitoes has the form $\mu_{\mathrm{v}} = 0.035 \sin(2\pi t/365 - 91(1.0172)) + 0.061$. The amplitude is chosen to be 0.035, the average of the range found in literature for the death rate of *A. aegypti* mosquitoes, and the vertical shift is chosen to be 0.061, the baseline value used in the first model with no seasonality. Similarly, the periodic death rates of *Wolbachia*-infected mosquitoes are given in the form $\mu_{\mathrm{vi}} = 0.055 \sin(2\pi t/365 - 91(1.0172)) + 0.068$, where the higher amplitude and vertical shift account for the higher death rate of *Wolbachia*-infected mosquitoes. As we can observe in Figure 1.15, in the months with higher temperatures, the conditions are good for the mosquitoes to thrive and multiply and the death rates are lower.

The temperature of the environment also affects how the immature stages of *A. aegypti* mosquitoes develop. The lower temperature threshold for immature stages to develop is 16°C, while 34°C is the upper limit.[54] The development time from the aquatic stage to adults was shorter at higher temperatures (30°C vs 21°C) and density and food availability played an important role. The time taken by larvae to complete their development was optimal at 32°C and that mortality was significant at 14°C and 38°C.[74] The highest temperature at which development fully occurred was 36°C. Immature stages were found to survive short exposure to temperatures up to 45°C.[75] Thus, it is important to assume that the transitioning rates from aquatic stage to adult are periodic.

We let γ_{wf} and γ_{wi} have the following form $\gamma_{\text{wf}} = \gamma_{\text{wi}} = 0.1\sin\left(2\pi(t-91)/365\right) + 0.12$.

As seen in Figure 1.15, the transitioning rates have a similar shape as the egg laying rates with an amplitude of 0.1, which is the average of the range of values for the transitioning rates found in literature. Even though *Wolbachia*-infected aquatic stage mosquitoes might take slightly longer to develop compared to wild mosquitoes, the change is so small that we allow both transitioning rates (wild and *Wolbachia*-infected) to have the same form.

When we allow for the egg laying rates, death rates, and transitioning rates, to depend explicitly on time, the model becomes nonautonomous. Nonautonomous models are harder to analyze theoretically, so we will only perform numerical simulations for the nonautonomous model. Our goal is to get some insight on how the seasonality of the rates affects *Wolbachia* spread and diseases dynamics when we simulate over a span of 3 years.

1.6.1 *Disease free simulations for nonautonomous model*

The nonautonomous model is simulated using periodicity for the egg laying rates, transitioning rates from aquatic stage to adults, and for death rates. All other parameters remain constant and the initial conditions stay the same as in Section 1.4.1.1. When the parameters that remain constant take the baseline values from Tables 1.3 and 1.4 and the periodic parameters take the form mentioned earlier, we illustrate the effect of the introduction of *Wolbachia*-infected mosquitoes through numerical simulations. The initial conditions used are the same as in Section 1.4.1.1 and listed in Table 1.6.

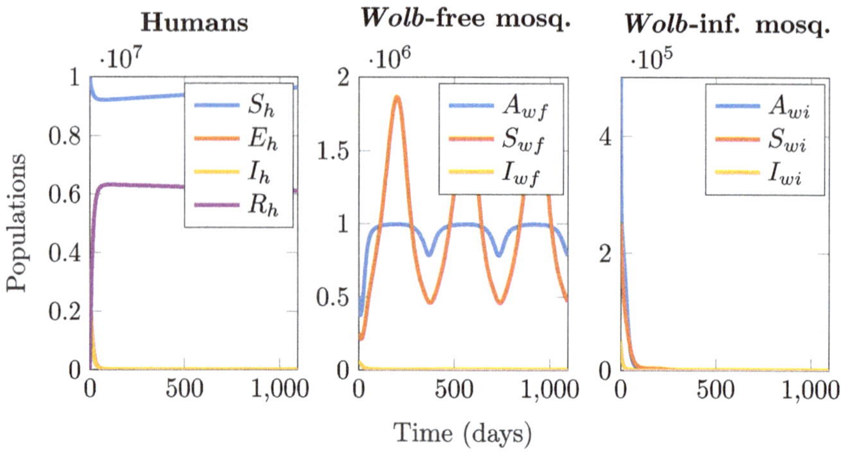

Fig. 1.16. Dominance of wild mosquitoes when initial conditions are the same as in Table 1.6.

1.6.1.1 *Zika takes longer to be eradicated when seasonality is included*

As we see in Figure 1.16, *Wolbachia*-free mosquitoes persist and the disease is eradicated in both humans and mosquitoes in approximately 350 days, a little longer than in the case where seasonality was not taken into account. *Wolbachia*-free female mosquitoes population oscillates between 500,000 and 1,750,000 compared to settling around 900,000 in the non-seasonal scenario.

1.6.1.2 *Wolbachia needs a higher threshold value to establish under seasonality*

Changing the initial conditions exactly like we did in Section 1.4.1.2 does not cause the *Wolbachia*-infected mosquitoes to persist. In the case where all parameters were constant, changing the initial amount of *Wolbachia*-infected susceptible female mosquitoes from $S_{\text{wi}}(0) = 250{,}000$ to $S_{\text{wi}}(0) = 850{,}000$ allowed the *Wolbachia* mosquitoes to persist. In the case where seasonality was taken into account, numerical simulations suggest that one would need to increase the *Wolbachia*-infected susceptible female mosquitoes to $S_{\text{wi}}(0) = 1.85$ millions to allow for *Wolbachia*-infected mosquitoes to persist as

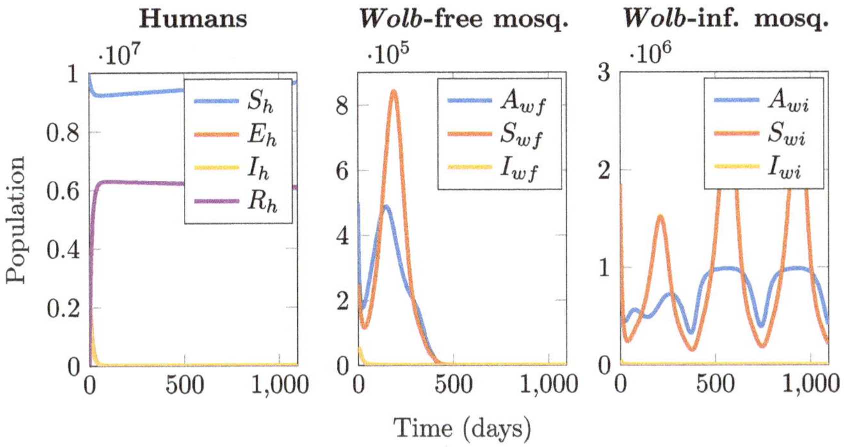

Fig. 1.17. Dominance of *Wolbachia*-infected mosquitoes when we start with 7.4 times as many *Wolbachia*-infected susceptible females compared to wild susceptible females.

observed in Figure 1.17. The elimination of wild mosquitoes takes around one year with seasonality taken into account and without seasonality. However, with seasonality, the number of *Wolbachia*-infected susceptible females oscillates yearly, with 2,750,000 females during the warm seasons compared to 800,000 *Wolbachia*-infected females without seasonality. Also, in the case of seasonality it takes a little longer for Zika to be eradicated in both humans and mosquitoes (around 300 days).

Simulations that involved releasing more *Wolbachia*-infected aquatic stage mosquitoes, *Wolbachia*-infected males, a combination of *Wolbachia*-infected aquatic stage mosquitoes and females and a combination of *Wolbachia*-infected aquatic stage mosquitoes and males in the periodic setting yield similar results. Persistence of *Wolbachia*-infected mosquitoes can be achieved when seasonality is taken into account using the same scenarios as in Sections 1.4.1.3–1.4.1.6, but it requires release of a higher number of *Wolbachia*-infected mosquitoes compared to wild mosquitoes. For example, one would need 3.01 million of *Wolbachia*-infected aquatic stage mosquitoes to allow for *Wolbachia*-infected mosquitoes to persist when seasonality is included compared to 1.85 millions when all parameters were constant as seen in Section 1.4.1.3. This is a little over six times as many *Wolbachia*-infected aquatic stage mosquitoes compared with *Wolbachia*-free aquatic stage mosquitoes.

1.6.2 *Disease present simulations for nonautonomous model*

Next we run numerical simulations using the initial conditions listed in Table 1.7 and baseline value for the parameters listed in Tables 1.3 and 1.4. Furthermore, we change the biting rate of the mosquitoes to $b = 1.25$ and the carrying capacity of the aquatic stage to $K = 10^9$.

1.6.2.1 *Zika takes longer to peak when Wolbachia infection is not established under seasonality*

As we can see in Figure 1.18, the wild mosquitoes dominate, *Wolbachia*-infected mosquitoes are eliminated in 300 days. The number of wild susceptible females mosquitoes oscillates between 460 millions and 1.8 billion. Zika-infected humans reach a peak of around 1.5 million in 50 days, which represents 15% of the total population. In the autonomous model, the peak was reached in 25 days and it has around 1.8 million infected humans (about 18% of the total population) as seen in Section 1.4.2.1.

The insets in Figure 1.18 show that the number of Zika-infected humans oscillates between 2,500 and 5,000. Zika-infected wild mosquitoes reach 100 millions and then oscillate approximately between 600,000 and 9.7 millions.

1.6.2.2 *Zika takes longer to be eradicated when Wolbachia is established with seasonality*

Changing only the initial conditions from $A_{\mathrm{wf}}(0) = 2{,}500{,}000$ to $A_{\mathrm{wf}}(0) = 500{,}000$ and $A_{\mathrm{wi}}(0) = 500{,}000$ to $A_{\mathrm{wi}}(0) = 1{,}500{,}000$, we see in Figure 1.19 that the *Wolbachia*-infected mosquitoes persist and the number of *Wolbachia*-infected susceptible females oscillates between 185 millions and 2.7 billions. Wild mosquitoes are eliminated in roughly one year. The disease is eradicated in humans in 400 days, compared to 300 days in the autonomous model Section 1.4.2.2, and in 422 days in *Wolbachia*-infected mosquitoes as observed in

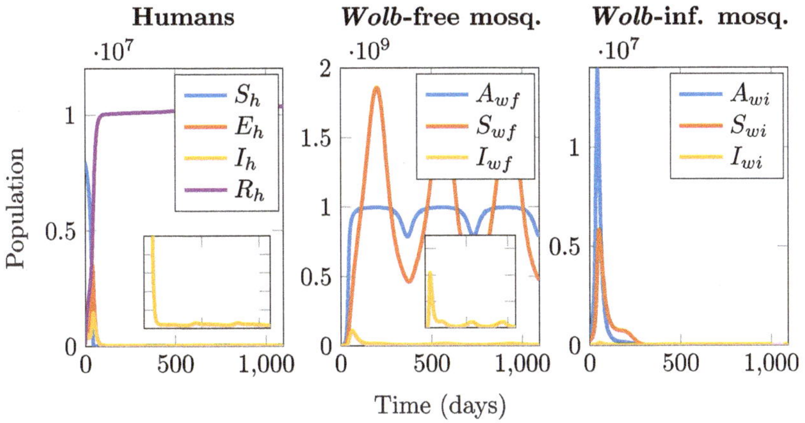

Fig. 1.18. Wild mosquitoes dominate and Zika is endemic in humans and wild mosquitoes. The peak of Zika-infected humans is reached in 50 days and it has around 1.5 million infected. Zika is endemic in humans and wild mosquitoes when seasonality is included.

Figure 1.19. This suggests that when seasonality is included, it takes longer for Zika to be eradicated in both humans and mosquitoes.

However, the number of Zika-infected humans first decline, and then reach a peak of around 300,000 infected in 100 days. This value is 3 times less than in the case where seasonality was not considered. Also, the number of total Zika-infected *Wolbachia* mosquitoes reaches a peak of 34 millions, compared to 14 millions. Therefore, even though the disease takes longer to be eliminated and there are more *Wolbachia*-infected mosquitoes, the number of Zika-infected humans is much lower in the case where seasonality is considered.

1.6.2.3 *The peak of Zika is smaller when seasonality is included and Wolbachia-infected vectors are more competent*

Next, we investigate the case where the *Wolbachia*-infected mosquitoes are less capable of blocking the virus. Increasing the competence of *Wolbachia*-infected mosquitoes allows for the disease to persist even when *Wolbachia*-infected mosquitoes dominate as seen in Figure 1.20. When seasonality is taken into account, we see that the number of Zika-infected humans reaches a peak of 1.3 million

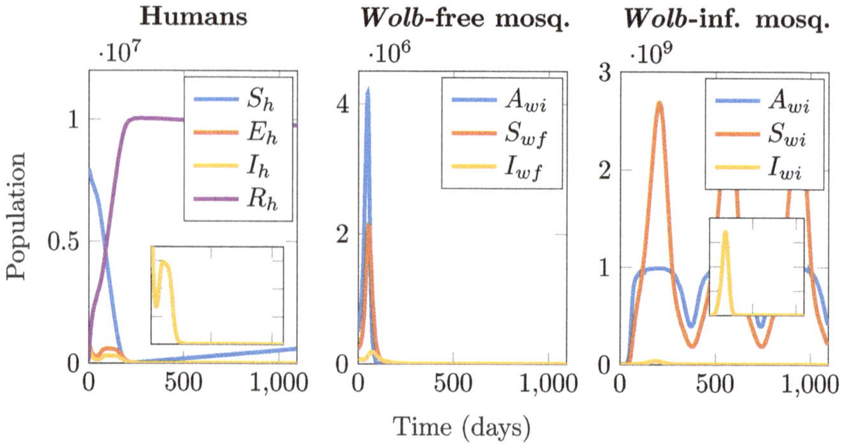

Fig. 1.19. *Wolbachia*-infected mosquitoes dominate and Zika takes longer to be eradicated under seasonality. Zika eradicated in humans in approximately 400 days when seasonality is included compared to 300 days in the autonomous model. Zika takes around 422 days to be eradicated in *Wolbachia*-infected mosquitoes under seasonality.

compared to 1.8 million in the autonomous model. Wild mosquitoes are eliminated in approximately one year and *Wolbachia*-infected susceptible females oscillate between 188 millions and 2.7 billions.

The insets in Figure 1.20 show that the Zika-infected human population goes down to a very small number (around 12) and then up to 30,000 infected. The peak of Zika-infected *Wolbachia*-infected mosquitoes is reached at around 70 millions and then oscillates between 1,300 and 30 millions as seen in Figure 1.20. In the autonomous model, the Zika-infected *Wolbachia*-infected mosquitoes reached a peak of 110 millions and then settled around 850,000.

Simulations for the nonautonomous model suggest that *Wolbachia* persistence can be achieved using similar release strategies but involve a larger number of initial *Wolbachia*-infected mosquitoes than in the case where no seasonality was taken into account. Also, in the nonautonomous case the disease takes longer to be eradicated in both humans and mosquitoes and in general there are many more mosquitoes present when seasonality is accounted for. Zika can stay endemic in the human and the dominant mosquito population over multiple years. Again, for realistic parameter values we were not able to observe coexistence of both mosquitoes.

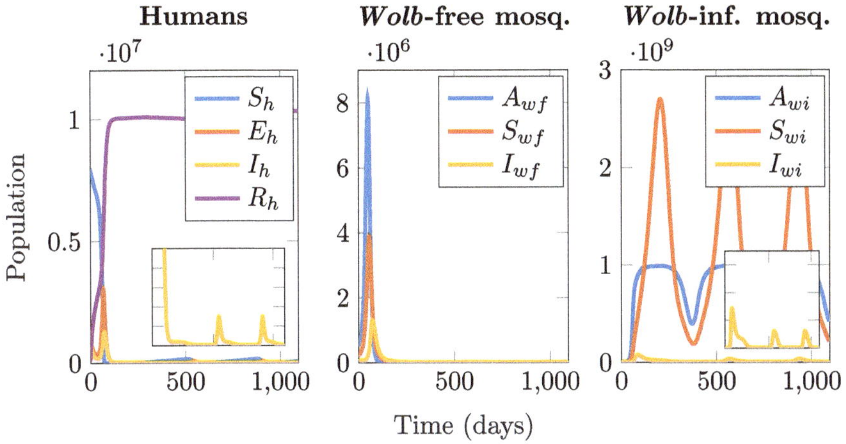

Fig. 1.20. *Wolbachia*-infected mosquitoes dominate and Zika is eradicated. Zika-infected humans reach a peak of 1.3 million compared to 1.8 million in the autonomous model. Zika is endemic in humans and *Wolbachia* mosquitoes when *Wolbachia* vectors are more competent.

1.7 Discussion

Wolbachia has been proven to be a powerful, environmentally friendly tool for controlling arboviral diseases including Zika. Mathematical models for transmission of Zika in presence of *Wolbachia* infection can be useful in providing better insights into the behavior of the disease and to help policymakers with the decision process regarding control and intervention strategies. A number of mathematical models for dengue transmission in presence of *Wolbachia*-infected mosquitoes exist, however, only a few mathematical models for Zika transmission with an underlying *Wolbachia*-infected mosquito population exist. We investigate a deterministic epidemic model of Zika with two types of vectors, wild mosquitoes and *Wolbachia*-infected mosquitoes. Similar to previous dengue-*Wolbachia* models, our model has an aquatic stage for each type of mosquito and an SEIR model for the human population. Unlike other models,[72] we chose an SI model for the female mosquitoes instead of SEI and included male mosquitoes. We also allow for the ratio between male and female mosquitoes to be different than a half. The model accounts for mating between mosquitoes, assumes complete cytoplasmic incompatibility, and allows for different parameters related to vector-borne transmission. Many of the dengue-*Wolbachia* models include the life cycle of the mosquitoes by having compartments for egg, larvae, and pupae grouped into one single compartment often denoted as an aquatic stage. The model in Ref. 40 has different compartments for eggs, pupae, and larvae, and female mosquitoes are split into two compartment: young immature females who do not mate and a mature female. They have separate models for *Wolbachia* infection within the mosquito population which includes complete cytoplasmic incompatibility and perfect maternal transmission of *Wolbachia* infection and then couple that model with the Dietz–Bailey model of dengue. Despite the extra compartments in Ref. 40, their results of the *Wolbachia* infection among the mosquitoes are very similar to ours. Two equilibria are asymptotically stable: an equilibrium where all the mosquito population is *Wolbachia*-free and an equilibria where all the mosquitoes are *Wolbachia*-infected. A third unstable equilibrium exists. This model however does not explore many numerical simulations or different release strategies nor does it look at the effects of seasonality.

The stability analysis of our model along with numerical simulations and sensitivity analysis show that once the *Wolbachia*-infected mosquitoes get established, they can ultimately dominate the mosquito population. The results show that the persistence of *Wolbachia* infection within the mosquito population can be achieved by multiple release strategies. The persistence of *Wolbachia*-infected mosquitoes can be achieved by releasing *Wolbachia*-infected females, *Wolbachia*-infected males, or a combination of both. This result agrees with other findings in the literature[43] that the *Wolbachia* infection persists if the ratio of the release rate of *Wolbachia*-infected mosquitoes over the number of wild mosquitoes at the initial state is large enough. The impact on Zika transmission in the human and mosquito population using the strategy of releasing *Wolbachia*-infected male mosquitoes was also explored in Ref. 44. Their findings that releasing *Wolbachia*-infected mosquitoes causes the disease to die out faster and the number of infected humans to decrease agree with our results. Also, in Ref. 44, numerical simulations show that introduction of *Wolbachia*-infected mosquitoes leads to a significant decrease in susceptible mosquito population and elimination of wild mosquitoes. The model in Ref. 44 does not have a compartment for *Wolbachia*-infected aquatic stage mosquitoes.

By including a compartment for *Wolbachia*-infected aquatic stage mosquitoes, we were able to investigate the release strategy that allows for investigating the effect of releasing *Wolbachia*-infected aquatic stage mosquitoes. Our findings suggest that persistence of *Wolbachia* infection can also be achieved by releasing *Wolbachia*-infected aquatic stage or by first reducing the existing *Wolbachia*-free aquatic stage and then releasing *Wolbachia*-infected aquatic stage mosquitoes.

Sensitivity analysis reveals that the basic reproduction number in presence of both types of mosquitoes is most sensitive to the death rate of adult mosquitoes and the direct transmission, results that match prior expectations as in Ref. 44. Moreover, we have concluded that the presence of *Wolbachia*-infected mosquitoes switches the control strategies. When no *Wolbachia*-infected mosquitoes are present, the control should be targeted toward decreasing the lifespan of the mosquitoes and when *Wolbachia*-infected mosquitoes are present, the efforts should be concentrated on treatment of humans to increase the recovery rate and sexual transmission control to decrease the direct transmission rate. Our findings agree with the conclusion in Ref. 44 that mitigation strategies should focus on both the mosquito-human transmission and direct transmission, but they are more specific to when each strategy is more efficient.

None of the previous Zika-*Wolbachia* models have studied the effect of environmental fluctuations of temperature. Since the transmission of Zika virus is affected by periodic seasonality, we explored that by including seasonal variation in the temperature-dependent parameters (egg laying rates, mosquito death rates, and transitioning rates). This lead to the model giving a more detailed outcome information about the mosquito population dynamics as temperature is assumed to vary. By including seasonality, we expect that the model would give a more practical guide for the case when the release process spans multiple seasons. When seasonality is considered, we find that the persistence of *Wolbachia* is still achieved by the same release strategies, but it requires a higher initial number of *Wolbachia*-infected released mosquitoes. From these results we can say that models that assume constant conditions produce threshold conditions that are less reliable when oscillations are taken into account and thus play an important role when releasing *Wolbachia* in seasonally varying climates. In both cases when seasonality is considered or not, numerical simulations suggest that if the *Wolbachia* strain does lose its strength to block virus proliferation within the mosquito, then even with *Wolbachia*-infected mosquitoes dominating, the disease can stay endemic.

Future work will include developing models for vector control methods that investigate the impact of releasing mosquitoes infected with two different *Wolbachia* strains (wAu and wMel) in a human population infected with Zika and explore how single or combined strategies will impact the disease dynamics. We are interested in understanding how different features of *Wolbachia* infection, such as non-induction of CI, high maintenance of the *Wolbachia* infection at high temperature (present in wAu strain), and loss of *Wolbachia* infection, imperfect maternal transmission (wMel) in mosquitoes could drive a reduction in Zika and other arboviral diseases transmission. These findings will further contribute to the effort of reducing or eliminating arboviral transmission throughout the world.

Acknowledgments

The authors would like to thank James Keesling for insights leading to the model, the reviewers for suggested improvements, Sergei Pilyugin for some corrections, and Nick Sizemore for assistance with the MATLAB code.

References

1. Enserink, M. (2008). A mosquito goes global. https://science.sciencemag.org/content/320/5878/864.
2. Schuffenecker, I., Iteman, I., Michault, A., Murri, S., Frangeul, L., Vaney, M.-C., Lavenir, R., Pardigon, N., Reynes, J.-M., Pettinelli, F., *et al.*, (2006). Genome microevolution of chikungunya viruses causing the Indian Ocean outbreak, *PLoS Med.* **3**, 7, e263.
3. Pattnaik, A., Sahoo, B. R., and Pattnaik, A. K. (2020). Current status of Zika virus vaccines: Successes and challenges, *Vaccines* **8**, 2, p. 266.
4. Dick, G., Kitchen, S., and Haddow, A. (1952). Zika virus (i). isolations and serological specificity, *Trans. Royal Soc. Trop. Med. Hyg.* **46**, 5, pp. 509–520.
5. Duffy, M. R., Chen, T.-H., Hancock, W. T., Powers, A. M., Kool, J. L., Lanciotti, R. S., Pretrick, M., Marfel, M., Holzbauer, S., Dubray, C.,

et al., (2009). Zika virus outbreak on yap island, federated states of micronesia, *N. Engl. J. Med.* **360**, 24, pp. 2536–2543.

6. Zanluca, C., Melo, V. C. A. D., Mosimann, A. L. P., Santos, G. I. V. D., Santos, C. N. D. D., and Luz, K. (2015). First report of autochthonous transmission of Zika virus in Brazil, *Mem. Inst. Oswaldo Cruz.* **110**, 4, pp. 569–572.
7. Centers for Disease Control and Prevention. (2020). 2020 case counts in the us. https://www.cdc.gov/zika/reporting/2020-case-counts.html. Accessed 22 February 2022.
8. Grubaugh, N. D., Ishtiaq, F., Setoh, Y. X., and Ko, A. I. (2019). Misperceived risks of Zika-related microcephaly in India, *Trends Microbiol.* **27**, 5, pp. 381–383.
9. de Oliveira, W. K., Cortez-Escalante, J., De Oliveira, W. T. G. H., Carmo, G. M. I. d., Henriques, C. M. P., Coelho, G. E., and de França, G. V. A. (2016). Increase in reported prevalence of microcephaly in infants born to women living in areas with confirmed Zika virus transmission during the first trimester of pregnancy — Brazil, 2015, *Morb. Mort. Weekly Rep.* **65**, 9, pp. 242–247.
10. Styczynski, A. R., Malta, J. M., Krow-Lucal, E. R., Percio, J., Nóbrega, M. E., Vargas, A., Lanzieri, T. M., Leite, P. L., Staples, J. E., Fischer, M. X., *et al.* (2017). Increased rates of Guillain-Barré syndrome associated with Zika virus outbreak in the Salvador metropolitan area, Brazil, *PLoS Neglect. Trop. Dis.* **11**, 8, e0005869.
11. Petersen, E. E., Meaney-Delman, D., Neblett-Fanfair, R., Havers, F., Oduyebo, T., Hills, S. L., Rabe, I. B., Lambert, A., Abercrombie, J., Martin, S. W., *et al.* (2016). Update: Interim guidance for preconception counseling and prevention of sexual transmission of Zika virus for persons with possible Zika virus exposure — United States, September 2016, *Morb. Mort. Weekly Rep.* **65**, 39, pp. 1077–1081.
12. Mead, P. S., Duggal, N. K., Hook, S. A., Delorey, M., Fischer, M., Olzenak McGuire, D., Becksted, H., Max, R. J., Anishchenko, M., Schwartz, A. M., *et al.* (2018). Zika virus shedding in semen of symptomatic infected men, *N. Engl. J. Med.* **378**, 15, pp. 1377–1385.
13. Brasil, P., Pereira, J. P. Jr., Moreira, M. E., Ribeiro Nogueira, R. M., Damasceno, L., Wakimoto, M., Rabello, R. S., Valderramos, S. G., Halai, U.-A., Salles, T. S., *et al.* (2016). Zika virus infection in pregnant women in Rio de Janeiro, *N. Eng. J. Med.* **375**, 24, pp. 2321–2334.
14. Pomar, L., Vouga, M., Lambert, V., Pomar, C., Hcini, N., Jolivet, A., Benoist, G., Rousset, D., Matheus, S., Malinger, G., *et al.* (2018). Maternal-fetal transmission and adverse perinatal outcomes in pregnant women infected with Zika virus: Prospective cohort study in French Guiana, *BMJ.* **363**, p. k4431.

15. Bloch, E. M., Ness, P. M., Tobian, A. A., and Sugarman, J. (2018). Revisiting blood safety practices given emerging data about Zika virus. *N. Engl. J. Med.*, **378**, 19, pp. 1837–1841.
16. Mann, T. Z., Haddad, L. B., Williams, T. R., Hills, S. L., Read, J. S., Dee, D. L., Dziuban, E. J., Pérez-Padilla, J., Jamieson, D. J., Honein, M. A., *et al.* (2018). Breast milk transmission of flaviviruses in the context of Zika virus: A systematic review, *Paediat. Perinat. Epidemiol.* **32**, 4, pp. 358–368.
17. Yen, J. H. and Barr, A. R. (1971). New hypothesis of the cause of cytoplasmic incompatibility in *Culex Pipiens* L., *Nature* **232**, 5313, pp. 657–658.
18. Jiggins, F. M. (2017). The spread of *Wolbachia* through mosquito populations, *PLoS Biol.* **15**, 6, p. e2002780.
19. Branca, A., Vavre, F., Silvain, J.-F., and Dupas, S. (2009). Maintenance of adaptive differentiation by *Wolbachia* induced bidirectional cytoplasmic incompatibility: The importance of sib-mating and genetic systems, *BMC Evolut. Biol.* **9**, 1, pp. 1–13.
20. Ross, P. A., Ritchie, S. A., Axford, J. K., and Hoffmann, A. A. (2019). Loss of cytoplasmic incompatibility, in *Wolbachia*-infected *Aedes aegypti* under field conditions, *PLoS Neglect. Trop. Dis.* **13**, 4, p. e0007357.
21. Tan, C. H., Wong, P. J., Li, M. I., Yang, H., Ng, L. C., and O'Neill, S. L. (2017). wMel limits Zika and chikungunya virus infection in a Singapore *Wolbachia*-introgressed *A. aegypti* strain, wMel-Sg, *PLoS Neglect. Trop. Dis.* **11**, 5, p. e0005496.
22. Ant, T. H., Herd, C. S., Geoghegan, V., Hoffmann, A. A., and Sinkins, S. P. (2018). The *Wolbachia* strain wAu provides highly efficient virus transmission blocking in *Aedes aegypti*, *PLoS Pathogens.* **14**, 1, p. e1006815.
23. Hoffmann, A. A., Ross, P. A., and Rašić, G. (2015). *Wolbachia* strains for disease control: Ecological and evolutionary considerations, *Evol. Appl.* **8**, 8, pp. 751–768.
24. Dickens, B. L., Yang, J., Cook, A. R., and Carrasco, L. R. (2016). Time to empower release of insects carrying a dominant lethal and *Wolbachia* against Zika, in *Open Forum Infectious Diseases*, Vol. 3, 2, Oxford University Press, Oxford, UK, p. ofw103.
25. Wiratsudakul, A., Suparit, P., and Modchang, C. (2018). Dynamics of Zika virus outbreaks: An overview of mathematical modeling approaches, *PeerJ.* **6**, p. e4526.
26. Funk, S., Kucharski, A. J., Camacho, A., Eggo, R. M., Yakob, L., Murray, L. M., and Edmunds, W. J. (2016). Comparative analysis of dengue and Zika outbreaks reveals differences by setting and virus, *PLoS Neglect. Trop. Dis.* **10**, 12, p. e0005173.

27. Champagne, C., Salthouse, D. G., Paul, R., Cao-Lormeau, V.-M., Roche, B., and Cazelles, B. (2016). Structure in the variability of the basic reproductive number (r0) for Zika epidemics in the Pacific Islands, *Elife* **5**, p. e19874.
28. Kucharski, A. J., Funk, S., Eggo, R. M., Mallet, H.-P., Edmunds, W. J., and Nilles, E. J. (2016). Transmission dynamics of Zika virus in island populations: A modelling analysis of the 2013–14 French polynesia outbreak, *PLoS Neglect. Trop. Dis.* **10**, 5, p. e0004726.
29. Gao, D., Lou, Y., He, D., Porco, T. C., Kuang, Y., Chowell, G., and Ruan, S. (2016). Prevention and control of Zika as a mosquito-borne and sexually transmitted disease: A mathematical modeling analysis, *Sci. Rep.* **6**, p. 28070.
30. Maxian, O., Neufeld, A., Talis, E. J., Childs, L. M., and Blackwood, J. C. (2017). Zika virus dynamics: When does sexual transmission matter? *Epidemics* **21**, pp. 48–55.
31. Tuncer, N., Marctheva, M., LaBarre, B., and Payoute, S. (2018). Structural and practical identifiability analysis of Zika epidemiological models, *Bull. Math. Biol.* **80**, 8, pp. 2209–2241.
32. Alzahrani, E. O., Ahmad, W., Khan, M. A., and Malebary, S. J. (1993). Optimal control strategies of Zika virus model with mutant, *Commun. Nonlinear Sci. Numer. Simulat.* **93**, p. 105532.
33. González-Parra, G., Díaz-Rodríguez, M., and Arenas, A. J. (2020). Optimization of the controls against the spread of Zika virus in populations, *Computation* **8**, 3, p. 76.
34. Caspari, E. and Watson, G. (1959). On the evolutionary importance of cytoplasmic sterility in mosquitoes, *Evolution* **13**, 4, pp. 568–570.
35. Turelli, M. and Hoffmann, A. A. (1991). Rapid spread of an inherited incompatibility factor in California Drosophila, *Nature* **353**, 6343, pp. 440–442.
36. Schofield, P. (2002). Spatially explicit models of Turelli-Hoffmann *Wolbachia* invasive wave fronts, *J. Theor. Biol.* **215**, 1, pp. 121–131.
37. Turelli, M. and Barton, N. H. (2017). Deploying dengue-suppressing *Wolbachia*: Robust models predict slow but effective spatial spread in *Aedes aegypti*, *Theoret. Popul. Biol.* **115**, pp. 45–60.
38. Hughes, H. and Britton, N. F. (2013). Modelling the use of *Wolbachia* to control dengue fever transmission, *Bull. Mathem. Biol.* **75**, 5, pp. 796–818.
39. Ndii, M. Z., Hickson, R. I., and Mercer, G. N. (2012). Modelling the introduction of *Wolbachia* into *Aedes aegypti* mosquitoes to reduce dengue transmission, *ANZIAM J.* **53**, 3, pp. 213–227.
40. Koiller, J., Da Silva, M., Souza, M., Codeço, C., Iggidr, A., and Sallet, G. (2014). Aedes, Wolbachia and dengue. PhD thesis, Inria Nancy-Grand Est, Villers-lès-Nancy, France.

41. Dorigatti, I., McCormack, C., Nedjati-Gilani, G., and Ferguson, N. M. (2018). Using *Wolbachia* for dengue control: Insights from modelling, *Trends Parasitol.* **34**, 2, pp. 102–113.
42. Ogunlade, S. T., Meehan, M. T., Adekunle, A. I., Rojas, D. P., Adegboye, O. A., and McBryde, E. S. (2021). A review: Aedes-borne arboviral infections, controls and *Wolbachia*-based strategies, *Vaccines* **9**, 1, p. 32.
43. Wang, L., Zhao, H., Oliva, S. M., and Zhu, H. (2017). Modeling the transmission and control of Zika in Brazil, *Sci. Rep.* **7**, 1, pp. 1–14.
44. Xue, L., Cao, X., and Wan, H. (2021). Releasing *Wolbachia*-infected mosquitoes to mitigate the transmission of Zika virus, *J. Math. Anal. Appl.* **496**, 1, p. 124804.
45. Gage, K. L., Burkot, T. R., Eisen, R. J., and Hayes, E. B. (2008). Climate and vectorborne diseases, *Am. J. Prev. Med.* **35**, 5, pp. 436–450.
46. Sutherst, R. W. (2004). Global change and human vulnerability to vector-borne diseases, *Clin. Microbiol. Rev.* **17**, 1, pp. 136–173.
47. Reeves, W. C., Hardy, J. L., Reisen, W. K., and Milby, M. M. (1994). Potential effect of global warming on mosquito-borne arboviruses, *J. Med. Entomol.* **31**, 3, pp. 323–332.
48. Reiter, P. (2001). Climate change and mosquito-borne disease, *Environ. Health Perspect.* **109**, suppl 1, pp. 141–161.
49. Reiter, P., Lathrop, S., Bunning, M., Biggerstaff, B., Singer, D., Tiwari, T., Baber, L., Amador, M., Thirion, J., Hayes, J. *et al.* (2003). Texas lifestyle limits transmission of dengue virus, *Emerg. Infect. Dis.* **9**, 1, p. 86.
50. Chretien, J.-P., Anyamba, A., Bedno, S. A., Breiman, R. F., Sang, R., Sergon, K., Powers, A. M., Onyango, C. O., Small, J., Tucker, C. J., *et al.* (2007). Drought-associated chikungunya emergence along coastal east Africa, *Am. J. Trop. Med. Hyg.* **76**, 3, pp. 405–407.
51. World Health Organization (2009). Dengue guidelines for diagnosis, treatment, prevention and control: New edition. World Health Organization. https://apps.who.int/iris/handle/10665/44188.
52. Marchoux, E., Salimbeni, A. T., and Simond, P.-L. (1903). La fièvre jaune: rapport de la mission française. Impr. Charair.
53. Connor, M. E. (1924). Suggestions for developing a campaign to control yellow fever1, *Am. J. Trop. Med. Hyg.* **1**, 3, pp. 277–307.
54. Christophers, S. *et al.* (1960). *Aedes aegypti (*L.*) The Yellow Fever Mosquito: Its Life History, Bionomics and Structure.* The Syndics of the Cambridge University Press, London.
55. C. for Disease Control, Prevention, *et al.* (2018). Estimated potential range of *A. aegypti* and aedes albopictus in the United States. https://www.cdc.gov/mosquitoes/mosquito-control/professionals/range.html.

56. Chang, C., Ortiz, K., Ansari, A., and Gershwin, M. E. (2016). The Zika outbreak of the 21st century, *J. Autoimmun.* **68**, pp. 1–13.
57. Foy, B. D., Kobylinski, K. C., Foy, J. L. C., Blitvich, B. J., da Rosa, A. T., Haddow, A. D., Lanciotti, R. S., and Tesh, R. B. (2011). Probable non–vector-borne transmission of Zika virus, Colorado, USA, *Emerg. Infect. Dis.* **17**, 5, p. 880.
58. Tabata, T., Petitt, M., Puerta-Guardo, H., Michlmayr, D., Wang, C., Fang-Hoover, J., Harris, E., and Pereira, L. (2016). Zika virus targets different primary human placental cells, suggesting two routes for vertical transmission, *Cell Host Microbe* **20**, 2, pp. 155–166.
59. Dutra, H. L. C., Rocha, M. N., Dias, F. B. S., Mansur, S. B., Caragata, E. P., and Moreira, L. A. (2016). *Wolbachia* blocks currently circulating Zika virus isolates in Brazilian *Aedes aegypti* mosquitoes, *Cell Host Microbe* **19**, 6, pp. 771–774.
60. McMeniman, C. J., Lane, R. V., Cass, B. N., Fong, A. W., Sidhu, M., Wang, Y.-F., and O'Neill, S. L. (2009). Stable introduction of a life-shortening *Wolbachia* infection into the mosquito *Aedes aegypti*, *Science* **323**, 5910, pp. 141–144.
61. WHO (2018). Zika virus. https://www.who.int/en/news-room/fact-sheets/detail/zika-virus. Accessed 15 September 2019.
62. Walker, T., Johnson, P., Moreira, L., Iturbe-Ormaetxe, I., Frentiu, F., McMeniman, C., Leong, Y. S., Dong, Y., Axford, J., Kriesner, P., *et al.* (2011). The wMel *Wolbachia* strain blocks dengue and invades caged *Aedes aegypti* populations, *Nature* **476**, 7361, pp. 450–453.
63. Styer, L. M., Minnick, S. L., Sun, A. K., and Scott, T. W. (2007). Mortality and reproductive dynamics of *Aedes aegypti* (Diptera: Culicidae) fed human blood, *Vector-Borne Zoonotic Dis.* **7**, 1, pp. 86–98.
64. Hoffmann, A. A., Iturbe-Ormaetxe, I., Callahan, A. G., Phillips, B. L., Billington, K., Axford, J. K., Montgomery, B., Turley, A. P., and O'Neill, S. L. (2014). Stability of the wMel wolbachia infection following invasion into *Aedes aegypti* populations, *PLoS Neglect. Trop. Dis.* **8**, 9, p. e3115.
65. Xue, L., Manore, C. A., Thongsripong, P., and Hyman, J. M. (2017). Two-sex mosquito model for the persistence of *Wolbachia*, *J. Biol. Dyn.* **11**, suppl 1, pp. 216–237.
66. Arrivillaga, J. and Barrera, R. (2004). Food as a limiting factor for *Aedes aegypti* in water-storage containers, *J. Vector Ecol.* **29**, pp. 11–20.
67. Van den Driessche, P. and Watmough, J. (2002). Reproduction numbers and sub-threshold endemic equilibria for compartmental models of disease transmission, *Mathe. Biosci.* **180**, 1–2, pp. 29–48.
68. Martcheva, M. (2019) Methods for deriving necessary and sufficient conditions for backward bifurcation, *J. Biol. Dyn.* **13**, 1, pp. 538–566.

69. Brady, O. J., Kharisma, D. D., Wilastonegoro, N. N., O'Reilly, K. M., Hendrickx, E., Bastos, L. S., Yakob, L., and Shepard, D. S. (2020). The cost-effectiveness of controlling dengue in Indonesia using wMel *Wolbachia* released at scale: A modelling study, *BMC Med.* **18**, 1, pp. 1–12.
70. Indriani, C., Tantowijoyo, W., Rancès, E., Andari, B., Prabowo, E., Yusdi, D., Ansari, M. R., Wardana, D. S., Supriyati, E., Nurhayati, I., *et al.* (2020). Reduced dengue incidence following deployments of *Wolbachia*-infected *Aedes aegypti* in Yogyakarta, Indonesia: A quasi-experimental trial using controlled interrupted time series analysis, *Gates Open Res.* **4**, 50, p. 50.
71. Qu, Z., Xue, L., and Hyman, J. M. (2018). Modeling the transmission of *Wolbachia* in mosquitoes for controlling mosquito-borne diseases, *SIAM J. Appl. Math.* **78**, 2, pp. 826–852.
72. Ndii, M. Z., Hickson, R. I., Allingham, D., and Mercer, G. (2015). Modelling the transmission dynamics of dengue in the presence of wolbachia, *Math. Biosci.* **262**, pp. 157–166.
73. Martcheva, M. (2015). *An Introduction to Mathematical Epidemiology*, Vol. 61. Springer, New York, NY.
74. Bar-Zeev, M. (1958). The effect of temperature on the growth rate and survival of the immature stages of *Aedes aegypti* (L.), *Bull. Entomol. Res.* **49**, 1, pp. 157–163.
75. Bar-Zeev, M. (1957). The effect of extreme temperatures on different stages of *Aedes aegypti* (L.), *Bull. Entomol. Res.* **48**, 3, pp. 593–599.

https://doi.org/10.1142/9789811263033_0002

Chapter 2

Dynamics and Optimal Control of HIV Infection and Opioid Addiction

Archana Neupane Timsina[*,‡] **and Necibe Tuncer**[†,§]

[*]*Population Health and Pathobiology Department, North Carolina State University*

[†]*Mathematical Sciences Department, Florida Atlantic University, Florida, USA*

[‡]*aneupan@ncsu.edu*

[§]*ntuncer@fau.edu*

On the basis of the growing association between opioid addiction and HIV infection, we develop a compartmental model to study dynamics and optimal control of two epidemics; opioid addiction and HIV infection. We show that the disease-free equilibrium is locally asymptotically stable when the basic reproduction number $\mathcal{R}_0 = \max(\mathcal{R}_0^u, \mathcal{R}_0^v) < 1$, here $\mathcal{R}_0^v$ is the reproduction number of the HIV infection, and $\mathcal{R}_0^u$ is the reproduction number of the opioid addiction. The addiction-only boundary equilibrium exists when $\mathcal{R}_0^u > 1$ and it is locally asymptotically stable when the invasion number of the opioid addiction is $\mathcal{R}_{inv}^u < 1$. Similarly, HIV-only boundary equilibrium exists when $\mathcal{R}_0^v > 1$ and it is locally asymptotically stable when the invasion number of the HIV infection is $\mathcal{R}_{\text{inv}}^v < 1$. We study structural identifiability of the parameters, estimate parameters employing yearly reported data from Central for Disease Control and Prevention (CDC), and practical identifiability of estimated parameters. We observe the basic reproduction number $\mathcal{R}_0$ using the parameters. Next, we introduce four distinct controls in the HIV — opioid model for the sake of control approach, including treatment for addictions, health care education about not sharing syringes, highly active anti-retroviral therapy (HAART), and rehab treatment for opiate addicts who are HIV infected. We apply single control in the beginning and observe the results, this provides us better understanding of the influence of each single control in the HIV–opioid model. We then apply those four controls together in the HIV–opioid model and compare their results using different control bounds and state variable weights. We conclude the result by presenting several graphs.

Keywords: Reproduction number, invasion number, identifiability, optimal control

2.1 Introduction

Human Immunodeficiency Virus (HIV) affects the human immune system by destroying essential cells that fight diseases and infection.[6] Blood transfusions, unprotected sex, needle sharing, and maternal-fetal transmission are all ways for this virus to spread. According to the Centers for Disease Control and Prevention (CDC), HIV can be controlled with good medical care, but it is not possible to completely cure HIV.[3] When HIV-positive patients do not receive therapy, they usually go through three stages of disease progression: acute HIV infection, chronic HIV infection, and Acquired Immunodeficiency Syndrome (AIDS).[3] People with the first acute HIV infection stage have a lot of HIV in their blood. They are quite contagious.[4] Some people may show flu like symptoms but some may not. The time period of this stage is 2–3 weeks.[3] The second, chronic HIV infection stage, is also called Asymptomatic HIV infection or clinical latency stage. In this stage, the virus still exists in the human body, but it reproduces at a very low rate.[4] People may not have any symptoms or get sick during this phase. The time period of this stage may last several years.[3] The third, AIDS is the most severe phase of HIV. AIDS attacks CD4 cells, which are active components of the immune system in the human body.[3] As the CD4 cell count decreases, immunity is gradually lost, and people become susceptible to other diseases.[40]

Opioids are a class of drugs including heroin, synthetic opioids, and pain relievers available legally by prescription. These can include drugs such as fentanyl, oxycodone, hydrocodone, codeine, and morphine.[7] Unlike viral infections, opioid addiction can be spread through the influence of other people, such as friends, family, and society. Opioid crises are detrimental to society, families, and even countries. This addiction can stop people from becoming useful members of society, as well as limit their ability to learn and work. The opioid epidemic is being addressed by varieties of organizations, teams, and hospitals all around the world.

HIV infection is often found alongside other diseases. In the past three decades, significant research has been conducted related to the co-infection of HIV with multiple diseases.[15,24,27,31,45] Among those studies, HIV infection along with other diseases has been a topic of interesting research. Many people are both HIV infected and addicted to opioids[45,46] because the virus can be transmitted through needle sharing and unprotected sex. Previous studies[16,45,46] have performed epidemiological modeling of HIV and drug use. On the basis of the review of these articles,[15,22,33,40,41] we determined that HIV infection and opioid addiction can be controlled to some extent.

In this study, we first use the reproduction disease-free equilibrium (DFE) number described in similar work[45] to study the local stability of the dynamics, and then examine addicted only equilibrium (AOE) and HIV only equilibrium (HOE) for stability by using invasion numbers as mentioned in the article by Xi-Chao Duana, Xue-Zhi Lib, and Maia Martcheva.[45] We focus not only on coincident HIV infection and opioid usage, but also on four control strategies. We use optimal control theory to mathematically study how controls should be applied to achieve the best outcomes of state variables.[19] The idea here is that the controls enter the system of ordinary differential equations and affect the dynamics. In this chapter, four controls such as group treatment for opioid use, precautions in needle sharing, highly active anti-retroviral therapy (HAART) for HIV positive people, and treatment for opioid with infected people are applied in the model. The articles by W.H. Fleming *et al.*, A.F. Filippovi, A.M. Steinberg *et al.*, M. Vidyasagar, A. Mallela *et al.*, S. Lanhart *et al.*, M. Martcheva, and O. Prosper *et al.*[9,10,15,29,30,38,42,44] are reviewed to demonstrate the existence and uniqueness of the control function solutions. Fellipo-Cesari theorem and Pontryagin's Maximum Principle are the primary tools used to demonstrate the sufficient and necessary conditions for the existence and uniqueness of control functions.[29] For the numerical analysis, parameter estimation and identifiability are important, so we use Mathematica and Monte Carlo simulation for structural and practical identifiability of the model. Similarly, we use curve fitting optimization[29] for parameter estimation. The CDC and National Center for Health Statistics[1,2] reports of cases of HIV, AIDS, and opioid use from 1999 to 2018 in the US are used for the parameter estimation. The Runge–Kutta algorithm forward and backward sweep method[42] are used to find the solutions to ordinary differential equations. We present different sets of results such as either by considering single control applied cases or all four controls applied cases and then discuss the solutions of state variables comparing the graphs with controls verses without controls.

The organization of this chapter is as follows; Section 2.2 is for mathematical formulation and Section 2.3 is for model analysis, which analyzes equilibrium points and stability. A model with four control strategies is introduced in Section 2.4 which describes optimal control problem set up with existence and necessary conditions. Next, Section 2.5 is for parameter estimation and identifiability with population data of United States available from CDC. Section 2.6 presents numerical results of the optimal control problem, and we discuss our results in Section 2.7. Section 2.8 is allocated for the Appendix, where the Mathematica code for identifiability, optimal control problem setup for single applied controls, and some results of optimal controls are given.

2.2 Mathematical Formulation

We propose a compartmental model to study the dynamics of HIV infection and opioid addiction (Model 2.1). The model is developed by splitting the total population N into six non-intersecting classes; individuals who are susceptible to both HIV-infection and opioid addiction $S(t)$, addicted individuals $U(t)$, HIV infected individuals $V(t)$, addicted *individuals* who are in the first stage of HIV infection $I_1(t)$, addicted *individuals* who are in the second stage of HIV infection $I_2(t)$, and *individuals* who have developed AIDS $A(t)$. Thus,

$$N = S + U + V + I_1 + I_2 + A.$$

The recruitment rate of the population is denoted by Λ and the natural death rate by μ. A susceptible individual can get in contact with an HIV-infected individual, become infectious and move to the infected class V, or can get in contact with an addicted individual, become addicted and move into the addiction class U. We denote the force of infection with λ_v and force of addiction with λ_u, given as

$$\text{Force of infection: } \lambda_v = \beta_v \frac{V + I_1 + I_2}{N},$$
$$\text{Force of addiction: } \lambda_u = \beta_u \frac{U + I_1 + I_2}{N}.$$

The addiction transmission rate β_u is the product of the population contact rate and the probability that a contact with an heroin user will result in addiction. Similarly, β_v denotes the product of the population contact rate and the probability that a contact with an HIV-infected individual will result in transmitting the disease. After therapy, an addicted individual goes back to the susceptible class. The number of addicted individuals going through therapy per unit of time is given by the term δU, where δ is the addiction treatment rate. The rate of change of a susceptible population is given, respectively, as follows:

$$S' = \Lambda - \lambda_u S - \lambda_v S - \mu S + \delta U.$$

We denote the death rate due to addiction and the death rate due to HIV infection as μ_u and μ_v, respectively. An addicted individual can become HIV-infected upon contact with a infected individual and moves to the co-affected class $I_1(t)$. The force of infection of a heroin user to HIV infection is denoted by $q_u\lambda_v$, where q_u is the coefficient to enrich or decrease the HIV transmission rate β_v for an addicted individual. Similarly, an infected individual becomes addicted upon contact with a heroin user and moves to the co-affected class $I_1(t)$. The force of addiction of an infected individual to addiction is denoted by $q_v\lambda_u$ where q_v decreases or increases the addiction transmission rate β_u for an infected individual. The parameters γ_v and γ_2 represent the transition into AIDS stage from V and I_2 classes, respectively. The duration of the first stage I_1 is short in comparison to the entire time span of HIV in the human body,[6] so we do not consider any addiction treatment in the first stage HIV-opioid infection. An addicted individual who is, respectively, in the second stage of the HIV infection (I_2) can recover from addiction and move to class V at rate δ_2. The rates of change for addicted and infected individuals are given by as follows:

$$U' = \lambda_u S - q_u \lambda_v U - (\mu + \mu_u + \delta)U$$
$$V' = \lambda_v S - q_v \lambda_u V - (\mu + \mu_v + \gamma_v)V + \delta_2 I_2.$$

The co-affected class I_1 decreases by natural death, by transition into the second stage of infection (I_2), or by death due to either HIV or opioid usage.[6] We denote the transition rate into I_2 as α and death rate due to addiction or infection as μ_1. Similarly, the death due to addiction or disease is μ_2 for the co-affected individuals who are in the second stage of HIV-infection. The rate of change of populations I_1 and I_2 are given, respectively, as follows:

$$I_1' = q_u \lambda_v U + q_v \lambda_u V - (\mu + \alpha + \mu_1)I_1$$
$$I_2' = \alpha I_1 - (\mu + \gamma_2 + \mu_2 + \delta_2)I_2.$$

Finally, the rate of change of population with AIDS A is given by

$$A' = \gamma_v V + \gamma_2 I_2 - (\mu + \mu_a)A,$$

where μ_a denotes the death rate due to AIDS.

Given the above considerations, the HIV-opioid model takes the following form:

$$\begin{aligned}
S' &= \Lambda - \lambda_u S - \lambda_v S - \mu S + \delta U \\
U' &= \lambda_u S - q_u \lambda_v U - (\mu + \mu_u + \delta)U \\
V' &= \lambda_v S - q_v \lambda_u V - (\mu + \mu_v + \gamma_v)V + \delta_2 I_2 \\
I_1' &= q_u \lambda_v U + q_v \lambda_u V - (\mu + \alpha + \mu_1)I_1 \\
I_2' &= \alpha I_1 - (\mu + \gamma_2 + \mu_2 + \delta_2)I_2 \\
A' &= \gamma_v V + \gamma_2 I_2 - (\mu + \mu_a)A.
\end{aligned} \tag{2.1}$$

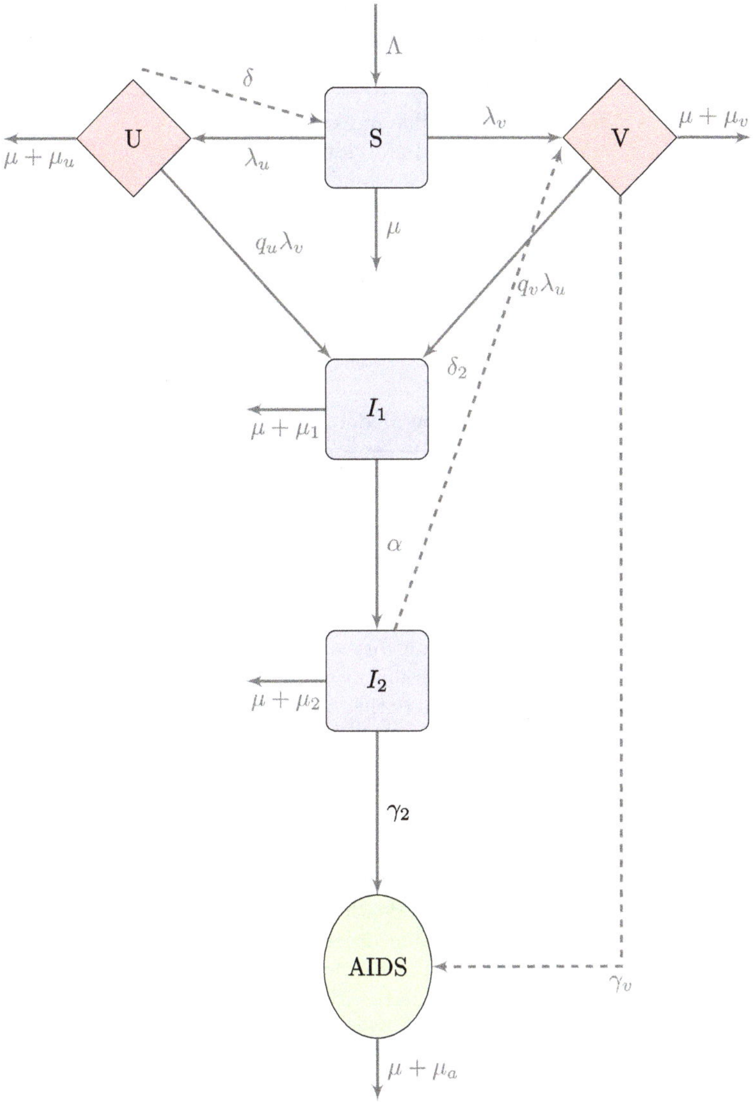

Fig. 2.1. Flow chart of the HIV–opioid model (2.2).

A schematic flow diagram of the HIV-opioid model (2.1) is given in a flowchart in Figure 2.1, and associated model variables and parameters are defined in Tables 2.1 and 2.2, respectively.

Table 2.1. State variables of model (2.2) and their definitions

Variables	Definition
$S(t)$	Number of susceptible individuals at time t
$U(t)$	Number of addicted individuals (opioid users) at time t
$V(t)$	Number of HIV-infected individuals at time t
$I_1(t)$	Number of addicted individuals in the first stage of HIV infection at time t
$I_2(t)$	Number of addicted individuals in the second stage of HIV infection at time t
$A(t)$	Number of individuals with AIDS at time t

Table 2.2. Parameters of the model 2. (2.1) and their definitions.

Variables	Definition
δ	Rate of recovery from addiction
δ_2	Rate of recovery from addiction for individuals in the $I_2(t)$ class
γ_v	Rate of transition to AIDS
β_u	Addiction transmission rate
β_v	HIV transmission rate
q_u	Coefficient to enrich or decrease the HIV-infection rate for addicted individuals
q_v	Coefficient to enrich or decrease the addiction transmission rate for HIV-infected individuals
α	Transition from the first to the second HIV stage
γ_2	Transition rate of I_2 to AIDS
μ	Natural death rate
μ_a	AIDS-induced death rate
μ_u	Addiction induced death rate
μ_v	HIV-induced death rate
μ_1	Death rate due to either HIV or opioid use
μ_2	Death rate due to either HIV or opioid use

2.3 Model Analysis

2.3.1 *Equilibria of HIV infection with opioid addiction dynamics*

In this section, we study the existence and stability of equilibria of the model (2.1). The equilibrium solutions of the model satisfy the following systems:

$$
\begin{aligned}
0 &= \Lambda - \beta_u \left(\frac{U + I_1 + I_2}{N}\right) S - \beta_v \left(\frac{V + I_1 + I_2}{N}\right) S - \mu S + \delta U \\
0 &= \beta_u \left(\frac{U + I_1 + I_2}{N}\right) S - q_u\beta_v \left(\frac{V + I_1 + I_2}{N}\right) U - (\mu + \mu_u + \delta)U \\
0 &= \beta_v \left(\frac{V + I_1 + I_2}{N}\right) S - q_v\beta_u \left(\frac{U + I_1 + I_2}{N}\right) V \\
&\quad - (\mu + \mu_v + \gamma_v)V + \delta_2 I_2 \\
0 &= q_u\beta_v \left(\frac{V + I_1 + I_2}{N}\right) U + q_v\beta_u \left(\frac{U + I_1 + I_2}{N}\right) V \\
&\quad -(\mu + \alpha + \mu_1)I_1 \\
0 &= \alpha I_1 - (\mu + \gamma_2 + \mu_2 + \delta_2)I_2 \\
0 &= \gamma_v V + \gamma_2 I_2 - (\mu + \mu_a)A. \qquad (2.2)
\end{aligned}
$$

2.3.2 *Disease free equilibrium (DFE)*

The system (2.2) has disease-free equilibrium $\mathcal{E}_0 = (S^*, 0, 0, 0, 0, 0) = \left(\frac{\Lambda}{\mu}, 0, 0, 0, 0, 0\right)$ in which both HIV and addiction are absent in the population. Linearizing the system (2.1) at DFE, we obtain the following Jacobian matrix:

$$
J(\mathcal{E}_0) = \begin{pmatrix}
-\mu & -\beta_u + \delta & -\beta_v & -\beta_u - \beta_v & -\beta_u - \beta_v & 0 \\
0 & \beta_u - (\mu + \mu_u + \delta) & & \beta_u & \beta_u & 0 \\
0 & 0 & \beta_v - (\mu + \mu_v + \gamma_v) & \beta_v & \beta_v + \delta_2 & 0 \\
0 & 0 & 0 & -(\mu + \alpha + \mu_1) & 0 & 0 \\
0 & 0 & 0 & \alpha & -(\mu + \gamma_2 + \mu_2 + \delta_2) & 0 \\
0 & 0 & \gamma_v & 0 & \gamma_2 & -(\mu + \mu_a)
\end{pmatrix}.
$$

The eigenvalues of $J(\mathcal{E}_0)$ are negative except for two, which are, $\lambda_4 = \beta_u - (\mu + \mu_u + \delta)$ and $\lambda_5 = \beta_v - (\mu + \mu_v + \gamma_v)$. Setting $\lambda_4 < 0$ and $\lambda_5 < 0$, we obtain the reproduction number for opioid addiction as

$$\mathcal{R}_0^u = \frac{\beta_u}{\mu + \mu_u + \delta} < 1$$

which represents the number of secondary addictions produced by one addicted individual in an entirely susceptible population during a life time of being addicted. The reproduction number for HIV infection is calculated as follows:

$$\mathcal{R}_0^v = \frac{\beta_v}{\mu + \mu_v + \gamma_v} < 1.$$

We then define the basic reproduction number as

$$\mathcal{R}_0 = \max\left(\mathcal{R}_0^u, \mathcal{R}_0^v\right).$$

Therefore, we proved the following Theorem 2.1 for the equilibrium $\mathcal{E}_0$.

Theorem 2.1. *The DFE of the model* (2.1), $\mathcal{E}_0 = \left(\frac{\Lambda}{\mu}, 0, 0, 0, 0, 0\right)$, *is locally, asymptotically stable when* $\mathcal{R}_0 < 1$.

2.3.3 *Addiction only equilibrium (AOE)*

Next, we consider the case when HIV infection dies out but addiction persists in the population. We denote such equilibrium as $\mathcal{E}_1 = (S^*, U^*, 0, 0, 0, 0)$. HIV–free equilibrium $\mathcal{E}_1$ satisfies the following system of equations:

$$\begin{aligned} \Lambda - \beta_u \frac{U^* S^*}{N^*} - \mu S^* + \delta U^* &= 0 \\ \beta_u \frac{U^* S^*}{N^*} - (\mu + \mu_u + \delta) U^* &= 0 \\ \Lambda - \mu N^* - \mu_u U^* &= 0, \end{aligned} \tag{2.3}$$

where $N^* = S^* + U^*$. The corresponding equilibrium point is

$$\mathcal{E}_1 = \left(\frac{\Lambda}{\mathcal{R}_0^u \mu + (\mathcal{R}_0^u - 1)\mu_u}, \frac{\Lambda(\mathcal{R}_0^u - 1)}{\mathcal{R}_0^u \mu + (\mathcal{R}_0^u - 1)\mu_u}, 0, 0, 0, 0\right).$$

Clearly, $\mathcal{E}_1$ exists if $\mathcal{R}_0^u > 1$. The Jacobian matrix at the equilibrium $\mathcal{E}_1$ is

$$J(\mathcal{E}_1) = \begin{pmatrix} A_1 & A_2 \\ 0 & A_4 \end{pmatrix},$$

where the major blocks of the matrix $J(\mathcal{E}_1)$ are

$$A_1 = \begin{pmatrix} -\beta_u \frac{U^*}{N^*} + \beta_u \frac{U^* S^*}{N^{*2}} - \mu & -\beta_u \frac{S^*}{N^*} + \beta_u \frac{U^* S^*}{N^{*2}} + \delta \\ \beta_u \frac{U^*}{N^*} - \beta_u \frac{U^* S^*}{N^{*2}} & -\beta_u \frac{U^* S^*}{N^{*2}} \end{pmatrix}$$

$$A_2 = \begin{pmatrix} \beta_u \frac{U^* S^*}{N^*} - \beta_v \frac{S^*}{N^*} & -\beta_u \left(\frac{S^*}{N^*} + \frac{U^* S^*}{N^{*2}}\right) - \beta_v \frac{S^*}{N^*} & -\beta_u \left(\frac{S^*}{N^*} + \frac{U^* S^*}{N^{*2}}\right) - \beta_v \frac{S^*}{N^*} & \beta_u \frac{U^* S^*}{N^{*2}} \\ -\beta_u \frac{U^* S^*}{N^{*2}} + q_u \beta_v \frac{U^*}{N^*} & \beta_u \left(\frac{S^*}{N^*} - \frac{U^* S^*}{N^{*2}}\right) - q_u \beta_v \frac{U^*}{N^*} & \beta_u \left(\frac{S^*}{N^*} - \frac{U^* S^*}{N^{*2}}\right) - q_u \beta_v \frac{U^*}{N^*} & -\beta_u \frac{U^* S^*}{N^{*2}} \end{pmatrix}$$

$$A_4 = \begin{pmatrix} \beta_v \frac{S^*}{N^*} - q_v \beta_u \frac{U^*}{N^*} - (\mu + \mu_v + \gamma_v) & \beta_v \frac{S^*}{N^*} & \beta_v \frac{S^*}{N^*} + \delta_2 & 0 \\ (q_u \beta_v + q_v \beta_u) \frac{U^*}{N^*} & q_u \beta_v \frac{U^*}{N^*} - (\mu + \alpha + \mu_1) & q_u \beta_v \frac{U^*}{N^*} & 0 \\ 0 & \alpha & -(\mu + \gamma_2 + \mu_2 + \delta_2) & 0 \\ \gamma_v & 0 & \gamma_2 & -(\mu + \mu_a) \end{pmatrix}$$

For the stability of the $\mathcal{E}_1$, we show that all eigenvalues are negative or have negative real part by using the Routh–Hurwitz criterion.[29] The block matrix A_1 has all negative eigenvalues, since

$$\text{Trace}(A_1) = -\beta_u \frac{U^*}{N^*} - \mu < 0,$$

$$\text{Det}(A_1) = \left(-\beta_u \frac{U^*}{N^*} + \beta_u \frac{U^* S^*}{N^{*2}} - \mu\right)\left(-\beta_u \frac{U^* S^*}{N^{*2}}\right) - \left(\beta_u \frac{U^*}{N^*} - \beta_u \frac{U^* S^*}{N^{*2}}\right)\left(-\beta_u \frac{S^*}{N^*} + \beta_u \frac{U^* S^*}{N^{*2}} + \delta\right)$$

$$= \mu\beta_u \frac{U^*S^*}{N^{*2}} + \beta_u^2 \frac{U^*S^*}{N^{*2}}\left(1 - \frac{S^*}{N^*}\right)$$

$$+ \delta\beta_u \frac{U^*}{N^*}\left(1 - \frac{S^*}{N^*}\right) > 0.$$

The block A_4 has eigenvalue $-(\mu + \mu_a) < 0$. For other eigenvalues, we set

$$E = \begin{pmatrix} \beta_v \frac{S^*}{N^*} - Q & \beta_v \frac{S^*}{N^*} & \beta_v \frac{S^*}{N^*} + \delta_2 \\ (q_u\beta_v + q_v\beta_u)\frac{U^*}{N^*} & q_u\beta_v \frac{U^*}{N^*} - Q_1 & q_u\beta_v \frac{U^*}{N^*} \\ 0 & \alpha & -Q_2 \end{pmatrix},$$

where $Q = q_v\beta_u \frac{U^*}{N^*} + (\mu+\mu_v+\gamma_v) = q_v(\mu+\mu_u+\delta)+(\mu+\mu_v+\gamma_v)$; $Q_1 = \mu + \alpha + \mu_1$ and $Q_2 = \mu + \gamma_2 + \mu_2 + \delta_2$. The characteristic polynomial of E takes the following form:

$$\lambda^3 + A\lambda^2 + B\lambda + C = 0, \tag{2.4}$$

where

$$A = Q_2 + \left(Q_1 - q_u\beta_v \frac{U^*}{N^*}\right) + \left(Q - \beta_v \frac{S^*}{N^*}\right)$$

$$B = Q_2\left(Q_1 - q_u\beta_v \frac{U^*}{N^*}\right) + Q_2\left(Q - \beta_v \frac{S^*}{N^*}\right) + \left(q_u\beta_v \frac{U^*}{N^*} - Q_1\right)$$
$$\times \left(\beta_v \frac{S^*}{N^*} - Q\right) - \alpha q_u\beta_v \frac{U^*}{N^*} - \beta_v \frac{S^*}{N^*}(q_u\beta_v + q_v\beta_u)\frac{U^*}{N^*}$$

$$C = \left(\beta_v \frac{S^*}{N^*} - Q\right)\left(q_u\beta_v \frac{U^*}{N^*} - Q_1\right) Q_2 - \beta_v \frac{S^*}{N^*} Q_2 (q_u\beta_v + q_v\beta_u)$$
$$\times \frac{U^*}{N^*} + \left(\beta_v \frac{S^*}{N^*} - Q\right)\alpha q_u\beta_v \frac{U^*}{N^*} - \alpha (q_u\beta_v + q_v\beta_u)$$
$$\times \frac{U^*}{N^*}\left(\beta_v \frac{S^*}{N^*} + \delta_2\right).$$

Setting $C > 0$, we obtain the following number:

$$\mathcal{R}_1^u = \frac{\beta_v \frac{S^*}{N^*}\left(Q_1Q_2 + \alpha q_v\beta_u \frac{U^*}{N^*} + Q_2 q_v\beta_u \frac{U^*}{N^*}\right) + q_u\beta_v \frac{U^*}{N^*} (QQ_2 + \alpha\delta_2 + \alpha Q) + q_v\beta_u \frac{U^*}{N^*}\alpha\delta_2}{QQ_1Q_2} < 1.$$

Here, $\dfrac{U^*}{N^*} = 1 - \dfrac{1}{\mathcal{R}_0^u}$ and $\dfrac{S^*}{N^*} = \dfrac{1}{\mathcal{R}_0^u}$
Note that when $\mathcal{R}_1^u < 1$, we have

$$\beta_v \frac{S^*}{N^* Q} < 1 \quad \text{and} \quad q_u \beta_v \frac{U^*}{N^* Q_1} < 1$$

which implies that $A = Q_2 + Q_1\left(1 - q_u\beta_v \frac{U^*}{N^*Q_1}\right) + Q\left(1 - \beta_v \frac{S^*}{N^*Q}\right) > 0$.
Next, we show that $AB - C > 0$, i.e.,

$$\begin{aligned}
AB = &\left(Q_2 + \left(Q_1 - q_u\beta_v \frac{U^*}{N^*}\right) + \left(Q - \beta_v \frac{S^*}{N^*}\right)\right)\\
&\times \left(Q_2\left(Q_1 - q_u\beta_v \frac{U^*}{N^*}\right) + Q_2\left(Q - \beta_v \frac{S^*}{N^*}\right)\right.\\
&+ \left(q_u\beta_v \frac{U^*}{N^*} - Q_1\right)\left(\beta_v \frac{S^*}{N^*} - Q\right) - \alpha q_u \beta_v \frac{U^*}{N^*}\\
&\left.-\beta_v \frac{S^*}{N^*}\left(q_u\beta_v + q_v\beta_u\right)\frac{U^*}{N^*}\right)\\
= &\left(1 + \frac{\left(Q_1 - q_u\beta_v \frac{U^*}{N^*}\right) + \left(Q - \beta_v \frac{S^*}{N^*}\right)}{Q_2}\right) Q_2\\
&\times \left(Q_2\left(Q_1 - q_u\beta_v \frac{U^*}{N^*}\right) + Q_2\left(Q - \beta_v \frac{S^*}{N^*}\right)\right.\\
&+ \left(q_u\beta_v \frac{U^*}{N^*} - Q_1\right)\left(\beta_v \frac{S^*}{N^*} - Q\right) - \alpha q_u \beta_v \frac{U^*}{N^*}\\
&\left.-\beta_v \frac{S^*}{N^*}\left(q_u\beta_v + q_v\beta_u\right)\frac{U^*}{N^*}\right).
\end{aligned}$$

Since $\left(1 + \frac{(Q_1 - q_u\beta_v \frac{U^*}{N^*}) + (Q - \beta_v \frac{S^*}{N^*})}{Q_2}\right) > 1$, we need to show that $Q_2 B > C$, hence,

$$\begin{aligned}
Q_2B - C = Q_2 &\left(Q_2\left(Q_1 - q_u\beta_v \frac{U^*}{N^*}\right) + Q_2\left(Q - \beta_v \frac{S^*}{N^*}\right)\right.\\
&+ \left(q_u\beta_v \frac{U^*}{N^*} - Q_1\right)\left(\beta_v \frac{S^*}{N^*} - Q\right)
\end{aligned}$$

$$
\begin{aligned}
&-\alpha q_u\beta_v\frac{U^*}{N^*}-\beta_v\frac{S^*}{N^*}\left(q_u\beta_v+q_v\beta_u\right)\frac{U^*}{N^*}\Bigg)\\
&-\left(\left(\beta_v\frac{S^*}{N^*}-Q\right)\left(q_u\beta_v\frac{U^*}{N^*}-Q_1\right)Q_2\right.\\
&-\beta_v\frac{S^*}{N^*}Q_2\left(q_u\beta_v+q_v\beta_u\right)\frac{U^*}{N^*}+\left(\beta_v\frac{S^*}{N^*}-Q\right)\alpha q_u\beta_v\frac{U^*}{N^*}\\
&\left.-\alpha\left(q_u\beta_v+q_v\beta_u\right)\frac{U^*}{N^*}\left(\beta_v\frac{S^*}{N^*}+\delta_2\right)\right).
\end{aligned}
$$

After cancellation, we obtain

$$
\begin{aligned}
&Q_2\left(Q_2\left(Q_1-q_u\beta_v\frac{U^*}{N^*}\right)+Q_2\left(Q-\beta_v\frac{S^*}{N^*}\right)-\alpha q_u\beta_v\frac{U^*}{N^*}\right)\\
&\quad-\left(\left(\beta_v\frac{S^*}{N^*}-Q\right)\alpha q_u\beta_v\frac{U^*}{N^*}-\alpha\left(q_u\beta_v+q_v\beta_u\right)\frac{U^*}{N^*}\right.\\
&\quad\left.\times\left(\beta_v\frac{S^*}{N^*}+\delta_2\right)\right)=Q_2\left(Q_2\left(Q_1-q_u\beta_v\frac{U^*}{N^*}\right)\right.\\
&\quad\left.+Q_2\left(Q-\beta_v\frac{S^*}{N^*}\right)-\alpha q_u\beta_v\frac{U^*}{N^*}\right)+\left(Q-\beta_v\frac{S^*}{N^*}\right)\alpha q_u\beta_v\frac{U^*}{N^*}\\
&\quad+\ \alpha\left(q_u\beta_v+q_v\beta_u\right)\frac{U^*}{N^*}\left(\beta_v\frac{S^*}{N^*}+\delta_2\right).
\end{aligned}
$$

Thus,

$$
\begin{aligned}
Q_2B-C=&\ Q_2\left(Q_2\left(Q_1-q_u\beta_v\frac{U^*}{N^*}\right)+Q_2\left(Q-\beta_v\frac{S^*}{N^*}\right)\right)\\
&+\left(Q-Q_2\right)\alpha q_u\beta_v\frac{U^*}{N^*}+\alpha q_u\beta_v\frac{U^*}{N^*}\delta_2\\
&+\alpha(q_v\beta_u)\frac{U^*}{N^*}\left(\beta_v\frac{S^*}{N^*}+\delta_2\right)
\end{aligned}
$$

this states that $Q_2B - C > 0$, when $Q > Q_2$ and $\mathcal{R}_1^u < 1$. We then define the invasion number as

$$\mathcal{R}_{\text{inv}}^u = \max\left\{\mathcal{R}_1^u, \frac{Q_2}{Q}\right\}.$$

Note that, since $A > 0$ and $C > 0$ and $AB > C$ implies that $B > 0$ as well, therefore, we prove the following Theorem 2.2 for the equilibrium $\mathcal{E}_1$.

Theorem 2.2. *The addiction-only boundary equilibrium $\mathcal{E}_1$ exists when $\mathcal{R}_0^u > 1$ and it is locally asymptotically stable when $\mathcal{R}_{\text{inv}}^u < 1$.*

2.3.4 *HIV-only equilibrium (HOE)*

Being free of opioid addiction entails the presence of HIV but no addicted people. The HIV-only equilibrium $\mathcal{E}_2 = (S^*, 0, V^*, 0, 0, A^*)$ satisfies the following equations:

$$\begin{aligned}
\Lambda - \beta_v \frac{V^* S^*}{N^*} - \mu S^* &= 0 \\
\left(\beta_v \frac{S^*}{N^*} - (\mu + \mu_v + \gamma_v)\right) V^* &= 0 \\
\gamma_v V^* - (\mu + \mu_a) A^* &= 0 \\
\Lambda - \mu N^* - \mu_v V^* - \mu_a A^* &= 0.
\end{aligned} \tag{2.5}$$

Solving the second and third equations in the above system (2.5), we obtain $A^* = \dfrac{\gamma_v}{\mu + \mu_a} V^*$ and $\dfrac{S^*}{N^*} = \dfrac{1}{\mathcal{R}_0^v}$, as $V^* \neq 0$. We have

$$\begin{aligned}
S^* + V^* + A^* &= N^* \\
\frac{V^*}{N^*} + \frac{A^*}{N^*} &= 1 - \frac{1}{\mathcal{R}_0^v} \\
\frac{V^*}{N^*} + \frac{\gamma_v}{\mu + \mu_a} \frac{V^*}{N^*} &= 1 - \frac{1}{\mathcal{R}_0^v} \\
\frac{V^*}{N^*} &= \left(1 - \frac{1}{\mathcal{R}_0^v}\right) \left(\frac{\mu + \mu_a}{\mu + \mu_a + \gamma_v}\right).
\end{aligned}$$

Substituting into the first equation, we obtain

$$\Lambda - \beta_v \frac{1}{\mathcal{R}_0^v} V^* - \frac{1}{\mathcal{R}_0^v}\left(\Lambda - \mu_v V^* - \mu_a \frac{\gamma_v}{\mu + \mu_a} V^*\right) = 0$$

this gives,

$$V^* = \frac{\Lambda\left(1 - \frac{1}{\mathcal{R}_0^v}\right)}{\mu + \mu_v\left(1 - \frac{1}{\mathcal{R}_0^v}\right) + \gamma_v\left(1 - \frac{\mu_a}{\mathcal{R}_0^v(\mu+\mu_a)}\right)}.$$

Clearly, the boundary equilibrium $\mathcal{E}_2 = (S^*, 0, V^*, 0, 0, A^*)$ exists when
$\mathcal{R}_0^v = \dfrac{\beta_v}{\mu + \mu_v + \gamma_v} > 1$. The Jacobian at the equilibrium $\mathcal{E}_2$ has the following form:

$$J(\mathcal{E}_2) = \begin{pmatrix} E_1 & E_2 \\ 0 & E_4 \end{pmatrix}$$

here

$$E_1 = \begin{pmatrix} -\beta_v \frac{V^*(V^*+A^*)}{N^{*2}} - \mu & \beta_v \frac{V^* S^*}{N^{*2}} & -\beta_v \frac{S^*(S^*+A^*)}{N^{*2}} \\ 0 & -P_4 & \gamma_v \\ \beta_v \frac{V^*(V^*+A^*)}{N^{*2}} & -\beta_v \frac{S^* V^*}{N^{*2}} & -\beta_v \frac{S^* V^*}{N^{*2}} \end{pmatrix}$$

$$E_2 = \begin{pmatrix} -\beta_v \frac{S^*}{N^*} - \beta_v \frac{S^*(S^*+A^*)}{N^{*2}} & -\beta_v \frac{S^*}{N^*} - \beta_v \frac{S^*}{N^*} + \beta_v \frac{V^* S^*}{N^{*2}} \\ 0 & \gamma_2 \\ \beta_v \frac{S^*}{N^*} - \beta_v \frac{S^* V^*}{N^{*2}} - q_v \beta_u \frac{V^*}{N^*} + \delta_1 & \beta_v \frac{S^*}{N^*} - \beta_v \frac{S^* V^*}{N^{*2}} - q_v \beta_u \frac{V^*}{N^*} \end{pmatrix.}$$

$$\begin{matrix} -\beta_u \frac{S^*}{N^*} + \beta_v \frac{V^* S^*}{N^{*2}} + \delta \\ 0 \\ -\beta_v \frac{V^* S^*}{N^{*2}} - q_v \beta_u \frac{V^*}{N^*} \end{matrix}\Bigg)$$

$$E_4 = \begin{pmatrix} q_v\beta_u\frac{V^*}{N^*} - P_2 & q_v\beta_u\frac{V^*}{N^*} & q_u\beta_v\frac{V^*}{N^*} + q_v\beta_u\frac{V^*}{N^*} \\ \alpha & -P_3 & 0 \\ \beta_u\frac{S^*}{N^*} & \beta_u\frac{S^*}{N^*} & \beta_u\frac{S^*}{N^*} - P_1 \end{pmatrix}.$$

and $P_1 = q_u\beta_v\frac{V^*}{N^*}+(\mu+\mu_u+\delta)$, $P_2 = \mu+\alpha+\mu_1$, $P_3 = \mu+\gamma_2+\mu_2+\delta_2$, $P_4 = \mu + \mu_a$. The characteristic polynomial of E_1 can be written as $\lambda^3 + A_1\lambda^2 + B_1\lambda + C_1 = 0$ here

$$\begin{aligned}
A_1 &= \left(\beta_v\frac{V^*(V^*+A^*)}{N^{*2}} + \mu\right) + \beta_v\frac{S^*V^*}{N^{*2}} + P_4 \\
B_1 &= \left(\beta_v\frac{V^*(V^*+A^*)}{N^{*2}} + \mu\right) P_4 + \left(\beta_v\frac{V^*(V^*+A^*)}{N^{*2}} + \mu\right) \\
&\quad \times \left(\beta_v\frac{S^*V^*}{N^{*2}}\right) + \left(\beta_v\frac{V^*(V^*+A^*)}{N^{*2}}\right)\left(\beta_v\frac{S^*V^*}{N^{*2}}\right) \\
&\quad + P_4\left(\beta_v\frac{S^*V^*}{N^{*2}}\right) + \left(\beta_v\frac{V^*(V^*+A^*)}{N^{*2}}\right)\left(\beta_v\frac{S^*(S^*+A^*)}{N^{*2}}\right) \\
C_1 &= P_4\left(\beta_v\frac{V^*(V^*+A^*)}{N^{*2}} + \mu\right)\left(\beta_v\frac{S^*V^*}{N^{*2}}\right) - \left(\beta_v\frac{V^*(V^*+A^*)}{N^{*2}}\right) \\
&\quad \times \left(\beta_v\frac{V^*S^*}{N^{*2}}\right)\gamma_v + \left(\beta_v\frac{V^*(V^*+A^*)}{N^{*2}} + \mu\right) \\
&\quad \times \left(\beta_v\frac{V^*(V^*+A^*)}{N^{*2}}\right)\left(\beta_v\frac{S^*(S^*+A^*)}{N^{*2}}\right) \\
&\quad + P_4\left(\beta_v\frac{V^*(V^*+A^*)}{N^{*2}}\right)\left(\beta_v\frac{S^*(S^*+A^*)}{N^{*2}}\right).
\end{aligned}$$

Clearly, $A_1 > 0$ and $B_1 > 0$. Note that, $C_1 > 0$ if $P_4 > \gamma_v$ or $\dfrac{\gamma_v}{\mu+\mu_a} < 1$. Next, we show $A_1B_1 > C_1$. In fact, we show that

$\left(\left(\beta_v \frac{V^*(V^*+A^*)}{N^{*2}} + \mu\right) + P_4\right) B_1 - C_1 > 0$ as follows,

$$
\begin{aligned}
&\left(\left(\beta_v \frac{V^*(V^*+A^*)}{N^{*2}} + \mu\right) + P_4\right) B_1 - C_1 \\
&\quad = \left(\beta_v \frac{V^*(V^*+A^*)}{N^{*2}} + \mu\right) P_4 \left(\beta_v \frac{V^*(V^*+A^*)}{N^{*2}} + \mu\right) \\
&\qquad + \left(\beta_v \frac{V^*(V^*+A^*)}{N^{*2}} + \mu\right) \left(\beta_v \frac{S^*V^*}{N^{*2}}\right) \left(\beta_v \frac{V^*(V^*+A^*)}{N^{*2}} + \mu\right) \\
&\qquad + \left(\beta_v \frac{V^*(V^*+A^*)}{N^{*2}}\right) \left(\beta_v \frac{S^*V^*}{N^{*2}}\right) \left(\beta_v \frac{V^*(V^*+A^*)}{N^{*2}} + \mu\right) \\
&\qquad + P_4 \left(\beta_v \frac{S^*V^*}{N^{*2}}\right) \left(\beta_v \frac{V^*(V^*+A^*)}{N^{*2}} + \mu\right) \\
&\qquad + \left(\beta_v \frac{V^*(V^*+A^*)}{N^{*2}} + \mu\right) P_4^2 + \left(\beta_v \frac{V^*(V^*+A^*)}{N^{*2}} + \mu\right) \\
&\qquad \times \left(\beta_v \frac{S^*V^*}{N^{*2}}\right) P_4 + P_4^2 \left(\beta_v \frac{S^*V^*}{N^{*2}}\right) + \left(\beta_v \frac{V^*(V^*+A^*)}{N^{*2}}\right) \\
&\qquad \times \left(\beta_v \frac{V^*S^*}{N^{*2}}\right) \gamma_v > 0.
\end{aligned}
$$

The characteristic polynomial of E_4 can be written as $\lambda^3 + A_4\lambda^2 + B_4\lambda + C_4 = 0$ here

$$
\begin{aligned}
A_4 &= \left(P_1 - \beta_u \frac{S^*}{N^*}\right) + \left(P_2 - q_v\beta_u \frac{V^*}{N^*}\right) + P_3, \\
B_4 &= -\beta_u \frac{S^*}{N^*} \left(q_u\beta_v \frac{V^*}{N^*} + q_v\beta_u \frac{V^*}{N^*}\right) + \left(\beta_u \frac{S^*}{N^*} - P_1\right) \\
&\quad \times \left(q_v\beta_u \frac{V^*}{N^*} - P_2\right) - \alpha q_v\beta_u \frac{V^*}{N^*} + P_3 \left(P_1 - \beta_u \frac{S^*}{N^*}\right) \\
&\quad + P_3 \left(P_2 - q_v\beta_u \frac{V^*}{N^*}\right),
\end{aligned}
$$

$$C_4 = -\alpha\beta_u \frac{S^*}{N^*}\left(q_u\beta_v\frac{V^*}{N^*} + q_v\beta_u\frac{V^*}{N^*}\right) + \left(\beta_u\frac{S^*}{N^*} - P_1\right)\alpha q_v\beta_u\frac{V^*}{N^*}$$
$$- P_3\beta_u\frac{S^*}{N^*}\left(q_u\beta_v\frac{V^*}{N^*} + q_v\beta_u\frac{V^*}{N^*}\right) + P_3\left(\beta_u\frac{S^*}{N^*} - P_1\right)$$
$$\times\left(q_v\beta_u\frac{V^*}{N^*} - P_2\right).$$

Setting $C_4 > 0$ gives the following requirement:

$$\mathcal{R}_1^v = \frac{\beta_u\frac{S^*}{N^*}\left(P_2P_3 + \alpha q_u\beta_v\frac{V^*}{N^*} + q_u\beta_v\frac{V^*}{N^*}P_3\right) + q_v\beta_u\frac{V^*}{N^*}(P_1P_3 + P_1\alpha)}{P_1P_2P_3}$$
$$< 1.$$

Note that when $\mathcal{R}_1^v < 1$, then $\frac{\beta_u\frac{S^*}{N^*}}{P_1} < 1$ and $\frac{q_v\beta_u\frac{V^*}{N^*}}{P_2} < 1$, which implies that $A_4 > 0$. Next we show that $A_4B_4 - C_4 > 0$. As before we proceed with showing that $P_3B_4 - C_4 > 0$, since $A_4 = (\frac{(P_1-\beta_u\frac{S^*}{N^*})+\left(P_2-q_v\beta_u\frac{V^*}{N^*}\right)}{P_3} + 1)P_3$ with $(\frac{\left(P_1-\beta_u\frac{S^*}{N^*}\right)+\left(P_2-q_v\beta_u\frac{V^*}{N^*}\right)}{P_3} + 1) > 1$

$$P_3B_4 - C_4 = -P_3\beta_u\frac{S^*}{N^*}\left(q_u\beta_v\frac{V^*}{N^*} + q_v\beta_u\frac{V^*}{N^*}\right) + P_3\left(\beta_u\frac{S^*}{N^*} - P_1\right)$$
$$\times\left(q_v\beta_u\frac{V^*}{N^*} - P_2\right) - P_3\alpha q_v\beta_u\frac{V^*}{N^*} + P_3^2\left(P_1 - \beta_u\frac{S^*}{N^*}\right)$$
$$+ P_3^2\left(P_2 - q_v\beta_u\frac{V^*}{N^*}\right) + \alpha\beta_u\frac{S^*}{N^*}\left(q_u\beta_v\frac{V^*}{N^*} + q_v\beta_u\frac{V^*}{N^*}\right)$$
$$-\left(\beta_u\frac{S^*}{N^*} - P_1\right)\alpha q_v\beta_u\frac{V^*}{N^*}$$
$$+ P_3\beta_u\frac{S^*}{N^*}\left(q_u\beta_v\frac{V^*}{N^*} + q_v\beta_u\frac{V^*}{N^*}\right) - P_3\left(\beta_u\frac{S^*}{N^*} - P_1\right)$$
$$\times\left(q_v\beta_u\frac{V^*}{N^*} - P_2\right).$$

$$\begin{aligned}P_3B_4 - C_4 = {} & P_3^2\left(P_1 - \beta_u\frac{S^*}{N^*}\right) + P_3^2\left(P_2 - q_v\beta_u\frac{V^*}{N^*}\right)\\ & + \alpha\beta_u\frac{S^*}{N^*}\left(q_u\beta_v\frac{V^*}{N^*} + q_v\beta_u\frac{V^*}{N^*}\right)\\ & - P_3\alpha q_v\beta_u\frac{V^*}{N^*} - \left(\beta_u\frac{S^*}{N^*} - P_1\right)\alpha q_v\beta_u\frac{V^*}{N^*}\\ P_3B_4 - C_4 = {} & P_3^2\left(P_1 - \beta_u\frac{S^*}{N^*}\right) + P_3^2\left(P_2 - q_v\beta_u\frac{V^*}{N^*}\right)\\ & + \alpha\beta_u\frac{S^*}{N^*}q_u\beta_v\frac{V^*}{N^*} + (P_1 - P_3)\alpha q_v\beta_u\frac{V^*}{N^*}.\end{aligned}$$

Hence, $P_3B_4 - C_4 > 0$ when $P_1 > P_3$. We then define the invasion number as

$$\mathcal{R}_{\text{inv}}^v = \max\left\{\mathcal{R}_1^v, \frac{P_3}{P_1}, \frac{\gamma_v}{\mu + \mu_a}\right\}.$$

Thus, we prove the following Theorem 2.3 for the equilibrium $\mathcal{E}_2$.

Theorem 2.3. *The HIV-only boundary equilibrium $\mathcal{E}_2$ exists when $\mathcal{R}_0^v > 1$ and it is locally asymptotically stable when $\mathcal{R}_{\text{inv}}^v < 1$.*

2.4 Model with Control Strategies

There is currently no effective cure for HIV.[3] However, through prevention and care, the number of HIV-infected individuals and opioid users can be reduced. We set up an optimal control problem to find the best control strategy that will minimize the numbers of HIV-infected individuals, drug users, and individuals with both HIV and opioid addiction while minimizing the cost of implementing such a strategy. We extend our model described by system (2.1) to include the HIV and opioid controls, which results in the following non-autonomous system;

$$
\begin{aligned}
S' &= \Lambda - \lambda_u S - \lambda_v S - \mu S + \delta u_1(t) U \\
U' &= \lambda_u S - q_u(1 - u_2(t))\lambda_v U - (\mu + \mu_u + \delta u_1(t))U \\
V' &= \lambda_v S - q_v \lambda_u V - (\mu + \mu_v + \gamma_v(1 - u_3(t)))V + \delta_2 u_4(t) I_2 \\
I_1' &= q_u(1 - u_2(t))\lambda_v U + q_v \lambda_u V - (\mu + \alpha + \mu_1) I_1 \\
I_2' &= \alpha I_1 - (\mu + \gamma_2(1 - u_3(t)) + \mu_2 + \delta_2 u_4(t)) I_2 \\
A' &= \gamma_v(1 - u_3(t))V + \gamma_2(1 - u_3(t)) I_2 - (\mu + \mu_a) A.
\end{aligned}
\tag{2.6}
$$

Opioid addiction can be treated and prevented by education and therapy. Furthermore, addicted individuals are treated with medicine in rehabilitation centers. We denote these combined addiction treatments as $u_1(t)$ and $u_4(t)$ for opioid addicted class, $U(t)$, and co-affected class $I_2(t)$, respectively. Injection is a method of opioid intake and needle sharing can cause HIV transmission from infected individual to non-infected addicted individual. Health care education about not sharing syringes, social protection, and rehabilitation, denoted by $u_2(t)$, are all aimed at reducing HIV-infection pathway from an infected to an addicted individual. Researchers have been working for more than three decades to find a cure for HIV. HAART[33] has been demonstrated to be highly effective in viral suppression, to prolonging the lives of infected people and to reducing the rate of HIV transmission. This class of control is denoted by $u_3(t)$. All controls are expressed as Lebesgue measurable functions and we denote the set of all admissible functions as Γ;

$$
\begin{aligned}
\Gamma = \{u = (u_1, u_2, u_3, u_4) : 0 \leq u_i(t) \leq u_{i_{\max}} : u_i(t) \in \mathcal{L}[0, T] \\
i = 1, 2, 3, 4\}.
\end{aligned}
$$

where T denotes the final time of interest, the controls u_i, $i = 1, 2, 3, 4$ are Lebesgue measurable functions ($\mathcal{L}[0, T]$) with upper bounds $u_{i_{\max}}$. It is clear that implementing all these control measures comes with a cost. We want to determine the control strategy which minimizes both the number of individuals with HIV and addictions and the cost of executing such a program. The objective function that we minimize is

$$\begin{aligned}\mathcal{J}(u) &= \int_0^T (a_1U + a_2V + a_3I_1 + a_4I_2 + a_5A + a_6u_1^2 + a_7u_2^2 \\ &\qquad + a_8u_3^2 + a_9u_4^2)\, dt \\ &= \int_0^T L(x,u)dt. \qquad (2.7)\end{aligned}$$

Here, the coefficients a_i, $i = 1,\ldots,5$ represent balancing weight of state variables which are minimizing and $a_i, i = 6,\ldots,9$ are weight of controls.

Thus, we seek optimal control $u^* = (u_1^*, u_2^*, u_3^*, u_4^*)$ such that

$$u^* = \min_{u\in\Gamma} \mathcal{J}(u) \qquad (2.8)$$

Using the Fellipo–Cesari theorem,[14] one can show that there exists a solution to the optimal control problem (2.6)–(2.8).

We continue with Pontryagin's Maximum Principle[42] to derive the necessary conditions that optimal control and corresponding state variables must satisfy. Pontryagin's Maximum Principle allows us to formulate an adjoint problem which will then transform the optimization problem into a problem of determining the pointwise minimum of the Hamiltonian. We present the Hamiltonian function H for the optimal control problem (2.8).

$$\begin{aligned}H &= a_1U + a_2V + a_3I_1 + a_4I_2 + a_5A + a_6u_1^2 + a_7u_2^2 + a_8u_3^2 + a_9u_4^2 \\ &\quad + \lambda_1(\Lambda - \lambda_uS - \lambda_vS - \mu S + \delta u_1(t)U) \\ &\quad + \lambda_2(\lambda_uS - q_u(1-u_2(t))\lambda_vU - (\mu + \mu_u + \delta u_1(t))U) \\ &\quad + \lambda_3(\lambda_vS - q_v\lambda_uV - (\mu + \mu_v + \gamma_v(1-u_3(t)))V \\ &\quad + \delta_2u_4(t)I_2t) + \lambda_4(q_u(1-u_2(t))\lambda_vU + q_v\lambda_uV \qquad (2.9) \\ &\quad - (\mu + \alpha + \mu_1)I_1) + \lambda_5(\alpha I_1 - (\mu + \gamma_2(1-u_3(t)) \\ &\quad + \mu_2 + \delta_2u_4(t))I_2) + \lambda_6((1-u_3)\gamma_v(t)V \\ &\quad + \gamma_2(1-u_3(t))I_2 - (\mu + \mu_a)A).\end{aligned}$$

Theorem 2.4. *Given an optimal control* $u^* = (u_1^*, u_2^*, u_3^*, u_4^*)$ *and corresponding state variables* $x^* = (S^*, U^*, V^*, I_1^*, I_2^*, A^*)$ *of the system (2.6), the adjoint functions* $\lambda_i(t)\ i = 1, \ldots, 6$ *satisfy the following system:*

$$
\begin{aligned}
\lambda_1'(t) &= (\lambda_1 - \lambda_2)\beta_u(U + I_1 + I_2)\frac{(N - S)}{N^2} + (\lambda_1 - \lambda_3) \\
&\quad \times \beta_v(V + I_1 + I_2)\frac{(N - S)}{N^2} + (\lambda_4 - \lambda_2)q_u(1 - u_2(t))\beta_v \\
&\quad \times \frac{(V + I_1 + I_2)}{N^2}U + (\lambda_4 - \lambda_3)q_v\beta_u\frac{(U + I_1 + I_2)}{N^2}V + \lambda_1\mu, \\
\lambda_2'(t) &= -a_1 + (\lambda_1 - \lambda_2)\beta_u S\frac{(S + V + A)}{N^2} + (\lambda_3 - \lambda_1)\beta_v S \\
&\quad \times \frac{(V + I_1 + I_2)}{N^2} + (\lambda_2 - \lambda_4)q_u(1 - u_2(t))\beta_v(V + I_1 + I_2) \\
&\quad \times \frac{(N - U)}{N^2} + (\lambda_3 - \lambda_4)q_v\beta_u V\frac{(S + V + A)}{N^2} \\
&\quad + (\lambda_2 - \lambda_1)\delta u_1(t) + \lambda_2(\mu + \mu_u), \\
\lambda_3'(t) &= -a_2 + (\lambda_2 - \lambda_1)\beta_u S\frac{(U + I_1 + I_2)}{N^2} + (\lambda_1 - \lambda_3)\beta_v S \\
&\quad \times \frac{(S + U + A)}{N^2} + (\lambda_2 - \lambda_4)q_u(1 - u_2(t))\beta_v U\frac{(S + U + A)}{N^2} \\
&\quad + (\lambda_3 - \lambda_4)q_v\beta_u(U + I_1 + I_2)\frac{N - V}{N^2} + (\lambda_3 - \lambda_6) \\
&\quad \times \gamma_v(1 - u_3(t)) + \lambda_3(\mu + \mu_v), \\
\lambda_4'(t) &= -a_3 + (\lambda_1 - \lambda_2)\beta_u S\frac{(S + V + A)}{N^2} + (\lambda_1 - \lambda_3)\beta_v S \\
&\quad \times \frac{(S + U + A)}{N^2} + (\lambda_2 - \lambda_4)q_u(1 - u_2(t))\beta_v U\frac{(S + U + A)}{N^2} \\
&\quad + (\lambda_3 - \lambda_4)q_v\beta_u V\frac{(S + V + A)}{N^2} + (\lambda_4 - \lambda_5)\alpha + \lambda_4(\mu + \mu_1), \\
\lambda_5'(t) &= -a_4 + (\lambda_1 - \lambda_2)\beta_u S\frac{(S + V + A)}{N^2} + (\lambda_1 - \lambda_3)\beta_v S
\end{aligned}
$$

$$\times \frac{(S+V+A)}{N^2} + (\lambda_2-\lambda_4)q_u(1-u_2(t))\beta_v U \frac{(S+V+A)}{N^2}$$
$$+ (\lambda_3-\lambda_4)q_v\beta_u V\frac{(S+V+A)}{N^2} + (\lambda_5-\lambda_3)\delta_2 u_4(t)$$
$$+ (\lambda_5-\lambda_6)\gamma_2(1-u_3(t)) + \lambda_5(\mu+\mu_2),$$
$$\lambda_6'(t) = -a_5 + (\lambda_2-\lambda_1)\beta_u S\frac{(U+I_1+I_2)}{N^2} + (\lambda_3-\lambda_1)\beta_v S$$
$$\times \frac{(V+I_1+I_2)}{N^2} + (\lambda_4-\lambda_2)q_u(1-u_2(t))\beta_v U\frac{(V+I_1+I_2)}{N^2}$$
$$+ (\lambda_4-\lambda_3)q_v\beta_u V\frac{(U+I_1+I_2)}{N^2} + \lambda_6(\mu+\mu_a). \qquad (2.10)$$

with $\lambda_i(T)=0$, for $i=1,\dots,6$.

And the optimal controls $u^* = (u_1^*, u_2^*, u_3^*, u_4^*)$ satisfy

$$u_1^* = \min\left(u_{1\max}, \max\left(0, \delta\frac{U^*(\lambda_2-\lambda_1)}{2a_6}\right)\right).$$
$$u_2^* = \min\left(u_{2\max}, \max\left(0, q_u\beta_v U^*\frac{(\lambda_4-\lambda_2)}{2a_7}\frac{(V^*+I_1^*+I_2^*)}{N^*}\right)\right).$$
$$u_3^* = \min\left(u_{3\max}, \max\left(0, \gamma_v\frac{V^*(\lambda_6-\lambda_3)}{2a_8} + \gamma_2\frac{(\lambda_6-\lambda_5)I_2^*}{2a_8}\right)\right).$$
$$u_4^* = \min\left(u_{4\max}, \max\left(0, \delta_2\frac{I_2^*(\lambda_5-\lambda_3)}{2a_9}\right)\right). \qquad (2.11)$$

Proof. Applying Pontryagin's Maximum Principle, we obtain the adjoint system from

$$\lambda_1' = -\frac{\partial H}{\partial S},\ \lambda_2' = -\frac{\partial H}{\partial U},\ \lambda_3' = -\frac{\partial H}{\partial V},\ \lambda_4' = -\frac{\partial H}{\partial I_1},$$
$$\lambda_5' = -\frac{\partial H}{\partial I_2},\ \lambda_6' = -\frac{\partial H}{\partial A}$$

with zero final time conditions. To obtain the system (2.11), we solve the following system for controls u^*:

$$\frac{\partial H}{\partial u_1} = 2a_6u_1 + \delta(\lambda_1 U - \lambda_2 U) = 0,$$

$$\frac{\partial H}{\partial u_2} = 2a_7u_2 + q_u\lambda_v(\lambda_2 U - \lambda_4 U) = 0 \tag{2.12}$$

$$\frac{\partial H}{\partial u_3} = 2a_8u_3 + \gamma_v(\lambda_3 V - \lambda_6 V) + \gamma_2(\lambda_5 I_2 - \lambda_6 I_2) = 0,$$

$$\frac{\partial H}{\partial u_4} = 2a_9u_4 + \delta_2(\lambda_3 I_2 - \lambda_5 I_2) = 0$$

and impose the bounds on the controls. □

2.5 Parameter Estimation

Data: Center for Disease Control (CDC) collects and reports surveillance data for US HIV infections since 2008. We obtained time series data of HIV — AIDS diagnoses, HIV deaths from the CDC website.[1] CDC defines the HIV diagnoses as the number of HIV infections, confirmed by laboratory or clinical evidence in one calendar year, regardless of the stage of HIV infection.[5] CDC divides the course of HIV infection into three stages; first stage is the initial acute HIV-infection phase, second stage is the chronic HIV-infection phase, and the third stage is AIDS. CDC reports HIV deaths data as the number of deaths of individuals who are diagnosed with HIV infection regardless of cause of death. Similarly, we obtained time series data of opioid deaths from the "Drug Overdose Deaths in the United-States 1999–2018" report by National Center for Health Statistics.[2] We present the data used to estimate the parameters of model (2.1) in Table 2.3.

Parameter estimation: In compact form, the model (2.1) can be rewritten as

$$\boldsymbol{x}'(t) = f(\boldsymbol{x}, \boldsymbol{p}), \quad \boldsymbol{x}(0) = \boldsymbol{x}_0, \quad \boldsymbol{y}(t) = h(\boldsymbol{x}, \boldsymbol{p}) \tag{2.13}$$

where $\boldsymbol{x}$ denotes the state variables of the system. The evolution of the system ($\boldsymbol{x}(t)$) depends on the parameters ($\boldsymbol{p}$) and the initial conditions ($\boldsymbol{x}_0$), while we observe the quantities ($\boldsymbol{y} = (y_1, y_2, y_3, y_4)$) such as HIV diagnoses, HIV deaths, AIDS diagnoses, and drug overdose deaths. All these observations are functions of the state variables and model parameters. The number of HIV cases per year is given in model (2.1) as $\beta_v\lambda_v S + q_u\lambda_v U$. CDC reports in the surveillance

Table 2.3. HIV cases, HIV deaths, AIDS cases, and drug overdose deaths.

Years	HIV diagnoses	HIV deaths	AIDS diagnoses	Opioid deaths
1999	—	—	—	8,050
2000	—	—	38,285	8,407
2001	—	—	36,922	9,496
2002	—	—	36,726	11,920
2003	—	—	37,317	12,940
2004	—	—	36,220	13,756
2005	—	—	34,261	14,916
2006	—	—	32,790	17,545
2007	—	—	31,984	18,516
2008	47,247	18,525	31,384	19,582
2009	44,716	18,043	30,187	20,422
2010	43,020	16,742	27,401	21,089
2011	41,265	16,300	25,620	22,784
2012	40,512	16,018	24,684	23,166
2013	39,232	15,908	23,656	25,052
2014	40,005	16,145	19,313	28,647
2015	39,817	15,860	18,590	33,091
2016	39,569	16,395	18,375	42,249
2017	38,351	16,358	17,749	47,600
2018	37,428	15,483	17,113	46,802

Note: Resources are Refs. 1 and 2.

data the HIV diagnoses regardless of the stage of the infection. That is the individuals who have been diagnosed with HIV infection for the first time and who have progressed into the third stage (AIDS) are reported in the HIV diagnoses category. So, HIV diagnoses as defined by CDC would be in the model (2.1) as

$$y_1(t) = \beta_v \lambda_v S + q_u \lambda_v U.$$

Similarly, the number of new AIDS cases per year is given in the model as

$$y_3(t) = \gamma_v V + \gamma_2 I_2.$$

Treatment with HAART delayed the progression into AIDS. Since the introduction of HAART, deaths due to AIDS have declined.[12] But another epidemic (opioid addiction) has emerged since then; and deaths due to drug overdose among people living with HIV have been rising. Authors reviewed the literature for the drug overdose epidemic among people living with HIV,[12] and they present that drug overdose mortality rate among HIV infected population varies from 0.7% to 31% (see Table 2.1 in Ref.[12]). In another study, researchers analyzed the drug-related deaths among people living with HIV in British Columbia between April 1996 and December 2017.[8,43] The analysis included 10,362 HIV-positive people who are 20 years of age or older, and 266 people (3%) who died of drug overdoses during the study period.[8,43] Based on this information, we split the disease-induced death rates of co-affected classes I_1 and I_2 as $q\%$ due to drug overdose and remaining $(1-q)\%$ due to HIV infection. So, we take the HIV and drug overdose deaths, respectively, as

$$\begin{aligned} y_2(t) &= \mu_v V + (1-q)\mu_1 I_1 + (1-q)\mu_2 I_2 + \mu_a A, \\ y_4(t) &= \mu_u U + q\mu_1 I_1 + q\mu_2 I_2. \end{aligned}$$

We estimate the model parameters by solving the following optimization problem using `fminsearhbnd` in MATLAB.

$$\begin{aligned} \hat{\boldsymbol{p}} = \min_{\boldsymbol{p}} \Bigg(& \sum_{i=1}^{n_1} \frac{|y_1(t_i) - Y_1^i|^2}{\hat{Y}_1} + \sum_{i=1}^{n_2} \frac{|y_2(t_i) - Y_2^i|^2}{\hat{Y}_2} + \sum_{i=1}^{n_3} \frac{|y_3(t_i) - Y_3^i|^2}{\hat{Y}_3} \\ & + \sum_{i=1}^{n_4} \frac{|y_4(t_i) - Y_4^i|^2}{\hat{Y}_4} \Bigg), \end{aligned} \tag{2.14}$$

where $Y_j,\ j = 1,2,3,4$ are the data presented in Table 2.3, and $\hat{Y}_j$ represents the average data value, which is derived as

$$\hat{Y}_j = \frac{1}{n_j} \sum_{i=1}^{n_j} Y_j^i.$$

Estimate parameter values are presented in Table 2.4.

Identifiability analysis: We observe structural and practical identifiability methodology for the model, since we have a nonlinear ODE model (2.1) of HIV and opioid. The structural identifiability is a theoretical method of determining identifiability for noise-free large enough data, while practical identifiability is the determination of identifiability using experimental data with noise.[34,39] If a unique parameter set leads to the unique output, the model is called globally structural identifiable. If an infinite number of parameter sets lead

Table 2.4. Results of parameter estimation problem (2.14).

β_u	β_v	δ	q_u	μ_u	q_v	α	μ_v	δ_2	μ_a	μ_1	μ_2
0.37	0.71	0.001	10.04	0.19	1	24.99	0.21	0.99	0.08	46.1	8.89
γ_v	γ_2	$\mathcal{R}_0^u$	$\mathcal{R}_0^v$	$\mathcal{R}_{\text{inv}}^u$	$\mathcal{R}_{\text{inv}}^v$						
0.499	0.99	1.77	0.98	12.2889	54.9602						

to the same observations, then the model is called structurally non-identifiable. If finite number of parameter sets lead to the same observational output, the model is called locally structural identifiable.[34] There are several methods for determining the identifiability of a nonlinear model. More information on the approaches can be found in the articles in Refs. 34, 39 and 23. We use Mathematica implementation of the probabilistic semi-numerical algorithm described in Refs. 23 and 39 naming `Identifiability Analysis` package to test identifiability of the model, as follows:

$$\boldsymbol{x}'(t) = f(\boldsymbol{x}, \boldsymbol{p}), \quad \boldsymbol{x}(0) = \boldsymbol{x}_0, \quad \boldsymbol{y}(t) = h(\boldsymbol{x}, \boldsymbol{p}). \tag{2.15}$$

The model is globally structural identifiable. Appendix 2.A contains the Mathematica code for the model for determining identifiability. The structural identifiability analysis serves as a theoretical foundation for the practical identifiability analysis. According to Ref. 39, a structurally identifiable model's practical identifiability is also unknown. As a result, the practical identifiability of a model from noisy data must be determined. For the practical identifiability analysis of the model (2.1), we use Monte Carlo simulation, a common simulation process. This approach employs random numbers and probability distributions as a sampling technique. More information on the Monte Carlo simulation can be found in Ref. 34. For the measurement or output model to have measurement errors from the observations such as HIV diagnoses, HIV deaths, AIDS diagnoses, and drug overdose deaths, the statistical model is as follows:

$$\boldsymbol{y}(t) = h(\boldsymbol{x}, \hat{\boldsymbol{p}}) + \epsilon(t), \tag{2.16}$$

where $\boldsymbol{p}$ denotes the true parameter values from curve fitting, $\boldsymbol{y}(t) = (y_1(t), y_2(t), y_3(t), y_4(t))$, and $\epsilon(t) = \{\epsilon_i, \quad i = 1, 2, 3, 4\}$ are the random variables that represent the measurement or observation error[39] in the discrete time $t = \{t_j\}_{j=1}^{n}$. The measurement error is considered to have the following form:

$$\epsilon_i = y_i(t)\mathcal{E}_i.$$

Here, $\mathcal{E}_i$ is standard deviation σ_0. The outlines of the procedure which we follow for the simulation are as follows:

(1) Determine the estimated true parameter values $\hat{\boldsymbol{p}}$ for simulation from curve fitting.
(2) Use the true parameter values $\hat{\boldsymbol{p}}$ to numerically solve the ODE model (2.1) to get the solution of the output $h(\boldsymbol{x}, \hat{\boldsymbol{p}})$ at the discrete data time points $\{t_i\}_{n_i=1}^{n}$.
(3) Generate $N = 1,000$ of simulated data from the measurement model (2.16) with measurement error using normal distribution with mean increasing errors gradually from $\sigma_0 = 0\%$, 1%, 5%, 10%, 20%, and to 30%.
(4) Fit the model (2.1) to each of the N simulated datasets to obtain parameter estimate $\hat{\boldsymbol{p}}_i, i = 1, \ldots, N$.
(5) Calculate the average relative estimation error (ARE) for each element of $\boldsymbol{p}$ as

$$\text{ARE} = 100\% \times \frac{1}{N}\sum_{i=1}^{N} \frac{|\hat{\boldsymbol{p}}^{(k)} - \hat{\boldsymbol{p}}_i^{(k)}|}{|\hat{\boldsymbol{p}}^{(k)}|},$$

where $\hat{\boldsymbol{p}}^{(k)}$ is the kth element of $\hat{\boldsymbol{p}}$ and $\hat{\boldsymbol{p}}_i^{(k)}$ is kth element of $\hat{\boldsymbol{p}}_i$.

We compute the ARE of the parameters to see if the estimates of parameters are suitable or not. The fact that a small increase in ARE of a parameter corresponds to an increase in noise level of σ_0 indicates that the parameter is essentially recognizable. The large increase in ARE, on the other hand, indicates that the parameter is sensitive and unidentifiable. For the purpose of observation, we use the CDC's natural death rate of $\mu = 1/79$ per year, the initial states vector from Figure (2.2), and the recruitment rate $\Lambda = S(0) * \mu$.

We observe two times the Monte Carlo Simulation of these estimated parameters from curve fitting. According to the ARE of parameters observed from the first time simulation presented in Table 2.5, parameters such as δ, q_u, δ_2, μ_a, μ_1, μ_2, α, γ_v, and γ_2 of the model (2.1) are observed practically unidentifiable since those

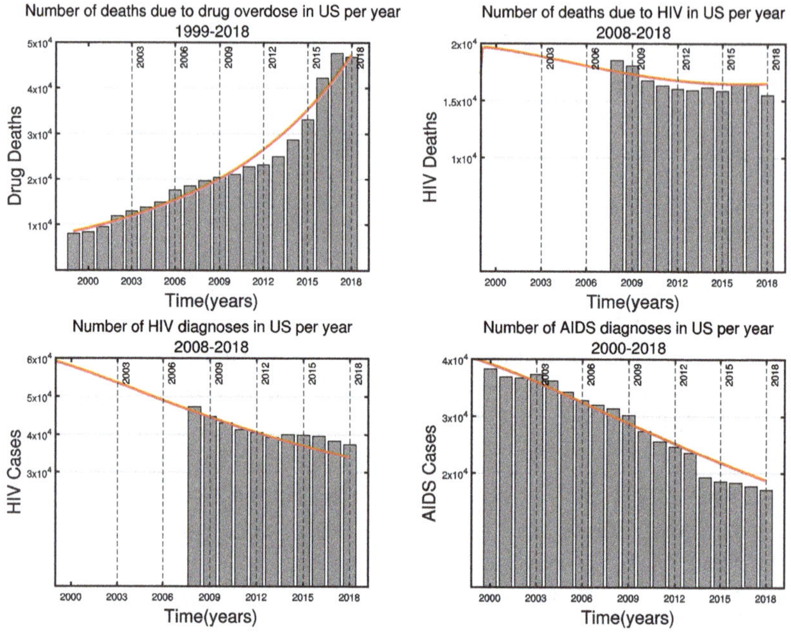

Fig. 2.2. This is the model fitting data for parameters taken from SSE (2.14). The vertical rectangles are for the deaths and cases data, and red curves are fitting modeling of corresponding data. In this case, initial values for susceptible $S(0) = 7779681$, Opioid addicted $U(0) = 42450$, HIV infected $V(0) = 42450$, opioid addicted first stage HIV infected $I_1(0) = 10$, opioid addicted second stage HIV infected $I_2(0) = 5$, and AIDS $A(0) = 31384$.

parameters show significant increase in ARE. Thus, we use the following definitions to fix some of the unidentifiable parameter values.

- $\gamma_v = 0.499$. The meaning of $\dfrac{1}{\gamma_v} \approx 2$ years is time duration to transfer HIV to AIDS. According to CDC page,[1] the duration is 2–10 years. We take low value of approx 2–3 years.
- $\alpha = 24.99$. $\dfrac{1}{\alpha} \approx 0.04$ year is time taken to transfer from I_1 to I_2 stage. CDC says acute stage infection takes 2–3 weeks to transfer into chronic stage infection. The duration is actually 0.04 years.
- Death rate μ_1 due to HIV is faster than death rate μ_2 due to opioid.
- q_u is high value as opioid user has more risk of transfer of HIV.

Table 2.5. These are Monte Carlo simulation results before fixing parameters.

Parameters	$\sigma_0 = 0\%$ ARE	$\sigma_0 = 1\%$ ARE	$\sigma_0 = 5\%$ ARE	$\sigma_0 = 10\%$ ARE	$\sigma_0 = 20\%$ ARE	$\sigma_0 = 30\%$ ARE
β_u	0	1.17	5.6	12.9	27.6	39.8
β_v	0	0.4	2.4	5.2	10.8	16.9
δ	0	437.9	2,430	4,611.3	8,042.3	10,511.1
q_u	0	6.8	39.5	84.2	174.3	220
μ_u	0	2	11.6	24.1	41.9	51.6
q_v	0	0.8	1.9	3.2	2.4	3.4
μ_v	0	2.7	11.4	24.2	55.6	89.5
δ_2	0	29.7	41.2	43	45.9	44.5
μ_a	0	20.5	76.6	127	161.8	194.3
μ_1	0	31.6	112.3	193.7	260.4	187.8
μ_2	0	37.3	132.6	241.7	222.5	347.2
α	0	15	23	25.1	26.1	25.3
γ_v	0	0.4	3.1	6.9	15	21.7
γ_2	0	15.5	19.3	20.1	22.3	24.1

Table 2.6. These are Monte Carlo simulation result after fixing some parameters.

Parameters	$\sigma_0 = 0\%$ ARE	$\sigma_0 = 1\%$ ARE	$\sigma_0 = 5\%$ ARE	$\sigma_0 = 10\%$ ARE	$\sigma_0 = 20\%$ ARE	$\sigma_0 = 30\%$ ARE
β_u	0	2	10.4	18.2	28.9	34.4
β_v	0	0.4	2	3.8	7.6	12.4
μ_u	0	4.8	25.7	45.2	69.8	81.3
q_v	0	3.5	18.3	20.6	16.5	7.4
μ_v	0	2.2	11.9	21.5	36.8	54.6

Fixing some unidentifiable parameters and then fitting rest of the set of parameters gives an idea to observe the Monte Carlo simulation. The second Monte Carlo simulation result with fixing some unidentifiable parameters from the first simulation is in Table 2.6. This shows that β_u, β_v, μ_u, q_v, and μ_v are practically identifiable. The parameters thus determined by fixing and fitting are presented in Table 2.4.

2.6 Optimal Control Numerical Analysis

In this section, we observe the influence of controls in the model (2.1) using optimal control theorem described in Section 2.4. To study the effect, we first apply the single control in model (2.1) and observe the scenario of their results obtained from individual control strategy. This study gives us an idea about the bounds of controls and weight of coefficients. Later we apply all four controls in model (2.1) and analyze the result.

While applying a control only at a time, models with a control thus formed for each control u_1, u_2, u_3 and u_4, are described in Appendix 2.B. To analyze numerically the solution of the model with controls and without controls, we use parameters from Table 2.4. The parameters value from Table 2.4 provides us the reproductive number $\mathcal{R}_0^u = 1.77$, which shows the presence of opioid in the general population, whereas $\mathcal{R}_0^v = 0.98$ means the HIV will disappear gradually in time. We also calculate the opioid invasion number $\mathcal{R}_{\text{inv}}^u = 12.289$ and the HIV invasion number $\mathcal{R}_{\text{inv}}^v = 54.9602$; these numbers indicate the presence of both diseases in the population with the corresponding parameters. Similarly, we obtain the initial values of susceptible $S(0) = 7,779,681$, opioid addicted $U(0) = 42,450$, HIV infected $V(0) = 42,450$, opioid addicted first stage HIV infected $I_1(0) = 10$, opioid addicted second stage HIV infected $I_2(0) = 5$ from the fitting, whereas the initial value of AIDS individuals $A(0) = 31,384$ is from the US CDC data.[1] We use the following algorithm to obtain the converged form of the optimal control function solutions.

(1) Take initial guesses for u_1, u_2, u_3, and u_4,
(2) Solve for the state function differential equations $x'(t) = f(t, x, u)$ by using the ode15s solver with taking initial value $x(0)$,
(3) Solve for the adjoint function differential equation $\lambda_i'(t) = -\frac{\partial H}{\partial x_i}$, $\lambda(T) = 0$ by using the ode15s solver,
(4) Update the new values of u_1, u_2, u_3, and u_4 by using (2.12),
(5) Iterate the same process until the following convergence criteria are met for every state variable, adjoint function, and control function,[42]

$$\frac{\|x^k - x^{k-1}\|}{\|x^k\|} > \kappa$$

for $\kappa > 0$.

Now we begin our study with applying a single control u_1, which is the recovery from addiction of opioid users, in model (2.1). The optimal control model (2.B.1) thus formed is in Appendix 2.B. Since the recovery rate δ from estimated parameters set is equal to 0.001, we take the bound of u_1 is in the interval [0,100] because the time taken for an individual to recover from opioid is $1/\delta = 100$ year. Similarly for coefficients, since u_1 is applied to control opioid users U, we take weight of U, V, I_1, and I_2 equally in objective functional (2.B.2) to minimize the infection and addiction. Thus, we take all coefficients $a_i = 1$ for all $i = 1, ..., 6$. We illustrate the dynamics of control u_1 only and state variables in Figure 2.3.

Next, we apply single control u_2, which is education for not sharing needles with opioid users, in the model HIV-opioid(2.1). The optimal control problem (2.B.3) thus developed is presented in Appendix 2.B. Since the bound of u_2 is applied in fraction form $(1 - u_2)$ in the model (2.B.3), we take bound of u_2 in the interval [0, 1]. In this case, $u_2 = 0$ denotes no control and $u_2 = 1$ denotes complete control. The control u_2 which is in fact several awareness programs, helps to prevent going from opioid users group to first-stage HIV infected I_1 class. While providing such activities at the highest level possible in society, opioid users are able to analyze and prevent needle sharing. In order to place more weight on minimizing HIV infected individuals, infected individuals, and AIDS people, we take $a_1 = 0.1$ for U and all other balancing coefficients a_2, a_3, a_4, a_5, and a_7 with equal weight 1 on objective functional (2.B.4). The dynamics of control u_2 only and state variables are in Figure 2.4.

Following above now, we apply the single control u_3, which is a HAART control in the HIV-opioid model (2.1). The HAART therapy treatment approach is described in detail in the Ref. 20. This therapy is effective in lowering the HIV viral load in body cells to the point that HIV becomes less active in the body, but it must be continued for a long period. This treatment does not completely

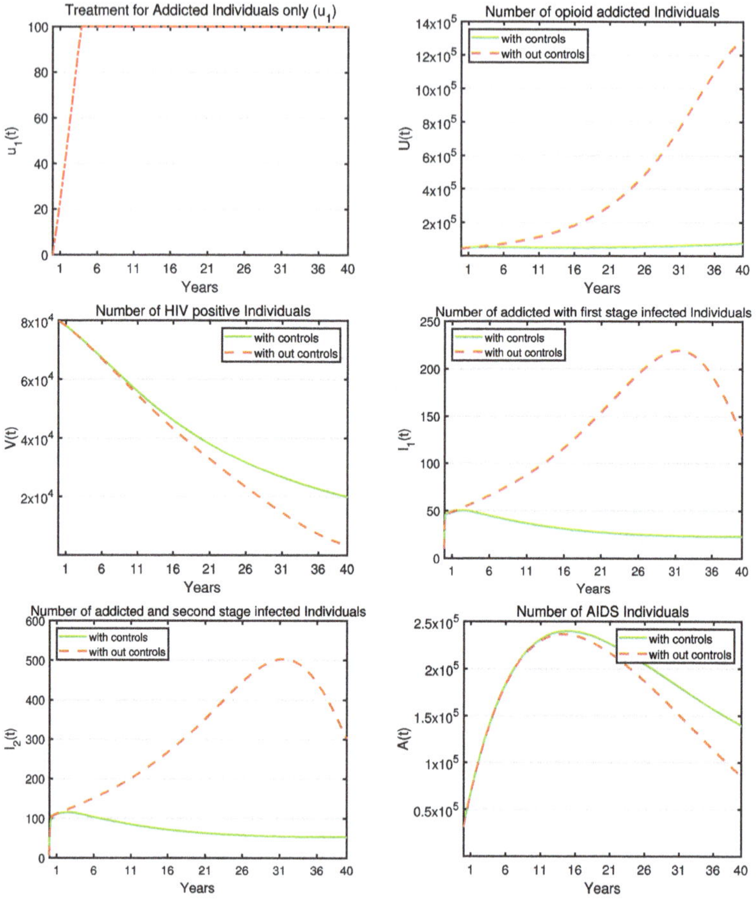

Fig. 2.3. These are computational analysis of optimal control problem with u_1 only (2.B.1) with the objective functional (2.B.2). We get this results with all weights a_1, a_2, a_3, a_4, a_5, as well as a_6 equal to 1. The boundary for u_1 is taken $[0, 100]$. The reproduction numbers for opioid is $\mathcal{R}_0^u = 1.77$ and so for HIV is $\mathcal{R}_0^u = 0.98$. Then the basic reproduction number of the model without control with these parameters is 1.77. The graphs visualize the dynamics behavior of the U, V, I_1, I_2, and A. After around the 5th year, the maximum control is seen applied in this 40-year time period for the u_1 control. When it comes to state variables, the red dash curve represents the solution with no control and the green curve represents the solution with control. This control reduces the opioid users, addicted with first stage infected individuals or second stage infected individuals, whereas number of HIV positive and AIDS individuals remain still slightly more than with no control.

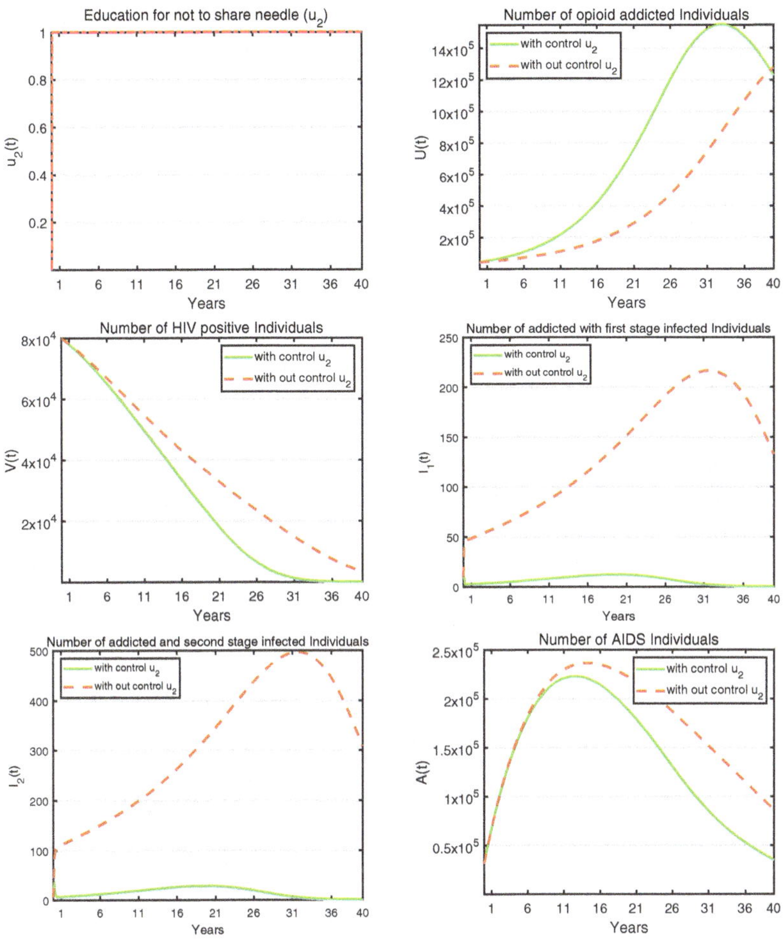

Fig. 2.4. Optimal control u_2 with bound $[0,1]$ is shown in the above graph. The control is applied maximum all the time. We take weight for U which is $a_1 = 0.1$ and all other weights a_2, a_3, a_4, a_5, and a_7 of state variables and control of (2.B.4) equal to 1. The basic reproduction number without control is $\mathcal{R}_0 = 1.77$. For the dynamics of state variables, the red dash curve represents the solution with no control and the green curve represents the solution with control. Even though the control u_2 increases the number of U individuals in the 40 years' duration, it reduces the numbers of V, I_1, I_2, and A individuals in the time interval. If people do not share the needle, then there is minimum chance of HIV spreading. So with this control opioid users increase their number but HIV positive, addicted with first stage infected, addicted with second stage infected, and AIDS are controlled to increase in number.

cure HIV patients but lowers the virus in infected people. The optimal problem setup formed by applying single control u_3 is presented in (2.B.5). The bound of u_3 is taken [0,1] as the control is applied in the form $(1-u_3)$. Once the viral load of infected people is reduced, the danger of HIV transmission is also lowered. Thus, considering the weight of implementation to minimize HIV and AIDS comparatively lower than the other state variables such as U, I_1, and I_2, we take $a_2 = 0.001, a_5 = 0.001$ on V and A, and all other coefficients a_1, a_3, and a_4 equal to 1. The dynamics of state variables are visualized in Figure 2.5. Similarly, we apply the rehab control u_4, which is for addicted with second stage HIV-infected people in the HIV-opioid model (2.1) . The corresponding optimal control model thus formed is presented in Appendix 2.B. Since the recovery rate is noted $\delta_2 = 0.99$, which is comparatively greater than δ, the time taken to recover from addiction in this case is $1/\delta_2$ which is around one to two years, so we take the bound of u_4 is $[0, 1]$. We consider the scenario in which relative costs are equal, and therefore take all coefficients are equal to 1. The dynamics of the state variables U, V, I_1, I_2, and A and u_4 are illustrated in Figure 2.6.

Finally, we analyze the case when all four controls — u_1, u_2, u_3, and u_4 — are applied together into the HIV-opioid model (2.1) to study the dynamics of U, V, I_1, I_2 and A. By observing the previous single control applied scenario, we note that the bound of controls and the weight of coefficients play a vital role in the optimal control problem. Keeping the importance of weight and bound in this four control case, we take different combination of bounds of control and coefficients. In the combination we can take the boundary of controls u_1 and u_4 more than 1, since u_1 and u_4 refer to the treatment of addicted people such as U and I_2 of model 2.6, whereas so of u_2 and u_3 in between 0 and 1 since these are in fraction. Similarly, the value of the coefficients a_i, $i = 1, ..., 5$ are taken according to the performance of state variables while minimizing the infection. The three combinations and their results are as follows:

(1) The bound of u_1 is $[0, 100]$, so of other controls are $[0, 1]$, and weight of all coefficients are 1. This is because of thinking that more force is to be given in the recovery of opioid by treatment

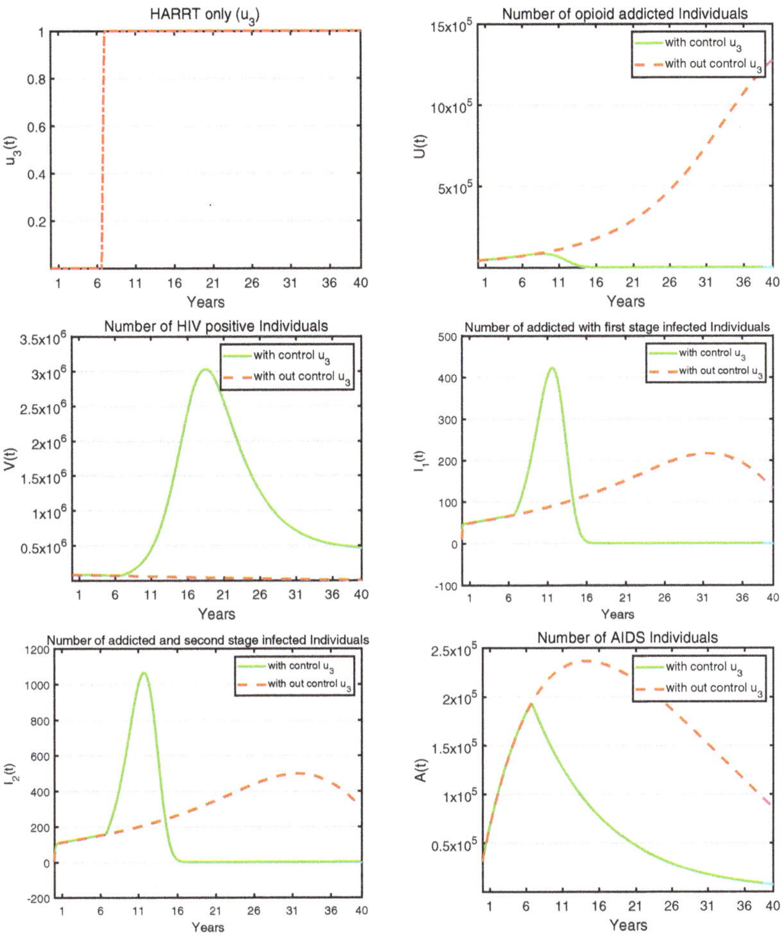

Fig. 2.5. u_3 has bound in $[0, 1]$. We take the weights $a_2 = 0.001$ and $a_5 = 0.001$ and all others a_1, a_3, and a_4 have equal weight 1 on (2.B.6). The basic reproduction number without control is $\mathcal{R}_0 = 1.77$. Like in other figures, the red dash curve represents the solution with no control and the green curve represents the solution with control. With these coefficients, the control starts to apply at highest level after 6 years and the number of U, I_1, I_2, and A individuals reduce to almost zero level, whereas number of V individuals increases sharply up to around 20th years, but later gets decreased continuously in the 40 years' time duration. This control helps to decrease viral load of HIV from body cells. So it prevents more people from being infected and to get into AIDS stage, but it helps to move into HIV positive but not infected group V. So increase in HIV group V is reasonable with this control.

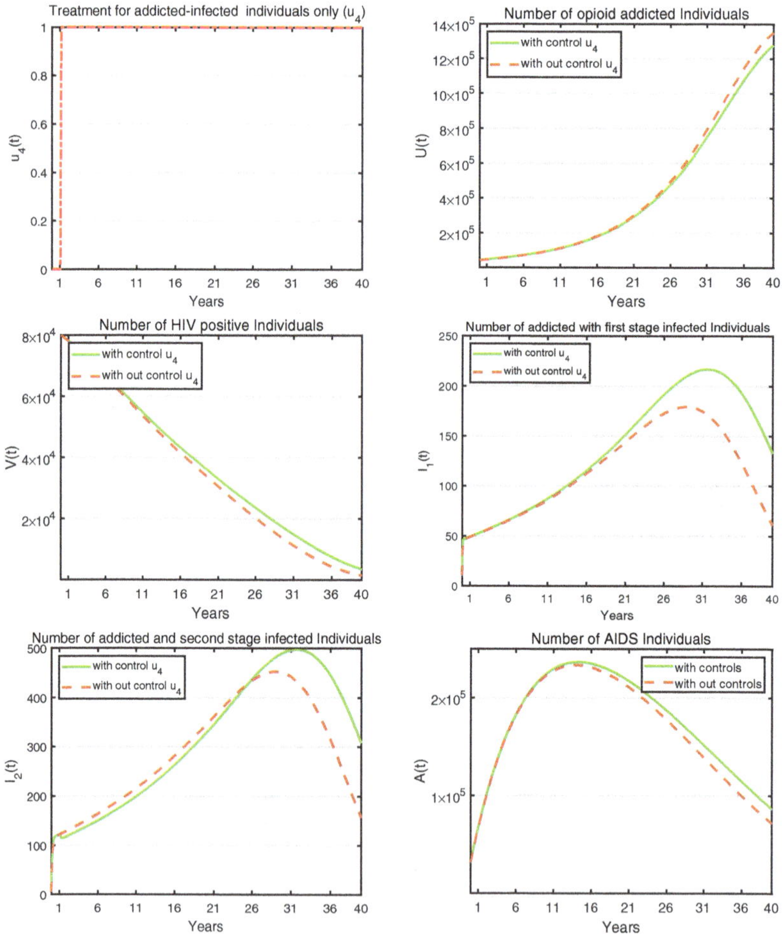

Fig. 2.6. With the bound [0, 1] of u_4, and having 1 weight for all a_1, a_2, a_3, a_4, a_5, and a_9, the control u_4 is seen applied in highest level after one year. The basic reproduction number without control is $\mathcal{R}_0 = 1.77$. In the case of state variables, the red dash curve represents the solution with no control and the green curve represents the solution with control. With this control, the number of opioid users U decrease whereas other number of individuals such as V, I_1, I_2, and A increase in the 40 years duration. This is the control which helps to get rid from opioid addiction of second stage infected people and replace the individuals into the HIV group V. So it is reasonable to say that this control helps to increase V, I_1, I_2, and A.

and similar level of focus to all other controls u_2, u_3, u_4, and equal weight to state variables U, V, I_1, I_2, and A. The dynamics of controls and state variables of this case are given in Figure 2.C.1.

(2) The bound of all controls are $[0, 1]$ and weight of a_2, a_5 is 0.001 as well as so of other coefficients is 1. This set is because of giving more focus to minimize U, I_1, and I_2. This is considered to be the case because once the HIV/AIDS disease invades into the body, the virus cannot be removed completely. However, the opioid addiction can be cured and viral load can be reduced from the body. The dynamics of controls and state variables are presented in Figure 2.C.2.

(3) The bound of u_1 is to be $[0, 10]$, so of other controls is $[0, 1]$, and weight of $a_2 = 0.01$, $a_5 = 0.89$ as well as so of other coefficients is 1. This set is for giving more range to opioid treatment and more weight to minimize U, I_1, and I_2. The solution of controls as well as dynamics of state variables of this case are visualized in Figures 2.7 and 2.8.

2.7 Discussion

We obtained from CDC information that one of the ways of HIV transmission is needle sharing. As people with drug addiction share needles to inject drugs like heroin, we considered these contagious needles help to spread HIV in the population. To study the intertwined spreading dynamics of these two diseases, we developed a HIV-opioid model (2.1). In the model, we verified that disease-free equilibrium existed locally if basic reproduction number $\mathcal{R}_0 < 1$, where $\mathcal{R}_0 = \max(\mathcal{R}_0^{\mathrm{u}}, \mathcal{R}_0^v)$. We determined that the invasion reproduction number for opioid addiction $\mathcal{R}_{\text{inv}}^u$ must be less than 1 for AOE to occur, with reproduction number for opioid addiction, $\mathcal{R}_0^u > 1$. Similarly, the invasion reproduction number $\mathcal{R}_{\text{inv}}^v$ should be less than 1 for HOE, with reproduction number for HIV, $\mathcal{R}_0^v > 1$.

To study this work with real life supported data, CDC surveillance data are employed for parameter estimation and identifiability of

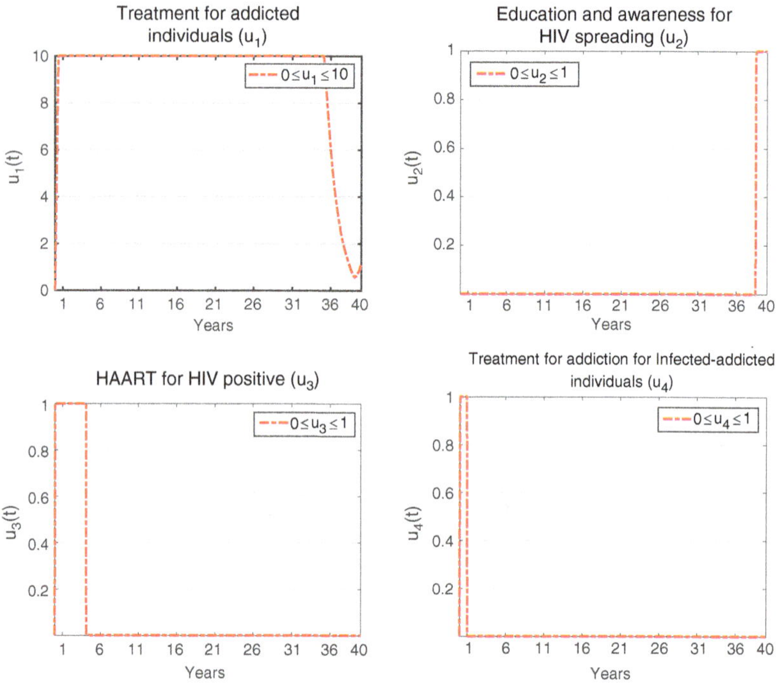

Fig. 2.7. These are solution graphs of controls u_1, u_2, u_3, and u_4 when we take bound of u_1 is $[0, 10]$, whereas $[0, 1]$ for all u_2, u_3, and u_4. The basic reproduction number without control is $\mathcal{R}_0 = 1.77$. To simulate these solutions of controls, we take $a_2 = 0.01$ and $a_5 = 0.89$, whereas all other weights of coefficients are 1. In this case, the u_1 is applied at highest level after around one year to around the 36th year. After then the control u_1 gets reduced to lower level. Next, the u_2 is seen applied to almost zero level until around the 38th year and then increase up to the highest level. Unlike u_2, the control u_3 is applied in the maximum level during beginning years up to around 5th year and then the control gets reduced to zero level. The fourth control u_4 is seen applied to the highest level up to first year and it gets decreased to around zero level all the way to 40 years. As a whole, it is reasonable to point out that u_1 is seen contributing more than other controls.

the HIV–opioid model. With the help of Mathematica and Monte Carlo simulation, the model is verified as structurally and practically identifiable.

The strength of this work is its examination of the significant effects on the solution of state variables after control functions

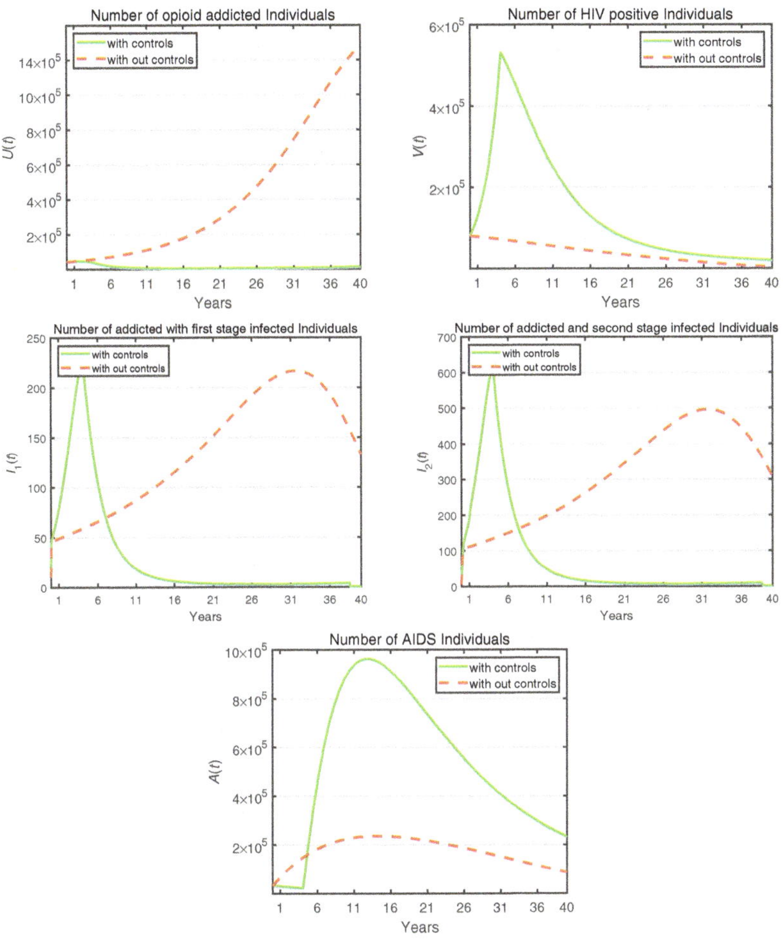

Fig. 2.8. These are dynamics of state variables in the model (2.6) with objective functional (2.8) when controls u_1 with bound $[0, 10]$ and u_2, u_3, u_4 with bound $[0, 1]$ are applied. In this case, the weights $a_2 = 0.01, a_5 = 0.89$, all other weights are taken as 1. In the above graphs, the red dash curve represents the solution with no control and the green curve represents the solution with control. The dynamics of U, I_1, I_2 are seen clearly reduced in the number during this 40 years, whereas V and A are a little bit higher than the no control case. While applying all the four controls at the same time with these bounds and weights, the opioid and addicted with both stage infected get almost zero level, whereas number of HIV and AIDS also become into lower number.

are applied. The novelty of this chapter is to study the dynamics of HIV infection and opioid addiction when four controls such as treatments for opioid addicted individuals, awareness for not sharing needles, HAART for HIV-infected people, and rehab for addicted with infected people are implemented at the same time in the population level. We used the Fellipo–Cesari theorem and Pontryagin's maximum principle to assess the sufficient and necessary conditions for the existence of control functions.

We began this study by observing the results with each single control applied in the HIV–opioid model (2.1). This single control approach is helpful to understand at least an insight of existence and significant effect of the control in the state variables. Those results are presented in Figures 2.3–2.6. We then applied all four controls together in the HIV–opioid model (2.1) and the model thus formed with four controls is given in equation (2.6). We used the comparison method to analyze the results obtained from optimal control analysis of (2.6) and (2.8). For that, we considered three sets of combinations of bounds of controls and weights of coefficients. The combinations are determined as follows:

- Bound of u_1 is [0, 100] and bound of all other three controls u_2, u_3, and u_4 is [0, 1]. Weight of all coefficients of U, V, I_1, I_2, and A from objective functional (2.8) is equally 1.
- Bound of all controls is [0, 1]. Weight of $a_2 = a_5 = 0.001$ and so of all other coefficients is equally 1.
- Bound of u_1 is [0, 10] and bound of all other three controls u_2, u_3, and u_4 is [0, 1]. Weight of $a_2 = 0.01$, $a_5 = 0.89$ and so of all other coefficients is equally 1.

The results of those set of problems are illustrated and described in Figures 2.7 and 2.8 and 2.C.1 and 2.C.2.

Observing these three cases, we preferred the case when bound of u_1 is $[0, 10]$, so of other controls is $[0, 1]$, and weight of $a_2 = 0.01$, $a_5 = 0.89$ as well as so of all other coefficients is 1. The dynamics of state variables and controls of this case are illustrated in Figures 2.7 and 2.8. By giving more bound for u_1 and more weight to the variables U, I_1, and I_2, we obtained that the number of individuals are decreased in the variables with controls. When decreasing the number of I_1 and I_2 with controls, obviously V and A get forced to increase because of the nature of the virus, which does not disappear from the body but might be lower in viral load. Once the viral load is decreased in the body, the transmission rate of virus is also reduced. As we see the graph of controls, the u_1, u_3, and u_4 are noted by applied to the maximum level in the beginning years but the nominal level in the later years, however u_2 is in nominal level in the beginning years but higher level after around 38th of the 40 years. This means cost of controls for u_1, u_3, and u_4 are going to be decreased after some years, whereas cost of u_2 is going to get to the maximum level once the other controls are nominal for giving the outcomes of state variables. Those are the reasons why we prefer the combination of bounds and weights.

As a conclusion, since no vaccinations and drugs are available as of now to remove HIV completely from the body, these controls are useful to reduce the number of HIV/AIDS patients from society so that the virus can be slowly eradicated from the world.

Note: All codes of the analysis are in this link: https://github.com/aneupanetims2016/HIV-opioid-code

Acknowledgments

Necibe Tuncer was partially supported by NSF DMS-1951626.

Appendix: 2.A Identifiability Analysis and Result

```
In[1]:= Clear all ["Global * "];
          Needs ["IdentifiabilityAnalysis"];
         m=1/79; L= 7779681*m;
      deq = {
x1'[t]==L - bu * (x2[t] + x4[t] +x5[t])*x1[t]/x7[t] - bv *(x3[t]+ x4[t] +x5[t])* x1[t] / x7[t]
- m*x1[t] +delta *x2[t],

x2'[t]== bu * (x2[t] + x4[t] +x5[t])*x1[t]/x7[t] - qu * bv * (x3[t]+ x4[t] +x5[t])* x2[t] / x7[t]
-(m +mu +delta) * x2[t],

x3'[t]== bv *(x3[t]+ x4[t] +x5[t])* x1[t] / x7[t] - qv * bu *  (x2[t] + x4[t] +x5[t])*x3[t] /x7[t]
- (m + mv + gammav)* x3[t] +delta2 *x5[t],

x4'[t]== qu*bv* (x3[t] + x4[t] + x5[t])*x2[t] / x7[t] + qv *bu *(x2[t] +x4[t] +x5[t]) * x3[t] / x7[t] -
(m +m1 + alpha)* x4[t],

x5'[t] == alpha * x4[t] - (m+m2+gamma2 +delta2)* x5[t],

x6'[t]== gammav * x3[t] + gamma2 *x5[t] - (m+ma) *x6[t],

x7'[t] == L - m* x7[t] -mu * x2[t] - mv * x3[t] - m1 * x4[t] - m2* x5[t] - ma * x6[t],

x1[0] == x10, x2[0] == x20, x3[0]== x30 , x4[0]==x40, x5[0]== x50, x6[0]== x60, x7[0] == x70
};
iad = IdentifiabilityAnalysis[{deq, mu*x2[t], mv* x3[t], gammav* x3[t] + gamma2 * x5[t],
bv * (x3[t] +x4[t]+ x5[t])* x1[t]/x7[t] +qu * bv * (x3[t] + x4[t] + x5[t])* x2[t] / x7[t]},
(x1, x2, x3, x4, x5, x6, x7},
{bu, bv, mu, delta, gammav, mv, qu, qv, gamma2, m2, m1, delta2, alpha, ma, x10, x20, x30,
x40, x50, x60, x70},t]

out[5] = IdentifiabilityAnalysisData [True, <>]
```

Appendix 2.B Optimal Control Problem Setup for Each of the Controls u_1, u_2, u_3, and u_4 Individually

2.B.1 *For u_1 only case*

$$
\begin{aligned}
S' &= \Lambda - \lambda_u S - \lambda_v S - \mu S + \delta u_1(t) U \\
U' &= \lambda_u S - q_u \lambda_v U - (\mu + \mu_u + \delta u_1(t)) U \\
V' &= \lambda_v S - q_v \lambda_u V - (\mu + \mu_v + \gamma_v V + \delta_2) I_2 \\
I_1' &= q_u \lambda_v U + q_v \lambda_u V - (\mu + \alpha + \mu_1) I_1 \\
I_2' &= \alpha I_1 - (\mu + \gamma_2 + \mu_2 + \delta_2) I_2 \\
A' &= \gamma_v V + \gamma_2 I_2 - (\mu + \mu_a) A.
\end{aligned}
\tag{2.B.1}
$$

The objective functional

$$
\mathcal{J}(u) = \int_0^T \left(a_1 U + a_2 V + a_3 I_1 + a_4 I_2 + a_5 A + a_6 u_1^2\right) dt. \tag{2.B.2}
$$

2.B.2 *For u_2 only case*

$$
\begin{aligned}
S' &= \Lambda - \lambda_u S - \lambda_v S - \mu S + \delta U \\
U' &= \lambda_u S - q_u (1 - u_2(t)) \lambda_v U - (\mu + \mu_u + \delta) U \\
V' &= \lambda_v S - q_v \lambda_u V - (\mu + \mu_v + \gamma_v) V + \delta_2 I_2 \\
I_1' &= q_u (1 - u_2(t)) \lambda_v U + q_v \lambda_u V - (\mu + \alpha + \mu_1) I_1 \\
I_2' &= \alpha I_1 - (\mu + \gamma_2 + \mu_2 + \delta_2) I_2 \\
A' &= \gamma_v V + \gamma_2 I_2 - (\mu + \mu_a) A.
\end{aligned}
\tag{2.B.3}
$$

The objective functional

$$
\mathcal{J}(u) = \int_0^T \left(a_1 U + a_2 V + a_3 I_1 + a_4 I_2 + a_5 A + a_7 u_2^2\right) dt. \tag{2.B.4}
$$

2.B.3 For u_3 only case

$$\begin{aligned}
S' &= \Lambda - \lambda_u S - \lambda_v S - \mu S + \delta U \\
U' &= \lambda_u S - q_u \lambda_v U - (\mu + \mu_u + \delta) U \\
V' &= \lambda_v S - q_v \lambda_u V - (\mu + \mu_v + \gamma_v(1 - u_3(t)))V + \delta_2 I_2 \\
I_1' &= q_u \lambda_v U + q_v \lambda_u V - (\mu + \alpha + \mu_1) I_1 \\
I_2' &= \alpha I_1 - (\mu + \gamma_2(1 - u_3(t)) + \mu_2 + \delta_2) I_2 \\
A' &= \gamma_v(1 - u_3(t))V + \gamma_2(1 - u_3(t)) I_2 - (\mu + \mu_a) A.
\end{aligned} \tag{2.B.5}$$

The objective functional

$$\mathcal{J}(u) = \int_0^T \left(a_1 U + a_2 V + a_3 I_1 + a_4 I_2 + a_5 A + a_8 u_3^2\right) dt. \tag{2.B.6}$$

2.B.4 For u_4 only case

$$\begin{aligned}
S' &= \Lambda - \lambda_u S - \lambda_v S - \mu S + \delta U \\
U' &= \lambda_u S - q_u \lambda_v U - (\mu + \mu_u + \delta) U \\
V' &= \lambda_v S - q_v \lambda_u V - (\mu + \mu_v + \gamma_v)V + \delta_2 u_4(t) I_2 \\
I_1' &= q_u \lambda_v U + q_v \lambda_u V - (\mu + \alpha + \mu_1) I_1 \\
I_2' &= \alpha I_1 - (\mu + \gamma_2 + \mu_2 + \delta_2 u_4(t)) I_2 \\
A' &= \gamma_v V + \gamma_2 I_2 - (\mu + \mu_a) A.
\end{aligned} \tag{2.B.7}$$

The objective functional

$$\mathcal{J}(u) = \int_0^T \left(a_1 U + a_2 V + a_3 I_1 + a_4 I_2 + a_5 A + a_9 u_4^2\right) dt. \tag{2.B.8}$$

Appendix 2.C Some Results of Optimal Control with Different Bounds of Controls and Weights

2.C.1 *Result of optimal controls with u_1 in $[0, 100]$ and all other controls have bound $[0, 1]$ and equal weight coefficients* 1 (Figure 2.C.1).

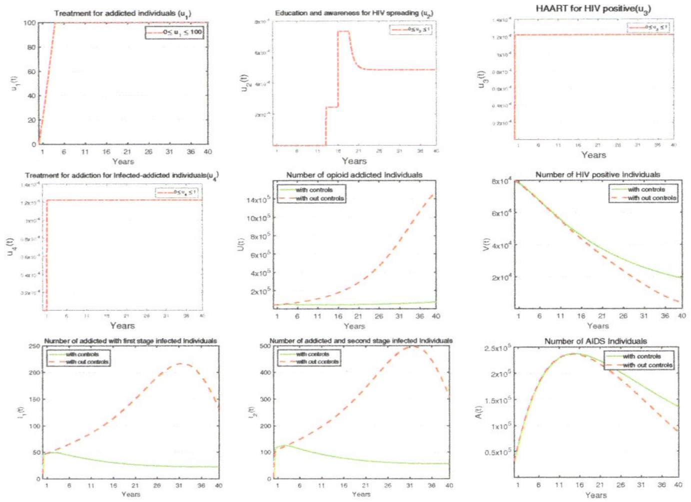

Fig. 2.C.1. These are graphs of solution of controls and state variables of the model (2.6) with objective functional (2.8) having controls as $0 \leq u_1 \leq 100$ and all other controls u_2, u_3, u_4 with boundaries $[0, 1]$. The weights of all coefficients of (2.7) are taken 1. The basic reproduction number without control is $\mathcal{R}_0 = 1.77$. As we see, in the application of controls, the u_1 is used in a maximum level after around 5th year to 40th year, whereas u_2, u_3, and u_4 are observed in a lower level in the duration. For state variables, the red dash curve represents the solution with no controls and the green curve represents the solution with controls. The number of opioid users U, addicted with first-stage infected I_1 and addicted with second-stage infected I_2 are getting reduced, whereas number with individuals with HIV V and AIDS A are slightly higher than in the no controls solution.

2.C.2 *Result of optimal controls with equal boundaries* $[0, 1]$ *for all controls and* $a_2 = a_5 = 0.001$ *with equal weight of other coefficients is* 1. (Figure 2.C.2).

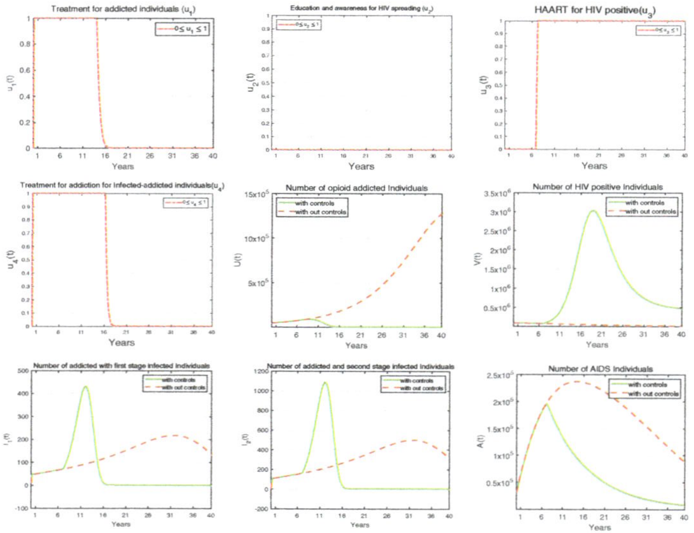

Fig. 2.C.2. These are graphs of solution of controls and state variables of the model (2.6) with objective functional (2.8) having controls with boundary of $[0, 1]$ as well as the weight of $a_2 = a_5 = 0.001$ and weight of all other coefficients being 1. The basic reproduction number without control is $\mathcal{R}_0 = 1.77$. To see application of controls, the u_1 and u_4 are getting reduced after some years, whereas u_3 is getting increased after 6th year but u_2 is applied nominally. Note that for state variables, the red dash curve represents the solution with no control and the green curve represents the solution with control. In this case, number of opioid individual U, addicted first-stage infected individuals I_1, addicted second-stage infected individuals I_2, and AIDS A are observed reduced with controls. But HIV positive cases are forced to increase with controls because of sharp decrease in I_1 and I_2 than with no controls case.

References

1. https://www.cdc.gov/nchhstp/atlas/index.htm.
2. https://www.cdc.gov/nchs/products/databriefs/db356.htm.
3. https://www.cdc.gov/hiv/basics/.
4. https://www.cdc.gov/hiv/basics/whatishiv.html.
5. https://www.cdc.gov/hiv/statistics/surveillance/terms.html.
6. https://www.cdc.gov/hiv/basics/index.html.
7. https://www.epbich.org/opioid.
8. https://www.aidsmap.com/news/jul-2020/overdose-deaths-have-reduced-survival-gains-among-people-hiv-british-columbia. Accessed March 1, 2021.
9. Steinberg, A. M. and Stalford, H. L. (1973). On existence of optimal control, *J. Optim. Theory Appl.*, **11**, 3, p. 266.
10. Filippovi, A. F. (1963). On the certain questions in the theory of optimal control, *J. SIAM Control* **1**, p. 76.
11. Arruda, E. F., Das, S. S., Dias, C. M., and Pastore, D. H. (2021). Modelling and optimal control of multi strain epidemics, with application to COVID-19, *PLoS One* **16**, 9, p. e0257512. https://doi.org/10.1371/journal.pone.0257512.
12. Manini, A. F. and Santos, C. D. (2011). The impact of the drug overdose epidemic on the mortality of persons with HIV/AIDS, *J. Clinic. Toxicol.*, **1**, S1, p. 4, doi:10.4172/2161-0495.S1-004.
13. Appa, A., Rowe, C., Hessol, N. A., and Coffin, P. (2020). Beyond overdose: Drug-related deaths in people with and without HIV in San Francisco 2007–2018. *Open Forum Infect. Dis.* **7**, 12, p. ofaa565. https://doi.org/10.1093/ofid/ofaa565.
14. Seierstad, A. and Sydsaster, K. (1986). *Optimal Control Theory with Economic Application*, Vol. 24, North Holland: Elsevier.
15. Mallela, A., Lenhart, S., and Vaidya, N. K. (2016). HIV-TB co-infection treatment: Modeling and optimal control theory perspectives, *J. Comput. Appl. Math.*, **307**, P. 143–161.
16. Fang, B., Li, X.-Z., Martcheva, M., and Cai, L.-M. (2015). Global asymptotic properties of a heroin epidemic model with treat-age, *Appl. Math. Comput.*, **263**, pp. 315–331.
17. Castilo-Chavez, C. and Feng, Z. (1997). To treat or not to treat: The case of tuberculosis, *Journal of Mathematical Biology*, **35**, 3/4, pp. 629–656.
18. Gao, D., Porco, T. C., and Ruan, S. (2016). Co-infection dynamics of two diseases in a single host population, *J. Math. Anal. Appl.*, **442**, pp. 171–188.

19. Krick, D. E. (2004). *Optimal Control Theory: An Introduction*, Dover Publications Inc., New York.
20. Granich, R., Crowley, S., Vitoria, M., Smyth, C., Kahn, J. G., Bennett, R., Lo, Y. R., Souteyrand, Y., and Williams, B. (2010). Highly active antiretroviral treatment as prevention of HIV transmission: review of scientific evidence and update. Curr. Opini. HIV and AIDS **5**, pp. 298–304. https://doi.org/10.1097/COH.0b013e32833a6c32.
21. Stein, E. and Sakarchi, R. (2005). *Real Analysis: Measure Theory, Integration and Hilbert Spaces*, Princeton University Press, Princeton, NJ.
22. Long, E. F., Vaidya, N. K., and Brandeau, M. L. (2008). Controlling co-Eepidemics: Analysis of HIV and tuberculosis infection dynamics, *Nat. Inst. Health*, **56**, pp. 1366–1381.
23. Karlsson, J., Anguelova, M., and Jirstrand, M. (2012). An efficient method for structural identifiability analysis of large dynamic systems. *IFAC Proc.*, **45**, 16, pp. 941–946. https://doi.org/10.3182/20120711-3-BE-2027.00381.
24. Blyuss, K. B. and Kyrychko, Y. N. (2005). On a basic model of a two-disease epidemic, *Appl. Math. Comput.* **160**, 1, pp. 177–187.
25. Ketcheson, D. I. (2021). Optimal control of an SIR epidemic through finite-time non-pharmaceutical intervention, *J. Math. Biol.*, **83**, 7, pp. 1–21. https://doi.org/10.1007/s00285-021-01628-9.
26. Lü, X., Hui, H.-W. and Liu, F.-F. *et al.* (2021). Stability and optimal control strategies for a novel epidemic model of COVID-19, *Nonlinear Dyn.* **106**, pp. 1491–1507. https://doi.org/10.1007/s11071-021-06524-x.
27. Roeger, L., Feng, Z., and Castillo-Chavez, C. (2009). Modeling TB and HIV co-infections, *Math. Biosci. Eng.* **6** 4, pp. 815–837.
28. A. M. Lerner and A. S. Fauci (2019). Opioid Injection in Rural Areas of the United States: A Potential Obstacle to Ending the HIV Epidemic. *JAMA* **322** 11, pp. 1041–1042, doi:10.1001/jama.2019.10657.
29. Martcheva, M. (2015). *An Introduction to Mathematical Epidemiology.* Chapter 9. Control Strategies. New York: Springer, pp. 215–244.
30. Vidyasagar, M. (1975). On the existence of optimal control, *J. Optim. Theory and Appl.*, **17**, 3/4, p. 273.
31. Martcheva, M. and Pilyugin, S. S. (2006). Pilyugin The role of coinfection in multi disease dynamics, *SIAM J. Appl. Math.*, **66**, pp. 843–872.
32. Bonnet, M., Baudin, E., Jani, I. V., Nunes, E., Verhoustraten, F., Calmy, A., Bastos, R., Bhatt, N. B., and Michon, C. (2013). Incidence of paradoxical tubercholosis-associated immune reconstitution inflammatory syndrome and impact on patient outcome, *PLoS One* **8**, 12, pp. 2577–2586.

33. Kretzchmar, M. and Wiessing, L. (2008). New challenges for mathematical and statistical modeling of HIV and hepatitis C virus in injecting drug users, *AIDS* **22**(13), pp. 1527–1537. ISSN 0269–9379.
34. Hongyu, M., Xia, X., Perelson, A. S., and Wu, H. (2011). On identifiability of nonlinear ODE models and applications in viral dynamics. *SIAM Review*, **53**, 1, pp. 3–39.
35. Tuncer, N., Torres, J., Martcheva, M., Barfield, M. and Robert, D. (2015). Dynamics of low and high pathogenic avian influenza in wild and domestic bird populations, *J. Biol. Dyn.*, **10**(1), pp. 104–139. ISSN: 1751–3758.
36. Neilan, R. M. and Lenhart, S. (2010). An introduction to optimal control with an application in disease modeling. In A. Gumel and S. Lenhart (eds.), *Modeling Paradigms and Analysis of Disease Transmission Models*, American Mathematical Society, Providence, RI, pp. 67–81.
37. NIDA (2021, May 21). Medication treatment of opioid use disorder can protect against overdose death. Retrieved from https://www.drugabuse.gov/news-events/nida-notes/2021/.05/medication-treatment-opioid-use-disorder-can-protect-against-overdose-death Accused 2021, May 28.
38. Prosper, O., Saucedo, O., Thompson, D., Torres-Garcia, G., Wang, X., and Castillo–Chavez, C. (2011). Modeling control strategies for concurrent epidemics of seasonal and pandemic H1N1 influenza. *Math. Biosci. Eng.*, **8**, pp. 141–170, doi: 10.3934/mbe.2011.8.141.
39. Tuncer, N., Martcheva, M., and LaBarre, B. *et al.* (2018). Structural and Practical Identifiability Analysis of Zika Epidemiological Models. *Bull. Math. Biol.*, **80**, pp. 2209–2241. https://doi.org/10.1007/s11538-018-0453-z
40. Nagina, P., Mbogo, R. W., and Luboobi, L. S. (2018). Modeling optimal control of In-host HIV dynamics using different control strategies, *Comuput. Math. Med. Methods Med.*, Article ID 9385080, p. 18.
41. Lynn, R. and Neilan, M. (2009). Optimal control applied to population and disease models, Doctorial dissertation, Tennessee Research and Creative Exchange.
42. Lanhart, S. and Workman, J. T. (2007). *Optimal Control Applied to Biological Models.* Chapter 4. Forward-Backward Sweep Method. Boca Raton: Champman & Hall/CRC, pp. 49–57.
43. St-Jean, M. *et al.* (2020). Drug overdoses are reducing the gains in life expectancy of people living with HIV (PLWH) in British Columbia. Canada. *23rd International AIDS Conference*, Abstract OAC0302.
44. Fleming, W. H. and Rishel, R. W. (1975). *Deterministic and Stochastic Optimal Control*, Springer-Verlag, New York.

45. Duan, X.-C., Lib, X.-Z., and Martcheva, M. (2020). Coinfection dynamics of heroin transmission and HIV infection in a single population, *J. Biol. Dyn.*, **14**, 1, pp. 116–142, https://doi.org/10.1080/17513758.2020.1726516.
46. Duan, X.-C., Li, X.-Z., and Martcheva, M. (2020). Qualitative analysis on a diffusive age-structured heroin transmission model, *Nonlinear Anal. Real World Appl.*, **54** pp. 103105.
47. Duan, X. C., Jung, I. H., Li, X. Z., and Martcheva, M. (2020). Dynamics and optimal control of an age-structured SIRVS epidemic model. *Math. Models Appl. Sci.*, **43**, Z, pp. 4239–4256.

https://doi.org/10.1142/9789811263033_0003

Chapter 3

The Effect of Lockdown on Mean Persistence Time of Highly Infectious Diseases: A Stochastic Model Based Study

Sk Shahid Nadim[*,†,‡,¶], **Bapi Saha**[*,§,||], **and Joydev Chattopadhyay**[‡,#]

[†]*Odum School of Ecology, University of Georgia, Athens, USA*
[§]*Govt. College of Engg. & Textile Technology, Berhampore, India*
[‡]*Agricultural and Ecological Research Unit, Indian Statistical Institute, India*
[¶]*nadimskshahid@gmail.com*
[||]*bapi.math@gmail.com*
[#]*joydev@isical.ac.in*

Infectious diseases have become a potential threat to public health over the last decade. This trend is possibly due to the emergence of highly pathogenic infections like Ebola, Influenza, West Nile virus, SARS, and very recently COVID-19, etc. These diseases are affecting the public health and have triggered significant economic damages worldwide. In this present study, we develop a stochastic epidemic model to study the effect of lockdown on infectious disease dynamics. The quarantined and un-quarantined susceptible and symptomatic and asymptomatic individuals are put into separate classes. The rate of susceptible to quarantine

[*]Sk Shahid Nadim and Bapi Saha share equal authorship.

and quarantine to susceptible are assumed to be functions of time to capture the non-uniformity in the said two rates during the lockdown phase and unlock phase. These two parameters are also assumed to be dependent on the period of complete lockdown. A new approach termed as maximum stability index is coined to see the effect of the period of complete lockdown on the mean persistence time, which is surprisingly difficult to achieve in case the dimension of the model is high. However, the mean persistence time of the total infected population including symptomatic and asymptomatic cases is obtained by taking the average of the observed time to extinctions based on simulation.

Keywords: Mathematical and Stochastic model, Lockdown, Discrete Time Markov Chain, Mean persistence time

3.1 Introduction

Infectious diseases are significant threats to modern society. Many new and re-emerging infectious diseases have endangered public health in the past few decades, posing new challenges to infectious disease experts throughout the world. The emergence of new zoonoses and the rapid spread of infectious diseases among humans are facilitated by the expansion of the human demographic, which results in a closer proximity to wildlife habitat and a massive urbanization process. Infectious diseases have played an undeniably significant role in human history, from plagues of biblical times to the current COVID-19 pandemic. Since prehistoric times, the continuous growth of human populations has resulted in successive invasions of the human population by a growing number of various pathogens. In recent years, COVID-19 has been one of humanity's greatest infectious issues, gaining pandemic status and having a huge effect on mortality rates in both rich and poor nations. Mathematical models of the transmission dynamics of infectious diseases are now omnipresent. These models play a significant role in the quantification of various ways for the prevention and mitigation of infectious diseases. An outbreak of a novel coronavirus pneumonia has been sweeping through China since December 2019.[1] The virus was named by the World Health Organization (WHO) as COVID-19 as of January 12; 2020. COVID-19 has recently spread to the large majority of countries such as the United States, India, France, Iran, Italy, Spain, etc. The epidemic of COVID-19 has become a major problem for public health in the medical community, as the virus spreads around the world. Symptoms of COVID-19 are most similar to severe acute respiratory syndrome (SARS) and MERS (breathing syndrome in the Middle East), like cough, nausea, shortness of breath, and difficulty.[2] The transmission routes include direct transmission, such as near contact and indirect air transmission through coughing and sneezing, even if any of the infected items are affected by virus particles. Given the large number of cases worldwide and low mortality compared with SARS and Middle Eastern Respiratory Syndrome (MERS), this virus is highly contagious and transmissible.

More than 22,000 healthcare workers in 52 countries are being screened positively for COVID-19 according to a regular monitoring report released by WHO.[3] There are many reports in India which clearly show that hospitals and quarantine centers are COVID-19 hotspots.[4,5] The physicians, nurses, and other health staffs are highly susceptible as they are closely associated to the COVID-19 patients.[6] Consequently, a large number of the susceptible population may be exposed to disease infection from patient interactions in hospitals and quarantine centers. However, recent experiences with the evolution of drug-resistant pathogens, the appearance of new diseases, and the problems of cost-effective treatment indicate that most diseases are still far from being eradicated. Recently, social distancing measure has been used successfully to monitor the spread of highly infectious disease although vaccines are available for many diseases. Governments from different countries ordered nationwide lockdowns restricting the movement of entire populations as a protective measure against the pandemic.[7] Many countries in the world implemented lockdown for several weeks to reduce the spread of infection. This situation demands the need for the study on the process of lockdown so that it can be implemented optimally.

The dynamics of highly infectious diseases are discussed in several mathematical models. There have been many modeling studies recently that provide details on various successful control measures to reduce disease transmission.[8–11,12] provided an effective lockdown and role of hospital-based COVID-19 transmission model. They also provided an effective lockdown strategy to regulate transmission of COVID-19 in India. They proposed cluster specific lockdown strategy for successful reduction of COVID-19 transmission in different parts of India.[13] It is observed that in almost all the work related to COVID-19, the parameters used to construct the models are assumed to be constant. We argue that the rate at which the susceptible population is becoming quarantined (l) and the rate at which the quarantined individuals enter into the susceptible compartment (ω) changes with time. It should be noted that it is quite natural that during the period of lockdown, the rate of making individuals quarantined is high in comparison to the period of unlocking. On the contrary, the rate at which the quarantined individuals are entering into the susceptible compartment is low in the lockdown period in comparison to the unlock period. This suggests that these two rates are not constant, but a sudden decrease in the rate of quarantine and sudden increase in the rate of unquarantine can be observed as soon as the lockdown is over. These rates are controlled by the restriction imposed by the Government and, in many cases, these are also influenced by the nature of the human population. All the remaining parameters are a model's inherent property and the population does not have direct control over it. Thus, in this chapter, we focus on these two parameters and consider these as functions of time and the functional forms are constructed with the help of Dirac delta function, which captures the required properties of these two rates.

Despite a potential number of studies in the deterministic model framework, study under stochasticity is limited though in real life stochastic perturbation is indispensable making stochastic modeling more appropriate. The literature of the stochastic epidemic model is old and vast and many of these works are related to mean persistence time of infection.[14] Motivated by this work, a stochastic model for infectious disease with hospital-based transmission is proposed and analyzed. We estimate the mean persistence time of the infected population in the system using the Discrete Time Markov Chain (DTMC) model although determining the time to extinction is surprisingly difficult. A series of papers came up regarding this issue.[14] To overcome the difficulty we utilize the 1,000 sample paths and take the average of the time to extinction of the infected population (both symptomatic and asymptomatic) of all the sample paths. The effect of the period of complete lockdown on the mean persistence time is discussed in detail. The analytical expressions for the mean time to extinction and the stability index of the infected individuals cannot be obtained due to the high value of the dimension of the model and very large value of the total population size. We propose a new index of stability which we call maximum stability index to have a feel about the effect of the period of lockdown and other relevant parameters on the stability index, which is the logarithm of the mean time to extinction of the infected population if the initial number of infected population is 1. We believe that this study will help to take proper management decisions to restrict the spread of infection.

3.2 Mathematical Model Utilized in the Creation of the Stochastic Model

In this study, an SEIR differential-equation model has been established to explain the transmission of highly infectious disease. Based on the known features of the disease pandemic, we assume that the total human population denoted by $N(t)$ is divided into seven mutually exclusive sub-populations such that each individual is in one of the following compartments:

- *Un-quarantined susceptible* (S_U): The un-quarantined individual who is not infected by the virus. For daily necessity, this community may go outside and get in contact with un-notified and hospitalized infected individuals.
- *Quarantined susceptible* (S_Q): The quarantined susceptible are vulnerable people who are in home isolation. They can become contaminated by un-notified infected individuals as un-quarantine individuals become infected without symptoms.
- *Exposed* (E): Population who are exposed are infected individuals but not infectious for the community. After becoming infected with the disease pathogen, the person is in the incubation phase and has no clear clinical signs.
- *Asymptomatic* (A): The individual infects others, but has not developed any clinical symptoms.
- *Symptomatic* (I): The individual infects others, begins to develop clinical symptoms, and the health authorities will identify and report when they arrive at the H compartments.
- *Hospitalized* (H): The individual is in hospital and is still able to infect others. A person passes to the recovered class at the end of this state.
- *Recovered* (R): The patient has survived the disease, is no longer contagious, and has acquired a natural immunity to the disease pathogen.

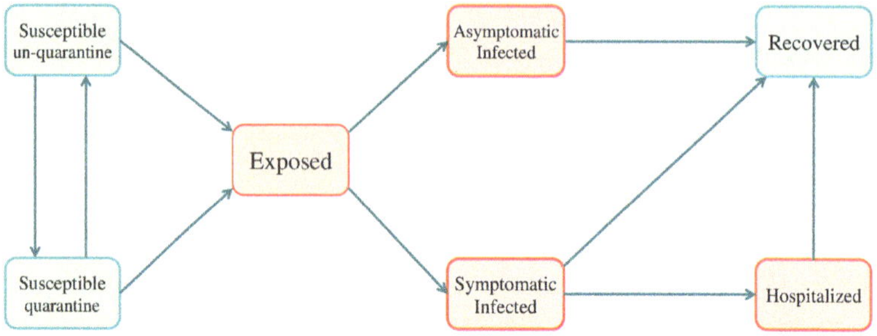

Fig. 3.1. Compartmental flow diagram of the proposed model (3.3).

where $N(t) = S_U(t) + S_Q(t) + E(t) + A(t) + I(t) + H(t) + R(t)$. The transmission mechanism of the COVID-19 diagram in India as seen in Figure 3.1. Susceptible individuals are generated by the recruitment of persons by birth or immigration to the community at a constant rate Π. Population decreases after infection and can be transmitted at a rate $\frac{\beta I}{N-S_Q}$, $\frac{\rho_1 \beta A}{N-S_Q}$, and $\frac{\rho_2 \beta H}{N-S_Q}$ by direct interaction between un-quarantined susceptible individuals with symptomatic infected, asymptomatic infected, and hospitalized individuals, respectively, and at a rate $\frac{\rho_3 \beta I}{N-S_U-H}$ and $\frac{\rho_4 \beta A}{N-S_U-H}$ by direct interaction between quarantined susceptible individuals with symptomatic infected and asymptomatic individuals, respectively. It is assumed that there is a recovery rate for asymptomatic, symptomatic, and hospitalized individuals, such as γ_1, γ_2, and γ_3, respectively. The model incorporates certain demographic impacts by estimating the disease-induced mortality rate δ of hospitalized persons and natural death rate μ of each of the seven sub-populations.

We focus our attention on separating the lock-down phase and unlock phase. We distinguish the two phases depending on the values of the parameters l and w. When lockdown phase gets over and the unlock phase starts, some restrictions are waived which is the surrogate of increase in the rate of population from quarantine to susceptible (ω). For the same reason, a decrease in the rate of susceptible to quarantine l can be observed. Thus, it is legitimate to assume that in the lockdown phase the value of the parameter l is high, whereas in the unlock phase it is low and vice versa for the parameter w. When the lockdown phase is over, a sudden decrease in l and a sudden increase in w are observed. Thus, we assume that l and w are functions of time. The sudden increase or decrease in l

and w, respectively, after lockdown are captured using delta function as follows:

$$\begin{aligned}\frac{\mathrm{d}l}{\mathrm{d}t} &= -\epsilon_1\delta(t-T),\\ \frac{\mathrm{d}w}{\mathrm{d}t} &= \epsilon_2\delta(t-T),\end{aligned} \tag{3.1}$$

where T is the period of complete or partial lockdown and the initial conditions stand as: $l(0) = l_0$ and $w(0) = w_0$. Under these condition, the solution for $l(t)$ and $w(t)$ can be written as

$$\begin{aligned}l(t) &= l_0 - \epsilon_1 H(t-T),\\ w(t) &= w_0 + \epsilon_2 H(t-T).\end{aligned} \tag{3.2}$$

Here, $\delta(t)$ is the Dirac delta function and $H(t)$ is the Heaviside function. Assuming all these, the following differential equations are considered to describe the disease transmission model.

$$\begin{aligned}
\frac{\mathrm{d}S_U}{\mathrm{d}t} &= \Pi + \omega(t)S_Q - \frac{\beta S_U I}{N-S_Q} - \frac{\rho_1\beta S_U A}{N-S_Q} - \frac{\rho_2\beta S_U H}{N-S_Q} - (\mu+l)S_U,\\
\frac{\mathrm{d}S_Q}{\mathrm{d}t} &= l(t)S_U - \frac{\rho_3\beta S_Q I}{N-S_U-H} - \frac{\rho_4\beta S_Q A}{N-S_U-H} - (\mu+\omega)S_Q,\\
\frac{\mathrm{d}E}{\mathrm{d}t} &= \frac{\beta S_U I}{N-S_Q} + \frac{\rho_1\beta S_U A}{N-S_Q} + \frac{\rho_2\beta S_U H}{N-S_Q}\\
&\quad + \frac{\rho_3\beta S_Q I}{N-S_U-H} + \frac{\rho_4\beta S_Q A}{N-S_U-H} - (\sigma+\mu)E,\\
\frac{\mathrm{d}A}{\mathrm{d}t} &= (1-\kappa)\sigma E - (\gamma_1+\mu)A\\
\frac{\mathrm{d}I}{\mathrm{d}t} &= \kappa\sigma E - (\tau+\mu+\gamma_2)I,\\
\frac{\mathrm{d}H}{\mathrm{d}t} &= \tau I - (\delta+\mu+\gamma_3)H,\\
\frac{\mathrm{d}R}{\mathrm{d}t} &= \gamma_1 A + \gamma_2 I + \gamma_3 H - \mu R.
\end{aligned} \tag{3.3}$$

All the parameters and their biological interpretation are given in Table 3.1, respectively.

Table 3.1. Parameters with their respective epidemiological explanation for the mechanistic mathematical model.

Parameters	Biological meaning	Value/Ranges	References
Π	Recruitment rate of human population	5.1234×10^4	—
$1/\mu$	Average life expectancy at birth	70.42 years	15
β	Transmission rate of infected individuals	1.7399	16
ρ_1	Modification factor	0.45	9
ρ_2	Modification factor	0.60	9
ρ_3	Modification factor	0.40	Assumed
ρ_4	Modification factor	0.35	Assumed
κ	Fraction of exposed population that become symptomatic infected	0.62 day^{-1}	13
σ	Rate of transition from exposed to infected class	0.1923 day^{-1}	1
l	Rate at which un-quarantined susceptible become quarantine	0.1041	13
τ	Rate at which symptomatic infected become hospitalized or notified	0.1632 day^{-1}	11
δ	Death rate of hospitalized or notified population	0.0162 day^{-1}	17
γ_1	Recovery rate for asymptomatic infected	0.73 day^{-1}	13
γ_2	Recovery rate for symptomatic infected	0.79 day^{-1}	13
ω	Rate at which quarantined susceptible become un-quarantine	0.0185 day^{-1}	13
γ_3	Recovery rate for hospitalized or notified individuals	0.7884 day^{-1}	17

Here the randomness in population growth rate is due to demographic stochasticity, which is the chance variation in the number of individual births and deaths and usually modeled using birth and death processes. It is customary to check some basic properties of a model, which are discussed in the next section.

3.2.1 *Basic properties*

3.2.1.1 *Positivity and boundedness of the solution for the model* (3.3)

This section is provided to prove the positivity and boundedness of solutions of the system (3.3) with initial conditions $(S_U(0), S_Q(0), E(0), A(0), I(0), H(0), R(0))^T \in \mathbb{R}_+^7$. We first state the following lemma.

Lemma 3.1. *Suppose* $\Omega \subset \mathbb{R} \times \mathbb{C}^n$ *is open,* $f_i \in C(\Omega, \mathbb{R}), i = 1, 2, 3, \ldots, n$. *If* $f_i|_{x_i(t)=0, X_t \in \mathbb{C}^n_{+0}} \geq 0$, $X_t = (x_{1t}, x_{2t}, \ldots, x_{1n})^T, i = 1, 2, 3, \ldots, n$, *then* $\mathbb{C}^n_{+0}\{\phi = (\phi_1, \ldots, \phi_n) : \phi \in \mathbb{C}([-\tau, 0], \mathbb{R}^n_{+0})\}$ *is the invariant domain of the following equations:*

$$\frac{\mathrm{d}x_i(t)}{\mathrm{d}t} = f_i(t, X_t), \quad t \geq \sigma, \quad i = 1, 2, 3, \ldots, n.$$

where $\mathbb{R}^n_{+0} = \{(x_1, \ldots, x_n) : x_i \geq 0, i = 1, \ldots, n\}$[18].

Proposition 3.1. *The system* (3.3) *is invariant in* $\mathbb{R}^7_+$.

Proof. By re-writing the system (3.3), we have

$$\frac{\mathrm{d}X}{\mathrm{d}t} = M(X(t)), \quad X(0) = X_0 \geq 0, \tag{3.4}$$

$M(X(t)) = (M_1(X), M_1(X), \ldots, M_6(X))^T$. We note that

$$\begin{aligned}
&\frac{\mathrm{d}S_U}{\mathrm{d}t}|_{S_U=0} = \Pi + \omega S_Q \geq 0, \quad \frac{\mathrm{d}S_Q}{\mathrm{d}t}|_{S_Q=0} = lS_U \geq 0,\\
&\frac{\mathrm{d}E}{\mathrm{d}t}|_{E=0} = \frac{\beta S_U I}{N - S_Q} + \frac{\rho_1 \beta S_U A}{N - S_Q} + \frac{\rho_2 \beta S_U H}{N - S_Q}\\
&\qquad\qquad + \frac{\rho_3 \beta S_Q I}{N - S_U - H} + \frac{\rho_4 \beta S_Q A}{N - S_U - H} \geq 0,\\
&\frac{\mathrm{d}A}{\mathrm{d}t}|_{A=0} = (1-\kappa)\sigma E \geq 0, \frac{\mathrm{d}I}{\mathrm{d}t}|_{I=0} = \kappa\sigma E \geq 0, \quad \frac{\mathrm{d}H}{\mathrm{d}t}|_{H=0} = \tau I \geq 0,\\
&\frac{\mathrm{d}R}{\mathrm{d}t}|_{R=0} = \gamma_1 A + \gamma_2 I + \gamma_3 H \geq 0.
\end{aligned}$$

Then it follows from the Lemma 3.1 that $\mathbb{R}^7_+$ is an invariant set. □

Lemma 3.2. *The system* (3.3) *is bounded in the region* $\Omega = \{(S_U, S_Q, E, A, I, H, R) \in \mathbb{R}^7_+ | S_U + S_Q + E + A + I + H + R \leq \frac{\Pi}{\mu}\}$

Proof. We observed from the system that

$$\frac{\mathrm{d}N}{\mathrm{d}t} = \Pi - \mu N - \delta H \leq \Pi - \mu N$$
$$\implies \lim_{t \to \infty} \sup N(t) \leq \frac{\Pi}{\mu}.$$

Hence, the system (3.3) is bounded. □

3.2.1.2 *Disease-free equilibrium and basic reproduction number*

The basic reproduction number R_0 is an epidemiologically significant threshold value, which determines the potential for an infectious disease to enter a population. The next generation matrix approach is used to obtain the basic reproduction number R_0 of the system (3.3). The system has a disease-free equilibrium given by

$$\varepsilon_0 = \left(\frac{\Pi(\mu + \omega)}{\mu(\mu + l + \omega)}, \frac{\Pi l}{\mu(\mu + l + \omega)}, 0, 0, 0, 0 \right).$$

For this, we assemble the compartments, which are infected from the system (3.3), and decompose the right-hand side as $\mathcal{F} - \mathcal{V}$, where $\mathcal{F}$ is the transmission part, expressing the production of new infection, and the transition part is $\mathcal{V}$, which describes the change in state.

$$\mathcal{F} = \begin{pmatrix} \frac{\beta S_U I}{N - S_Q} + \frac{\rho_1 \beta S_U A}{N - S_Q} + \frac{\rho_2 \beta S_U H}{N - S_Q} + \frac{\rho_3 \beta S_Q I}{N - S_U - H} + \frac{\rho_4 \beta S_Q A}{N - S_U - H} \\ 0 \\ 0 \\ 0 \end{pmatrix},$$

$$\mathcal{V} = \begin{pmatrix} (\sigma + \mu)E \\ -(1 - \kappa)\sigma E + (\gamma_1 + \mu)A \\ -\kappa\sigma E + (\tau + \mu + \gamma_2)I \\ -\tau I + (\delta + \mu + \gamma_3)H \end{pmatrix}.$$

Now we calculate the Jacobian of $\mathcal{F}$ and $\mathcal{V}$ at DFE ε_0. Following the next generation matrix method, the matrix F of the transmission terms and the matrix V of the transition terms calculated at ε_0 are

$$F = \frac{\partial \mathcal{F}}{\partial X} = \begin{pmatrix} 0 & \frac{\rho_1 \beta S_U^0}{N^0 - S_Q^0} + \frac{\rho_4 \beta S_Q^0}{N^0 - S_U^0} & \frac{\beta S_U^0}{N^0 - S_Q^0} + \frac{\rho_3 \beta S_Q^0}{N^0 - S_U^0} & \frac{\rho_2 \beta S_U^0}{N^0 - S_Q^0} \\ 0 & 0 & 0 & 0 \\ 0 & 0 & 0 & 0 \\ 0 & 0 & 0 & 0 \end{pmatrix},$$

$$V = \frac{\partial \mathcal{V}}{\partial X} = \begin{pmatrix} \sigma + \mu & 0 & 0 & 0 \\ -(1-\kappa)\sigma & \gamma_1 + \mu & 0 & 0 \\ -\kappa\sigma & 0 & \tau + \mu + \gamma_2 & 0 \\ 0 & 0 & -\tau & \delta + \mu + \gamma_3 \end{pmatrix}.$$

Calculating the spectral radius of the next generation matrix FV^{-1}, we obtain the basic reproductive number as follows[19, 20]:

$$R_0 = \frac{\kappa\sigma\tau\rho_2\beta}{(\sigma+\mu)(\tau+\mu+\gamma_2)(\delta+\mu+\gamma_3)} + \frac{\kappa\sigma\beta(1+\rho_3)}{(\sigma+\mu)(\tau+\mu+\gamma_2)} + \frac{(1-\kappa)\sigma\beta(\rho_1+\rho_4)}{(\sigma+\mu)(\gamma_1+\mu)}. \tag{3.5}$$

The basic reproduction number R_0 is defined as the expected number of secondary cases generated by one infected individual during their lifespan as infectious in a fully susceptible population. The basic reproduction number R_0 of (3.3) is given in 3.5.

Using Theorem 2 in Ref. 20, the following result is established.

Lemma 3.3. *The disease-free equilibrium ε_0 of system (3.3) is locally asymptotically stable whenever $R_0 < 1$, and unstable whenever $R_0 > 1$.*

3.3 Stochastic Model

In this section, we give the stochastic counterpart of the deterministic model to capture the randomness in the spread of the disease. We include demographic stochasticity with the help of birth and death processes. Suppose $S_U(t)$, $S_Q(t)$, $E(t)$, $A(t)$, $I(t)$, $H(t)$, and $R(t)$ are the random variables which denote the number of susceptible, home quarantined, exposed, infected asymptomatic, infected symptomatic, hospitalized, and recovery population, respectively, at time t. The values of the said random variables belong to the state space $(1, 2, 3, \ldots M)$. Here M is the maximum possible population size in the system. A strict upper limit for the maximum population size is not easy to determine. One standard way is to consider M to be the largest possible value for which the computation remains feasible.[21] In our present case, we consider M to be the total population size N of the system. Note that $S_U(t) + S_Q(t) + E(t) + A(t) + I(t) + H(t) + R(t) = N$. In the next section, we give the process of formation of joint probability distribution, which is useful to analyze the nature of the spread of the disease.

3.3.1 *Derivation of joint probability distribution*

The state of the system at time t can be characterized by the probability $p_{sqeaihr}(t)$ of having s number of susceptible, q no. of quarantine, e no. of exposed, a no. of asymptomatic, i no. of symptomatic infected, h no of hospitalized, and r no. of recovery individuals.

Suppose, $p_{sqeaihr}(t) = \mathbb{P}[S_U(t) = s, S_Q(t) = q, E(t) = e, A(t) = a, I(t) = i, H(t) = h, R(t) = r]$. The possible changes in the states of the different class of population and the corresponding probabilities are given in Table 3.2.

Note that the total time interval is divided in to small time intervals $(0, \Delta t, 2\Delta t, \ldots, n\Delta t)$. Here Δt is taken sufficiently small so that $\sum p_i \leq 1$.

Now,

$$\begin{aligned}
p_{sqeaihr}(t + \Delta t) = {} & p_{s-1qeaihr}(t)\pi\Delta t + p_{s-1q+1eaihr}(t)w(q+1)\Delta t \\
& + p_{s+1q-1eaihr}(t)l(s+1)\Delta t \\
& + p_{s+1qe-1aihr}(t)\frac{\beta(s+1)i}{N-q}\Delta t \\
& + p_{s+1qe-1aihr}(t)\frac{\rho_1\beta(s+1)a}{N-q}\Delta t \\
& + p_{s+1qe-1aihr}(t)\frac{\rho_2\beta(s+1)h}{N-q}\Delta t \\
& + p_{s+1qeaihr}(t)\mu(s+1)\Delta t \\
& + p_{sq+1e-1aihr}(t)\frac{\beta(q+1)(\rho_3 i + \rho_4 a)}{N-s-h}\Delta t
\end{aligned}$$

Table 3.2. Transitions of different states and corresponding probabilities.

State at time t	State after transition during Δt	Transition probability (p_i)
S_U, S_Q, E, A, I, H, R	$(S_U+1, S_Q, E, A, I, H, R)$	$\pi\Delta t = p_1$
	$(S_U+1, S_Q-1, E, A, I, H, R)$	$wS_Q\Delta t = p_2$
	$(S_U-1, S_Q+1, E, A, I, H, R)$	$lS_U\Delta t = p_3$
	$(S_U-1, S_Q, E+1, A, I, H, R)$	$\frac{\beta S_U I\Delta t}{N-S_Q} = p_4$
	$(S_U-1, S_Q, E+1, A, I, H, R)$	$\frac{\rho_1\beta S_U A\Delta t}{N-S_Q} = p_5$
	$(S_U-1, S_Q, E+1, A, I, H, R)$	$\frac{\rho_2\beta S_U H\Delta t}{N-S_Q} = p_6$
	$(S_U-1, S_Q, E, A, I, H, R)$	$\mu S_U\Delta t = p_7$
	$(S_U, S_Q-1, E+1, A, I, H, R)$	$\frac{\rho_3\beta S_Q I\Delta t}{N-S_U-H} = p_8$
	$(S_U, S_Q-1, E+1, A, I, H, R)$	$\frac{\rho 4\beta S_Q A}{N-S_U-H}\Delta t = p_9$
	$(S_U, S_Q-1, E, A, I, H, R)$	$\mu S_q\Delta t = p_{10}$
	$(S_U, S_Q, E-1, A+1, I, H, R)$	$(1-\kappa)\sigma E\Delta t = p_{11}$
	$(S_U, S_Q, E-1, A, I+1, H, R)$	$\sigma E\Delta t = p_{12}$
	$(S_U, S_Q, E-1, A, I, H, R)$	$\mu E\Delta t = p_{13}$
	$(S_U, S_Q, E, A-1, I, H, R+1)$	$\gamma_1 A\Delta t = p_{14}$
	$(S_U, S_Q, E, A-1, I, H, R)$	$\mu A\Delta t = p_{15}$
	$(S_U, S_Q, E, A, I-1, H+1, R)$	$\tau I\Delta t = p_{16}$
	$(S_U, S_Q, E, A, I-1, H, R+1)$	$\gamma_2 I\Delta t = p_{17}$
	$(S_U, S_Q, E, A, I-1, H, R)$	$\mu I\Delta t = p_{18}$
	$(S_U, S_Q, E, A, I, H-1, R+1)$	$\gamma_3 H\Delta t = p_{19}$
	$(S_U, S_Q, E, A, I, H-1, R)$	$(\mu+\delta)H\Delta t = p_{20}$
	$(S_U, S_Q, E, A, I, H, R-1)$	$\mu R\Delta t = p_{16}$
	$(S_U, S_Q, E, A, I, H, R)$	$1-\sum p_i,\ i=1,\ldots,21$

$$
\begin{aligned}
&+p_{sq+1eaihr}(t)\mu\,(q+1)\,\Delta t\\
&+p_{sqe+1a-1ihr}(t)(1-\kappa)\sigma\,(e+1)\,\Delta t\\
&+p_{sqe+1ai+1hr}\kappa\sigma\,(e+1)\,\Delta t+p_{sqea+1isr-1}(t)\gamma_1 a\Delta t\\
&+p_{sqea+1ihr}(t)\mu a\Delta t+p_{sqeai+1h-1r}(t)\tau\,(i+1)\,\Delta t\\
&+p_{sqeai+1hr-1}(t)\gamma_2\,(i+1)\,\Delta t+p_{sqeai+1hr}(t)\mu i\Delta t\\
&+p_{sqeaih+1r-1}(t)\gamma_3 h\Delta t\\
&+p_{sqeaih+1r}(t)\,(\mu+\delta)\,(h+1)\,\Delta t\\
&+p_{sqeaihr+1}(t)\mu r\Delta t+p_{sqeaihr}(t)\,(1-\lambda\Delta t)\,.
\end{aligned}
$$

Here, $\lambda = \sum p_i(t)$. This equation is the forward Kolmogorov equation. This will give the following differential equation as $\Delta t \to 0$.

$$
\begin{aligned}
\frac{p_{sqeaihr}}{dt} &= p_{s-1qeaihr}(t)\pi + p_{s-1q+1eaihr}(t)w(q+1)\\
&\quad + p_{s+1q-1eaihr}(t)l(s+1) + p_{s+1qe-1aihr}(t)\frac{\beta(s+1)i}{N-q}\\
&\quad + p_{s+1qe-1aihr}(t)\frac{\rho_1\beta(s+1)a}{N-q}\\
&\quad + p_{s+1qe-1aihr}(t)\frac{\rho_2\beta(s+1)h}{N-q}\\
&\quad + p_{s+1qeaihr}(t)\mu(s+1)\\
&\quad + p_{sq+1e-1aihr}(t)\frac{\beta(q+1)(\rho_3 i+\rho_4 a)}{N-s-h}\\
&\quad + p_{sq+1eaihr}(t)\mu\,(q+1) + p_{sqe+1a-1ihr}(t)(1-\kappa)\sigma\,(e+1)\\
&\quad + p_{sqe+1ai+1hr}\kappa\sigma\,(e+1) + p_{sqea+1isr-1}(t)\gamma_1 a\Delta t\\
&\quad + p_{sqea+1ihr}(t)\mu a + p_{sqeai+1h-1r}(t)\tau\,(i+1)\\
&\quad + p_{sqeai+1hr-1}(t)\gamma_2\,(i+1) + p_{sqeai+1hr}(t)\mu i\\
&\quad + p_{sqeaih+1r-1}(t)\gamma_3 h + p_{sqeaih+1r}(t)\,(\mu+\delta)\,(h+1)\\
&\quad + p_{sqeaihr+1}(t)\mu r + \lambda p_{sqeaihr}(t).
\end{aligned}
\tag{3.6}
$$

A matrix representation of Eq. (3.6) is useful to study the long-term behavior of the distribution which will help us anticipate pre-cautionary measures to prevent the spread of the disease. To find a matrix representation, we need to map the quantities $p_{sqeaihr}$ into one-dimensional vector p. A possible way of achieving this is to assign the probability $p_{sqeaihr}$ as the $r(N+1)^6 + h(N+1)^5 + i(N+1)^4 + a(N+1)^3 + e(N+1)^2 + q(N+1) + s + 1$ element. Thus, the differential equation (3.6) can be written as

$$\frac{\mathrm{d}p}{\mathrm{d}t} = Qp \tag{3.7}$$

where Q is the matrix consisting of the transition rates between the states. The matrix Q is a Sparse matrix since most of the elements of Q are zero. The solution of the vector differential equation (3.7) can be written as

$$p(t) = p(0)\exp(Qt) \tag{3.8}$$

The eigenvalues of the matrix Q are important in order to determine the nature of the distribution after a large time. Despite this fact, for the present study, the explicit expression of matrix representation cannot be obtained due to a very large value of N. A common way to tackle the problem is to generate a large number of sample paths, which will help to give some ideas about the long-term process. To generate sample paths, the following method will be used.

The deterministic model can be written in a vector form as

$$\dot{x}(t) = \mu(x(t)) \tag{3.9}$$

where x denotes the vector $(S_U, S_Q, E, A, I, H, R)$ and "." represents the derivative w.r.t time. If we incorporate the stochasticity, the above deterministic model takes the form

$$\dot{x}(t) = \mu(x(t)) + b(x(t))\frac{\mathrm{d}W(t)}{\mathrm{d}t} \tag{3.10}$$

Here, $V(x(t)) = b(x(t))b(x(t))^T$ represent the variance–co-variance matrix of the components of the vector x at time t and is given by

$$\begin{pmatrix} \mathcal{F}_1 & \mathcal{F}_2 \\ \mathcal{F}_3 & \mathcal{F}_4 \end{pmatrix},$$

where

$$\mathcal{F}_1 = \begin{pmatrix} v_{11} & -lS_U - wS_Q \\ -lS_U - wS_Q & v_{22} \\ -\frac{\beta S_U I}{N-S_Q} - \frac{\rho_1\beta S_U A}{N-S_Q} - \frac{\rho_2\beta S_U H}{N-S_Q} & -\frac{\rho_3\beta S_Q I}{N-S_U-H} - \frac{\rho_4\beta S_Q A}{N-S_U-H} \end{pmatrix},$$

$$= \begin{pmatrix} -\frac{\beta S_U I}{N-S_Q} - \frac{\rho_1\beta S_U A}{N-S_Q} - \frac{\rho_2\beta S_U H}{N-S_Q} \\ -\frac{\rho_3\beta S_Q I}{N-S_U-H} - \frac{\rho_4\beta S_Q A}{N-S_U-H} \\ v_{33} \end{pmatrix},$$

$$\mathcal{F}_2 = \begin{pmatrix} 0 & 0 & 0 & 0 \\ 0 & 0 & 0 & 0 \\ -(1-\kappa)\sigma E & -\kappa\sigma E & 0 & 0 \end{pmatrix} \mathcal{F}_3 = \begin{pmatrix} 0 & 0 & -(1-\kappa)\sigma E \\ 0 & 0 & -\kappa\sigma E \\ 0 & 0 & 0 \\ 0 & 0 & 0 \end{pmatrix},$$

$$\mathcal{F}_4 = \begin{pmatrix} v_{44} & 0 & 0 & 0 \\ 0 & v_{55} & -\tau I & -\gamma_2 I \\ 0 & -\tau I & v_{66} & -\gamma_3 H \\ 0 & 0 & -\gamma_3 H & v_{77} \end{pmatrix},$$

$$v_{11} = \Pi + wS_Q + \frac{\beta S_U(I + \rho_1 A + \rho_2 H)}{N - S_Q} + (\mu + l)S_U,$$

$$v_{22} = lS_U + \frac{\rho_3\beta S_Q I}{N - S_U - H} + \frac{\rho_4\beta S_Q A}{N - S_U - H} + (\mu + w)S_Q,$$

$$v_{33} = \frac{\beta S_U(I + \rho_1 A + \rho_2 H)}{N - S_Q} + \frac{\rho_3\beta S_Q I}{N - S_U - H} + \frac{\rho_4\beta S_Q A}{N - S_U - H} + (\sigma + \mu)\,E,$$

$$v_{44} = (1-\kappa)\sigma E + (\gamma_1 + \mu)A,$$

$$v_{55} = \kappa\sigma E + (\tau + \mu + \gamma_2)I,$$

$$v_{66} = \tau I + (\delta + \mu + \gamma_3)H,$$

$$v_{77} = \gamma_1 A + \gamma_2 I + \gamma_3 H + \mu R$$

$V = (v_{ij})_{7\times 7}$. Here, $v_{11} = \Pi + wS_Q + \frac{\beta S_U(I + \rho_1 A + \rho_2 H)}{N - S_Q} + (\mu + l)S_U$.

The equivalent difference equation of Eq. 3.10 can be written as

$$x(t+\mathrm{d}t) = x(t) + \mu(x(t))\mathrm{d}t + b(x(t))\mathrm{d}W(t) \tag{3.11}$$

Here $\mathrm{d}W$ is the standard Weiner process with mean 0 and variance $\mathrm{d}t$. Equation (3.11) will be used to generate the sample paths. From these sample paths, one can measure the approximate mean persistence time by observing the time to reach the zero population size of infected individuals for each sample path followed by taking the average of all these times to extinction. As we mentioned earlier about the difficulties of the analytic expression for the mean persistence time, we can use the stability index,[22] which is the logarithm of the mean persistence time of the infection if the process starts with 1 infected individual, in place of mean persistence time. But the stability index is also very difficult to determine for a model with higher dimensions. In the present study, we formulate an upper bound of the stability index and term it as "maximum stability index" which, we believe, can put some light on the stability index of the infected class (symptomatic and asymptomatic).

3.3.2 *Mean persistence time and maximum stability index*

The mathematical complexity increases with the increase in the dimension of the model. This problem is well addressed in the work of.[21] The analysis of extinction related issues like probability of extinction, time to extinction for any particular category of individual are challenging issues. Thus, it is difficult to determine the effect of complete lockdown period T on the stability of the infected population or on the mean persistence time of the infection analytically. In order to have some clue about the effect of complete lockdown period and other relevant parameters on the mean persistence time of infection, we formulate the maximum stability index as follows:

Here, we consider the mean persistence time of the total infected class $X(t) = A(t) + I(t)$ because the extinction of the total infected class will ensure the extinction of infected population both symptomatic and asymptomatic instead of considering the symptomatic and asymptomatic cases separately.

Now,

$$\begin{aligned}\frac{\mathrm{d}A}{\mathrm{d}t}+\frac{\mathrm{d}I}{\mathrm{d}t}&=\sigma E-(\gamma_1+\mu)\,A-(\tau+\mu+\gamma_2)\,I,\\ \frac{\mathrm{d}X}{\mathrm{d}t}&\le \sigma N-\xi X,\end{aligned}\tag{3.12}$$

where $\xi=\min\{(\gamma_1+\mu),(\tau+\mu+\gamma_2)\}$. Thus, we have the following relation.

$$\begin{aligned}\frac{\mathrm{d}A}{\mathrm{d}t}+\frac{\mathrm{d}I}{\mathrm{d}t}&=\sigma E-(\gamma_1+\mu)\,A-(\tau+\mu+\gamma_2)\,I,\\ \frac{\mathrm{d}X}{\mathrm{d}t}&\le \sigma N-\xi X.\end{aligned}\tag{3.13}$$

Let $p_{x\cdot}(t)=\mathbb{P}\{X(t)=x\}$ and $p_x^{ji}=\mathbb{P}\{X(t+\Delta t)=j|X(t)=i\}$. Suppose T_x is the random variable representing the time to extinction of the total infected population if the process starts with x number of total infected individuals and τ_x is the mean persistence time that is $E(T_x)=\tau_x$. For a Discrete time Markov Chain model (DTMC), τ_x satisfies the following difference equation[23]:

$$\begin{aligned}\tau_x&=b(x)\Delta t\,(\tau_{x+1}+\Delta t)+d(x)\Delta t\,(\tau_{x-1}+\Delta t)\\ &\quad+(1-(b(x)+d(x))\,\Delta t)\,,\end{aligned}\tag{3.14}$$

where $x=0,1,2,\ldots,N$ and $b(x)$ and $d(x)$ with $b(N)=0$ are birth and death functions of the variable x. The difference equation (3.14) can be put in the simplified form

$$d(x)\tau_{x-1}-(b(x)+d(x))\,\tau_x+b(x)\tau_{x+1}=-1,\tag{3.15}$$

where $x=0,1,2,\ldots,N$. The solutions of the difference equation (3.15) can be written as,[22]

$$\begin{aligned}\tau_x&=\frac{1}{d(1)}+\sum_{i=2}^{\infty}\frac{b(1)\ldots b(i-1)}{d(1)\ldots d(i)}\qquad x=1\\ &=\tau_1+\sum_{s=1}^{y-1}\left[\frac{d(1)\ldots d(s)}{b(1)\ldots b(s)}\sum_{i=s+1}^{\infty}\frac{b(1)\ldots b(i-1)}{d(1)\ldots d(i)}\right]\qquad x=2,\ldots,N.\end{aligned}\tag{3.16}$$

To find the time to extinction of the infected population for a given number of the infected population is a cumbersome job and to avoid it we consider τ_1, i.e., the mean persistence time of the infected population if the process starts with 1 infected population.[22] In order to get the solution for τ_1, we need the birth function $b(x)$ and death function $d(x)$. The exact forms of $b(X)$ and $d(X)$ cannot be determined due to the complexity of the model and its dimension being large. In this case, we will use an upper bound of the birth function $b(x)$ and a lower bound of the death function $d(x)$. This will give the upper bound of τ_1 and hence an upper bound of the stability index defined as $\log(\tau_1)$.[22] Although the value of this bound is not so effective due to its large value, it can give a feel about its dependence on other parameters. We have,

$$\begin{aligned}\tau_1 &= \sum_{i=1}^{\infty} \frac{b(1)b(2)\ldots b(i-1)}{d(1)d(2)\ldots d(i)} \\ &\leq \sum_{i=1}^{\infty} \frac{(\sigma N)^{i-1}}{(\xi^i i!)}. \end{aligned} \tag{3.17}$$

Now,

$$\begin{aligned}\sum_{i=1}^{\infty} \frac{(\sigma N)^{i-1}}{(\xi^i i!)} &= \frac{1}{\sigma N} \sum_{i=1}^{\infty} \frac{(\sigma N)^{i}}{(\xi^i i!)} \\ &= \frac{1}{\sigma N} \sum_{i=1}^{\infty} \frac{\left(\frac{\sigma N}{\xi}\right)^i}{i!} \\ &= \frac{1}{\sigma N}\left(e^{\frac{\sigma n}{\xi}} - 1\right) \\ &\cong \frac{1}{\sigma N} e^{\frac{\sigma n}{\xi}} \end{aligned}$$

when N is sufficiently large. Therefore, the stability index $\log(\tau_1)$ satisfies the inequality, stability index $\leq \frac{\sigma N}{\xi} - \log(\sigma N)$.

As mentioned earlier, $\frac{\sigma N}{\xi} - \log(\sigma N)$ loses its practical importance due to its very large value. Moreover, this expression does not involve relevant parameters except σ, N, γ_1, γ_2, τ, and μ. To see the effect of important parameters like period of complete lockdown (T) we define the maximum stability index at time t as follows:

$$\begin{aligned}\phi_t &= \frac{\max\{E(t)\}\sigma}{\xi} - \log\left(\max\{E(t)\}\sigma\right)\\ &= \frac{\max\{E(t)\}\sigma}{\xi} - \log\left(\max\{E(t)\}\sigma\right).\end{aligned}$$

Note that in the above definition we replace the total population size N by the $\max\{E(t)\}$ in the expression $\frac{\sigma N}{\xi} - \log(\sigma N)$.

It is also very difficult to get the exact analytic expression of $\max\{E(t)\}$. For numerical simulation, we use a suitable estimator for it. We generate 1,000 sample path to have a sample of size 1,000 at time t for the parameter $\max\{E(t)\}$. We use 1,000th order statistics as an estimator for the parameter $\max\{E(t)\}$. This will give an estimate $\hat{\phi}_t$ of ϕ_t at time t. To see the effect of a relevant parameter, we plot this $\hat{\phi}_t$ against the parameter under consideration.

3.4 Results and Discussion

Recently, since the emergence of the new virus COVID-19, the scientific community is engaged to ward off the rapid spread of the said virus with the aid of a vaccine or some alternate way out. Initially, when the success in discovering the vaccine was not yet achieved, lockdown was proposed as an immediate remedy to reduce the spread of this infectious virus worldwide. In some countries, the lockdown came into effect after a potential number of people were infected. Even today when vaccination is going on, some countries specially where population size is large like India are implementing partial or full lockdown. Thus, there may be some debates and cross debates regarding the timing of the start of the lockdown and the optimal duration of the lockdown. Here we construct a model where the period of lockdown is incorporated explicitly. In the present study, we consider the effect of lockdown in the form of the rate of susceptible to quarantine (l) and the rate of quarantine to susceptible ω in our proposed model where the infected asymptomatic and symptomatic classes are considered separately. As lockdown is proposed to be one of the effective methods to reduce the spread of the infection, the optimal duration of complete lockdown is naturally an important factor and it demands more research work on it. On the contrary, the work toward this direction is found to be limited in stochastic set up. Here we pay attention to this direction and show the effect of the duration of complete lockdown (T) on mean persistence time and on the maximum stability index of the total infected population based on the simulated sample paths.

The values of the parameters associated with the model 3.3 are taken from the literature. The parameter values are depicted in Table 3.1.

First, we focus our attention on the basic reproduction number R_0. Our observation as depicted in Figure 3.2 shows that if ρ_2, the fraction of individuals who comes to exposed class after being contaminated with infected individuals, increases (>1), keeping ρ_1 fixed, R_0 increases. This increase may result in the destabilization of the disease-free equilibrium. Thus, the spread of infection in the quarantined class of individuals should be restricted with more care. On the contrary, the effect of ρ_1 when ρ_2 is fixed is found to be not so significant.

We consider the stochastic counterpart of the deterministic model to capture the demographic variation in the system through birth–death processes. The mean persistence time of the infection in the system is an important parameter to take appropriate management decisions and to take precautionary measures well in advance. To assess the mean persistence time we have to rely on the estimate of

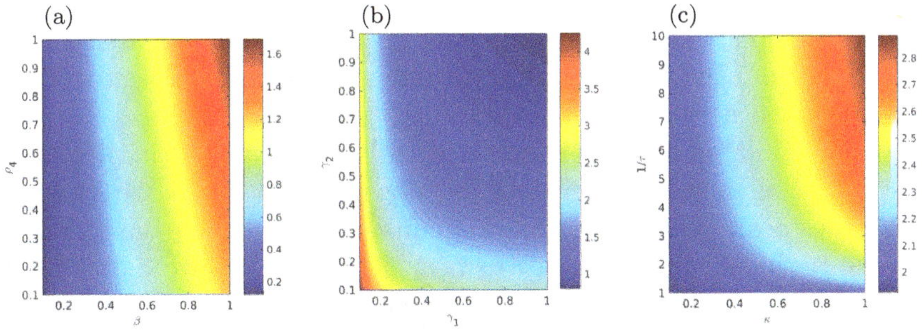

Fig. 3.2. Contour plot of R_0, with respect to (a) β (transmission rate) and ρ_4 (modification factor), (b) γ_1 (recovery rate of asymptomatic individuals) and γ_2 (recovery rate of symptomatic individuals) and (c) κ (fraction of exposed individuals) and $\frac{1}{\tau}$ (average days until hospitalization). All the parameters are taken from Table 3.1.

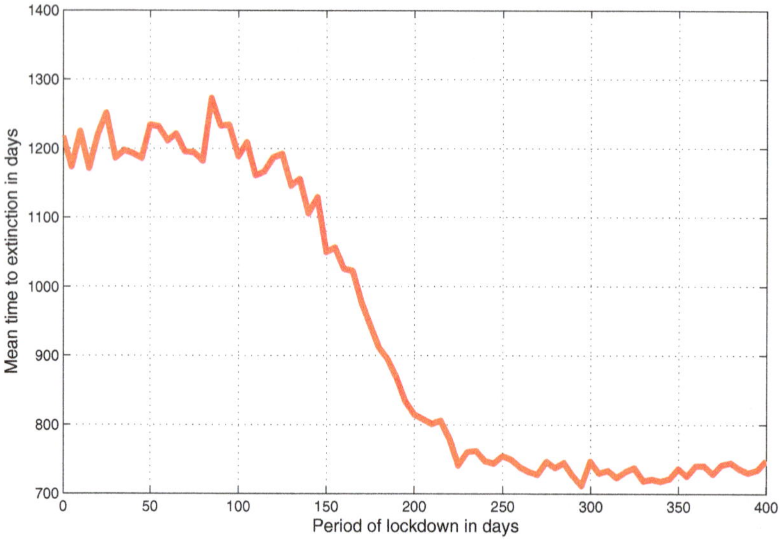

Fig. 3.3. The mean persistence time (in days) of infection vs the period of complete lockdown (T). Up to a certain days of complete lockdown, the mean persistence time remains constant and then it starts decreasing for further increase in the complete lockdown period.

the mean time to extinction based on 1,000 sample paths as it is very difficult to determine analytically (Figure 3.3). So we consider 1,000 sample paths and take the mean of the observed time to extinction. We coin the term "maximum stability index" as an alternate measure instead of the stability index, which we believe will give some clues about the stability of a particular class of a dynamical system when the dimension of the underlying model is pretty high. The maximum stability index for the symptomatic and asymptomatic class together is considered here.

The estimated mean persistence time based on 1,000 sample paths is plotted against the period of lockdown (T) in (3.3). We observe that the mean persistence time remains unaltered with the increase in the period of lockdown up to near 100 days (Figure 3.3). If the period of lockdown is increased further, the mean persistence time starts decreasing and continues to decrease until it reaches near a threshold value. The mean persistence time remains almost constant despite a further increase in the period of lockdown. A similar effect can be observed for the maximum stability index and this supports

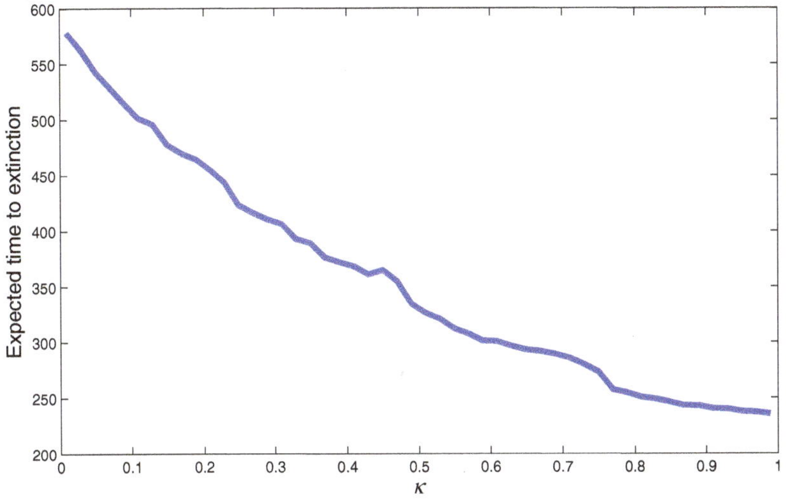

Fig. 3.4. The mean persistence time (in days) of infection vs fraction of symptomatic individuals (κ) getting infected per day.

the observed relationship between the mean persistence time and the period of lockdown. The maximum stability index (ϕ_t) of the infected class (X), which is highest at 5 days of lockdown period, starts decreasing with the increase in the period of lockdown. A sudden fall in the max. stability index can be observed and continues to decrease up to near a threshold value and then remains almost constant despite a further increase in the complete lockdown period (Figure 3.5). This observation gives us a clue about the optimal choice of the complete period of lockdown for which the mean persistence time of the infection can be controlled. In addition, this shows that the increased lockdown period does not allow the stability index to increase further, which results in the increase of the chance of extinction of the total infected population.

The fractions of symptomatic (κ) and asymptomatic cases among the infected individuals play an important role in the process of the spread of the disease. We observe from Figure 3.4 that the mean persistence time of the infection in the system decreases with the increase in the fraction κ. This observation clearly indicates that if the number of asymptomatic cases among the infected individuals increases, the infection will persist for a longer duration. This result is quite natural because the movement of the asymptomatic individuals

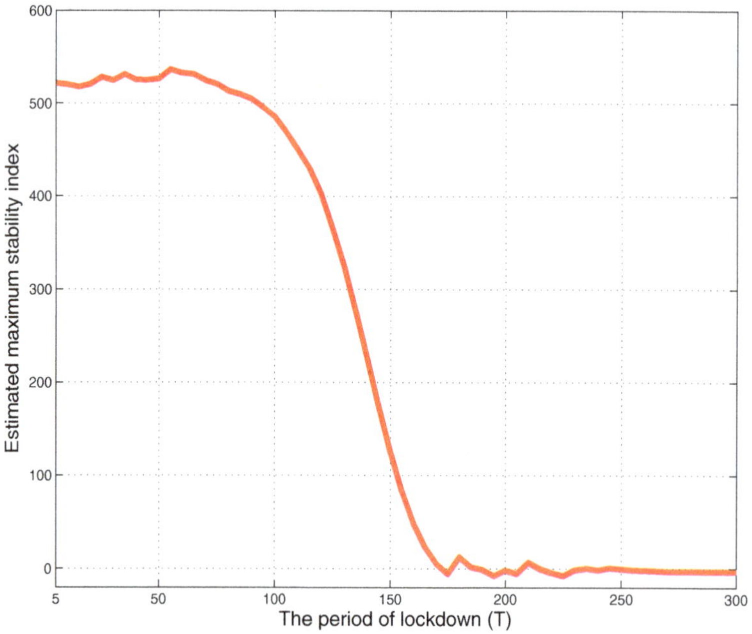

Fig. 3.5. This figure depicts the change in maximum stability index vs period of complete lockdown (T). This figure reveals that the maximum stability index decreases with the increase in the complete lockdown period (T).

are not restricted as they do not have any symptoms and so the infection is spread by them very easily. In addition, our proposed model is capable of capturing the fact that if the recovery rate (γ_1, γ_2) of infected population increases, the system will take less time to become disease-free. In Figures 3.6 and 3.7, it is observed that if the recovery rates increases the maximum stability index of the infected class decreases which supports our claim.

To capture the distributional aspect, we construct a histogram of the infected population based on 500 sample paths. In the simulation process we use the parameter values given in Table 3.1. We took 100 initial infected population and total population size around 1,300,000,000. The histograms in panel 1 of Figures 3.8 and 3.9 show that the distributions of symptomatic cases and asymptomatic cases at the 200th day form the first case reported and these distributions are symmetric around the mode and have a clear resemblance with normal distribution. This distribution reveals that the number

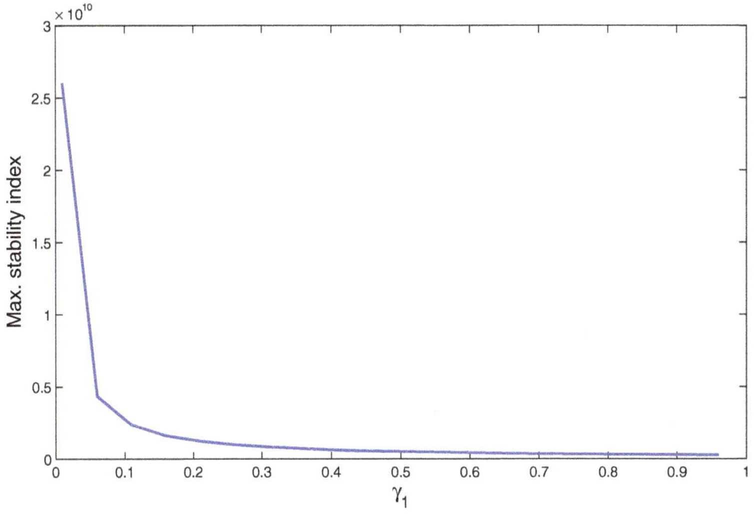

Fig. 3.6. Effect of recovery rate of asymptomatic individuals (γ_1) on the maximum stability index.

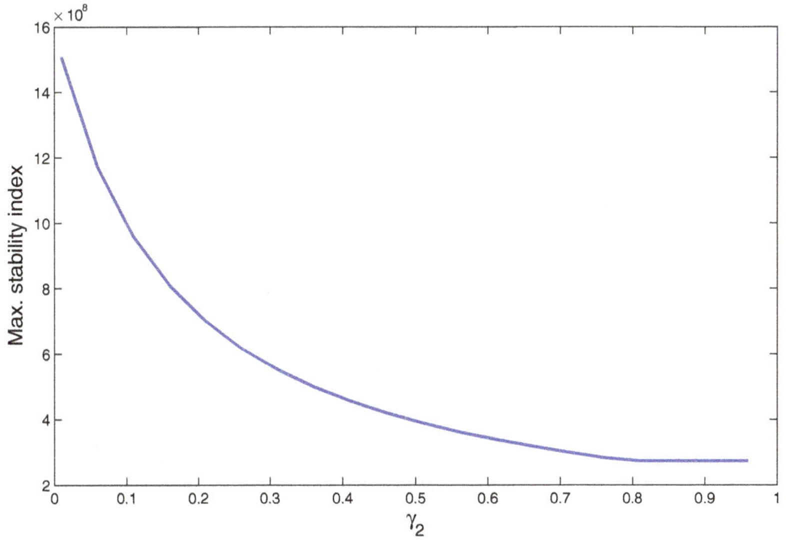

Fig. 3.7. Effect of recovery rate of symptomatic individuals (γ_2) on the maximum stability index.

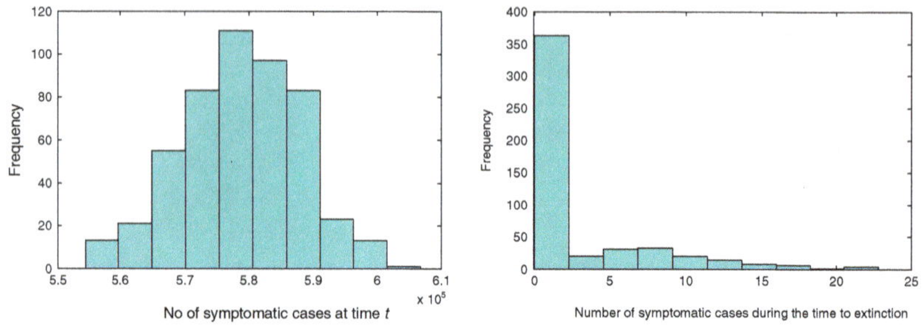

Fig. 3.8. The histogram of symptomatic infected cases based on 500 sample paths at 200th day (panel 1) and at the time when the infection is likely to be absent in the system (panel 2). The histogram in panel 1 shows a symmetric distribution with mode near the size 5.1×10^5, whereas that in panel 2 shows a skewed distribution with mode near 0.

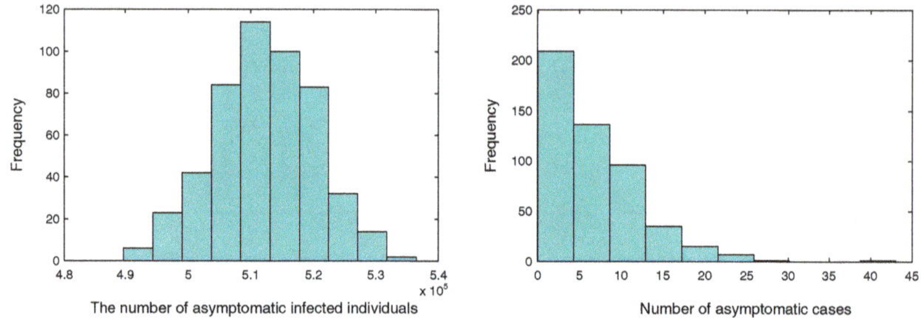

Fig. 3.9. The histogram of asymptomatic infected cases based on 500 sample paths at 200th day (panel 1) and at the time when the infection is likely to be absent in the system (panel 2). The histogram in panel 1 shows a symmetric distribution with mode near the size 5.8×10^5, whereas that in panel 2 shows a skewed distribution with mode near 0.

of asymptomatic and symptomatic cases are most likely to lie near 5.1×10^5 and 5.8×10^5, respectively, at around the 200th day. The distribution loses its symmetry as the time progresses to the situation when the infection is likely to cease and the distribution becomes skewed (see panel 2 of Figures 3.8 and 3.9).

To find an optimal lockdown period so that the duration of the persistence of the pandemic can be minimized is a challenging issue. Thus, it is difficult to frame a policy on up to what extend the restriction can be retained or up to what extend it can be waived so that a balance can be maintained between the normal routine life and control on the pandemic situation, when lockdown gets over and the unlock phase starts. This is one of our main focuses in this study.

We argue that the policy for unlock phases can be framed depending on the values of the parameters ϵ_1 and ϵ_2 since these two parameters represent the sudden change in the rate of quarantine l and the rate at which the people move from the quarantine class to susceptible class ω when unlock phase starts.

We also performed a global sensitivity analysis, which gives us a clear understanding of the important parameters, to observe the effect of the most important parameters that have a significant effect on the basic reproduction number. The PRCCs of R_0 for these parameters suggest that COVID-19 transmission coefficient, β, has a maximum positive correlation with R_0 and the recovery rate for symptomatic infected individuals, γ_2, has a significant maximum negative correlation with R_0. Although there is a strong positive correlation with the modification factor with susceptible quarantine and asymptomatic individuals, ρ_4, and the fraction of the exposed individuals being symptomatic infected, κ, with the basic reproductive number R_0. Further, we draw the contour plot of R_0 with respect to the six sensitive parameters β, ρ_4, γ_1, γ_2, κ, and τ for the model to explore the influence of the most sensitive parameters on R_0, which we have already discussed.

3.4.1 *Sensitivity analysis and contour plot*

A global sensitivity analysis was carried out using techniques described in Marino *et al.*[24] to identify the most important parameters that have a major impact on the significant output variable of the system (3.3). In order to investigate the sensitivity, we calculate the Partial Rank Correlation Coefficients (PRCC) of the parameters $\beta, \rho_1, \rho_2, \rho_3, \rho_4, \gamma_1, \gamma_2, \gamma_3, \kappa$, and τ with respect to the basic reproduction number R_0 as the output. The nonlinear and monotone relations between the response variable and the input parameters are observed, which is a prerequisite for computing the PRCCs. The Latin Hypercube Sampling (LHS) has been used to draw 1,000 samples of the above parameters. The bar diagram of the PRCC values of R_0 against those parameters is shown in Figure 3.10. The PRCC values with the response of these parameters imply that the COVID-19 transmission coefficient, β, has a maximum positive correlation with R_0. The recovery rate for symptomatic infected individuals (γ_2) has a significant maximum negative correlation with R_0. Besides these, there is a strong positive correlation with the modification factor

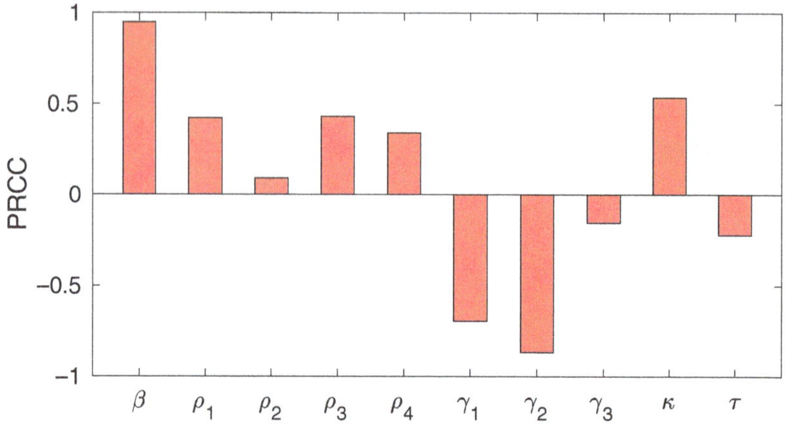

Fig. 3.10. Effect of uncertainty of the model (3.3) on basic reproduction number (R_0).

with susceptible quarantine and asymptomatic individuals (ρ_4) and the fraction of the exposed individuals being symptomatic infected (κ) with the basic reproductive number (R_0). Whereas, the recovery rate of asymptomatic infected individuals (γ_1) and the rate of hospitalization of symptomatic infected individuals (τ) have a significant negative correlation with the basic reproduction number (R_0).

In addition, we draw the contour plot of R_0 with respect to the six sensitive parameters β, ρ_4, γ_1, γ_2, κ, and τ for the model (3.3), in Figure 3.2, to explore the influence of the most sensitive parameters on R_0. We draw the contour plots of R_0 in $\beta - \rho_4$ plane (first figure), $\gamma_1 - \gamma_2$ plane (second figure), and $\kappa - \frac{1}{\tau}$ plane (third figure). Here we fix all the other parameters as shown in Table 3.1. It is evident from the first figure that, if the rate of transmission of infectious individuals and the modification factor between asymptomatic and susceptible quarantined individuals decreases, the basic reproduction number decreases and for a certain range of β and ρ_4 can be less than unity. The second figure reveals that the basic reproduction number declines by increasing the recovery rate of asymptomatic and symptomatic individuals. In the third figure, the axes of these plots are given as average days from the starting of symptoms to hospitalization ($1/\tau$) and fraction of the exposed population that become symptomatic infected (κ). We have observed that the basic reproduction number increases as the number of average days before hospitalization after symptoms increases. This suggests that more testing should be done and the isolation of infected individuals should be carried out effectively to reduce the basic reproduction number and hence the disease burden. Finally, we have shown that there is still a major role in the percentage of exposed individuals who become symptomatically infected. By the value of κ, the basic reproduction number is increased.

3.5 Conclusion

A major concern about the highly infectious diseases is the number of asymptomatic cases. In this work we consider the symptomatic and asymptomatic cases separately. The process of lockdown is thought to be the most effective way to ward off the spread of infection and it has come into force in almost all the countries in case of COVID-19. But the optimal period of complete lockdown for which the mean time to extinction of the infection is minimum is not easy to achieve. Moreover, the determination of mean time to extinction analytically is a challenging task as the dimension of the model is high as is the largeness of population size. In this chapter, we primarily focus on these two issues.

We incorporate the period of complete lockdown (T) through the parameters l and w assuming these two parameters to be functions of time. To avoid the complexity of determining the mean persistence time, we propose an alternate strategy termed as maximum stability index which can help to have a feel about the effect of T on mean persistence time as well as the stability of the infected population in the system. We also estimate the mean persistence time by taking the average of the times to extinctions for 1,000 sample paths and observe the effect of different relevant parameters on it.

The essence of our study is that, if the period of lockdown is increased, the mean persistence time of both the infected population, symptomatic and asymptomatic, will decrease. We observe that if the fraction of asymptomatic population $(1 - \kappa)$ decreases, the mean time to extinction diminishes. (see 3.4). Thus, if the number of infected populations who are asymptomatic increases, the infection will persist in the system longer. This finding indicates that we should focus on the number of asymptomatic cases and to accomplish it the number of testing should be increased. Further, our proposed model can give a clue about the optimal period of lockdown and the optimal strategy to implement unlock process under the constraint that the desired balance can be maintained between the spread of the infectious disease and the economic condition of a country, which is our future endeavor.

References

1. Li, Q., Guan, X., Wu, P., Wang, X., Zhou, L., Tong, Y., Ren, R., Leung, K. S., Lau, E. H., Wong, J. Y., *et al.* (2020). Early transmission dynamics in Wuhan, China, of novel coronavirus–infected pneumonia, *N. Engl. J. Med.* **382**, 1199–1207.
2. Wu, P., Hao, X., Lau, E. H., Wong, J. Y., Leung, K. S., Wu, J. T., Cowling, B. J., and Leung, G. M. (2020). Real-time tentative assessment of the epidemiological characteristics of novel coronavirus infections in Wuhan, China, as at 22 January 2020, *Eurosurveillance* **25**, 3, p. 2000044.
3. Health Economic Times (2020). Who says over 22,000 healthcare workers across 52 countries infected by COVID-19. https://health.economictimes.indiatimes.com/news/industry/who-says-over-22000-healthcare-workers-across-52-countries-infected-by-covid-19/75107238. Accessed 3 April 2020.
4. The Print (2020a). 89 health dept members, including doctors and ias officers, test positive for COVID-19 in mp. https://theprint.in/health/89-health-dept-members-including-doctors-ias-officers-test-positive- for-covid-19-in-mp/403102/. Accessed 3 April 2020.
5. Economics Times (2020c). COVID-19 impact: Over 40 nurses test positive in Mumbai hospital. https://economictimes.indiatimes.com/news/politics-and-nation/two-private-hospitals-in-mumbai-sealed-after-staff-test-positive/articleshow/75009359.cms. Accessed 3 April 2020.
6. NDTV (2020b). 90 health workers infected with covid-19, total cases over 8,000 in India. https://www.ndtv.com/india-news/coronavirus-india-coronavirus-cases-in-india-cross-8-000-mark-34-dead-in-24-hours-2210282. Accessed 3 April 2020.
7. New York Times (2020). Modi orders 3-week total lockdown for all 1.3 billion Indians. https://www.nytimes.com/2020/03/24/world/asia/india-coronavirus-lockdown.html. Accessed 3 April 2020.

8. Moghadas, S. M., Shoukat, A., Fitzpatrick, M. C., Wells, C. R., Sah, P., Pandey, A., Sachs, J. D., Wang, Z., Meyers, L. A., Singer, B. H., *et al.* (2020). Projecting hospital utilization during the COVID-19 outbreaks in the United States, *Proc. Natl. Acad. Sci.* **117**, 16, pp. 9122–9126.
9. Nadim, S. S., Ghosh, I., and Chattopadhyay, J. (2020). Short-term predictions and prevention strategies for COVID-2019: A model based study, arXiv preprint arXiv:2003.08150 .
10. Tang, B., Wang, X., Li, Q., Bragazzi, N. L., Tang, S., Xiao, Y., and Wu, J. (2020). Estimation of the transmission risk of the 2019-ncov and its implication for public health interventions, *J. Clin. Med.* **9**, 2, p. 462.
11. Nadim, S. S. and Chattopadhyay, J. (2020). Occurrence of backward bifurcation and prediction of disease transmission with imperfect lockdown: A case study on COVID-19, *Chaos Solitons Fractals* **140**, p. 110163.
12. Sardar, T. and Rana, S. (2020). Effective lockdown and role of hospital-based COVID-19 transmission in some Indian states: An outbreak risk analysis, arXiv preprint arXiv:2005.01149.
13. Sardar, T., Nadim, S. S., Rana, S., and Chattopadhyay, J. (2020). Assessment of lockdown effect in some states and overall India: A predictive mathematical study on COVID-19 outbreak, *Chaos Solitons Fractals*, **139**, p. 110078.
14. Nasell, I. (1999). On the time to extinction in recurrent epidemics, *J. Roy. Stat. Soci. Ser. B.* **61**, 2, pp. 309–330.
15. NITI (2020). Life expectancy at birth. https://niti.gov.in/content/life-expectancy. Accessed 3 April 2020.
16. Senapati, A., Rana, S., Das, T., and Chattopadhyay, J. (2021). Impact of intervention on the spread of COVID-19 in India: A model based study, *J. Theor. Biol.* **523**, p. 110711.
17. COVID-19 (2020). India COVID-19 tracker. https://www.covid19india.org/. Accessed 3 April 2020.
18. Yang, X., Chen, L., and Chen, J. (1996). Permanence and positive periodic solution for the single-species nonautonomous delay diffusive models, *Comput. Mathe. Appl.* **32**, 4, pp. 109–116.
19. Diekmann, O., Heesterbeek, J. A. P., and Metz, J. A. (1990). On the definition and the computation of the basic reproduction ratio R0 in models for infectious diseases in heterogeneous populations, *J. Math. Biol.* **28**, 4, pp. 365–382.

20. Van den Driessche, P. and Watmough, J. (2002). Reproduction numbers and sub-threshold endemic equilibria for compartmental models of disease transmission, *Mathe. Biosci.* **180**, 1–2, pp. 29–48.
21. Keeling, M. J. and Ross, J. V. (2008). On methods for studying stochastic disease dynamics, *J. R. Soc. Interface* **5**, pp. 171–181.
22. Renshaw, E. (1993). *Modelling Biological Populations in Space and Time*, Vol. 11, Cambridge University Press, England, UK.
23. Allen, L. J. and Allen, E. J. (2003). A comparison of three different stochastic population models with regard to persistence time, *Theor. Popul. Biol.* **64**, 4, pp. 439–449.
24. Marino, S., Hogue, I. B., Ray, C. J., and Kirschner, D. E. (2008). A methodology for performing global uncertainty and sensitivity analysis in systems biology, *J. Theor. Biol.* **254**, 1, pp. 178–196.

https://doi.org/10.1142/9789811263033_0004

Chapter 4

A Model of Virus Infection with Immune Responses Supports Boosting CTL Response to Balance Antibody Response

Tyler Meadows[*,‡] **and Elissa J. Schwartz**[†,§]

[*]*Department of Mathematics,*
University of Idaho, Moscow, Idaho, USA

[†]*Department of Mathematics and Statistics,*
and School of Biological Sciences, Washington State University,
Pullman, Washington, USA

[‡]*tyler.meadows@queensu.ca*

[§]*ejs@wsu.edu*

We analyze a within-host model of virus infection with antibody and $CD8^+$ cytotoxic T lymphocyte (CTL) responses proposed by Schwartz *et al.*[22] The goal of this work is to gain an overview of the stability of the biologically-relevant equilibria as a function of the model's immune response parameters. We show that the equilibria undergo at most two forward transcritical bifurcations. The model is also explored numerically and results are applied to equine infectious anemia virus (EIAV) infection. In order to arrive at stability of the biologically-relevant endemic equilibrium characterized by coexistence of antibody and CTL responses, the parameters promoting CTL responses need to be boosted over parameters promoting antibody production. This result may seem counter intuitive (in that a weaker antibody response is better) but can be understood in terms of a balance between CTL and antibody responses that is needed to permit existence of CTLs. In conclusion, an intervention such as a vaccine that is intended to control a persistent viral infection with both immune responses should moderate the antibody response to allow for stimulation of the CTL response.

Keywords: virus dynamics, transcritical bifurcations, immune system dynamics, equine infectious anemia virus infection

4.1 Introduction

Equine infectious anemia virus (EIAV) is an infection in horses that is similar to human immunodeficiency virus (HIV) in structure, genome, and life cycle.[1] However, horses infected with EIAV do not develop AIDS, as do HIV-infected individuals without treatment. Instead, EIAV-infected horses produce immune responses that control the infection.[2] This control has been shown to be mediated by both CD8$^+$ cytotoxic T lymphocytes (CTLs) and antibody responses, which ultimately limit virus replication and prevent symptoms in long-term infected animals, even though the virus infection is not cleared.[3] Specifically, T cell epitopes that indicate a broadening of CTL responses have been identified to persist in long-term infection.[4–6] Furthermore, evolving broadly neutralizing antibodies have been found to be needed to maintain asymptomatic disease.[7–9] Thus, EIAV infection has been the subject of many controlled experiments to investigate the immune response to infection.[10–14] The more we understand how CTLs and antibodies control EIAV infection, the more insight will be gained on how best to develop effective interventions that control other similar viral infections.

The standard model of virus infection[15,16] is a system of three ordinary differential equations (ODEs) that can account for many experimental observations in the stages of both HIV and EIAV infection.[12,17–20] This model depicts the concentrations of uninfected cells, infected cells, and virus particles, and represents a virus infecting a target cell population, with the infected cells then producing more viral particles. This model does not include additional equations that explicitly model immune responses. Since immune responses are dynamic as well, further realism can be gained by explicitly including equations representing immune responses.

Nowak and Bangham[15] developed a model consisting of a system of four ODEs that includes the dynamics of one component of the immune system: the population of CTLs that kill infected cells. A subsequent model by Wodarz[21] presents a system of five equations to model an infection by hepatitis C virus; this model explicitly includes immune responses given by populations of both CTLs and antibodies.

In 2013, Schwartz *et al.*[22] published a five-equation model of EIAV infection that also includes two populations of immune responses, one for CTLs and one for antibodies. This mathematical model differs from that of Wodarz[21] in the equation describing the antibody response. Schwartz *et al.*[22] depict antibody production as proportional to the concentration of virus, rather than proportional to the interactions between viruses and pre-existing antibodies. Other authors have built more complexity into this equation, such as by including B cell dynamics and differentiation into antibody producing cells,[23] but this approach necessarily relies upon the addition of more parameters with unknown values in the case of EIAV infection. The Schwartz *et al.*[22] model takes a step back and uses a more straightforward approach, in which antibody production is modeled as first order in V. This choice still captures the essence of the biology, given that antibody production is correlated with the quantity of virus.[2,24,25]

In the notation of Schwartz *et al.*,[22] the system of equations is

$$\dot{M} = \lambda - \rho M - \beta M V, \tag{4.1a}$$

$$\dot{I} = \beta M V - \delta I - k I C, \tag{4.1b}$$

$$\dot{V} = b I - \gamma V - f V A, \tag{4.1c}$$

$$\dot{C} = \psi I C - \omega C, \tag{4.1d}$$

$$\dot{A} = \alpha V - \mu A. \tag{4.1e}$$

The state variables are the concentrations of uninfected cells M (in this infection, uninfected cells are macrophages, a type of white blood cell that is the target cell of EIAV), infected cells I, virus V, CTLs C, and antibodies A. Solutions of the EIAV model are only biologically relevant when all of the state variables are non-negative. (We use "biological" as a synonym for "non-negative.") The lower-case Greek and Latin letters denote 12 parameters, which are assumed to be positive. These equations are interpreted as follows: Eq. (4.1a) describes uninfected cells introduced at rate λ, removed at rate ρM, and infected by virus particles at rate βMV; Eq. (4.1b) describes infected cells produced at rate βMV, removed at rate δI, and killed by CTLs at rate kIC; Eq. (4.1c) describes virus particles (measured in viral RNA, vRNA) produced by infected cells at rate bI, removed at rate γV, and neutralized by interaction with antibodies at rate fVA; Eq. (4.1d) describes CTLs produced at rate ψIC and removed at rate ωC, as in Nowak and Bangham[15] and Wodarz[21]; and finally, Eq. (4.1e) describes antibody molecules produced at rate αV and removed at rate μA.

The model has five equilibria, yet only three of these can have non-negative values for all of the state variables and thus be biologically relevant. These equilibrium states are (1) the infection-free equilibrium (IFE) $\mathbf{E}_0$, (2) the antibody-only equilibrium $\mathbf{E}_1$ describing an infection limited by an antibody response but not a CTL response, and (3) the coexistence equilibrium $\mathbf{E}_3$ describing an infection limited by both an antibody response and a CTL response.

Schwartz *et al.*[22] showed that the existence of the biologically relevant equilibria could be determined by the basic reproduction number R_0 and a second threshold R_1. They identified example parameter sets corresponding to scenarios where each equilibrium is stable. However, they did not show that the thresholds determine the stability of the equilibria. In this chapter, we build upon the results of Schwartz *et al.*[22] by investigating the bifurcation structure of system (4.1), and by showing that there are at most two forward transcritical bifurcations. We use a combination of standard techniques to analyze the dynamics of (4.1) in general and verify that the thresholds are associated with bifurcation points.

We show that the IFE $\mathbf{E}_0$ is globally asymptotically stable when $R_0 < 1$ and unstable when $R_0 > 1$. The antibody-only equilibrium $\mathbf{E}_1$ is locally asymptotically stable when $R_1 < 1 < R_0$, indicating that there are bifurcations when $R_0 = 1$ and $R_1 = 1$. In particular, we use the next generation matrix method[26–28] to show that the $R_1 = 1$ bifurcation is a forward transcritical bifurcation between $\mathbf{E}_1$ and $\mathbf{E}_3$. Our analysis is supported by bifurcation diagrams that can also be used to help identify parameter ranges associated with the stability of these equilibria.

The biological goal of our analysis is to identify the key parameters that correspond to the stability of each equilibrium. In particular, since EIAV infection is controlled by both antibodies and

CTLs, we aim to determine which parameter ranges lead the system to the coexistence equilibrium $\mathbf{E}_3$. We can then use this information to determine the specific parameters to target with an intervention such as a vaccine. For example, using the CTL production rate (ψ) or antibody production rate (α) as control parameters, we can identify which ranges would be needed by a vaccine that stimulates CTL production or antibody production sufficiently to drive the system to $\mathbf{E}_3$. Such advances in our understanding of the modes of action of the immune system that control EIAV could indicate the targets for a potential vaccine to control other infections, like HIV.

4.2 Analytical Results

Proposition 4.1. *If* (M, I, V, C, A) *is a solution to the EIAV model* (4.1) *with biological initial conditions, then* (M, I, V, C, A) *remains biological for all time. Furthermore,* (M, I, V, C, A) *is bounded for all time.*

Proof. First, suppose that $M(t)$ becomes negative, then there exists $t_0 \geq 0$ such that $M(t_0) = 0$. At $t = t_0$, $\dot{M} = \lambda > 0$, and so there exists $\epsilon > 0$ such that $M(t) > 0$ for all $t \in (t_0, t_0 + \epsilon)$. Thus $M(t)$ cannot become negative. Suppose now that $C(t)$ becomes negative, then there exists t_0 such that $C(t_0) = 0$. At $t = t_0$ $\dot{C} = 0$, and therefore $C(t) = 0$ for all t by the Picard–Lindelöf theorem. Thus, if $C(0) > 0$, then $C(t) > 0$ for all t.

Suppose that one of I or V becomes negative. Then there exists $t^* \geq 0$ such that $\min\{I(t^*), V(t^*)\} = 0$. If $I(t^*) = 0$ and $V(t^*) > 0$, then $\dot{I}(t^*) = \beta M(t^*)V(t^*) > 0$, and thus there exists $\epsilon > 0$ such that $I(t) > 0$ for all $t \in (t^*, t^* + \epsilon)$. If $I(t^*) > 0$ and $V(t^*) = 0$, then $\dot{V}(t^*) = bI(t^*) > 0$, and therefore there exists $\epsilon > 0$ such that $V(t) > 0$ for all $t \in (t^*, t^* + \epsilon)$. Finally, if $I(t^*) = V(t^*) = 0$, then $\dot{I}(t^*) = \dot{V}(t^*) = 0$, and thus $I(t) = V(t) = 0$ for all $t > 0$. It follows that $I(t) \geq 0$ and $V(t) \geq 0$ for all $t \geq 0$.

Finally, suppose that $A(t)$ becomes negative, then there exists $t_0 \geq 0$ such that $A(t_0) = 0$. At $t = t_0$, $\dot{A}(t_0) = \alpha V(t_0) \geq 0$. It follows that $A(t) > 0$ for all $t \geq 0$.

To show that solutions are bounded, consider $\Sigma(t) := M(t) + I(t) + \frac{k}{\psi}C(t)$, which satisfies

$$\begin{aligned}\dot{\sum}(t) &= \lambda - \rho M(t) - \delta I(t) - \frac{\omega k}{\psi}C(t),\\ &\leq \lambda - \kappa \sum(t),\end{aligned} \tag{4.2}$$

where $\kappa = \min\{\rho, \delta, \omega\}$. By multiplying Eq. (4.2) by $e^{\kappa t}$, rearranging, and integrating, we find

$$\begin{aligned}\sum(t) &\leq \frac{\lambda}{\kappa} + \sum(0)e^{-\kappa t} - \frac{\lambda}{\kappa}e^{-\kappa t},\\ &\leq \max\left\{\frac{\lambda}{\kappa}, \sum(0)\right\}.\end{aligned}$$

Since $M(t)$, $I(t)$, and $C(t)$ are positive, it follows that they are bounded in forward time. Since $I(t)$ is bounded above, there exists $T > 0$ such that

$$\begin{aligned}\dot{V}(t) &\leq T - \gamma V(t) - fV(t)A(t),\\ &\leq T - \gamma V(t),\end{aligned}$$

where the second inequality follows from the fact that $fV(t)A(t) \geq 0$. It similarly follows that $V(t)$ is bounded for all $t \geq 0$. Finally, since $I(t)$ is bounded above, there exists $Q > 0$ such that

$$\dot{A}(t) \leq Q - \mu A(t),$$

and a similar argument shows that $A(t)$ is bounded for all $t \geq 0$. □

4.2.1 *Equilibria of the EIAV model*

The equilibria of the EIAV model, (4.1), are derived in Schwartz *et al.*[22] We report the non-negative ones here for convenience.

The IFE is given by

$$\mathbf{E}_0 = (M_0, I_0, V_0, C_0, A_0) = (\lambda/\rho, 0, 0, 0, 0). \tag{4.3}$$

Only uninfected cells are present.

The boundary equilibrium is given by $\mathbf{E}_1 = (M_1, I_1, V_1, C_1, A_1)$ where

$$M_1 = \frac{\lambda}{(\rho + \beta V_1)}, \tag{4.4a}$$

$$I_1 = \frac{\lambda}{\delta} \cdot \frac{\beta V_1}{\rho + \beta V_1}, \tag{4.4b}$$

$$V_1 = \frac{-(\alpha f \rho + \beta\gamma\mu) + \sqrt{(\alpha f \rho - \beta\gamma\mu)^2 + \frac{4\alpha b \beta^2 f \mu \lambda}{\delta}}}{2\alpha\beta f}, \tag{4.4c}$$

$$C_1 = 0, \tag{4.4d}$$

$$A_1 = \frac{\alpha}{\mu} V_1. \tag{4.4e}$$

The boundary equilibrium describes an infection — with both virus particles and infected cells present — that elicits an antibody response but not a CTL response.

Finally, the endemic equilibrium is given by $\mathbf{E}_3 = (M_3, I_3, V_3, C_3, A_3)$ where

$$M_3 = \frac{\lambda}{\rho + \beta V_3}, \tag{4.5a}$$

$$I_3 = \omega/\psi, \tag{4.5b}$$

$$V_3 = \frac{-\gamma\mu\psi + \sqrt{(\gamma\mu\psi)^2 + 4\alpha b f \mu \psi \omega}}{2\alpha f \psi}, \tag{4.5c}$$

$$C_3 = \frac{\lambda\psi}{k\omega} \cdot \frac{\beta V_3}{\rho + \beta V_3} - \frac{\delta}{k}, \tag{4.5d}$$

$$A_3 = \frac{\alpha}{\mu} V_3. \tag{4.5e}$$

The endemic equilibrium describes an infection that elicits both antibody and CTL responses from the immune system. Note that this equilibrium is only non-negative when $C_3 \geq 0$.

4.2.2 *Stability analysis*

Schwartz *et al.*[22] made use of the next generation matrix method[28] to analyze the linear stability of the IFE. We combine their results in the statement of Theorem 4.1 that follows. If an infected cell is introduced into an IFE, the basic reproduction number R_0 is roughly the average number of infected cells produced. The number is a threshold value. That is, if $R_0 < 1$, then the infection will die out and if $R_0 > 1$, then the infection will grow.

Theorem 4.1. *Consider the EIAV model with positive parameter values. The basic reproduction number is*

$$R_0 = \frac{b\beta\lambda}{\delta\gamma\rho}. \tag{4.6}$$

Further, the infection-free equilibrium $\mathbf{E}_0$ *is globally asymptotically stable for* $R_0 < 1$ *and unstable for* $R_0 > 1$.

Proof. Linear stability results for $\mathbf{E}_0$ and a calculation of R_0 using the next generation matrix method were done in Schwartz *et al.*[22]

Suppose that $R_0 < 1$, then there exists $\epsilon > 0$ such that $R_0 + \epsilon \leq 1$. To show global stability of $\mathbf{E}_0$, define the Lyapunov function

$$U_\epsilon = M(t) - M_0 \log\left(\frac{M(t)}{M_0}\right) + I(t) + \frac{\delta}{\beta}V(t) + \frac{k}{\psi}C(t) + \epsilon\frac{\gamma\delta}{b\alpha}A(t),$$

so that, for any $\epsilon > 0$, $U_\epsilon \geq 0$ with $U_\epsilon = 0$ only when $(M, I, V, C, A) = \mathbf{E}_0$. The derivative of U_ϵ along trajectories is

$$\begin{aligned}\dot{U}_\epsilon = &-\frac{\rho}{M(t)}(M_0 - M(t))^2 - \frac{\gamma\delta}{b}(1 - R_0 - \epsilon)V(t)\\ &- \gamma\delta f V(t)A(t) - \frac{k\omega}{\psi}C(t) - \epsilon\frac{\gamma\delta\mu}{b\alpha}A(t) \leq 0.\end{aligned}$$

By LaSalle's invariance principle, solutions converge to the largest compact invariant set of (4.1) that is contained in $\{(M(t), I(t), V(t), C(t), A(t)) : \dot{U}_\epsilon = 0\} = \{(M_0, I(t), 0, 0, 0)\}$. That is $\mathbf{E}_0$. □

Note that V_1 may be written in terms of R_0 as

$$V_1 = \frac{-(\alpha f\rho + \beta\gamma\mu) + \sqrt{(\alpha f\rho + \beta\gamma\mu)^2 + 4\alpha f\rho\beta\gamma\mu(R_0 - 1)}}{2\alpha\beta f}, \tag{4.7}$$

so that when $R_0 = 1$, $V_1 = 0$. It is easy to see from this that $\mathbf{E}_0 = \mathbf{E}_1$ when $R_0 = 1$.

The antibody-only equilibrium $\mathbf{E}_1$ represents a viral infection with an antibody response, $A_1 > 0$, but no CTL response, $C_1 = 0$. We may treat the CTL response as an active variable, similar to the use of "infectious variable" in the terminology of van den Driessche and Watmough.[28] From this new point of view the boundary equilibrium $\mathbf{E}_1$ is analogous to an IFE, in that it is free of the active variable C. Thus, we can obtain an expression for the threshold for CTLs using the next generation matrix method. The term that introduces new CTLs into the system is $\mathcal{F}_1 = \psi IC$ and the term that eliminates them is $\mathcal{V}_1 = \omega C$. Let $\mathcal{F} = (\frac{\partial \mathcal{F}_1}{\partial C})(\mathbf{E}_1) = \psi I_1$ and $\mathcal{V} = (\frac{\partial \mathcal{V}_1}{\partial C})(\mathbf{E}_1) = \omega$. Then, by the next generation matrix method,

$$\hat{R}_1 = \mathcal{F}\mathcal{V}^{-1} = \frac{\lambda\psi}{\delta\omega} \cdot \frac{\beta V_1}{\rho + \beta V_1}. \tag{4.8}$$

Alternatively, following Schwartz *et al.*,[22] we write the $\mathbf{E}_3$ CTL response (given by Eq. (4.5d)) in the form $C_3 = (\delta/k)(R_1 - 1)$ where

$$R_1 = \frac{\lambda\psi}{\delta\omega} \cdot \frac{\beta V_3}{\rho + \beta V_3} \tag{4.9}$$

Those authors show that $R_1 < R_0$ (Ref. 22, Theorem 3). $\mathbf{E}_3$ is biological only when $R_1 \geq 1$.

When $\mathbf{E}_1 = \mathbf{E}_3$, we have $V_1 = V_3$, which implies $R_1 = \hat{R}_1$, and $C_3 = C_1 = 0$, which implies $R_1 = 1$. On the other hand, if $R_1 = \hat{R}_1$, it follows that $V_1 = V_3$ and, by virtue of the equations that V_1 and V_3 solve, i.e.,

$$\frac{f\alpha}{\mu} V_1^2 + \gamma V_1 - bI_1 = 0, \tag{4.10}$$

$$\frac{f\alpha}{\mu} V_3^2 + \gamma V_3 - bI_3 = 0, \tag{4.11}$$

that $I_1 = I_3$. Since $I_1 = \frac{\psi}{\omega}\hat{R}_1 = I_3\hat{R}_1$ we obtain $\hat{R}_1 = 1$, and therefore that $\mathbf{E}_1 = \mathbf{E}_3$. Finally, if $\hat{R}_1 = 1$, then $I_1 = I_3$ and by Eqs. (4.10) and (4.11), $V_1 = V_3$. Therefore $R_1 = \hat{R}_1$. Thus, the following three statements are equivalent:

(1) $R_1 = \hat{R}_1$,
(2) $\mathbf{E}_1 = \mathbf{E}_3$,
(3) $\hat{R}_1 = 1$.

Therefore, both R_1 and $\hat{R}_1$ may be used as a threshold to determine the existence (and stability) of $\mathbf{E}_3$, although only $\hat{R}_1$ should be thought of as a basic reproduction number.

Lemma 4.1. *If V_1 depends on a parameter, that dependence is strictly monotone.*

Proof. By letting $x = \mu/(2\alpha f)$, $y = \rho/(2\beta)$, and $z = \lambda b/(2\delta)$, we can write

$$V_1 = -\gamma x - y + \sqrt{(\gamma x + y)^2 + 4x(z - y\gamma)}.$$

Straightforward calculations show that, for positive parameter values,

$$\frac{\partial V_1}{\partial y} = -\frac{x\gamma - y + \sqrt{(x\gamma - y)^2 + 4xz}}{\sqrt{(x\gamma - y)^2 + 4xz}} < 0,$$

$$\frac{\partial V_1}{\partial z} = \frac{2x}{\sqrt{(x\gamma - y)^2 + 4xz}} > 0,$$

$$\frac{\partial V_1}{\partial \gamma} = \frac{x(x\gamma - y - \sqrt{(x\gamma - y)^2 + 4xz})}{\sqrt{(x\gamma - y)^2 + 4xz}} < 0,$$

$$\frac{\partial V_1}{\partial x} = \frac{2z\gamma V_1}{\gamma(x\gamma - y + \sqrt{(x\gamma - y)^2 + 4xz})\sqrt{(x\gamma - y)^2 + 4xz}}.$$

The sign of $\frac{\partial V_1}{\partial x}$ is determined by the sign of V_1, which is determined by the sign of $R_0 = z/(y\gamma)$, which is independent of x. Applying the chain rule gives us the desired result. □

Since the dependence of $\hat{R}_1$ on V_1 is strictly monotone, a corollary to Lemma 4.1 is that if $\hat{R}_1$ depends on a parameter, then that dependence is also strictly monotone.

Theorem 4.2. *Consider (4.1) with positive parameter values. The curves of equilibria $\mathbf{E}_0$ and $\mathbf{E}_1$ intersect in a forward transcritical bifurcation when $R_0 = 1$. The IFE $\mathbf{E}_0$ is globally asymptotically stable for $R_0 < 1$ and unstable for $R_0 > 1$. The antibody-only equilibrium $\mathbf{E}_1$ is non-biological for $R_0 < 1$, is locally asymptotically stable for $\hat{R}_1 < 1 < R_0$, and is unstable if $\hat{R}_1 > 1$.*

Proof. The stability results properties of $\mathbf{E}_0$ follow directly from Theorem 4.1. If $R_0 < 1$, then it follows from (4.7) that $V_1 < 0$, and thus $\mathbf{E}_1$ is non-biological. By Lemma 4.1, the intersection between the curves of equilibria $\mathbf{E}_0$ and $\mathbf{E}_1$ is transverse. When $R_0 > 1$, the antibody-only equilibrium $\mathbf{E}_1$ is biological.

By writing the equation for CTLs first, the Jacobian at $\mathbf{E}_1$ may be written as

$$D\mathbf{g}(\mathbf{E}_1) = \begin{pmatrix} \omega\hat{R}_1 - \omega & 0 & 0 & 0 & 0 \\ 0 & -\rho - \beta V_1 & 0 & -\beta M_1 & 0 \\ -\frac{k\omega}{\psi}\hat{R}_1 & \beta V_1 & -\delta & \beta M_1 & 0 \\ 0 & 0 & b & -\gamma - fA_1 & -fV_1 \\ 0 & 0 & 0 & \alpha & -\mu \end{pmatrix}, \tag{4.12}$$

which has a lower-block-triangular form. The eigenvalues of $D\mathbf{g}(\mathbf{E}_1)$ are $\omega(\hat{R}_1 - 1)$ and the eigenvalues of the lower 4×4 matrix. Thus if $\hat{R}_1 > 1$, then $D\mathbf{g}(\mathbf{E}_1)$ has a positive eigenvalue and $\mathbf{E}_1$ is unstable.

When $R_0 > 1$ the characteristic equation of the lower 4×4 matrix satisfies the Routh–Hurwitz criteria (see, for example, Ref. 29), and therefore has negative real part (for details see the supplementary material). When $R_0 = 1$, there is a transcritical bifurcation that is "forward" in the sense that $\mathbf{E}_1$ is locally asymptotically stable for R_0 in a neighborhood of the form $(1, 1 + \epsilon)$, for ϵ sufficiently small. □

Theorem 4.3. *Consider the EIAV model with positive parameter values and* $\lambda\psi > \delta\omega$. *The curves of the equilibria* $\mathbf{E}_1$ *and* $\mathbf{E}_3$ *intersect in a forward transcritical bifurcation when* $\hat{R}_1 = 1$. *The coexistence equilibrium* $\mathbf{E}_3$ *is non-biological if* $\hat{R}_1 < 1$ *and is locally asymptotically stable for* $\hat{R}_1$ *in a neighborhood of form* $(1, 1+\epsilon)$ *for sufficiently small* $\epsilon > 0$.

Proof. It is easy to see that the Jacobian at $\mathbf{E}_1$ — given by Eq. (4.12) — has a simple zero eigenvalue when $\hat{R}_1 = 1$. The right and left null vectors of $D\mathbf{g}(\mathbf{E}_1)|_{\hat{R}_1=1}$ are $w = (1, w_2, w_3, w_4, w_5)^T$ and

$v = (1, 0, 0, 0, 0)$, respectively, where

$$w_3 = -\frac{k\lambda(2\alpha f\delta\rho\omega + \beta\gamma\mu(\lambda\psi - \delta\omega)}{\delta^2(\beta\gamma\mu(\lambda\psi - \delta\omega) + f\alpha\rho(\lambda\psi + \delta\omega))}, \tag{4.13}$$

which is negative since $\lambda\psi > \delta\omega$. Because of the exact structure of $D\mathbf{g}(\mathbf{E}_1)|_{\hat{R}_1=1}$, the expressions for w_2, w_4, and w_5 are not important, but for completion they are shown in the supplementary material. From here, we follow van den Driessche and Watmough.[28] Let p be one of the parameters that defines $\hat{R}_1$. By Lemma 4.1, $\frac{\partial \hat{R}_1}{\partial p} \neq 0$, and so the transversality condition,

$$v \cdot D_{xp}\mathbf{g}(\mathbf{E}_1)|_{\hat{R}_1=1} \cdot w = \omega\frac{\partial \hat{R}_1}{\partial p} \neq 0, \tag{4.14}$$

is satisfied. Checking the nondegeneracy condition, we have

$$v \cdot D_{xx}\mathbf{g}(\mathbf{E}_1)|_{\hat{R}_1=1} \cdot w^2 = \psi w_3 < 0, \tag{4.15}$$

and thus the bifurcation at $\hat{R}_1 = 1$ is a transcritical bifurcation. The sign of $\partial\hat{R}_1/\partial p$ corresponds to the stability of $\mathbf{E}_1$ for $p < p^*$, where p^* is the value of p such that $\hat{R}_1\Big|_{p=p^*} = 1$. If $\partial\hat{R}_1/\partial p > 0$, then $\mathbf{E}_1$ is stable for $p < p^*$ ($\hat{R}_1 < 1$), and if $\partial\hat{R}_1/\partial p < 0$, then $\mathbf{E}_1$ is unstable for $p < p^*$ ($\hat{R}_1 > 1$). Thus, the transcritical bifurcation is "forward" in the sense that $\mathbf{E}_3$ is biological and locally asymptotically stable for $\hat{R}_1$ in a neighborhood of the form $(1, 1+\epsilon)$ with $\epsilon > 0$ sufficiently small. □

Remark. Theorem 4.3 only guarantees stability of $\mathbf{E}_3$ for $\hat{R}_1$ in a neighborhood of the form $(1, 1+\epsilon)$ for $\epsilon > 0$ sufficiently small. The characteristic equation of the Jacobian at $\mathbf{E}_3$ is a fifth-order polynomial to which we were unable to apply the Routh–Hurwitz criterion. However, we can see from the determinant of the Jacobian,

$$\det(D\mathbf{g}(\mathbf{E}_3)) = -(\rho + \beta V_3)(R_1 - 1)(2V_3\alpha f + \gamma\mu), \tag{4.16}$$

that there is only a zero eigenvalue if $R_1 = 1$, and thus that $\mathbf{E}_3$ does not undergo any further transcritical or saddle-node bifurcations. However, it is not clear that $\mathbf{E}_3$ does not lose stability in a Hopf bifurcation or in some other, more exotic way. We did not observe any Hopf bifurcations or more complex dynamics in our numerical exploration.

4.3 Numerical Results and Application to EIAV Infection

In this section, we use bifurcation analysis to explore the EIAV system numerically and determine which parameters play key roles in the system. Values of immune system parameters k, f, ψ, ω, and μ were obtained from Schwartz *et al.*,[22] while the other parameters were taken from a simplified model fitted to data from horses experimentally infected with EIAV[12] (Table 4.1). Equilibrium $\mathbf{E}_0$ corresponds to the IFE in which infection does not persist. Boundary equilibrium $\mathbf{E}_1$ corresponds to EIAV infection with only an antibody response, which have not been observed among infected horses. Interior endemic equilibrium $\mathbf{E}_3$ corresponds to an infection that persists but is controlled at a manageable level by both CTLs and antibodies, which is what is observed in horses.[1] For this set of parameters, we have $\lambda\psi > \delta\omega$ and Theorem 4.3 applies.

Figure 4.1 uses the viral infectivity β as the control parameter, and the virus particle concentration V is shown on the vertical axis. The curve of IFE $\mathbf{E}_0$ intersects the curve of boundary equilibrium $\mathbf{E}_1$ when the basic reproduction number $R_0 = 1$. This occurs when $\beta = \beta_0 \approx 1.50 \times 10^{-5}\mu\text{L}/(\text{vRNA}\cdot\text{day})$. There is a forward transcritical bifurcation at the intersection; the IFE $\mathbf{E}_0$ is stable for $\beta < \beta_0$ and unstable for $\beta > \beta_0$. The antibody-only equilibrium

Table 4.1. Parameter values used in numerical results.

Symbol	Definition	Value	Units
α	Antibody production rate	15	molecules/(vRNA·day)
β	Infectivity rate	0.000325	μL/(vRNA·day)
γ	Virus clearance rate	6.73	1/day
δ	Infected cell death rate	0.0476	1/day
λ	Uninfected cell arrival rate	2.019	cells/(μL·day)
μ	Antibody clearance rate	20	1/day
ρ	Uninfected cell death rate	0.0476	1/day
ψ	CTL production rate	0.75	μL/(cell·day)
ω	CTL death rate	5	1/day
b	Virus production rate	505	vRNA/(cell·day)
f	Antibody neutralization rate	3	μL/(molecule·day)
k	Rate of killing by CTLs	0.01	μL/(cell·day)

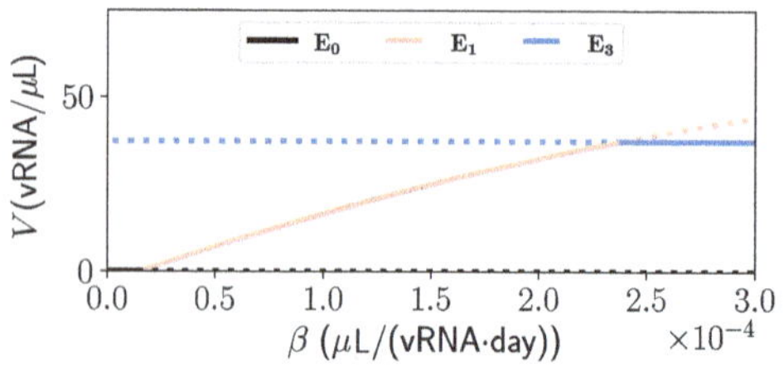

Fig. 4.1. Equilibrium values of the virus concentration V as a function of the viral infectivity β. The solid lines indicate when an equilibrium is stable, while the dotted lines indicate when the equilibrium is unstable. When $\mathbf{E}_3$ is unstable, it is also non-biological, and when $\mathbf{E}_3$ is stable, it is biological. Parameter values (except for β) are given in Table 4.1.

$\mathbf{E}_1$ is non-biological for $\beta < \beta_0$ and is stable for $\beta > \beta_0$. As the infectivity β increases beyond β_0, the equilibrium virus concentration V increases. Another forward transcritical bifurcation takes place when $\beta = \beta_1 \approx 2.38 \times 10^{-4}\mu$L/(vRNA·day). The antibody-only equilibrium $\mathbf{E}_1$ is stable for $\beta < \beta_1$ and unstable for $\beta > \beta_1$. The coexistence equilibrium $\mathbf{E}_3$, in which the antibody and CTL responses coexist, is non-biological for $\beta < \beta_1$ and is stable for $\beta > \beta_1$. As the infectivity β increases beyond β_1, the equilibrium virus concentration V remains steady.

Similarly, Figure 4.2 shows the equilibrium virus concentration using the virus production rate b as the control parameter. There is a forward transcritical bifurcation at the intersection of the IFE $\mathbf{E}_0$ and boundary equilibrium $\mathbf{E}_1$ curves, which occurs when $b = b_0 \approx 20$ vRNA/(cell·day). Another forward transcritical bifurcation occurs at the intersection of the boundary equilibrium $\mathbf{E}_1$ and interior equilibrium $\mathbf{E}_3$ curves, which occurs at $b = b_1 \approx 213$ vRNA/(cell·day).

Figures 4.1 and 4.2 show how different values of parameters β and b drive the system to different equilibria. Lower viral infectivity (β) and lower production of virus (b) correspond to lower levels of virus V and stability of antibody-only equilibrium $\mathbf{E}_1$. Alternatively, greater infectivity (β) and greater virus production (b) give higher virus levels V and stability of the coexistence equilibrium $\mathbf{E}_3$.

Figure 4.3 shows a two-parameter bifurcation diagram using viral infectivity β and the virus production rate b as bifurcation parameters. The first transcritical bifurcation occurs when $R_0 = 1$. Since R_0 depends on both β and b, it is possible to solve for b as a function of β, and the bifurcation appears as a decreasing curve in the

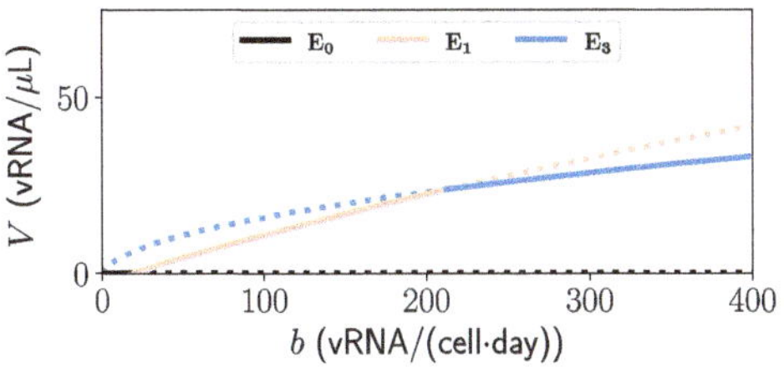

Fig. 4.2. Equilibrium values of the virus concentration V as a function of the virus production rate b. The solid lines indicate when an equilibrium is stable, while the dotted lines indicate when the equilibrium is unstable. When $\mathbf{E}_3$ is unstable, it is also non-biological. Parameter values (except for b) are given in Table 4.1.

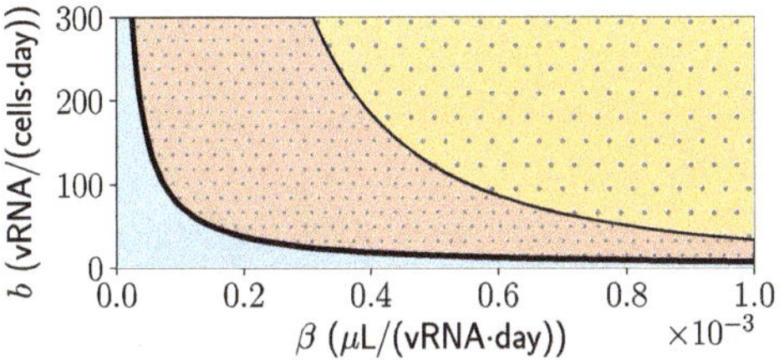

Fig. 4.3. Two-parameter bifurcation diagram showing how the critical values of the virus production rate (b) depend on infectivity (β). The black curve on the left represents the first transcritical bifurcation, when $R_0 = 1$. In the blue area to the left of and below this line, the IFE $\mathbf{E}_0$ is stable. The black curve on the right represents the second transcritical bifurcation, when $R_1 = 1$. In the tightly dotted red region between these two curves, the boundary equilibrium $\mathbf{E}_1$ is stable, and in the loosely dotted yellow region to the right of this curve, the endemic equilibrium $\mathbf{E}_3$ is stable. Parameter values (other than β and b) are taken from Table 4.1.

(β, b) plane, separating the region where $\mathbf{E}_0$ is stable from the region where $\mathbf{E}_1$ is stable. The second transcritical bifurcation occurs when $R_1 = 1$ and separates the region where $\mathbf{E}_1$ is stable from the region where $\mathbf{E}_3$ is stable. Note that if the value of the infectivity (β) or virus production rate (b) is low enough, then IFE $\mathbf{E}_0$ can be reached with a broad range of values in the other parameter.

The implication of the results shown in Figures 4.1–4.3 is that modification of the infectivity β or virus production rate b, such as by using antiretroviral therapies (ARTs) that block infection of cells or inhibit production of virus, respectively, can theoretically shift the

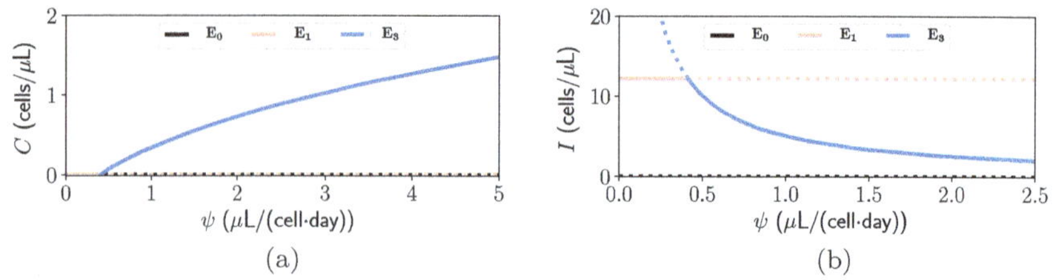

Fig. 4.4. (a) Equilibrium values of the CTL concentration C as a function of the CTL production coefficient ψ. (b) Equilibrium values of the infected cell concentration I as a function of the CTL production coefficient ψ. The solid lines indicate when an equilibrium is stable, while the dotted lines indicate when the equilibrium is unstable. When $\mathbf{E_3}$ is unstable, it is also non-biological. Parameter values (except for ψ) are given in Table 4.1.

system to a more preferred state, such as the IFE $\mathbf{E_0}$. ART, however, is not used in practice for treating EIAV-infected horses, whose immune systems manage the persistent infection without symptoms throughout most of their lives. Thus, we next investigate how to shift the system's equilibria by modifying the immune system parameters. In general, immune responses can be boosted by vaccination. Consequently, we examine the CTL production rate ψ and the antibody production rate α.

Figure 4.4 shows the $\mathbf{E_1}$–$\mathbf{E_3}$ bifurcation with CTL production rate ψ as the control parameter. The antibody-only equilibrium $\mathbf{E_1}$ is stable for $\psi < \psi_1$ and is unstable for $\psi > \psi_1$, with $\psi_1 \approx 0.48$ μL/(cell·day). The coexistence equilibrium $\mathbf{E_3}$ is non-biological for $\psi < \psi_1$ and is stable for $\psi > \psi_1$. As the value of ψ increases above ψ_1, the equilibrium CTL concentration C increases above zero (Figure 4.4a) and the equilibrium infected cell concentration I decreases (Figure 4.4b). This is consistent with the known function of CTLs, whose role is killing infected cells. In other words, the greater the production of CTLs, the higher the CTL level and the lower the number of infected cells. Equilibrium $\mathbf{E_3}$ is characterized by the presence of both CTLs and antibodies, which is the condition that gives rise to control of virus infection in EIAV-infected horses.

Figure 4.5 shows the $\mathbf{E_1}$–$\mathbf{E_3}$ bifurcation with antibody production rate α as the control parameter and virus particle concentration V (Figure 4.5a) and antibody concentration A (Figure 4.5b) on the vertical axes. When the antibody production rate (α) takes on

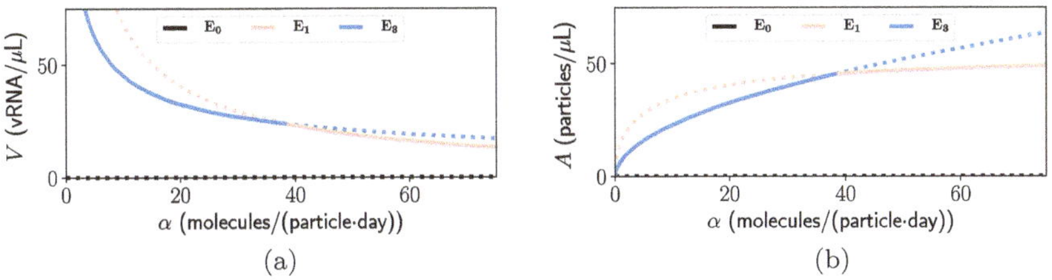

Fig. 4.5. (a) Equilibrium values of the virus concentration V as a function of the antibody production rate α. (b) Equilibrium values of the antibody concentration A as a function of the antibody production rate α. The solid lines indicate when an equilibrium is stable, while the dotted lines indicate when the equilibrium is unstable. When $\mathbf{E}_3$ is unstable, it is also non-biological. Parameter values (except for α) are given in Table 4.1.

lower values (less than ≈ 38 molecules/(vRNA·day)), the equilibrium viral load V is higher (Figure 4.5a), the equilibrium antibody level A is lower (Figure 4.5b), and the system is driven to stability of $\mathbf{E}_3$ (characterized by the existence not only of antibodies but also CTLs). When antibody production α takes on higher values (greater than $\approx$38 molecules/(vRNA·day)), the equilibrium viral load V is lower (Figure 4.5a), the equilibrium antibody level A plateaus (Figure 4.5b), and the antibody-only equilibrium $\mathbf{E}_1$ is stable. This result suggests that an antibody response that moderately reduces, but does not strongly reduce, the virus, is consistent with stability of coexistence equilibrium $\mathbf{E}_3$.

Figure 4.6 shows a two-parameter bifurcation diagram using the antibody production rate α and the CTL production rate ψ as bifurcation parameters. The second transcritical bifurcation, occurring when $R_1 = 1$, is an increasing curve in the α–ψ plane, separating the region where $\mathbf{E}_1$ is stable from the region where $\mathbf{E}_3$ is stable. Since R_1 depends on both ψ and α, it is possible to solve for ψ as a function of α, appearing almost as an inverse relationship, where higher values of α require even higher values of ψ for the stability of $\mathbf{E}_3$.

This result suggests that for stability of coexistence equilibrium $\mathbf{E}_3$, the required value of the CTL production rate ψ increases with increasing antibody production rate α. In other words, any increase in antibody production should be coupled with an increase in CTL production; otherwise the system is driven toward $\mathbf{E}_1$, with antibody responses and no CTLs. In summary, these numerical results may

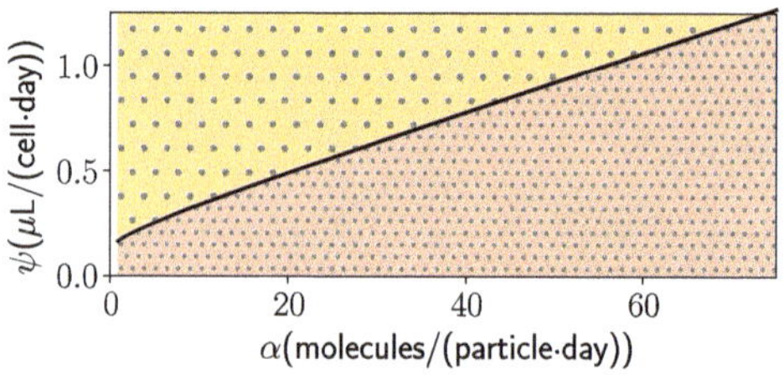

Fig. 4.6. Two parameter bifurcation diagram showing how the critical value of the CTL production rate ψ depends on the antibody production rate α. The curve represents the second transcritical bifurcation, when $R_1 = 1$. In the tightly dotted red area below the curve, the boundary equilibrium $\mathbf{E}_1$ is stable, and in the loosely dotted yellow area above this curve, the endemic equilibrium $\mathbf{E}_3$ is stable. Parameter values (other than α and ψ) are taken from Table 4.1.

seem somewhat counter-intuitive: To obtain stability of $\mathbf{E}_3$, characterized by the coexistence of both antibody and CTL responses, the desired immune response has lower α (i.e., less production of antibodies), and greater ψ (i.e., greater production of CTLs); a very strong antibody production (i.e., high α), associated with $\psi < \psi_1$, however, correlates with an absence of CTLs.

4.4 Discussion

In this chapter, we analyze the equilibrium states of a virus infection model with immune system responses in the form of antibodies and CTLs. Using a standard Lyapunov function argument, we show that the IFE $\mathbf{E}_0$ is globally asymptotically stable when the basic reproduction number R_0 is less than one. When $R_0 = 1$, there is a forward transcritical bifurcation where the IFE loses stability to the boundary equilibrium $\mathbf{E}_1$, which describes an infection that is controlled by antibodies but not CTLs. Using the next generation matrix method by Diekmann *et al.*,[26] van den Driessche and Watmough,[28] and van den Driessche,[27] we derive a reproduction number for CTLs, $\hat{R}_1$, and show that $\mathbf{E}_1$ is locally asymptotically stable when $\hat{R}_1 < 1 < R_0$. When $\hat{R}_1 = 1$, there is a second forward transcritical bifurcation where the boundary equilibrium loses stability to the endemic equilibrium $\mathbf{E}_3$, which describes an infection that is controlled by both antibodies and CTLs. We are unable to show that $\mathbf{E}_3$ remains locally stable as $\hat{R}_1$ increases, but our numerical analysis suggests that this is the case.

Our results are similar to those of Gómez-Acevedo *et al.*,[30] who examine a three-equation model of infection by Human T cell Leukemia/Lymphoma virus, HTLV. They obtain a basic reproduction number R_0 for infected cells and a second threshold R_1, which they also interpret as a basic reproduction number for CTLs. Their Theorem 3.1 corresponds to our results relating the stability of the equilibria to the basic reproduction numbers R_0 and $\hat{R}_1$, consistent with our analysis (of the bifurcations between $\mathbf{E}_0$, $\mathbf{E}_1$, and $\mathbf{E}_3$) showing that $\hat{R}_1$ is precisely the basic reproduction number of CTLs.

The numerical results presented here offer insights into the potential relationships between immune response parameters in EIAV infection. Furthermore, such insights have implications for vaccine development. For instance, an antibody response that moderately, but not strongly, reduces virus is consistent with stability of the coexistence equilibrium $\mathbf{E}_3$ (i.e., control of virus infection with CTLs and antibodies). A vaccine, therefore, that aims to stimulate antibody production modestly would drive the system to this state of control of infection. This would lead to a lower antibody concentration and a higher virus concentration, which may seem counter intuitive, since conventional wisdom would presume that a vaccine that stimulates more antibody production would lead to greater reduction of virus and more control. However, the results shown here describe how the antibody response must be tempered in order to allow for the coexistence of a CTL response. Thus, a potential vaccine that stimulates the production of antibodies would need to stimulate the production of CTLs as well. Overstimulation of antibodies would drive the system to the equilibrium state devoid of CTLs. Consequently, a vaccine intended to control virus infection by stimulating both immune responses would aim to favor the CTL response in order to balance the antibody response. An ideal vaccine would accomplish this balance.

Some limitations of the work presented here should be discussed. In this work, we are motivated by the finding that EIAV infection (unlike HIV) is controlled by the host adaptive immune response, and this control is mediated by both antibody and CTL responses. Thus our goal here is to use the knowledge that both responses are crucial for control as the basis to explore the asymptotic behavior of a model that considers each response's dynamics separately. However, other models more explicitly describe the clonal expansion of the antibody response as well as the kinetics of CTL growth.[31] Future work that addresses these modeling hurdles will help understand the role of complex immune responses. Another caveat of this work is a reliance on deterministic population dynamics; stochastic interactions are not taken into account in this model. In addition, this study does not consider spatial heterogeneity or diffusion.

Other studies in the literature do consider stochasticity in within-host dynamics,[32,33] as well as spatial heterogeneity and diffusion.[32,34] Gibelli *et al.*[32] expands upon the basic model[15,35] by including population heterogeneity (in this case, in the age-distributed time of cell death and variation in the timing of eclipse phase dynamics). Bellomo *et al.*[34] takes into account spatial effects of the three populations of the basic model (i.e., uninfected cells, infected cells, and virus), particularly the contributions of diffusion and movement by chemotaxis. While research on stochastic dynamical systems shows that the deterministic structure of a model is often still strongly apparent with the addition of stochasticity,[36] future studies that use hybrid models that include stochastic and deterministic dynamics, and modeling that considers heterogeneity, will advance the field by leading to more precise depictions of the biological scenarios being modeled. Our model and analysis presented here may form the foundation of such future work.

Acknowledgments

We would like to thank Mark Schumaker for substantial input on an earlier draft of the chapter. We would also like to thank Christina Cobbold, Adriana Dawes, Abba Gumel, Fabio Milner, Stacey Smith? [sic], Rebecca Tyson, and Gail Wolkowicz for their suggestions on the mathematical and numerical analyses and conversations about this work. We also thank two anonymous reviewers for their suggestions, which improved the chapter.

This work was partially supported by the Simons Foundation and partially supported by the National Institute of General Medical Sciences of the National Institutes of Health under Award Number P20GM104420. The content is solely the responsibility of the authors and does not necessarily represent the official views of the National Institutes of Health.

Appendix 4: Supplementary Material

This section briefly describes the supplementary materials, which are Jupyter notebooks that fill in some of the details of the proofs in this paper. The DOI provides the link to this material online.

Notebook 1: In this notebook, we develop the Routh–Hurwitz conditions for the stability of a fourth-order polynomial by constructing the table described by Meinsma.[29] https://doi.org/10.7273/000002580.

Notebook 2: In this notebook, we calculate the characteristic polynomial associated with the Jacobian matrix $D\mathbf{g}(\mathbf{E}_1)$ given by Eq. (4.12). We cast the coefficients of the characteristic polynomial in forms that are manifestly positive, and verify the Routh–Hurwitz criterion. The third-order Routh–Hurwitz criterion comes out as a sum of 172 terms, of which two are negative. We show that squares can be completed, combining the negative terms with other terms in a form that is positive. https://doi.org/10.7273/000002581.

Notebook 3: In this notebook, we calculate the left and right null-vectors of the Jacobian matrix $D\mathbf{g}(\mathbf{E}_1)$ given by Eq. (4.12). We then cast the nondegeneracy condition in a form that is manifestly negative. https://doi.org/10.7273/000002582.

References

1. Leroux, C., Cadoré, J.-L., and Montelaro, R. C. (2004). Equine infectious anemia virus (EIAV): What has HIV's country cousin got to tell us? *Vet. Res.* **35**, pp. 485–512.
2. Craigo, J. K. and Montelaro, R. C. (2013). Lessons in AIDS vaccine development learned from studies of equine infectious anemia virus infection and immunity, *Viruses* **5**, pp. 2963–2976, doi:10.3390/v5122963.
3. Cook, R. F., Leroux, C., and Issel, C. J. (2013). Equine infectious anemia and equine infectious anemia virus in 2013: A review, *Vet. Microbiol.* **167**, pp. 181–204, doi:10.1016/j.vetmic.2013.09.031.

4. Issel, C. J., Cook, R. F., Mealey, R. H., and Horohov, D. W. (2014). Equine infectious anemia in 2014: Live with it or eradicate it? *Vet. Clin. Equine* **30**, pp. 561–577, doi:10.1016/j.cveq.2014.08.002.
5. McGuire, T., Leib, S., Lonning, S., Zhang, W., Byrne, K., and Mealey, R. (2000). Equine infectious anaemia virus proteins with epitopes most frequently recognized by cytotoxic T-lymphocytes from infected horses, *J. Gen. Virol.* **81**, pp. 2735–2739.
6. Tagmyer, T., Craigo, J., Cook, S., Issel, C., and Montelaro, R. (2007). Envelope-specific T-helper and cytotoxic T-lymphocyte responses associated with protective immunity to equine infectious anemia virus, *J. Gen. Virol.* **88**, pp. 1324–1336.
7. Craigo, J. K., Durkin, S., Sturgeon, T. J., Tagmyer, T., Cook, S. J., Issel, C. J., and Montelaro, R. C. (2007). Immune suppression of challenged vaccinates as a rigorous assessment of sterile protection by lentiviral vaccines, *Vaccine* **25**, 5, pp. 834–845.
8. Hammond, S. A., Cook, S. J., Lichtenstein, D. L., and Issel, C. J. (1997). Maturation of the cellular and humoral responses to persistent infection in horses by equine infectious anemia virus is a complex and lengthy process, *J. Virol.* **71**, 5, pp. 3840–3852.
9. Sponseller, B., Sparks, W., Wannemuehler, Y., Li, Y., Antons, A., Oaks, J., and Carpenter, S. (2007). Immune selection of equine infectious anemia virus env variants during the long-term inapparent stage of disease, *Virology* **363**, pp. 156–165.
10. Mealey, R. H., Littke, M. H., Leib, S. R., Davis, W. C., and McGuire, T. C. (2008). Failure of low-dose recombinant human IL-2 to support the survival of virus-specific CTL clones infused into severe combined immunodeficient foals: Lack of correlation between in vitro activity and in vivo efficacy, *Vet. Immunol. Immunop.* **121**, pp. 8–22, doi:10.1016/j.vetimm.2007.07.011.
11. Schwartz, E. J., Nanda, S., and Mealey, R. H. (2015). Antibody escape kinetics of equine infectious anemia virus infection of horses, *J. Virol.* **89**, pp. 6945–6951, doi:10.1128/JVI.00137-15.
12. Schwartz, E. J., Vaidya, N. K., Dorman, K., Carpenter, S., and Mealey, R. H. (2018). Dynamics of lentiviral infection in vivo in the absence of adaptive immune responses, *Virology* **513**, pp. 108–113, doi:10.1016/j.virol.2017.09.023.
13. Taylor, S. D., Leib, S. R., Carpenter, S., and Mealey, R. H. (2010). Selection of a rare neutralization-resistant variant following passive transfer of convalescent immune plasma in equine infectious anemia virus-challenged SCID horses, *J. Virol.* **84**, 13, pp. 6536–6548, doi: 10.1128/JVI.00218-10.

14. Taylor, S. D., Leib, S. R., Wu, W., Nelson, R., Carpenter, S., and Mealey, R. H. (2011). Protective effects of broadly neutralizing immunoglobulin against homologous and heterologous equine infectious anemia virus infection in horses with severe combined immunodeficiency, *J. Virol.* **85**, 13, pp. 6814–6818, doi:10.1128/JVI.00077-11.
15. Nowak, M. A. and Bangham, C. R. M. (1996). Population dynamics of immune responses to persistent viruses, *Science* **272**, pp. 74–79.
16. Perelson, A. S., Neumann, A. U., Markowitz, M., Leonard, J. M., and Ho, D. D. (1996). HIV-1 dynamics in vivo: Virion clearance rate, infected cell life-span, and viral generation time, *Science* **271**, 5255, pp. 1582–1586, doi:10.1126/science.271.5255.1582.
17. Noecker, C., Schaefer, K., Zaccheo, K., Yang, Y., Day, J., and Ganusov, V. V. (2015). Simple mathematical models do not accurately predict early SIV dynamics, *Viruses* **7**, pp. 1189–1217, doi:10.3390/v7031189.
18. Perelson, A. S. and Ribeiro, R. M. (2013). Modeling the within-host dynamics of HIV infection, *BMC Biol.* **11**, p. 96, doi:10.1186/1741-7007-11-96.
19. Phillips, A. N. (1996). Reduction of HIV concentration during acute infection: independence from a specific immune response, *Science* **271**, pp. 497–499.
20. Stafford, M. A., Corey, L., Cao, Y., Daar, E. S., Ho, D. D., and Perelson, A. S. (2000). Modeling plasma virus concentration during primary HIV infection, *J. Theor. Biol.* **203**, pp. 285–301, doi:10.1006/jtbi.2000.1076.
21. Wodarz, D. (2003). Hepatitis C virus dynamics and pathology: The role of CTL and antibody responses, *J. Gen. Virol.* **84**, pp. 1743–1750.
22. Schwartz, E. J., Pawelek, K. A., Harrington, K., Cangelosi, R., and Madrid, S. (2013). Immune control of equine infectious anemia virus infection by cell-mediated and humoral responses, *Appl. Math.* **4**, pp. 171–177, doi:10.4236/am.2013.48A023.
23. Le, D., Miller, J. D., and Ganusov, V. V. (2015). Mathematical modeling provides kinetic details of the human immune response to vaccination, *Front. Cell. Infect. Microbiol.* **4**, p. 177, doi:10.3389/fcimb.2014.00177.
24. Koopman, G., Mooij, P., Dekking, L., Mortier, D., Nieuwenhuis, I. G., van Heteren, M., Kuipers, H., Remarque, E. J., Radošević, K., and Bogers, W. M. J. M. (2016). Correlation between virus replication and antibody responses in macaques following infection with pandemic influenza A virus, *J. Virol.* **90**, 2, pp. 1023–1033, doi:10.1128/JVI.02757-15.
25. Sajadi, M. M., Guan, Y., DeVico, A. L., Seaman, M. S., Hossain, M., Lewis, G. K., and Redfield, R. R. (2011). Correlation between circulating HIV-1 RNA and broad HIV-1 neutralizing antibody activity,

J. Acquir. Immune. Defic. Syndr. **57**, 1, pp. 9–15, doi:10.1097/QAI.0b013e3182100c1b.

26. Diekmann, O., Heesterbeek, J. A. P., and Metz, J. A. J. (1990). On the definition and the computation of the basic reproduction ratio R_0 in models for infectious diseases in heterogeneous populations, *J. Math. Bio.* **28**, 4, pp. 365–382, doi:10.1007/BF00178324.
27. van den Driessche, P. (2017). Reproduction numbers of infectious disease models, *Infect. Dis. Model.* **2**, pp. 288–303.
28. van den Driessche, P. and Watmough, J. (2002). Reproduction numbers and sub-threshold equilibria for compartmental models of disease transmission, *Math. Biosci.* **180**, pp. 29–48.
29. Meinsma, G. (1995). Elementary proof of the Routh-Hurwitz test, *Syst. Control Lett.* **25**, pp. 237–242.
30. Gómez-Acevedo, H., Li, M. Y., and Jacobson, S. (2010). Multistability in a model for CTL response to HTLV infection and its implications to HAM/TSP development and prevention, *Bull. Math. Bio.* **72**, 3, pp. 681–696, doi:10.1007/s11538-009-9465-z.
31. Antia, R., Ganusov, V. V., and Ahmed, R. (2005). The role of models in understanding CD8+ T-cell memory, *Nat. Rev. Immunol.* **5**, 2, pp. 101–111.
32. Gibelli, L., Elaiw, A., and Alghamdi, M. (2017). Heterogeneous population dynamics of active particles: Progression, mutations, and selection dynamics, *Math. Mod. Meth. Appl. Sci.* **27**, pp. 617–640.
33. Schwartz, E. J., Yang, O. O., Cumberland, W. G., and de Pillis, L. G. (2013). Computational model of HIV-1 escape from the cytotoxic T lymphocyte response, *Canad. Appl. Math. Quart.* **21**, 2, pp. 261–279.
34. Bellomo, N., Painter, K., Tao, Y., and Winkler, M. (2019). Occurrence versus absence of taxis-driven instabilities in a May-Nowak model for virus infection, *SIAM J. Appl. Math.* **79**, 5, pp. 1990–2010.
35. Perelson, A. S. (2002). Modeling viral and immune system dynamics, *Nat. Rev. Immunol.* **2**, pp. 28–36, doi:10.1038/nri700.
36. Abbott, K. C. and Nolting, B. C. (2017). Alternative (un)stable states in a stochastic predator-prey model, *Ecol. Complex.* **32**, pp. 181–195.

https://doi.org/10.1142/9789811263033_0005

Chapter 5

Structural and Practical Identifiability Analysis of a Multiscale Immuno-Epidemiological Model

Laura Nemeth[*,§], **Necibe Tuncer**[†,¶], **and Maia Martcheva**[‡,||]

[*]*Department of Mathematical Sciences, Florida Atlantic University, Boca Raton, Florida, USA*

[†]*Department of Mathematical Sciences, Florida Atlantic University, Boca Raton, Florida, USA*

[‡]*Department of Mathematics, University of Florida, 358 Little Hall, PO Box 118105, Gainesville, Florida, USA*

[§]*knemeth1@my.fau.edu*
[¶]*ntuncer@fau.edu*
[||]*maia@ufl.edu*

In this chapter, we perform the identifiability analysis of a multiscale model of seasonal influenza with multiscale data. We show that the well studied target cell limited within-host model is not structurally identifiable. So, we reformulate the model and work with a scaled within-host model which is structurally identifiable. We find that the scaled within-host model is practically identifiable with respect to two distinct viremia datasets while fitting with weighted or unweighted least squares. We introduce a methodology on how to study the structural identifiability of multiscale epidemic models, specifically nested immuno-epidemiological models. All parameters of the multiscale model are practically identifiable. Furthermore, we find that the practical identifiability of the multiscale model is significantly better when fitted to viremia and incidence data as opposed to when fitted to viremia and cumulative incidence data. Comparing first and second order numerical methods for solving the partial differential equations suggests that using a higher order numerical method does not affect the identifiability of the parameters. Further simulations suggest that the choice of the linking functions has some impact on identifiability when viremia and incidences are fitted, but no impact when viremia and cumulative incidences are fitted.

5.1 Introduction

There are many types of multiscale models for infectious diseases which have been reviewed on several occasions (see Refs. 7, 13). Nested multiscale immuno-epidemiological models are one of the many multiscale infectious disease models. They were first proposed by Gilchrist and Sasaki[8] in 2002. A salient strength of these models is that they allow the use of multiscale data. Data on the within-host level, which are typically collected in days post infection, can be connected via nested immuno-epidemiological models to data on the between-host level, which are collected in chronological time. Linking multiscale data in turn may allow us to study infectious disease processes through well-validated multiscale models.

Fitting epidemic or immune models to time series data with the goal of inferring the parameters is not new and it is a typical first step before some biological question(s) are addressed. This process has become fairly common when single scale models are used. When we fit to derive the parameters of the model, it becomes clear that we need to seek identifiability to strengthen the reliability of parameters and, consequently, model outputs. In recent years, structural/practical identifiability of ODE models as often been addressed because of its importance for inferring the parameters.[6,22] The mathematical tools to study structural and practical identifiability of ODE models are well developed. In contrast, for the partial differential equation models (PDEs) these tools and knowledge are not developed at all. One of the difficulties of the identifiability analysis for PDEs is that the parameters of these models are not constants but functions which define an infinite dimensional space to be identified. However, this drawback is not present in nested immuno-epidemiological models. Even though they are PDE age-since-infection models, once the functions linking the within-host scale to the between-host scale have been decided on, the parameter set to be determined from output data is a vector of constants. That makes the problem somewhat more approachable but raises the question whether the form of the linking functions has impact on the identifiability of the parameters.

In this chapter, we study the question of identifiability of multiscale nested immuno-epidemiological models. Nested models can be fit to data in two ways: (1) sequentially, by fitting first the within-host model to within-host data, fixing its parameters to the estimated values, and then fitting the between host models to population-level data; (2) simultaneously, by fitting the full model to data from both scales. In our previous work[20] we found that the simultaneous fitting improves the practical identifiability of the parameters. Thus, in this chapter we fit simultaneously and address other factors that may influence the parameter identifiability. In particular, we are interested in (1) how the choice of the linking functions influences the identifiability, (2) how the choice of the numerical method affects the identifiability, and last but not least, (3) how the choice of data affects identifiability.

In the next section, we introduce a nested immuno-epidemiological model of influenza. In Section 5.3, we study the structural and practical identifiability of the within-host model relative to two within-host datasets of viremia. In Section 5.4, we study the structural and practical identifiability of the multiscale model relative to viremia data and incidences/cumulative incidences of influenza. We formulate two different numerical methods for solving the PDE model, and two different transmission rate linking functions. In Section 5.5, we summarize our results and observations from the simulations.

5.2 Immuno-Epidemiological Model

5.2.1 *Within-host model*

We model the within-host dynamics of the Influenza virus with the following well-known target cell limited model (5.1) (Table 5.1). The within-host model (5.1) describes the dynamics of uninfected target

Table 5.1. Definition of the variables and the parameters in the within-host model (5.1).

Variable	Meaning
$V(\tau)$	Viremia level at time-since infection τ
$T(\tau)$	Number of target cells at time-since infection τ
$T_i(\tau)$	Number of infected target cells at time-since infection τ
Parameter	**Meaning**
β_T	Infection rate of the target cells
δ	Clearance rate of infected target cells
π	The viral release rate
c	The viral clearance rate

cells $T(\tau)$, infected target cells $T_i(\tau)$, and virus $V(\tau)$. Such models have been studied extensively for human immunodeficiency virus (HIV), West Nile virus (WNV), and Influenza virus infections.[2, 18, 19] The process of virus infection within a cell is modeled by the mass action term $\beta_T VT$, which states that the rate at which the virus finds a target cell, successfully binds to the surface of the cell, and/or enters the target cell, is assumed to be proportional to the product of the number of target cells and free virus. The infected (virus-producing) target cells, with average life span of $1/\delta$, produce free virus at rate π. Free virus is cleared by non-specific mechanisms (innate and adaptive immune response) at rate c. The initial viral load and the initial number of target cells are denoted by V_0 and T_0, respectively.

$$\textbf{Within-host model:} \begin{cases} \dfrac{dT}{d\tau} & = -\beta_T VT, \\ \dfrac{dT_i}{d\tau} & = \beta_T VT - \delta T_i, \\ \dfrac{dV}{d\tau} & = \pi T_i - cV, \end{cases} \tag{5.1}$$

5.2.2 *Between-host model*

To model the Influenza H1N1 outbreak, we use a time-since infection structured SIR outbreak model. That is, we divide the total population $N(t)$ into three non intersecting classes: susceptible, infected, and recovered classes. The number of susceptible individuals at time

Table 5.2. Definition of the variables in the between-host model (5.2).

Variable	Meaning
$S(t)$	Number of susceptible individuals at time t
$i(\tau, t)$	Density of infected individuals at time-since-infection τ and at time t
$R(t)$	Number of recovered individuals at time t

t is denoted by $S(t)$ and the number of recovered individuals at time t is denoted by $R(t)$. The density of infected individuals, denoted by $i(t, \tau)$, is structured by time-since infection parameter τ to introduce heterogeneity to the host population (Table 5.2). The incidences are modeled by standard incidence term, with transmission rate $\beta(\tau)$ varying with the infection age (τ) of the infected individual. Thus, the susceptible individuals who move to infected compartment per unit of time is given by

$$\frac{S}{N}\int_0^{\tau^*} \beta(\tau)i(\tau,t)\mathrm{d}\tau.$$

Infected individuals recover at a rate $\gamma(\tau)$ and move to the recovered class. The transmission and recovery rates of the epidemiological model are linked to the immune response of the host and depend on the within-host model state variables. Let the chronological time be denoted by $t > 0$ and time-since-infection $\tau \in [0, \tau^*]$ with $\tau^* = \infty$, then the between-host model is an SIR model with the following form:

$$\textbf{Between-host model:} \begin{cases} \dfrac{\mathrm{d}S}{\mathrm{d}t} &= -\dfrac{S}{N}\displaystyle\int_0^{\tau^*} \beta(\tau)i(\tau,t)\mathrm{d}\tau, \\ \dfrac{\partial i}{\partial t} + \dfrac{\partial i}{\partial \tau} &= -\gamma(\tau)i(\tau,t), \\ i(0,t) &= \dfrac{S}{N}\displaystyle\int_0^{\tau^*} \beta(\tau)i(\tau,t)\mathrm{d}\tau, \\ \dfrac{\mathrm{d}R}{\mathrm{d}t} &= \displaystyle\int_0^{\tau^*} \gamma(\tau)i(\tau,t)\mathrm{d}\tau, \end{cases} \tag{5.2}$$

which is equipped with initial conditions $S(0) = S_0$, $i(\tau, 0) = i_0(\tau)$ $R(0) = 0$. The probability of still being infectious at time τ

is given by

$$\pi(\tau) = e^{-\int_0^\tau \gamma(s)\mathrm{d}s}.$$

Integrating with respect to τ the density of infected population $i(\tau, t)$, we obtain the total number of infected individuals at time t, $I(t)$, given by

$$I(t) = \int_0^{\tau^*} i(\tau, t)\mathrm{d}\tau.$$

Since τ^* is the maximum time-since infection, the density of infected individuals vanishes for $\tau \geq \tau^*$. Hence, $\lim_{\tau \to \tau^*} i(\tau, t) = 0$. The total population $N(t) = S(t) + I(t) + R(t)$ satisfies the equation $N' = 0$, so we set $N(t) = N$ where

$$N = S(0) + \int_0^\infty i_0(\tau)\mathrm{d}\tau + R(0).$$

To link the epidemiological parameters to within-host immune dynamics, we formulate the recovery rate $\gamma(\tau)$ as a function of the pathogen load of the infected host, $V(\tau)$. We assume that $\gamma(\tau)$ is inversely related to the viremia level of the host as follows:

$$\gamma(\tau) = \frac{\kappa}{V(\tau) + \epsilon},$$

where κ is the proportionality constant and ϵ is a small number. Similarly, the transmission rate $\beta(\tau)$ is also dependent on the infected host's pathogen load, and we assume that transmission rate of the infected host is directly proportional to the viremia level as follows:

$$\beta(\tau) = c_1 V(\tau).$$

The goal of this study is to examine the structural and practical identifiability of the nested epidemiological models, hence, we study the effects of the linking relationships on the identifiability of the model parameters, so we also consider another linking relationship of the transmission rate to the pathogen load (Table 5.3). For the second linking, we formulate $\beta(\tau)$ as

$$\beta(\tau) = \frac{c_1 V(\tau)}{C_0 + V(\tau)}.$$

Table 5.3. Definition of the parameters in the between-host model (5.2) and their linking-relation to immune response dynamics.

Parameter	Meaning	Linking relation
$\beta(\tau)$	Transmission rate	$\beta(\tau) = c_1 V(\tau)$ or $\beta(\tau) = \dfrac{c_1 V(\tau)}{C_0 + V(\tau)}$,
$\gamma(\tau)$	Recovery rate	$\gamma(\tau) = \dfrac{\kappa}{V(\tau) + \epsilon}$

5.2.3 *Finite difference method for the immune epidemiological model*

To approximate solutions of the between-host model (5.2), we perform two different finite difference methods, which converge with first-order and second-order accuracies. The parameters of the epidemiological model are linked to the infected host's immune response. Hence, we combine the finite difference schemes with a numerical ODE solver and incorporate solutions of the immunological model into the epidemiological parameters: transmission and recovery rates.

Each finite difference method discretizes both the time variable t and the time-since infection variable τ. Let Δt be both the time step and the time-since infection step, then the discrete time points are $\{t_n\}_{n=0}^{N_T}$, where $t_n = n\Delta t$ for $n = 0, 1, 2, \ldots, N_T$. Similarly, let $\{\tau_k\}_{k=0}^{M}$, where $\tau_k = k\Delta t$ for $k = 0, 1, 2, \ldots, M$. Clearly, the final time of interest is $T = N_T \Delta t$ and the final time of infection is $A = M\Delta t$. We denote the approximate solutions of the state variables and the linked epidemiological parameters at (t_n, τ_k) as $S(t_n) \approx S^n$, $i(\tau_k, t_n) \approx i_k^n$, and $\beta(\tau_k) = \beta_k$, $\gamma(\tau_k) = \gamma_k$. The derivatives of the state variables, S and R, are approximated by the backward Euler method and the integrals are approximated by the right end rule. The nonlinear term $\dfrac{S}{N}\displaystyle\int_0^{\mathcal{T}*} \beta(\tau) i(\tau, t) \mathrm{d}\tau$ is treated with single Picard iteration. For instance, the approximate solution of the nonlinear term $\dfrac{S^{n+1}}{N}\displaystyle\int_0^{\mathcal{T}*} \beta(\tau) i(\tau, t_{n+1}) \mathrm{d}\tau$ is linearized with single Picard iteration as $\dfrac{S^{n+1}}{N}\displaystyle\int_0^{\mathcal{T}*} \beta(\tau) i(\tau, t_n) \mathrm{d}\tau$. Then, we use the following backward

Euler method to approximate the state variables S and R in (5.2), which takes the following form:

$$\textbf{Backward Euler method:} \begin{cases} S^{n+1} = \dfrac{S^n}{1 + \frac{\Delta t}{N}\sum_{k=1}^{M} \Delta t \beta_k i_k^n} \\ R^{n+1} = R^n + \Delta t \sum_{k=1}^{M} \Delta t \gamma_k i_k^n. \end{cases} \tag{5.3}$$

The first finite difference we perform is the backward difference approximation of the characteristic derivative which converges with a first-order accuracy.[12] That is the first numerical method is given as

$$\textbf{Numerical method 1:} \begin{cases} \dfrac{i_{k+1}^{n+1} - i_k^n}{\Delta t} = -\gamma_{k+1} i_{k+1}^{n+1}, \\ i_0^{n+1} \qquad = \dfrac{S^{n+1}}{N} \displaystyle\sum_{k=1}^{M} \Delta t \beta_k i_k^n \\ i_k^0 \qquad = i_0(\tau_k). \end{cases} \tag{5.4}$$

The second finite difference we perform is the central difference approximation of the characteristic derivative which converges with a second-order accuracy.[9] That is the second numerical method is given as

$$\textbf{Numerical method 2:} \begin{cases} \dfrac{i_{k+1}^{n+1} - i_k^n}{\Delta t} = -\gamma_{k+1/2} \dfrac{i_{k+1}^{n+1} + i_k^n}{2}, \\ i_0^{n+1} \qquad = \dfrac{S^{n+1}}{N} \displaystyle\sum_{k=1}^{M} \Delta t \beta_k i_k^n \\ i_k^0 \qquad = i_0(\tau_k). \end{cases} \tag{5.5}$$

The epidemiological parameters $\beta(\tau)$ and $\gamma(\tau)$ are evaluated by solving the within-host model using the `MATLAB ODE` solver `ode45`.

5.3 Identifiabilty Analysis of the Within-Host Model

5.3.1 *Structural identifiability of the within-host model*

Structural identifiability of the dynamical models represented by system of ODEs addresses whether the parameters of the model can be determined from the noise-free data. If the parameters of the model can be estimated uniquely or have finitely many solutions, then the ODE model is said to be structurally identifiable or structurally locally identifiable, respectively. If infinitely many parameter sets can be obtained from the same noise-free data, then the model is said to be unidentifiable. Structural identifiability is a characteristic of the ODE model for the ideal noise-free data. Hence, it is crucial to check the structural identifiably of the model before estimating the parameters numerically.

Structural identifiability analysis tries to answer whether the model is structured to identify its parameters for perfect experimental data. Suppose $g(\boldsymbol{x}(\tau), \boldsymbol{p})$ is the perfect noise-free output function, where $\boldsymbol{x}(\tau)$ is the array of state variables and $\boldsymbol{p}$ are the parameters of the ODE model. The noise-free experimental data be on the trajectory given by $g(\boldsymbol{x}(\tau), \boldsymbol{p})$. The definition of the structural identifiability is given in the literature such as Refs. 6 and 22.

Definition 5.1. A parameter set $\boldsymbol{p}$ is called *structurally globally (or uniquely) identifiable* if for every $\boldsymbol{q}$ in the parameter space the equation

$$g(\boldsymbol{x}(\tau), \boldsymbol{p}) = g(\boldsymbol{x}(\tau), \boldsymbol{q}) \implies \boldsymbol{p} = \boldsymbol{q}.$$

That is, for any distinct parameter set, the noise-free data are distinct. There are many mathematical methods to study the structural identifiability of a dynamic system of ODEs. Among these methods are the differential algebra approach, Taylor series approach, generating series approach, and a method based on the Implicit Function

Theorem.[3,6,10,14,17] For details about these methods, we refer the reader to the excellent reviews written on the topic.[5,14] In this chapter, we will use differential algebra approach to study the structural identifiability of the immunological model (5.1). For the within-host model, we are interested in evaluating immunological parameters such as infection rate of the target cells, β_T, clearance rate of infected cells, δ, the viral release rate of infected cells, π, and the viral clearance rate, c. Since the data are the viremia levels, the initial viremia level can often be determined from the data. We study whether the immunological model (5.1) is structured to reveal its immunological parameters from noise-free viremia levels using the differential algebra approach. We will briefly summarize the differential algebra approach here, but for more details the reader is referred to Refs. 3 and 14.

The first step in differential algebra approach is to find the input–output equations of the dynamical ODE model. This is done by reducing the model to its *characteristic set* via Ritt's algorithm. The characteristic set is then used to derive the input–output equations which contain all the identifiability information of the model. Input–output equations, simply put, are a subset of the characteristic set. The characteristic set of a dynamical ODE model is not unique, hence, it depends on the chosen ranking of the state variables. However, the coefficients of the input–output equation can be fixed uniquely by normalizing the equations to make them monic.

For example, with the ranking $V < T < T_i$, the model (5.1) yields to the following input–output equation:

$$\begin{aligned} &V'''V - V''V' - \delta c V'V - (\delta + c)(V')^2 + \beta_T V''V^2 \\ &+\beta_T \delta c V^3 + \beta_T(\delta + c)V'V^2 + \delta c V'V + (\delta + c)V''V = 0. \end{aligned} \tag{5.6}$$

The input–output equation is monic because the coefficient in front of the highest derivative is one. The input–output equation is a differential algebraic equation of the output, which in this case is the viremia level, V, and its derivatives, with coefficients being the parameters of the model. In other words, the viremia level V obtained by solving the within-host model (5.1) is the same as the solution of the input–output equation (5.6). The dynamical model is said to be globally structurally identifiable if the map from the parameter space to the coefficients of the input–output equations is injective.[6] Within the differential algebra approach, the definition of the structural identifiability becomes:

Definition 5.2. The dynamical model is globally identifiable if and only if

$$c(\boldsymbol{p}) = c(\boldsymbol{q}) \implies \boldsymbol{p} = \boldsymbol{q},$$

where $c(\boldsymbol{p})$ are the coefficients of the normalized input–output equation.

Suppose, to the contrary, that another set of the parameters $\boldsymbol{q} = [\hat{\beta}_T, \hat{\delta}, \hat{\pi}, \hat{c}]$ produced the same observed state $g(\boldsymbol{x}(\tau), \boldsymbol{p})$, then equaling the coefficients of the differential algebraic equation (5.6), we obtain

$$\beta_T = \hat{\beta}_T, \quad \delta c = \hat{\delta}\hat{c}, \quad \delta + c = \hat{\delta} + \hat{c}. \tag{5.7}$$

Solving the nonlinear equation system (5.7) results in two sets of solutions,

$$\beta_T = \hat{\beta}_T,\ \delta = \hat{\delta},\ c = \hat{c} \quad \text{or} \quad \beta_T = \hat{\beta}_T,\ \delta = \hat{c},\ c = \hat{\delta}\,.$$

Hence, the model parameters δ and c are only locally identifiable from viral load observations. On the other hand, the parameter π does not appear in the input–output equation (5.6). The reason for that can be easily seen by scaling the unobserved state variables by any $\sigma > 0$. Hence, both (T, T_i, V) and $(\sigma T, \sigma T_i, V)$ solve the system (5.1), but now π is replaced with π/σ, and there is no information on how to determine σ. Hence, π is not identifiable. We summarized the results in the following proposition.

Proposition 5.1. *The within host model* (5.1) *is not structurally identifiable from the viremia observations. Only the parameter β_T is identifiable, parameters δ and c are only locally identifiable and the parameter π is not identifiable, from the viremia observations.*

Since (5.1) is not structurally identifiable, we scale the unobserved state variables, target cells and infected target cell with the unidentifiable parameter, π, to obtain a structurally locally identifiable model,

$$\textbf{Scaled within-host model (SM}_1\textbf{):}\quad \begin{cases} \hat{T}' &= -\beta_T V \hat{T}, \\ \hat{T}_i' &= \beta_T V \hat{T} - \delta \hat{T}_i, \\ V' &= \hat{T}_i - cV, \end{cases} \tag{5.8}$$

where $\hat{T} = \pi T$ and $\hat{T}_i = \pi T_i$.

Proposition 5.2. *The scaled within-host model* (5.8) *is locally identifiable from the viremia observations.*

5.3.2 *Parameter estimation problem*

Let the within-host model (5.1) be given in the following compact form:

$$\boldsymbol{x}' = f(\boldsymbol{x}(\tau), \boldsymbol{p}) \quad \boldsymbol{x}(0) = \boldsymbol{x}_0, \tag{5.9}$$

where $\boldsymbol{p}$ denotes the parameters of the within-host model, $\boldsymbol{x}(\tau)$ denotes the state variables, and $\boldsymbol{x}_0$ is the initial values. The parameter estimation problem is to determine the parameters of the mathematical model from given datasets. In an ideal setting, when the measurements are not contaminated with errors and the model is the correct model for the process, datasets are the discrete points of the output function $g(\boldsymbol{x}(\tau_i), \boldsymbol{p})$, which can be a function of the state variables $\boldsymbol{x}(\tau_i)$. Here, $\{\tau_i\}_{i=1}^n$ represents the discrete time points at which some output of the within-host model is measured to obtain the data. We then define the statistical model as

$$y_i = g(\boldsymbol{x}(\tau_i), \hat{\boldsymbol{p}}) + E_i, \tag{5.10}$$

where $\hat{\boldsymbol{p}}$ denotes the true parameters that generate the observations $\{y_i\}_{i=1}^n$ and E_i are the random variables that represent the observation or measurement error which cause the observations not to fall exactly on the points $g(\boldsymbol{x}(\tau_i), \hat{\boldsymbol{p}})$ of the smooth path $g(\boldsymbol{x}(\tau), \hat{\boldsymbol{p}})$.[1]

In a general setting, the measurement errors are assumed to have the following form:

$$E_i = g(\boldsymbol{x}(\tau_i), \hat{\boldsymbol{p}})^{\xi} \epsilon_i, \tag{5.11}$$

where $\xi \geq 0$ and ϵ_i are independent and normally distributed with mean zero and constant variance σ_0^2. That is $\epsilon_i \sim$ Normal $(0; \sigma_0)$, and as a consequence, the random variables y_i have mean $\mathbb{E}(y_i) = g(\boldsymbol{x}(\tau_i), \hat{\boldsymbol{p}})$ and variances $\text{Var}(y_i) = g(\boldsymbol{x}(\tau_i), \hat{\boldsymbol{p}})^{2\xi}\sigma_0^2$. Varying ξ allows for varying error scales in the measurements. We use the *relative error* model, that is $\xi = 1$ in (5.11), and use *generalized least squares* in the parameter estimation problem. Parameter estimation problem in the least squares sense solves the following optimization problem;

$$\hat{\boldsymbol{p}} = \min_{\boldsymbol{p}} \sum_{i=1}^{n} w_i \left(y_i - g(\boldsymbol{x}(\tau_i), \boldsymbol{p})\right)^2 . \tag{5.12}$$

Here, w_i denotes the weights and we set the weights as $w_i = \frac{1}{g(\boldsymbol{x}(\tau_i), \hat{\boldsymbol{p}})^{2\xi}}$.

5.3.3 *Within-host data*

To estimate the parameters of the within-host model, we use two viral load data, both are published.[4,24] First dataset is published at Ref. 4. Authors in Ref. 4 reviewed a total of 56 published studies where the total of 1280 participants were challenged with human influenza virus. For the within-host data, we used the viremia levels of type A H1N1 virus infected individuals. The authors presented the average viremia levels (log scale) of 116 participants who are shedding the H1N1 influenza virus in several studies (see Figure 5.2 in Ref. 4). We used the published averaged viral load data[4] to estimate the parameters of the scaled within-host model (5.8) by (5.12). The output measured in the laboratory settings for the infected individual's within-host pathogen dynamics is the viremia level. The viremia levels in the assays taken from infected individuals at discrete time points are recorded. Thus, $g(\boldsymbol{x}(\tau_i), \boldsymbol{p}) = \log_{10} V(\tau_i; \boldsymbol{p})$. We fitted the scaled within host model (5.8) using weighted least squares. The result of fitting the scaled within host model (5.8) to the viral load in Ref. 4 is presented in Figure 5.1 (left). For the second dataset, we used the H1N1 viral load data published at Ref. 24

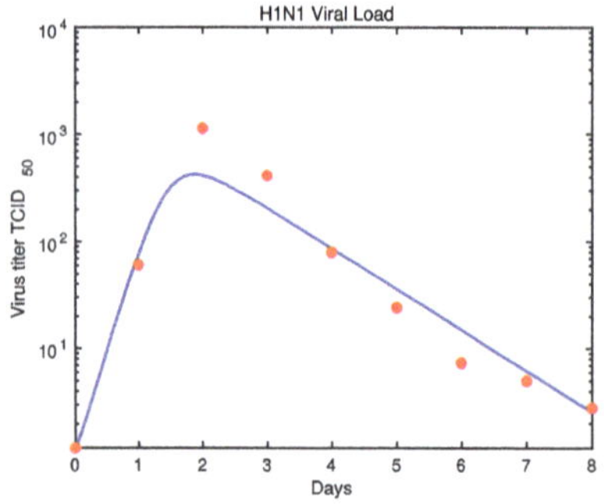

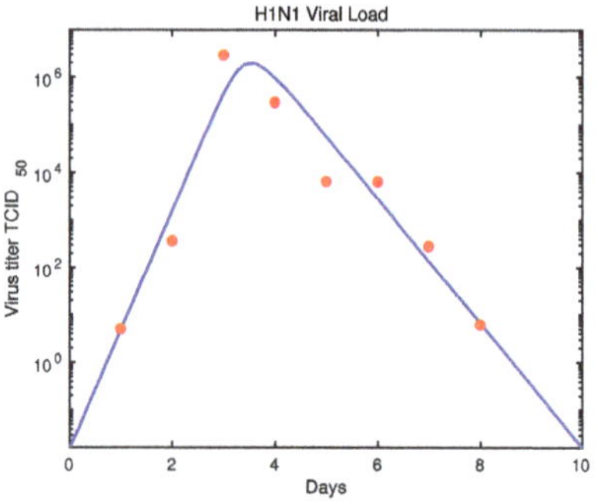

Fig. 5.1. Fitting of structurally locally identifiable model (5.8) to the $\log_{10}$ H1N1 viremia load data. (left figure) The estimated parameters are $\hat{\boldsymbol{p}} = [\beta_T, \delta, c] = [0.0076, 41.0697, 0.8910]$. Initial values are $\hat{T}_0 = 10^{4.5}$, $\hat{T}_{i0} = 0$, $V_0 = 1.1901$. Blue curve is the model prediction and the red dots are the average H1N1 influenza viremia levels given in Ref. 4. (right figure) The estimated parameters are $\hat{\boldsymbol{p}} = [\beta_T, \delta, c, V_0] = [1.4 \times 10^{-6}, 51.56, 3.9115, 0.017]$. Initial values are set to $\hat{T}_0 = 4 \times 10^8$, $\hat{T}_{i0} = 0$. Blue curve is the model prediction and the red dots are the average H1N1 influenza viremia levels given in Ref. 24.

(see Figure 5.3 in Ref. 24), where the data collection was performed by GlaxoSmithKline. We also fitted the scaled within host model (5.8) to the H1N1 viral load in Ref. 24 (given in log scale), using weighted least squares. That is in (5.12), we set $w_i = \frac{1}{g(\boldsymbol{x}(\tau_i), \hat{\boldsymbol{p}})^{2\xi}}$ where $g(\boldsymbol{x}(\tau_i), \boldsymbol{p}) = \log_{10} V(\tau_i; \boldsymbol{p})$. For this second data set initial viremia level is not given (see Figure 5.1, right), so we also fitted the initial viremia level V_0, our result was in agreement with the result in Ref. 24. The results of fitting are presented in Figure 5.1 (right).

5.3.4 *Practical identifiability of the within-host models*

To further analyze the identifiability of the within-host model, we perform Monte Carlo simulations which have been widely used for practical identifiability of ODE models.[14] Let's call the estimated parameters of the scaled model (5.8) from the within-host data as true parameters. We generate 1,000 synthetic datasets using the true parameter set $\hat{\boldsymbol{p}}$ and adding noise at increasing levels. We outline the Monte Carlo simulations in the following steps:

(1) Solve the within-host model, SM_1, numerically with the true parameters $\hat{\boldsymbol{p}}$ and obtain the output vector $\boldsymbol{g}(\boldsymbol{x}(\tau), \hat{\boldsymbol{p}})$ at the discrete data time points $\{\tau_i\}_{i=1}^n$.

(2) Generate $M = 1,000$ datasets from the statistical model (5.10) with a given measurement error. Datasets are drawn from a normal distribution whose mean is the output vector obtained in step (1) and standard deviation is the $\sigma_0\%$ of the mean. That is, we set $\xi = 1$ in the error structure given in (5.10)

$$y_i = g(\boldsymbol{x}(\tau_i), \hat{\boldsymbol{p}}) + g(\boldsymbol{x}(\tau_i), \hat{\boldsymbol{p}})\epsilon_i \quad i = 1, 2, \ldots, n.$$

where ϵ_i are drawn from a normal distribution with mean 0 and $\text{Var}(\epsilon_i) = \sigma_0^2$, that is $\epsilon_i \sim \text{Normal}\ (0, \sigma_o)$. Hence, as a consequence, the random variables y_i have mean $\mathbb{E}(y_i) = g(\boldsymbol{x}(\tau_i), \hat{\boldsymbol{p}})$ and variances $\text{Var}(y_i) = g(\boldsymbol{x}(\tau_i), \hat{\boldsymbol{p}})^2\sigma_0^2$.

(3) Fit the immunological model SM_1 to each of the M simulated data-sets to estimate the parameter set $\boldsymbol{p}_j$ for $j = 1, 2, \ldots, M$. That is

$$\boldsymbol{p}_j = \min_{\boldsymbol{p}} \sum_{i=1}^{n} w_i \left(y_i - g(\boldsymbol{x}(\tau_i), \boldsymbol{p})\right)^2, \quad j = 1, 2, \ldots, M,$$

where $w_i = \frac{1}{g(\boldsymbol{x}(\tau_i), \hat{\boldsymbol{p}})^{2\xi}}$.

(4) Calculate the average relative estimation error for each parameter in the set $\boldsymbol{p}$ by[14]

$$\text{ARE}(p^{(k)}) = 100\% \frac{1}{M} \sum_{j=1}^{M} \frac{|\hat{p}^{(k)} - p_j^{(k)}|}{\hat{p}^{(k)}},$$

where $p^{(k)}$ is the kth parameter in the set $\boldsymbol{p}$, $\hat{p}^{(k)}$ is the kth parameter in the true parameter set $\hat{\boldsymbol{p}}$, and $p_j^{(k)}$ is the kth parameter in the set $\boldsymbol{p}_j$.

(5) Repeat steps 1 through 4 with increasing level of noise, that is take $\sigma_0 = 0, 1, 5, 10,\ 20\%$.

Table 5.4. Monte Carlo simulations: Average relative estimation error (ARE) for parameters of the scaled within-host model SM_1 when fitted to data in Ref. 4.

Noise (σ_0) (%)	Infection rate β_T	Infected cells clearance rate δ	Clearance rate c
0	0	0	0
1	0.5	0.5	0.2
5	2.3	2.7	1.0
10	4.6	5.4	2.0
20	9.4	11.0	4.0

Note: The true parameter is set to $\hat{\boldsymbol{p}} = [\beta_T, \delta, c] = [0.0076, 41.0697, 0.8910]$. which is obtained by validating SM_1 from data in Ref. 4. We use weighted least squares. The initial conditions are the same as the left figure in Figure 5.1. $\tau^* = 15$ days.

Table 5.5. Monte Carlo simulations: Average relative estimation error (ARE) for parameters of the scaled within-host model SM_1 when fitted to data in Ref. 24.

Noise (σ_0) (%)	Infection rate β_T	Infected cells clearance rate δ	Clearance rate c
0	0	0	0
1	0.5	0.6	0.1
5	2.3	2.8	0.6
10	4.6	5.6	1.2
20	9.4	11.3	2.4

Note: The true parameter is set to $\hat{\boldsymbol{p}} = [\beta_T, \delta, c] = [1.4 \times 10^{-6}, 51.56, 3.9115]$, which are obtained by fitting to data in Ref. 24. Here we use weighted least squares. The initial conditions are set to $\hat{T}_0 = 4 \times 10^8$, $\hat{T}_{i0} = 0$, $V_0 = 0.017$. $\tau^* = 15$ days.

We perform Monte Carlo simulations, and then report the relative estimation errors (ARE) for each parameter in the scaled within-host model SM_1 (5.8) in Tables 5.4 and 5.5. The ARE gives the practical identifiability of the parameters. When $\sigma_0 = 0$, that is when there is no noise in the data, the ARE of the parameters of the structurally identifiable model should be 0 or very close to 0. The ARE of the model parameters increase as we increase noise in the data. If a parameter is not practically identifiable, then the ARE of that parameter will be significantly high even for a reasonable level of measurement error. To be specific, we say that if the ARE of the parameter is higher than the measurement error σ_0, then we say that the parameter is practically unidentifiable. However, it is worth mentioning that this is a very strict rule for practical identifiability. As seen in Tables 5.4 and 5.5, the AREs for each parameter remain below the measurement noise level. Hence, we conclude that the structurally locally identifiable model SM_1 is also practically identifiable.

5.4 Identifiability Analysis of the Between-Host Model

Investigating the structural identifiability of PDE models is a very undeveloped area, but it is necessary in order to put model fitting and parameter estimation of multiscale models on solid ground. To the best of our knowledge, only two references address identifiability of age-structured models,[15,16] which analyze simpler single-scale models. In this chapter, we are developing structural and practical identifiability techniques for multiscale immuno-epidemiological models. These nested models allow us to link immunological data with epidemiological data and as a consequence using these multiple datasets improves the identifiability of the parameters.[20]

5.4.1 *Parameter estimation of immuno-epidemiological model*

Parameter estimation of age-structured models have been studied.[15,16] However, estimating parameters of multiscale nested models with multi-scale data has been addressed for the first time in Ref. 20. In this study, for the first time we will study the structural identifiability of the multiscale model (5.2) where the epidemiologically relevant parameters such as the transmission and the recovery rate are linked to the within-host model state variables. Hence, we would like to determine epidemiological parameters such as transmission and recovery rates from the multiscale data. At the individual scale the data are the infected host's viremia level. At the population scale, we consider two types of data, incidences or cumulative incidences (Figure 5.2). For instance, if the population scale data are incidences,

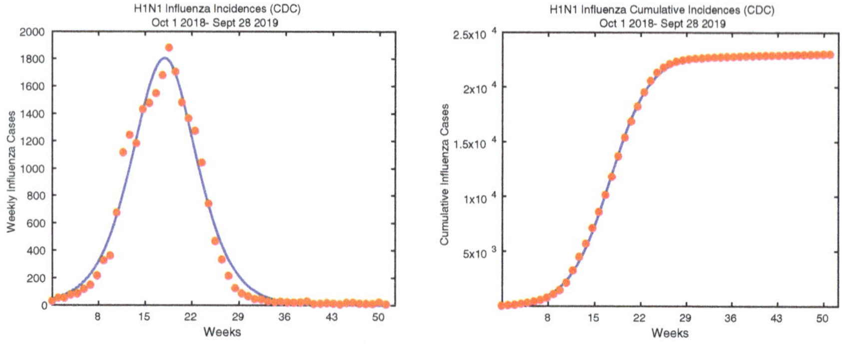

Fig. 5.2. Multiscale fitting results: Left figure: Weekly influenza cases as reported by CDC are linked to the viremia levels in patients (reported by Glaxo Data). Blue curve is the multiscale model (5.2) predictions and red dots are the weekly incidences data. Right figure: Weekly cumulative influenza incidences are linked to the viremia levels in patients. Blue curve is the model (5.2) predictions and red dots are the weekly cumulative incidences.

then incidences for the ideal noise-free situation lie on the smooth curve given by

$$G(t;\boldsymbol{p}_e) = \frac{S(t,\boldsymbol{p}_e)}{N}\int_0^{\mathcal{T}*}\beta(\tau;\boldsymbol{p}_e)i(\tau,t,\boldsymbol{p}_e)\mathrm{d}\tau \quad \text{for the time interval } t\in[0,T].$$

Similarly, when the population scale data are cumulative incidences, then in the ideal noise-free situation cumulative incidences lie on the smooth curve given by

$$C(t;\boldsymbol{p}_e) = \int_0^t \frac{S(s,\boldsymbol{p}_e)}{N}\int_0^{\mathcal{T}*}\beta(\tau;\boldsymbol{p}_e)i(\tau,s,\boldsymbol{p}_e)\mathrm{d}\tau\mathrm{d}s.$$

Here $\boldsymbol{p}_e$ denotes the vector of parameters to be estimated such as c_1, κ, ϵ, or C_0.

We will continue with the case when incidences are considered for the population scale data. Thus, $\{G(t_j;\boldsymbol{p}_e)\}_{j=1}^m$ are the discrete time points at which number of new cases are given. As in the within-host model, we assume that the observations, Y_j, are contaminated with measurement errors, E_j. That is

$$Y_j = G(t_j; \boldsymbol{p}_e) + E_j, \tag{5.13}$$

where E_j are the random variables with the following relative error form:

$$E_j = G(t_j; \boldsymbol{p}_e)\epsilon_j$$

with ϵ_j being independent and normally distributed with mean zero and constant variance.

Fitting multiscale data to a multiscale model comes with its challenges. Data at the within-host scale and between-host scale have different magnitudes. Furthermore, for the within-host data there are only $n = 8$ data points, while for the epidemiological data there are $m = 52$ data points. In such a situation, any iterative numerical algorithm used to estimate the multi-scale model parameters by minimizing a cost function, as in the case of least squares, tends to minimize the data with the highest magnitude at the price of deviating from the small magnitude data. To overcome this issue, we normalize with the average data value. So, to estimate the parameters of the multi-scale model (5.2), we minimize the following cost function where it normalizes the multi-scale data with different scales. Let $\hat{Y}$ denote the average of the epidemiological data, $\hat{y}$ the average of the immune data, that is

$$\hat{Y} = \frac{1}{m}\sum_{j=1}^{m} Y_j \qquad \hat{y} = \frac{1}{n}\sum_{i=1}^{n} y_i.$$

We then estimate the parameters of the multi-scale model by minimizing the following cost functional:

$$\hat{\boldsymbol{p}}_{\mathrm{s}} = \min_{\boldsymbol{p}_{\mathrm{s}}} \left(\frac{1}{m}\sum_{j=1}^{m} \frac{\left(Y_j - G(t_j, \boldsymbol{p}_e)\right)^2}{\hat{Y}^2} + \frac{1}{n}\sum_{i=1}^{n} \frac{\left(y_i - g(\boldsymbol{x}(\tau_i), \boldsymbol{p})\right)^2}{\hat{y}^2} \right). \tag{5.14}$$

We denote the parameters of the immunological model by $\boldsymbol{p}$, the epidemiological model by $\boldsymbol{p}_e$, and $\boldsymbol{p}_s = [\boldsymbol{p}_e, \boldsymbol{p}]$.

5.4.2 *Structural identifiability of the multiscale model*

Structural identifiability of the PDE epidemic models, especially nested immuno-epidemiological models, has not been studied before. Here, we give an outline on how to study the structural identifiability of such models. We link the between-host model (5.2) to the structurally identifiable within-host model (5.8) and study the structural identifiability of the nested immuno-epidemiological model (5.2). The methodology we introduce here is similar to the identifiability analysis of ODE models, in the sense that we will first derive the input–output equation and then define what structural identifiability means for the nested immuno-epidemiological models.

We will study the structural identifiability of the between-host model (5.2) when the observations are the incidences. The observations in the parameter estimation problem are the nonlocal boundary conditions for the between-host model (5.2). Namely,

$$G(t) = \frac{S(t)}{N}\int_0^{T*} \beta(\tau) i(\tau, t)\mathrm{d}\tau. \tag{5.15}$$

Integrating the density of infected population along the characteristics, we obtain

$$i(\tau,t) = \begin{cases} i_0(\tau - t)\dfrac{\pi(\tau)}{\pi(\tau - t)}, & \tau \geq t. \\ G(t-\tau)\pi(\tau), & \tau < t, \end{cases} \tag{5.16}$$

Substituting (5.16) into (5.15), we obtain the following nonlinear Volterra integral of the second kind,

$$G(t) = \frac{S(t)}{N}\int_0^t G(t-\tau)\beta(\tau)\pi(\tau)\mathrm{d}\tau + \frac{S(t)}{N}\int_0^\infty \beta(\tau+t)\frac{\pi(\tau+t)}{\pi(\tau)} i_0(\tau)\mathrm{d}\tau. \tag{5.17}$$

The nonlinear Volterra integral equation (5.17) has a unique solution.[23] Substituting

$$S(t) = S_0 - \int_0^t G(s)\mathrm{d}s$$

into (5.17), we obtain

$$G(t) = \frac{1}{N}\left(S_0 - \int_0^t G(s)\mathrm{d}s\right)\int_0^t G(t-\tau)K(\tau)\mathrm{d}\tau$$
$$+\frac{1}{N}\left(S_0 - \int_0^t G(s)\mathrm{d}s\right)F(t), \tag{5.18}$$

where

$$K(\tau) = \beta(\tau)\pi(\tau) \quad \text{and} \quad F(t) = \int_0^\infty \beta(\tau+t)\frac{\pi(\tau+t)}{\pi(\tau)}i_0(\tau)\mathrm{d}\tau.$$

Equation (5.18) is the input–output equation for the nested immuno-epidemiological model. The inputs are the parameters of the nested model, such as $S_0, \beta(\tau), \gamma(\tau)$ and $i_0(\tau)$, and the output is the new incidences, $G(t)$. Solving the immuno-epidemiological model and obtaining the new incidences is equivalent to solving the input–output equation (5.18) for $G(t)$. We are interested in studying the structural identifiability for epidemiologically important parameters such as transmission and recovery rates and assume that the initial conditions S_0 and $i_0(\tau)$ are known. Thus, we define the structural identifiability for the nested immuno-epidemiological model (5.2) as follows:

Definition 5.3. A parameter set $\boldsymbol{p}_e$ is called *structurally globally (or uniquely) identifiable* if for every $\boldsymbol{q}_e$ in the parameter space the equation

$$G(t;\boldsymbol{p}_e) = G(t;\boldsymbol{q}_e) \implies \boldsymbol{p}_e = \boldsymbol{q}_e.$$

For simplicity, let's assume that the transmission rate $\beta(\tau)$ is directly proportional to the infected individual's pathogen load with the proportionality constant c_1. Similarly, the recovery rate $\gamma(\tau)$ is inversely proportional to the pathogen load with the proportionality constant κ. We add ϵ which is set as a small number to avoid dividing by zero when the pathogen load reaches zero in the infected host. That is for the structural identifiability analysis of the immuno-epidemiological model, we will continue with the following linking relations:

$$\beta(\tau) = c_1 V(\tau) \quad \text{and} \quad \gamma(\tau) = \frac{\kappa}{\epsilon + V(\tau)}.$$

We analyze whether the input–output equation for the between host model (5.18) is one-to-one with respect to the epidemiological parameters $\boldsymbol{p}_e = [c_1, \kappa, \epsilon]$. Recall that the pathogen load in infected host is determined by a structurally identifiable model, that is $V(\tau, \boldsymbol{p}) = V(\tau, \boldsymbol{q})$ implies $\boldsymbol{p} = \boldsymbol{q}$. We first show that the probability of survival rate is one-to-one with respect to epidemiological parameters.

Proposition 5.3. *The probability of survival rate $\pi(\tau; \boldsymbol{p}_e)$ is one-to-one with respect to epidemiological parameters $\boldsymbol{p}_e$. That is*

$$\pi(\tau; \boldsymbol{p}_e) = \pi(\tau; \boldsymbol{q}_e) \implies \boldsymbol{p}_e = \boldsymbol{q}_e.$$

for any $\boldsymbol{p}_e, \boldsymbol{q}_e$ in the parameter space.

Proof. Suppose $\pi(\tau; \boldsymbol{p}_e) = \pi(\tau; \boldsymbol{q}_e)$, that is

$$e^{-\int_0^\tau \frac{\kappa}{\epsilon + V(s)} ds} = e^{-\int_0^\tau \frac{\hat{\kappa}}{\hat{\epsilon} + V(s)} ds}.$$

which implies

$$\int_0^\tau \frac{\kappa}{\epsilon + V(s)} ds = \int_0^\tau \frac{\hat{\kappa}}{\hat{\epsilon} + V(s)} ds.$$

Differentiating both sides, we obtain $\frac{\kappa}{\epsilon + V(s)} = \frac{\hat{\kappa}}{\hat{\epsilon} + V(s)}$, which implies $\kappa = \hat{\kappa}$ and $\epsilon = \hat{\epsilon}$. □

Proposition 5.4. *The kernel $K(\tau; \boldsymbol{p}_e) = \beta(\tau; \boldsymbol{p}_e)\pi(\tau; \boldsymbol{p}_e)$ of the nonlinear Volterra integral equation (5.18) is one-to-one with respect to epidemiological parameters $\boldsymbol{p}_e$. That is*

$$K(\tau; \boldsymbol{p}_e) = K(\tau; \boldsymbol{q}_e) \implies \boldsymbol{p}_e = \boldsymbol{q}_e,$$

for any $\boldsymbol{p}_e, \boldsymbol{q}_e$ in the parameter space.

Proof. We first show that the transmission rate $\beta(\tau) = c_1 V(\tau)$ and recovery rate $\gamma(\tau) = \dfrac{\kappa}{\epsilon + V(\tau)}$ are both one-to-one with respect to parameters c_1, κ, ϵ. Set $\boldsymbol{p}_e = [c_1, \kappa, \epsilon]$ and $\boldsymbol{q}_e = [\hat{c}_1, \hat{\kappa}, \hat{\epsilon}]$. Note that $\beta(\tau, \boldsymbol{p}_e) = \beta(\tau, \boldsymbol{q}_e)$, that is $c_1 V(\tau) = \hat{c}_1 V(\tau)$ implies $c_1 = \hat{c}_1$. Similarly, setting $\gamma(\tau, \boldsymbol{p}_e) = \gamma(\tau, \boldsymbol{q}_e)$, gives $\dfrac{\kappa}{\epsilon + V(\tau)} = \dfrac{\hat{\kappa}}{\hat{\epsilon} + V(\tau)}$. We then obtain

$$\kappa\hat{\epsilon} + \kappa V(\tau) = \hat{\kappa}\epsilon + \hat{\kappa} V(\tau),$$

which implies $\kappa = \hat{\kappa}$ and $\epsilon = \hat{\epsilon}$. Next, we set $K(\tau; \boldsymbol{p}_e) = K(\tau; \boldsymbol{q}_e)$, that is

$$c_1 V(\tau) e^{-\int_0^\tau \frac{\kappa}{\epsilon + V(s)} \mathrm{d}s} = \hat{c}_1 V(\tau) e^{-\int_0^\tau \frac{\hat{\kappa}}{\hat{\epsilon} + V(s)} \mathrm{d}s}.$$

For $\tau = 0$, we obtain $c_1 = \hat{c}_1$. Then, we have $\pi(\tau; \boldsymbol{p}_e) = \pi(\tau; \boldsymbol{q}_e)$ which implies $\kappa = \hat{\kappa}$ and $\epsilon = \hat{\epsilon}$, by Proposition 5.3. □

Next, we state the structural identifiability condition for the immuno-epidemiological model.

Theorem 5.1. *If $F(t; \boldsymbol{p}_e) = F(t; \boldsymbol{q}_e)$ holds, then the immuno-epidemiological model (5.2) is structurally identifiable from the incidence observations.*

Proof. To prove that the immuno-epidemiological model (5.2) is structurally identifiable, we need to show that

$$G(t; \boldsymbol{p}_e) = G(t; \boldsymbol{q}_e) \implies \boldsymbol{p}_e = \boldsymbol{q}_e.$$

We rewrite the input–output equation (5.18) in the following equivalent form:

$$\int_0^t G(t - \tau) K(\tau) \mathrm{d}\tau + F(t) - \frac{N G(t)}{S_0 - \displaystyle\int_0^t G(s) \mathrm{d}s} = 0.$$

Clearly, when $G(t; \boldsymbol{p}_e) = G(t; \boldsymbol{q}_e)$, we have

$$\frac{N G(t; \boldsymbol{p}_e)}{S_0 - \displaystyle\int_0^t G(s; \boldsymbol{p}_e) \mathrm{d}s} = \frac{N G(t; \boldsymbol{q}_e)}{S_0 - \displaystyle\int_0^t G(s; \boldsymbol{q}_e) \mathrm{d}s}.$$

Thus, we obtain,

$$\int_0^t G(t-\tau;\boldsymbol{p}_e)K(\tau;\boldsymbol{p}_e)d\tau + F(t;\boldsymbol{p}_e)$$
$$= \int_0^t G(t-\tau;\boldsymbol{q}_e)K(\tau;\boldsymbol{q}_e)d\tau + F(t;\boldsymbol{q}_e).$$

Setting $t = 0$, we obtain $F(0;\boldsymbol{p}_e) = F(0;\boldsymbol{q}_e)$, which implies

$$F(0;\boldsymbol{p}_e) = \int_0^\infty \beta(\tau;\boldsymbol{p}_e)i_0(\tau)\mathrm{d}\tau = \int_0^\infty \beta(\tau;\boldsymbol{q}_e)i_0(\tau)\mathrm{d}\tau = F(0;\boldsymbol{q}_e).$$

Since $\beta(\tau,\boldsymbol{p}_e) = c_1V(\tau)$ and $\beta(\tau,\boldsymbol{q}_e) = \hat{c}_1V(\tau)$, we have

$$\int_0^\infty c_1V(\tau)i_0(\tau)\mathrm{d}\tau = \int_0^\infty \hat{c}_1V(\tau)i_0(\tau)\mathrm{d}\tau,$$

which implies that $c_1 = \hat{c}_1$. Thus, $\beta(\tau,\boldsymbol{p}_e) = \beta(\tau,\boldsymbol{q}_e)$. Since $G(t;\boldsymbol{p}_e) = G(t;\boldsymbol{q}_e)$, we obtain

$$\int_0^t G(t-\tau;\boldsymbol{p}_e)\left(K(\tau;\boldsymbol{p}_e) - K(\tau;\boldsymbol{q}_e)\right)\mathrm{d}\tau + \left(F(t;\boldsymbol{p}_e) - F(t;\boldsymbol{q}_e)\right) = 0. \tag{5.19}$$

Assuming, $F(t;\boldsymbol{p}_e) = F(t;\boldsymbol{q}_e)$ and taking the Laplace transform, we would get

$$\mathcal{L}\{G(t-\tau;\boldsymbol{p}_e)\}\mathcal{L}\{\beta(\tau,\boldsymbol{p}_e)\pi(\tau,\boldsymbol{p}_e) - \beta(\tau,\boldsymbol{q}_e)\pi(\tau,\boldsymbol{q}_e)\} = 0.$$

Clearly, Laplace transform of new incidences is not zero, thus we set $\mathcal{L}\{\beta(\tau,\boldsymbol{p}_e)\pi(\tau,\boldsymbol{p}_e) - \beta(\tau,\boldsymbol{q}_e)\pi(\tau,\boldsymbol{q}_e)\} = 0$, which implies that

$$\beta(\tau,\boldsymbol{p}_e)\pi(\tau,\boldsymbol{p}_e) - \beta(\tau,\boldsymbol{q}_e)\pi(\tau,\boldsymbol{q}_e) = 0.$$

We already showed that $\beta(\tau,\boldsymbol{p}_e) = \beta(\tau,\boldsymbol{q}_e)$, thus we have

$$\pi(\tau,\boldsymbol{p}_e) = \pi(\tau,\boldsymbol{q}_e).$$

By Proposition 5.3, we have $\boldsymbol{p}_e = \boldsymbol{q}_e$. □

Theorem 5.1 states that the between-host model is structurally identifiable only if the requirement $F(t;\boldsymbol{p}_e) = F(t;\boldsymbol{q}_e)$ is met. In other words, structural identifiability will follow if $F(t;\boldsymbol{p}_e) = F(t;\boldsymbol{q}_e)$ can be derived from (5.19).

5.4.3 *Practical identifiability analysis of multiscale model with multiscale data*

Our prior investigations of a nested vector–host model of Rift Valley Fever suggested that the practical identifiability of the nested immuno-epidemiological models improves when within-host and between-host models are fitted simultaneously to immunological and epidemiological data sets.[20] We perform the Monte Carlo simulations where we adjust it for fitting within-host and between-host data simultaneously. As before, let's call the estimated values of the scaled within-host model (5.8) true parameter and denote it by $\hat{\boldsymbol{p}}$, and similarly the estimated values of the between host model as $\hat{\boldsymbol{p}}_e$. With the following steps, we outline the Monte Carlo simulations when the population scale data are incidences. Same steps are taken when the cumulative incidences are used at the population scale for the Monte Carlo simulations.

(1) Solve the within-host model, SM_1, numerically with the true parameters $\hat{\boldsymbol{p}}$ and obtain the immune data by computing the output vector $g(\boldsymbol{x}(\tau), \hat{\boldsymbol{p}})$ at the discrete data time points $\{\tau_i\}_{i=1}^n$. Solve the between-host model (5.2) with true parameters $\hat{\boldsymbol{p}}_e$ and obtain the epidemiological data by computing $G(t, \hat{\boldsymbol{p}}_e)$ at the discrete data time points $\{t_j\}_{j=1}^m$.

(2) Generate $M = 1,000$ datasets from the statistical models (5.10) and (5.13) with a given measurement error. Datasets are drawn from a normal distribution whose mean is the output vector obtained in step (1) and standard deviation is the $\sigma_0\%$ of the mean.

$$y_i = g(\boldsymbol{x}(\tau_i), \hat{\boldsymbol{p}}) + g(\boldsymbol{x}(\tau_i), \hat{\boldsymbol{p}})\epsilon_i \quad i = 1, 2, \ldots, n$$

immune data

$$Y_j = G(t_j, \hat{\boldsymbol{p}}_e) + G(t_j, \hat{\boldsymbol{p}}_e)\epsilon_i \quad j = 1, 2, \ldots, m$$

epidemiological data

Here $\epsilon \sim \text{Normal}(0; \sigma_0)$, as a consequence the random variables $\{y_i\}_{i=1}^n$ and $\{Y_j\}_{j=1}^m$ have mean $\mathbb{E}(y_i) = g(\boldsymbol{x}(\tau_i), \hat{\boldsymbol{p}})$, $\mathbb{E}(Y_j) = G(t_j, \hat{\boldsymbol{p}}_e)$ and variances $\text{Var}(y_i) = g(\boldsymbol{x}(\tau_i), \hat{\boldsymbol{p}})^2\sigma_0^2$, $\text{Var}(Y_j) = G(t_j, \hat{\boldsymbol{p}}_e)^2\sigma_0^2$, respectively.

Note that in the case of the cumulative incidences,

$$Y_j = C(t_j, \hat{\boldsymbol{p}}_e) + C(t_j, \hat{\boldsymbol{p}}_e)\epsilon_i \quad j = 1, 2, \ldots, m$$

epidemiological data

(3) Fit the multiscale model (within-host linked to between-host model) to each of the M simulated datasets to estimate the parameter set $\boldsymbol{p}_l$ for $l = 1, 2, \ldots, M$. That is,

$$\boldsymbol{p}_l = \min_{[\boldsymbol{p}_e, \boldsymbol{p}]} \left(\frac{1}{m} \sum_{j=1}^{m} \frac{(Y_j - G(t_j, \boldsymbol{p}_e))^2}{\hat{Y}^2} + \frac{1}{n} \sum_{i=1}^{n} \frac{(y_i - g(\boldsymbol{x}(\tau_i), \boldsymbol{p}))^2}{\hat{y}^2} \right).$$

$l = 1, 2, \ldots, 1,000.$

(4) Calculate the average relative estimation error for each parameter in the set $\boldsymbol{p}$ and $\boldsymbol{p}_e$ by[14]

$$\text{ARE}(p^{(k)}) = 100\% \frac{1}{M} \sum_{l=1}^{1,000} \frac{|\hat{p}^{(k)} - p_l^{(k)}|}{\hat{p}^{(k)}},$$

where $p^{(k)}$ is the kth parameter in the set $\boldsymbol{p}$ (or $\boldsymbol{p}_e$), $\hat{p}^{(k)}$ is the kth parameter in the true parameter set $\hat{\boldsymbol{p}}$ (or $\hat{\boldsymbol{p}}_e$), and $p_l^{(k)}$ is the kth parameter in the set $\boldsymbol{p}_l$.

(5) Repeat steps (1) through (4) with increasing level of noise, that is take $\sigma_0 = 0, 1, 5, 10, 20\%$.

5.4.4 *Results of the simulations*

We perform simulations with the two numerical methods and with the two linking functions for $\beta(\tau)$. We fit in some simulations incidences and other cumulative incidences. We seek to understand the impact of the rate of convergence of the numerical method, the different linking functions, and the type of epidemic data used on the practical identifiability of the multiscale model.

Comparing Tables 5.6 and 5.7 where we fit the weekly incidence data with a first-order numerical method but vary the linking function $\beta(\tau)$, we see that with the simpler $\beta(\tau) = c_1 V(\tau)$ the AREs

Table 5.6. Monte Carlo simulations: Average relative estimation error (ARE) for parameters of the immuno-epidemiological model (5.2).

Noise (%)	Infection rate β_T	Infected cell clearance rate δ	Clearance rate c	Transmission rate linking constant c_1	Recovery rate linking constant κ
0	0	0	0	0	0
1	0.5	0.5	0.3	0.3	0.3
5	2.5	2.4	1.3	1.4	1.4
10	5.2	5.0	2.9	3.0	2.8
20	12.1	12.3	7.7	9.8	6.5

Note: The between-host model is linked to the within-host model SM_1 with true parameters $\beta_T = 2.2 \times 10^{-6}$, $\delta = 88.94$, and $c = 3.21$. The linking relations are set as $\beta(\tau) = c_1 V(\tau)$ and $\gamma(\tau) = \dfrac{\kappa}{\epsilon_0 + V(\tau)}$, where $c_1 = 2.2 \times 10^{-5}$ and $\kappa = 0.75$. ϵ_0 is fixed at 10^{-2}. First-order finite difference method is used to solve the multiscale model. At within-host level, GlaxoSmithKline data are used, and at between-host level, weekly influenza incidences are used.

Table 5.7. Monte Carlo simulations: Average relative estimation error (ARE) for parameters of the immuno-epidemiological model (5.2).

Noise (%)	Infection rate β_T	Infected cell clearance rate δ	Clearance rate c	Transmission rate linking constant c_1	Recovery rate linking constant κ
0	0	0	0	0	0
1	3.5	3.9	0.9	0.9	0.6
5	8.8	9.7	3.7	3.5	2.7
10	13.9	15.3	6.7	6.1	5.2
20	23.1	25.2	14.1	12.5	10.4

Note: The between-host model is linked to the within-host model SM_1 with true parameters $\beta_T = 2.9 \times 10^{-6}$, $\delta = 125.09$, and $c = 3.08$. The linking relation is used as $\beta(\tau) = \frac{c_1 V(\tau)}{C_0 + V(\tau)}$ and $\gamma(\tau) = \frac{\kappa}{\epsilon_0 + V(\tau)}$ where $c_1 = 10.97$, $\kappa = 0.82$, and $C_0 = 1 \times 10^4$. ϵ_0 is fixed at 10^{-2}. First-order finite difference method is used to solve the multiscale model. At within-host level GlaxoSmithKline data are used; at between-host level, weekly incidence is used.

are about twice smaller than with the function $\beta(\tau) = \frac{c_1 V(\tau)}{C_0 + V(\tau)}$. The reason for that may be that the simpler linking function has only one parameter and simpler dependence of the within-host viral load. While in Table 5.6 all AREs are below the measurement error, and

Table 5.8. Monte Carlo simulations: Average relative estimation error (ARE) for parameters of the immuno-epidemiological model (5.2).

Noise (%)	Infection rate β_T	Infected cell clearance rate δ	Clearance rate c	Transmission rate linking constant c_1	Recovery rate linking constant κ
0	0	0	0	0	0
1	1.4	1.6	0.8	2.4	1.3
5	8.2	9.1	4.1	14.2	6.6
10	15.4	17.2	8.2	29.2	13.6
20	34.7	40.8	18.4	87.5	26.3

Note: The between-host model is linked to the within-host model SM_1 with true parameters $\beta_T = 2.7\times 10^{-6}$, $\delta = 113.61$, and $c = 3.19$. The linking relations used are $\beta(\tau) = c_1 V(\tau)$ and $\gamma(\tau) = \frac{\kappa}{\epsilon_0+V(\tau)}$, where $c_1 = 3.6\times 10^{-6}$ and $\kappa = 0.21$. ϵ_0 is fixed at 10^{-2}. First-order finite difference method is used to solve the multiscale model. At within-host level, GlaxoSmithKline data are used; at between-host level, weekly cumulative incidence is used.

Table 5.9. Monte Carlo simulations: Average relative estimation error (ARE) for parameters of the immuno-epidemiological model (5.2).

Noise (%)	Infection rate β_T	Infected cell clearance rate δ	Clearance rate c	Transmission rate linking constant c_1	Recovery rate linking constant κ
0	0	0	0	0	0
1	1.8	2	0.8	1.1	0.8
5	9.5	10.5	3.9	5.8	4.2
10	20.7	23.2	8.1	13.6	9.6
20	49.2	54.9	18.6	31.6	19.9

Note: The between-host model is linked to the within-host model SM_1 with true parameters $\beta_T = 2.5\times 10^{-6}$, $\delta = 101.79$, and $c = 2.41$. The linking relation is used as $\beta(\tau) = \frac{c_1 V(\tau)}{C_0+V(\tau)}$ and $\gamma(\tau) = \frac{\kappa}{\epsilon_0+V(\tau)}$, where $c_1 = 1.9$ and $\kappa = 0.42$. ϵ_0 is fixed at 10^{-2}. First-order finite difference method is used to solve the multiscale model. At within-host level, GlaxoSmithKline data are used. And at between-host level, weekly cumulative incidences are used.

the model is identifiable, in Table 5.7 some AREs are above the measurement, although they are slightly above.

In Tables 5.6 and 5.8, we use the simpler $\beta(\tau)$ and the first-order numerical method to test practical identifiability when we fit weekly incidences as opposed to when we fit cumulative incidences.

Table 5.10. Monte Carlo simulations: Average relative estimation error (ARE) for parameters of the immuno-epidemiological model (5.2).

Noise (%)	Infection rate β_T	Infected cell clearance rate δ	Clearance rate c	Transmission rate linking constant c_1	Recovery rate linking constant κ
0	0	0	0	0	0
1	1.2	1.3	0.8	0.9	0.2
5	5.7	6.3	3.3	4.9	1.1
10	9.5	10.4	6.1	7.5	2.1
20	15.8	16.9	11.1	13.3	4.2

Note: The between-host model is linked to the within-host model SM_1 with true parameters $\beta_T = 2.5 \times 10^{-6}$, $\delta = 104.06$, and $c = 3.27$. The linking relation is used as $\beta(\tau) = c_1 V(\tau)$ and $\gamma(\tau) = \frac{\kappa}{\epsilon_0 + V(\tau)}$, where $c_1 = 2.7 \times 10^{-5}$ and $\kappa = 0.36$. ϵ_0 is fixed at 10^{-2}. Second-order finite difference method is used to solve the multiscale model. At within-host level, GlaxoSmithKline data are used. And at between-host level, weekly incidences are used.

In Table 5.6 the AREs are below measurement error, but in Table 5.8 they are not for most parameters. We conclude that in principle fitting incidences leads to better identifiable parameters compared to fitting cumulative incidences.

In Tables 5.8 and 5.9, we compare the impact of the linking function $\beta(\tau)$ when we fit cumulative incidences. When cumulative incidences are fitted, there is no clear linking function for which the AREs are smaller across all parameters. For instance, $\beta(\tau) = c_1 V(\tau)$ leads to smaller AREs in β_T and δ but higher in c_1 and k.

Finally, we wish to understand how the accuracy of the numerical method reflects on the identifiability of the parameters. Because $\beta(\tau) = c_1 V(\tau)$ is superior when incidences are fitted, and fitting incidences leads to smaller AREs, we fit $\beta(\tau) = c_1 V(\tau)$ to the weekly incidences, and we vary the numerical method used. The results with the first-order numerical method are given in Table 5.6, while the results with the second-order method are given in Table 5.10. Comparing the results in Tables 5.6 and 5.10, we see that the higher order numerical method does not give lower AREs. The AREs from both numerical methods are quite comparable. We conclude that for nested immuno-epidemiological models using a first-order semi-implicit method is sufficient, and there is no advantage in the identifiability results to using a second-order method.

5.5 Discussion

In this chapter, we consider a nested multiscale model of influenza. The within-host model is a target cell limited model typically used in the literature for modeling influenza, and in modified version — HIV and HCV. The between-host model is an SIR outbreak model. We use two different linking functions for the epidemiological transmission rate — a simple function in which $\beta(\tau)$ is proportional to the within-host viral load, and a Holling II functional response function which takes into account the fact that the probability of transmission is bounded by one.[11] Our goal is to study the identifiability of the model under different modeling and simulational choices. We summarize all simulations in Table 5.11.

First we study the identifiability of the within-host model with respect to two different datasets. The investigation of the structural identifiability of the within-host model reveals that the within-host model as originally developed is not structurally identifiable. Thus, we consider a re-scaled model that is structurally identifiable. We fit the scaled model to two distinct within-host data sets with unweighted and weighted least squares. We conclude that regardless of the data set or the fitting method, all parameters of the scaled within-host model are practically identifiable.

To study the identifiability of the multi-scale model, we chose to fit the within-host and between-host models simultaneously, as our previous research suggests that this approach leads to better identifiability of all parameters.[20] For the within-host data, we chose the data collected by GlaxoSmithKline (termed Glaxo data). We

Table 5.11. Summary of simulations for practical identifiability analysis of multiscale model with multiscale data.

Tables	Transmission rate linking form	Order of finite difference method	Population level data type
5.6	$c_1 V(\tau)$	1st	Incidences
5.7	$\dfrac{c_1 V(\tau)}{C_0 + V(\tau)}$	1st	Incidences
5.8	$c_1 V(\tau)$	1st	Cumulative incidences
5.9	$\dfrac{c_1 V(\tau)}{C_0 + V(\tau)}$	1st	Cumulative incidences
5.10	$c_1 V(\tau)$	2nd	Incidences

study the impact of the numerical method used to solve the PDE between-host model on the identifiability. We compare the identifiability results of a first-order method as opposed to a second-order method. We find no significant difference in the identifiability of the parameters between the two methods with the first-order method being slightly better. We surmise that this is due to the fact that higher order methods require higher order differentiability of the solutions to be convergent, which may not be present in this PDE model.

Next, we explore the impact of fitting the multiscale model to incidences of influenza as opposed to fitting to cumulative incidences. Here the results show that fitting to incidence gives smaller AREs and better identifiability of the parameters compared to fitting to cumulative incidences. Fitting to cumulative incidences gives on average 2–3 times larger AREs. Our conclusion is that fitting to incidences should be preferred and used if possible. This result is consistent with our prior investigation of ODE models where we found out that fitting to prevalence makes the model identifiable while fitting to cumulative incidences makes the model practically unidentifiable.[21]

We also explore the choice of a linking function on the identifiability of the model parameters. We vary the transmission rate $\beta(\tau)$ for which we explore two possible choices. We compare the performance of the two linking functions both when we fit incidences and when we fit cumulative incidences. When we fit incidences, the simpler transmission rate performs somewhat better, but when we fit cumulative incidences, the two functions perform comparably.

In summary, we find the multi-scale model identifiable from viremia data and incidence data when N is fixed. The identifiability of the parameters declines when viremia and cumulative incidence data are used. The choice of the numerical method does not have significant impact on the identifiability of the parameters.

References

1. Banks, H. T., Hu, S., and Thompson, W. C. (2014). *Modeling and Inverse Problems in the Presence of Uncertainty*, CRC Press, Boca Raton, FL, 2014.
2. Banerjee, S., Guedj, J., Ribeiro, R. M., Moses, M., and Perelson, A. S. (2016). Estimating biologically relevant parameters under uncertainty for experimental within-host murine West Nile virus infection, *J. R. Soc. Interface* **13**, 117, doi:10.1098/rsif.2016.0130.

3. Bellu, G., Saccomani, M. P., Audoly, S., and D'Angio, DAISY L. (2007). A new software tool to test global identifiability of biological and physiological systems, *Comput. Methods Programs Biomed.* **88**, 1, pp. 52–61.
4. Carrat, F., Vergu, E., Ferguson, N. M., Lemaitre, M., Cauchemez, S., Leach, S., and Valleron, A.-J. (2008). Time lines of infection and disease in human influenza: A review of volunteer challenge studies, *Am. J. Epidemiol.* **167**, 7, 1, pp. 775–785.
5. Chris, O. T., Banga, J. R., and Balsa-Canto, E. Structural identifiability of systems biology models: A critical comparison of methods, *PLoS One* **6**, 11, e27755.
6. Eisenberg, M., Robertson, S., and Tien, J. (2013). Identifiability and estimation of multiple transmission pathways in cholera and waterborne disease, *JTB* **324**, pp. 84–102.
7. Garira, W. (2017). A complete categorization of multiscale models of infectious disease systems, *J. Biol. Dynamics* **11**, 1, pp. 378–435.
8. Gilchrist, M. A. and Sasaki, A. (2002). Modeling host-parasite coevolution: a nested approach based on mechanistic models, *J. Theor. Biol.* **218**, 3, pp. 289–308.
9. Milner, F. A. and Rabbiolo, G. (1992). Rapidly converging numerical algorithms for models of population dynamics, *J. Math. Biol.* **30**, pp. 733–753.
10. Ljung, L. and Glad, T. (1994). On the global identifiability of arbitrary model parametrizations, *Automotica* **30**, pp. 265–276.
11. Li, X. Z., Yang, J. Y., and Martchev, M. (2020). *Age Structured Epidemic Modeling*, Springer, New York.
12. Martcheva, M. (2015). *An Introduction to Mathematical Epidemiology*, Springer, New York.
13. Martcheva, M., Tuncer, N., and St Mary, C. (2015). Coupling within-host and between-host infectious diseases models, *BIOMATH* **4**, 2, p. 1510091.
14. Miao, H., Xia, X., Perelson, A. S., and Wu, H. (2011). On identifiability of nonlinear ODE models and applications in in viral dynamics, *SIAM Review* **53**, 1, pp. 3–39.
15. Perasso, A., Laroche, B., and Touzeau, S. (2010). Identifiability analysis of an epidemiological PDE model, *Proceedings of the 19th International Symposium on Mathematical Theory of Networks and Systems*, pp. 2077–2082.
16. Perasso, A. and Razafison, U. (2015). Identifiability problem for recovering the mortality rate in an age-structured population dynamics model, *Inverse Problems Sci. Eng.* **24**, 4, pp. 711–728.

17. Pohjanpalo, H. (1978). System identifiability based on power-series expansion of solution, *Math. Biosci.* **41**, pp. 21–33.
18. Perelson, A. S. (2002). Modelling viral and immune system dynamics, *Nat. Rev. Immunol.* **2**, 1, pp. 28–36.
19. Smith, A. (2018). Host-pathogen kinetics during influenza infection and coinfection: Insights from predictive modeling, *Immunol. Rev.* **285**, pp. 97–112.
20. Tuncer, N., Gulbudak, H., Cannataro, V., and Martcheva, M. (2016). Structural and practical identifiability issues of immuno-epidemiological vector-host models with application to Rift Valley fever, *Bull. Mathe. Bio.* **78**, 9, pp. 1769–1827.
21. Tuncer, N. and Le, T. (2018). Structural and practical identifiability analysis of outbreak models, *Math. Biosci.* **299**, pp. 1–18.
22. Tuncer, N., Martcheva, M., LaBarre, B., and Payoute, S. (2018). Structural and practical identifiability analysis of Zika epidemiological models, *Bull. Math. Biol.* **80**, pp. 2209–2241.
23. Thieme, R. H. (2003). *Mathematics in Population Biology*, Princeton University Press.
24. Hadjichrysanthou, C., Cauët, E., Lawrence, E., Vegvari, C., de Wolf F., and Anderson, R. M. Understanding the within-host dynamics of influenza A virus: From theory to clinical implications, *J. R. Soci. Interface.* **13**, 20160289. http://doi.org/10.1098/rsif.2016.0289.

https://doi.org/10.1142/9789811263033_0006

Chapter 6

Multiple Dimensions of *Aedes aegypti* Population Growth: Modeling the Impacts of Resource Dependence on Mass and Age at Emergence

Melody Walker[*,§,¶], **Michael A. Robert**[†,‡,††,||], **and Lauren M. Childs**[‡,††,#]

**Department of Mathematics, Virginia Tech, Blacksburg, Virginia, USA*

[†]Department of Mathematics and Applied Mathematics, Virginia Commonwealth University, Richmond, Virginia, USA; Center for Emerging Zoonotic and Arthropod-Borne Pathogens, Virginia Tech, Blacksburg, Virginia, USA

[‡]Center for Emerging Zoonotic and Arthropod-Borne Pathogens, Virginia Tech, Blacksburg, Virginia, USA

[§]Laboratory for Systems Medicine, University of Florida Health, Gainesville, Florida, USA

[††]Department of Mathematics, Virginia Tech, Blacksburg, Virginia, USA

[¶]Melody.walker@medicine.ufl.edu

[||]robertma@vcu.edu

[#]lchilds@vt.edu

Mosquitoes are responsible for the transmission of many diseases which lead to a large burden on public health. The age and size of adult mosquitoes impacts their ability to transmit disease. Older mosquitoes are more likely to have acquired a pathogen, and larger mosquitoes typically have higher fitness. In this chapter, we examine how larval resources affect the age and mass distributions of adult mosquitoes. We develop a partial differential equation model of juvenile and adult mosquitoes across time, age, and mass, and we incorporate resource dependence in the juvenile growth and death functions to determine differential effects of these on mosquito population dynamics. We find that the resource-dependent growth shows much larger changes in population dynamics than resource-dependent death. Furthermore, we predict that longer oscillations in resource lead to more extreme swings in population size than shorter oscillations in resources or decaying resources. We discuss our results in the context of mitigation strategies for mosquito and mosquito-borne disease control.

Keywords: *Aedes aegypti* (*Ae. aegypti*), Dengue, Mosquito-borne disease, Body size, Larval development, Resource-dependent (resource dependence), Sinko and Streifer, von Bertalanffy, Mass distribution, Age distribution, Difference method

6.1 Introduction

Mosquitoes are responsible for the death of millions of people each year.[1,2] Mosquito-borne diseases such as dengue and malaria are a significant global health burden, and dengue in particular is among the most important mosquito-borne viruses, threatening over 96 million people across 129 countries each year.[3,4] In 2013, the estimated global cost due to dengue was almost $9 billion, including direct and indirect costs such as those associated with health care and loss of work time.[5] The mosquito species *Aedes aegypti* is the principle vector of dengue as well as other pathogens such as those responsible for Zika virus, Mayaro virus, chikungunya, yellow fever, West Nile virus, and Rift Valley fever.[6–10] Reducing the global burden of disease will require an integrated approach that includes the development of vaccines and treatments as well as mosquito population reduction strategies. In fact, in some cases, vector control is the only available method to prevent mosquito-borne disease spread.[11]

Many current strategies to eliminate the vectors of dengue are centered around use of insecticides. Approaches such as indoor residual spraying and insecticide-treated bed nets focus on applying insecticides in individual homes.[11] Novel methods of mosquito control are being considered including those that would release biologically or genetically modified mosquitoes into wild populations. These modified mosquitoes are altered to be less fit mates, produce fewer or no offspring, or are less capable of transmitting pathogens.[12,13] Importantly, these novel control methods primarily influence the early stages of the mosquito life cycle. Many of these approaches would result in changes in the reproductive cycle such as blocking females from laying viable eggs or by preventing survival of larvae to adulthood, both of which would lead to changes in larval population dynamics.[12–14] Variation in the larval population size alters competition for resources and can have effects on mosquito body size, development time, survival, and immune system development.[15,16] All of these can ultimately affect mosquito population size and the spread of diseases they transmit. Thus, understanding larval dynamics is important for accurately predicting the effectiveness of these measures.

Ae. aegypti mosquitoes lay their eggs above small pockets of water, often in artificial containers. Once the conditions are appropriate, i.e. sufficient water is present, temperatures are warm enough, and daylight is long enough, eggs hatch and go through several stages of molting in their aquatic environment before becoming adults. A juvenile mosquito transitions through four larval stages called instars, often denoted by their order, such that L1, L2, L3, and L4 represent the first, second, third, and fourth instar larvae. Larvae consume microorganisms and organic debris,[17] which they find by browsing nearby leaves or carcasses.[18,19] At each stage the larva molts and gets larger. Finally, once the larva has reached a critical mass and sufficient age, it molts to become a pupa.[20,21] After time as a pupa, it emerges as an adult. Development time from hatching of an egg to emergence of an adult mosquito varies, but on average in suitable conditions it is 6 days for *Ae. aegypti*.[22,23]

Once adults emerge, female mosquitoes need to mate in order to produce and lay viable eggs. Adult mosquitoes gain nutrients from plant nectar and honeydew; however, female mosquitoes also take blood meals to assist in egg development.[24] Once a female is old enough, she seeks a host for a blood meal. Mating does not need to occur before taking a blood meal, but mated females seek hosts at greater rates than virgin females.[25] Once the female has a sufficient amount of blood, she rests to digest the blood and produce eggs. After resting for about a day, she then finds a suitable place to oviposit (i.e., lay eggs). Once the female has laid all her eggs, she repeats this cycle until she dies. The cycle, which includes seeking a host, resting, and laying eggs, is known as the gonotrophic cycle. Here, we do not focus on the details of each life cycle stage but separate mosquitoes into juveniles and adults.

Recently, there has been improved understanding of the role that larval density plays in determining adult body size and the influence of mosquito body size on disease transmission. In particular, body size is known to affect adult mosquito behavior, fecundity, and longevity. Mosquito behavior is differentiated by size as large mosquitoes spend more time trying to bite a single host, while smaller mosquitoes bite multiple hosts with greater frequency and travel further to feed.[26–28] In addition, body size has been shown to affect fecundity and longevity in harsh environments, and larger females are generally considered more fit than smaller females.[26,29–35]

Although the primary focus of this work is on the impacts of mass, longevity of adults also has an impact on population dynamics and disease spread.[36–39] As a mosquito needs to become infected via a blood meal and survive the incubation period prior to transmitting a pathogen, a longer lifespan increases the likelihood for disease spread.[36,37] Furthermore, both survival and fecundity of *Ae. aegypti* are age-dependent, with younger mosquitoes having higher daily survival than older mosquitoes,[38,40,41] and average fecundity oscillating with age as mosquitoes progress through gonotrophic cycles,[40] although fecundity ultimately decreases with age.[42]

In mosquito larval development, resources affect both mortality and growth. Without sufficient resources, larvae stop growing and die, but the details of changes in mortality vary by larval instar stage. Young mosquitoes are known to have long periods of resistance to starvation, especially if resources are available for the first stages of development.[43] However, there is a switch in behavior based on larval stage. Lack of resources in young larval stages (L1 and L2) leads to greater death, but in later stages (early L4), limited food merely results in a delay in pupation.[44] Both starvation in early stages and delayed pupation lead to a decrease in the number of emerging adults, which can then impact mosquito-borne disease dynamics.

To date, little modeling work has focused on the role of mosquito body size. In one of the first modeling studies involving mosquito body size, Gilpin and McClelland created a system of ordinary differential equations to describe changes in size that had separate descriptions for average mass, food resource, and variance of mass.[45] The Gilpin and McClelland model used mass and age to determine when a juvenile mosquito develops into an adult. In their work, they discussed that a lack of reserves led to starvation and found an empirical linear relationship between reserves and body weight. However, they did not model energy reserves directly. The Gilpin and McClelland model of growth and maturation has been incorporated into larger and more complicated models.[46–48] More recently, Padmanabha *et al.* emphasized the importance of reserves over weight alone.[20] In their work, their data showed that growth occurs in phases: early stages, intense growth stage, and a late fourth instar stage in which larvae are committed to pupate. The early stages in their model were marked by more deaths due to lack of food.[44,45] Padmanabha *et al.* concluded that the inclusion of reserves modeled the data best, and that using a threshold value of growth instead of monotonic growth as in Gilpin and McClelland was better. In Ref. 49, Aznar *et al.* included food-dependent and food-independent stages to separate the changes in behavior for early and late stage larvae. In previous work,[50] we considered a discrete time model that separated mass into three groups: small, medium, and large. We showed that the proportion which grew increased with available food. A model considering mortality in *Anopheles* used adult body size, temperature, and humidity as independent variables.[51] They found their model including mass predicted mortality better than without mass.[52]

Although modeling work devoted to body size in mosquitoes is limited,[20,45,49–51,53] there is a large amount of modeling literature considering body size (or mass as a proxy) for other organisms. In these models, mass is often used as an indicator of when an individual matures to adulthood, which results in a shift of energy use toward reproduction. von Bertalanffy growth is one of the most common methods used, especially in fishery models, to describe growth in mass.[54–61] von Bertalanffy made the assumption that growth in mass is proportional to a power of current mass.[62] The mosquito models including body size discussed above all use an adapted von Bertalanffy growth form.[20,45,49] However, there is criticism over the use of mass alone to model growth, as is done by von Bertalanffy. This is because not all of an individual's mass is usable for functions such as reproduction,[63,64] rather, some mass is required for structural components. Furthermore, the von Bertalanffy growth curve was shown to be negative for some values of the growth constant.[65] The criticism on distribution of mass for various purposes is rectified with Dynamic Energy Budget (DEB) theory, which argues that it is important to separate growth into usable energy (reserves) and structural energy.[66] Separating mass into two distinct components adds at least one extra equation into the models. Thus, it is common practice to use a proportion of mass as a proxy for reserves instead of tracking reserves and structural weight separately. DEB models are often agent-based and/or discrete, because of their complicated structure.[67,68]

Other models, however, use differential equations to describe mosquito dynamics incorporating mass dependence, i.e. Aznar *et al.*[49] Such models consider mass as a separate equation, but little modeling work incorporates mass as an independent variable. Sinko and Streifer developed a partial differential equation to describe population density as a function of time, age, and mass, and demonstrated its utility through modeling a population of *Daphnia*.[69,70] As mass is an attribute of the larvae, the inclusion of it as an independent variable more accurately describes the distribution of mass across a population.

In this work, we are interested in describing the impacts of competition for resources during the *Aedes* larval stages on characteristics of adults. To that end, we build upon Sinko and Streifer's model[69,70] to develop a system of two partial differential equations (PDEs) to model juvenile and adult mosquitoes. We couple these two PDEs at birth and maturation. We then consider how resources affect mass distribution and age distribution. We include the effects of resources in two ways: impacting growth and death. First, we consider decaying resources incorporated into the growth function and compare this to published data where resources are only provided at the outset of the experiments.[50,71] Once we determine parameters for a realistic setting with decaying resource, we simulate cyclic resources, which is meant to reflect a simplified field setting. For example, resources decay as larvae eat, but occasionally detritus falls into the larval environment restoring the resource level. We find that details of larval growth and death and their resource dependence impact age and mass distributions of adults.

6.2 Methods

In the following, we describe the development of our model and its numerical implementation.

6.2.1 *General PDE formulation*

We consider two biological stages of mosquitoes: adults (A) and juveniles (J), the latter of which consist of all aquatic phases: egg, larvae, and pupae. We assume that population density of adults $A(t, y, m)$ and juveniles $J(t, a, m)$ are functions of time (t), mass (m), and age (represented by y for adults and a for juveniles). The death rate for adults is an age-dependent function, $d_a(y)$, and death rate for juveniles is a time-dependent function, $d_j(t)$. Birth rate, $b(y, m)$, is a function of the mother's age y and mass m. We assume that all juveniles are the same mass, $m_0 > 0$, when born. Thus, for any $m_* < m_0$, $J(t, a, m_*) = 0$. This assumption arises as all juveniles start out as eggs, which are roughly the same size. Furthermore, the size differences as eggs are negligible compared to variation in larval sizes. The boundary condition for the juvenile compartment is given by the renewal equation,

$$J(t, 0, m) = \begin{cases} \int_0^\infty \int_0^\infty b(\nu, \eta) A(t, \nu, \eta) \mathrm{d}\nu \; \mathrm{d}\eta, & m = m_0, \\ 0, & m \neq m_0. \end{cases}$$

Juveniles leave via mortality or by maturing to become an adult with the rate of development $w(a, m)$, which is a function of age and mass. The maturation function is incorporated into the boundary condition for the adult population as

$$A(t, 0, m) = \int_0^\infty w(\nu, m) J(t, \nu, m) \mathrm{d}\nu.$$

We assume that growth as an adult is given by the constant, γ_a. For juveniles, growth in mass is a function of time and mass, $g(t, m)$. Thus, our system of partial differential equations is

$$\frac{\partial J}{\partial t} + \frac{\partial J}{\partial a} + \frac{\partial}{\partial m}\Big(g(t,m)J(t,a,m)\Big) + \Big(d_j(t) + w(a,m)\Big)$$

$$\times\ J(t,a,m) = 0,$$

$$\frac{\partial A}{\partial t} + \frac{\partial A}{\partial y} + \gamma_a \frac{\partial A}{\partial m} + d_a(y)A(t,y,m) = 0,$$

with initial conditions:

$$J(0,a,m) = h_j(a,m),$$
$$A(0,y,m) = h_a(y,m),$$

where $h_j(a,m)$ and $h_a(y,m)$ are distributions in age and mass.

6.2.1.1 *Details of functions*

We choose the transition function, $w(a,m)$, with the assumption that juveniles need to be a certain mass and age before they are able to emerge as adults. The maturation function is

$$w(a,m) = \Big(1 - \exp\big(-\alpha_1 a^{n_1}\big)\Big)\Big(1 - \exp\big(-\alpha_2 m^{n_2}\big)\Big),$$

where α_i and n_i are constants that determine the inflection point and steepness of the curve, respectively. With the defined function, the rate of maturation increases as age and mass increase, has a very low rate when a and m are small, and is smooth.

In the relatively few mosquito models that include mass-dependent growth, this idea of requiring a minimum mass or weight is often included.[20,45,46,48,49] Our chosen maturation function is initially flat with a very small value and then increases steeply before reaching an asymptote. The parameters n_1 and n_2 determine how steep the change in maturation is by age and mass, respectively. The parameters α_1 and α_2 shift the value when the maximum steepness occurs. The function does allow a relatively small number of mosquitoes to emerge as adults in situations that are biologically unlikely, e.g., very old, small mosquitoes or young, large ones. Despite this, we prefer this function compared to a piecewise function because its smoothness allows for the PDE to be solvable and leads to fewer numerical issues.

The birth function, $b(y, m)$, is a piecewise function, so that young adults are unable to give birth, and is given by

$$b(y, m) = \begin{cases} \beta_0 + \beta_1 m, & y \geq y_*, \\ 0, & y < y_*. \end{cases}$$

As the birth function only appears in the boundary condition for the juvenile population, there is no restriction on the smoothness of the function. Our chosen function incorporates the generally accepted assumption that increased body mass in mosquitoes leads to increase female fecundity. While it is likely that the mechanism is more complicated than our simple functional form, the use of a line can be thought of as a general Taylor expansion of some unknown function, $b(y, m) = \beta_0 + \beta_1 m + \mathcal{O}(m^2)$.

Mass is incorporated into the growth function, via a modification of the von Bertalanffy growth function. von Bertalanffy calls his implementation the surface rule, where growth is proportional to the surface area of the individual, which is roughly two-thirds of the mass.[62] This choice was validated for *Ae. aegypti* in a model incorporating data.[20,49] Our modification of the von Bertalanffy growth function excludes term for loss due to catabolism (breaking down due to sustaining a current mass) which appears in the original von Bertalanffy function. We neglect this component because Day and Taylor conclude that the catabolism rate term makes the function decrease with respect to the growth rate parameter (γ_j) at large masses, which is contradictory.[65] They argued that most biologically relevant uses of Bertalanffy growth fit this piece to zero or practically zero. Thus, our growth function for juveniles is

$$g(t, m) = \gamma_j m^{2/3}.$$

Many modeling papers argue that assuming a constant death rate in adults leads to incorrect results, and one important factor in adult mortality is age.[38,72] We assume a simple age-dependent function for adult death, given by

$$d_a(a) = \delta_0 + \delta_a a.$$

As with birth, this function can represent a Taylor approximation of death as a function of age. The juvenile death is a constant value $d_J(t) = \delta_j$.

As our goal is to compare the impact of resources on juvenile death vs their impact on juvenile growth and how these differential effects impact the population, we incorporate resource dependence in our juvenile death and growth functions. Without enough food, juveniles die because low resource levels should increase mortality. Furthermore, it is known that juvenile mosquitoes are affected by density, at least in part because of the need to share resources, which limits their growth. Thus, we employ a resource function which increases death, or limits growth, when resources are scarce.

6.2.1.2 *Incorporation of resources*

We consider two forms for resources:

$$r_1(t) = \mathrm{e}^{-\rho t},$$

$$r_2(t) = \frac{1}{2}\left(\cos\left(\frac{2\pi t}{\theta}\right) + 1\right).$$

The first function, r_1, represents resources decaying over time at a rate ρ; the second function, r_2, represents cyclic resources where the amount of resources oscillates repeatedly with a period of θ. Both of these functions have a maximum of one and a minimum of zero. We choose both functions to have the same bounds to be comparable. In reality, resources should be a function of density and maybe even a history of density. Additionally, resources might fluctuate for other reasons outside of density, such as rain water flushing out resources or bringing in new resources. We do not consider density or other complicated dependencies here as their inclusion would make the PDE nonlinear. We assume resources are simply a function of time and investigate how this effects the population.

Option 1: Growth affected by resource

We let the growth function be given by

$$g(t, m) = \gamma_j m^{2/3} r_i(t),$$

where $r_i(t)$ represents resources available at time t and is defined in the previous section. As the resource function has a maximum value of one, it only serves to limit the growth function.

Option 2: Death affected by resource

We let the death function of juveniles be given by

$$d_j(t,m) = c_1 \frac{(1-r_i(t))^5}{(1-r_i(t))^5 + c_3^5} \frac{m_{\max} - m}{m_{\max}} + c_2,$$

where $m_{\max}$ is the maximum mass considered in the discretization, c_1 influences the maximum death rate, c_2 is the minimum death rate, c_3 dictates what resource value is the half maximal rate, and $r_i(t)$ is the resource function. By convention, $m < m_{\max}$ so that $d_j(t,m) >= c_2$. The death rate is maximized when resources are low with a value of $\frac{c_1 m_{\max} - m}{c_3^5 m_{\max}} + c_2$. Young and malnourished larvae (juvenile mosquitoes) are more susceptible to starvation. While biologically it is known that both mass and age are important, we assume death is only a function of mass. This disproportionately affects larvae that are smaller in size, which also tend to be younger. This function is sigmoidal with respect to resource, so that from a maximal resource level, small reductions in resources negligibly increase death rate. Increases in mass shift the upper bound, such that the death rate of small juveniles is greater than that of large juveniles.

6.2.2 *Numerical solution*

In order to find solutions to this system of partial differential equations, we introduce the numerical schemes we use. Our boundary conditions are integrals of the solution to the PDEs. In order to find the integrals, we use a numerical integration technique known as Simpson's rule of $\frac{1}{3}$. Simpson's rule divides up the area into smaller pieces, approximates the smaller pieces using quadratic interpolation of the end points, and finds the integral of the quadratic approximation. We let Δa and Δm be the width of each piece or the step size. The approximation for the boundaries are

$$\begin{aligned} J(t,0,m_0) &= \int_0^\infty \int_0^\infty b(\nu,\eta) A(t,\nu,\eta)\ \mathrm{d}\nu\ \mathrm{d}\eta, \\ &\approx \frac{\Delta a \Delta m}{9} \sum_{k=0}^{m_{\max}} \sum_{j=1}^{a_{\max}} b(j\Delta a, m_0 + k\Delta m) W_{j,k} \\ &\quad \times A(t, j\Delta a, m_0 + k\Delta m), \end{aligned}$$

$$A(t,0,m_0+k\Delta m) = \int_0^\infty w(\nu, m_0+k\Delta m)J(t,\nu,m_0+k\Delta m)\,\mathrm{d}\nu,$$
$$\approx \frac{\Delta a}{3}\sum_{j=1}^{a_{\max}} W_{j,1}w(j\Delta a, m_0+k\Delta m)$$
$$\times J(t, j\Delta a, m_0+k\Delta m),$$

where W are the weights for implementing Simpson's rule.

Notice that we cut off the numerical approximation at some finite point given by $a_{\max}$ in age and $m_{\max}$ in mass. This is reasonable as mosquitoes will reach some finite maximum age and mass. In the approximation, we do not consider the left end of age or when age is zero. This does not alter the approximation as we do not allow individuals of age zero to mature or adults of age zero to produce offspring. For the boundary conditions, we integrate the necessary pieces in the analytic solution using the same numerical method.

6.2.2.1 *Discretization by difference method*

We use the forward time-backwards space difference scheme to solve the system of PDEs where we consider mass and age to be space. For notation, the superscripts represent the step in time with step size given by Δt; subscripts represent a step in age and mass with step size indicated by Δa and Δm, respectively. The value $J^n_{j,k}$ represents $J(t_0+n\Delta t, a_0+j\Delta a, m_0+k\Delta m) = J(n\Delta t, j\Delta a, m_0+k\Delta m)$ with n steps from the initial time (0), j steps from the initial age (0), and k steps from the initial mass (m_0). This is similar for the adult equation, where $A^n_{j,k} = A(n\Delta t, j\Delta a, m_0+k\Delta m)$. We discretize each partial derivative as

$$A^{n+1}_{j,k} = A^n_{j,k} - \Delta t\left(\frac{A^n_{j,k}-A^n_{j-1,k}}{\Delta a} + \gamma_a\frac{A^n_{j,k}-A^n_{j,k-1}}{\Delta m} + d_a(j\Delta a)A^n_{j,k}\right),$$

$$J^{n+1}_{j,k} = J^n_{j,k} - \Delta t\left(\frac{J^n_{j,k}-J^n_{j-1,k}}{\Delta a} + g(m_0+k\Delta m)\frac{J^n_{j,k}-J^n_{j,k-1}}{\Delta m}\right.$$
$$\left.+\left(d_j(n\Delta t) + w(j\Delta a, m_0+k\Delta m) + \frac{\partial g}{\partial m}\right)J^n_{j,k}\right).$$

First we define the initial condition that describes $A^0_{j,k}$ and $J^0_{j,k}$ for all $j>0$ and k. Then we find when $j=0$ by using the boundary condition, as this is when age is zero. Note that this is also the initial condition, but the boundary and initial condition must be equal when they overlap. We use Simpson's rule to find the integrals in the boundary condition. The newborns and new adults at time zero are given as

$$\begin{aligned}
J_{0,0}^0 = J(0,0,m_0) &= \int_0^\infty \int_0^\infty b(\nu,\eta)A(0,\nu,\eta)\ \mathrm{d}\nu\ \mathrm{d}\eta, \\
&\approx \frac{\Delta a \Delta m}{9} \sum_{k=0}^{m_{\max}} \sum_{j=0}^{a_{\max}} b(j\Delta a, m_0 + k\Delta m) W_{j,k} A_{j,k}^0. \\
A_{0,k}^0 = A(0,0,m_0 + k\Delta m) &= \int_0^\infty w(\nu,m)J(0,\nu,m)\ \mathrm{d}\nu, \\
&\approx \frac{\Delta a}{3} \sum_{j=0}^{a_{\max}} w(j\Delta a, m_0 + k\Delta m) W_{j,1} J_{j,k}^0.
\end{aligned}$$

The approximation explicitly includes the term when $j = 0$, meaning age is zero, but there is no contribution as $b(0,m) = 0$ and $w(0,m) = 0$. Then using the difference scheme, we find $A_{j,k}^1$ for all $j, k > 0$ as follows:

$$A_{j,k}^1 = A_{j,k}^0 - \Delta t \left(\frac{A_{j,k}^0 - A_{j-1,k}^0}{\Delta a} + \gamma_a \frac{A_{j,k}^0 - A_{j,k-1}^0}{\Delta m} + d_a(j\Delta a) A_{j,k}^0 \right),$$

$$\begin{aligned}
J_{j,k}^1 = J_{j,k}^0 - \Delta t \Bigg(& \frac{J_{j,k}^0 - J_{j-1,k}^0}{\Delta a} + g(m_0 + k\Delta m) \frac{J_{j,k}^0 - J_{j,k-1}^0}{\Delta m} \\
& + \left(d_j(n\Delta t) + w(j\Delta a, m_0 + k\Delta m) + \frac{\partial g}{\partial m} \right) J_{j,k}^0 \Bigg).
\end{aligned}$$

When $k = 0$ and $j \geq 1$, for consistency, we set $A_{j,-1}^0, J_{j,-1}^0 = 0$, so that the difference equation becomes

$$A_{j,0}^1 = A_{j,0}^0 - \Delta t \left(\frac{A_{j,0}^0 - A_{j-1,0}^0}{\Delta a} + \gamma_a \frac{A_{j,0}^0}{\Delta m} + d_a(j\Delta a) A_{j,0}^0 \right),$$

$$\begin{aligned}
J_{j,0}^1 = J_{j,0}^0 - \Delta t \Bigg(& \frac{J_{j,0}^0 - J_{j-1,0}^0}{\Delta a} + g(m_0) \frac{J_{j,0}^0}{\Delta m} \\
& + \left(d_j(n\Delta t) + w(j\Delta a, m_0) + \frac{\partial g}{\partial m} \right) J_{j,0}^0 \Bigg).
\end{aligned}$$

Finally, we find $A_{0,k}^{n+1}$, $J_{0,k}^{n+1}$, i.e. when age is zero. This comes from the boundary condition as shown previously and is given by

$$J_{0,k}^{n+1} = \frac{\Delta a \Delta m}{9} \sum_{k=0}^{m_{\max}} \sum_{j=1}^{a_{\max}} b(j\Delta a, m_0 + k\Delta m) W_{j,k} A_{j,k}^{n},$$

$$A_{0,k}^{n+1} = \frac{\Delta a}{3} \sum_{j=1}^{a_{\max}} w(j\Delta a, m_0 + k\Delta m) W_{j,1} J_{j,k}^{n}.$$

During each time step, we must determine new adults and newborns last because the integrals require information from the current time step. We then repeat this process until we reach our maximum time desired.

Full Numerical Scheme

For each time step $n+1$, we first find when $j \geq 1$ and $k \geq 1$ by

$$A_{j,k}^{n+1} = A_{j,k}^{n} - \Delta t \left(\frac{A_{j,k}^{n} - A_{j-1,k}^{n}}{\Delta a} + \gamma_a \frac{A_{j,k}^{n} - A_{j,k-1}^{n}}{\Delta m} + d_a(j\Delta a) A_{j,k}^{n} \right), \tag{6.1}$$

$$J_{j,k}^{n+1} = J_{j,k}^{n} - \Delta t \left(\frac{J_{j,k}^{n} - J_{j-1,k}^{n}}{\Delta a} + g(m_0 + k\Delta m) \frac{J_{j,k}^{n} - J_{j,k-1}^{n}}{\Delta m} + \left(d_j(n\Delta t) + w(j\Delta a, m_0 + k\Delta m) + \frac{\partial g}{\partial m} \right) J_{j,k}^{n} \right). \tag{6.2}$$

Then we find all $j \geq 1$ when $k = 0$, by

$$A_{j,0}^{n+1} = A_{j,0}^{n} - \Delta t \left(\frac{A_{j,0}^{n} - A_{j-1,0}^{n}}{\Delta a} + \gamma_a \frac{A_{j,0}^{n}}{\Delta m} + d_a(j\Delta a) A_{j,0}^{n} \right), \tag{6.3}$$

$$J_{j,0}^{n+1} = J_{j,0}^{n} - \Delta t \left(\frac{J_{j,0}^{n} - J_{j-1,0}^{n}}{\Delta a} + g(m_0) \frac{J_{j,0}^{n}}{\Delta m} + \left(d_j(n\Delta t) + w(j\Delta a, m_0) + \frac{\partial g}{\partial m} \right) J_{j,0}^{n} \right). \tag{6.4}$$

The last step is when $j = 0$, for all k, by

$$J_{0,k}^{n+1} = \frac{\Delta a \Delta m}{9} \sum_{k=0}^{m_{\max}} \sum_{j=1}^{a_{\max}} b(j\Delta a, m_0 + k\Delta m) W_{j,k} A_{j,k}^n, \quad (6.5)$$

$$A_{0,k}^{n+1} = \frac{\Delta a}{3} \sum_{j=1}^{a_{\max}} w(j\Delta a, m_0 + k\Delta m) W_{j,1} J_{j,k}^n. \quad (6.6)$$

In Section 6.3, we show the consistency and stability of our full numerical scheme. Note that we use the following functions $g(t,m), d_j(a)$, and $w(a,m)$ as described in the text, and we plug in the age and mass at the discretized point. Specifically, $g(t,m) = g(n\Delta t, d_j(j\Delta a)$, and $w(j\Delta a, m_0 + k\Delta m)$. Similarly, we find $\frac{\partial g}{\partial m}$ and plug in the particular discretized point.

$$\frac{\partial g}{\partial m}(t,m) = \frac{2}{3}\gamma_j m^{-1/3} r_i(t),$$

$$\frac{\partial g}{\partial m}(n\Delta t, m_0 + k\Delta m) = \frac{2}{3}\gamma_j (m_0 + k\Delta m)^{-1/3} r_i(n\Delta t).$$

6.2.3 *Data and fitting*

6.2.3.1 *Data*

We use publicly available data that was collected from *Ae. aegypti*.[50,71] Two types of experimental conditions were performed: one with low density (26 larvae) and one with high density (78 larvae). A pre-determined amount of resources was present at the beginning of the experiments, and no additional resources were added. All experiments began with mosquitoes in the L1 stage, and mosquitoes were weighed upon emergence as adults. For each individual, the mass and development time were recorded. For both low and high density conditions, we used the histogram function in MATLAB to group the data into gridded bins based on development time and mass. Unless otherwise specified, MATLAB automatically groups data into uniform bin sizes that fit the distribution of the data. For the high density data, development time was binned by 2 days from 6 to 22 days, and for the low density data, development time was binned by one day from 5 to 12 days. We shifted all development times by adding three to account for the assumed time of 3 days for eggs to hatch. The mass for both low and high density was binned by groups of length 0.3 mg, with high density from 0.6 to 2.4 mg and low density from 1.8 to 3.6 mg.

6.2.3.2 *Fitting*

Initially, we ran the model assuming decaying resource, $r_1(t)$, in the growth function. This condition reflects the experimental setup where there was an initial amount of food which is reduced by larval use over time. We fixed parameters for the juvenile death rate, γ_j, adult death (δ_0 and δ_a), minimum age adults can reproduce, y_*, adult growth rate, γ_a, and birth rate (β_0 and β_1), as found in Table 6.1. As this work focuses on the impact of changes in larval nutrition, we work under the assumption that the change in body mass during the larval stages is significantly more important than changes in mass as an adult. While mass does fluctuate as an adult, these changes are based on many things that we do not directly incorporate in the model such as time since the most recent blood meal or deposition of eggs as well as hydration. From the model output, we determine the number of adults of each mass and age group with the same discretization as the data. We compared the values from the model and the data, grouped by mass and age, with the following error:

$$\left(\sum_k \left(\text{Model}(k) - \text{Data}(k)\right)^2\right)^{1/2},$$

where k is at the kth grid box. Using this error function, we fit six parameters: four relating to the maturation function $(\alpha_1, \alpha_2, n_1, n_2)$,

Table 6.1. Fixed parameters.

Symbol	Description	Value (units)	Source
δ_j	Juvenile death rate	0.05 (d^{-1})	73
δ_0	Constant adult death rate	0.05 (d^{-1})	38
δ_a	Age-dependent adult death coefficient	0.025 $(\text{age} \cdot d)^{-1}$	38
y_*	Minimum female reproduction age	8 (d)	36
γ_a	Adult growth rate	0 $(\text{mg} \cdot d^{-1})$	See 6.2.3.2
β_0	Baseline birth rate	10 eggs $(a \cdot d)^{-1}$	40
β_1	Relative birth rate increase with mass	1 egg $(a \cdot d \cdot \text{mg})^{-1}$	29

Note: Parameters fixed during fitting to the data. d = day; a = adult.

Table 6.2. Fitted parameters.

Symbol	Description	Low	High	Combined	Units
α_1	Coefficient of age maturation	2e-3	6.2e-7	6e-4	$(\text{day}^{n_1})^{-1}$
α_2	Coefficient of mass maturation	0.038	1.2	0.6	$(\text{mg}^{n_2})^{-1}$
n_1	Exponent of age maturation	2.8	6	3	None
n_2	Exponent of mass maturation	3.6	16	10.8	None
γ_j	Juvenile growth rate	0.4	0.3	0.33	day^{-1}
ρ	Resource decay rate	0.028	0.035	0.05	day^{-1}

Note: Parameter values fitted to the data in the low density setting, high density setting, and with the datasets combined.

the maximal juvenile growth rate γ_j, and the resource decay rate, ρ. Fitted parameters are found in Table 6.2.

6.2.3.3 *Sensitivity analyses*

To evaluate our parameter choices, we considered alternative parameter sets with values 20% below and above the standard for each parameter of interest, varied uni-variately. We then compared the error with results from the revised parameter set to those from the standard parameter set. We only considered a uni-variate comparison of each parameter as each run requires substantial computation time.

6.2.4 *Simulations*

All simulations were performed in MATLAB 2020a, and all results shown use the numerical scheme described above. All code is available at the GitHub Repository: https://github.com/melody289/Mosquito-Body-Size-Dynamics-PDE.

The initial values for the PDEs were found by initializing with 100 eggs and running iterations of the difference method with resources set to one. Once a stable mass distribution was reached, we used the resulting age and mass values, re-scaled as described in what follows, for both juveniles and adults at time zero. We re-scaled the initial values to start with 50 adults distributed across the age and mass ranges accordingly. For juveniles, we scaled the amount to be proportional to adults at the stable mass distribution.

Once initial values for the PDEs were found, we ran four types of simulations with cyclic or decaying resources in each of the growth and death functions. For decaying resources, we considered three values for the decay rate, ρ: 0.02, 0.05, and 0.2. For cyclic resources, we varied the period to be 7, 13, and 19 days. We also looked at periods from 1 to 31 days by increments of two for cyclic resources. Parameters for the maturation function were chosen from fitting the data, and other parameters were fixed at determined biological values.

6.3 Analytical Results

Here, we show the consistency and stability of our numerical implementation of the PDE.

6.3.1 *Consistency of difference scheme*

In order for the difference method to converge to the PDE, it must be consistent. We show the difference between the PDE and the difference scheme approaches zero pointwise as the step sizes go to zero.[74]

Theorem 6.1. *The numerical scheme given in Eqs.* (6.1)–(6.6) *is consistent.*

Proof. Let the difference scheme be written as $P_\Delta A = f_1$ and $Q_\Delta J = f_2$ and the partial differential equation as $PA = f_1$ and $QJ = f_2$ for adults and juvenile equations, respectively, and $f_1, f_2 = 0$. Thus, $PA = P_\Delta A$ and $QJ = Q_\Delta J$ such that

$$\begin{aligned} PA &= A_t(n\Delta t, j\Delta a, m_0 + k\Delta) + A_a(n\Delta t, j\Delta a, m_0 + k\Delta) \\ &\quad + \gamma_a A_m(n\Delta t, j\Delta a, m_0 + k\Delta) + d_a(a)A^n_{j,k}, \\ QJ &= J_t(n\Delta t, j\Delta a, m_0 + k\Delta) + J_a(n\Delta t, j\Delta a, m_0 + k\Delta) \\ &\quad + g(t,m)J_m(n\Delta t, j\Delta a, m_0 + k\Delta) \\ &\quad + \left(d_j(t) + w(a,m) + \frac{\partial g}{\partial m}\right) J^n_{j,k}, \end{aligned}$$

where A_t, A_a, and A_m represent the partial derivative of A with respect to time, age, and mass, respectively. Similarly, J_t, J_a, and

J_m are partial derivatives for J. For consistency, we need

$$P_\Delta A - PA \to 0 \text{ as } \Delta t,\ \Delta a,\ \Delta m \to 0$$
$$Q_\Delta J - QJ \to 0 \text{ as } \Delta t,\ \Delta a,\ \Delta m \to 0.$$

We use Taylor series expansions of the functions A and J about the point $(n\Delta t, j\Delta a, m_0 + k\Delta)$ and input these into the difference scheme. For the sake of clarity, we hide the arguments for the partial derivatives, such that $A_t = A_t(n\Delta t, j\Delta a, m_0 + k\Delta)$, $A_a = A_a(n\Delta t, j\Delta a, m_0 + k\Delta)$, and $A_m = A_m(n\Delta t, j\Delta a, m_0 + k\Delta)$. Similarly, for juveniles, J_t, J_a, and J_m. The Taylor expansion are

$$\begin{aligned}
A_{j,k}^{n+1} &= A_{j,k}^n + \ \Delta t \cdot A_t + (\Delta t)^2 K_1,\\
A_{j-1,k}^n &= A_{j,k}^n - \Delta a \cdot A_a + (\Delta a)^2 K_2,\\
A_{j,k-1}^n &= A_{j,k}^n - \Delta m \cdot A_m + (\Delta m)^2 K_3,\\
J_{j,k}^{n+1} &= J_{j,k}^n + \ \Delta t \cdot J_t + (\Delta t)^2 K_4,\\
J_{j-1,k}^n &= J_{j,k}^n - \Delta a \cdot J_a + (\Delta a)^2 K_5,\\
J_{j,k-1}^n &= J_{j,k}^n - \Delta m \cdot J_m + (\Delta m)^2 K_6.
\end{aligned}$$

where $K_1, \ldots, K_6$ are constant values which are known to exist through Taylor remainder theorem.

Adult equations. To begin, we focus on the adult equations. The difference equation for the adults (Eq. (6.1)), after replacing the function values for $A_{j,k}^{n+1}$, $A_{j-1,k}^n$, and A_{k-1}^n with the Taylor expansions shown above, is

$$\begin{aligned}
P_\Delta A &= \frac{A_{j,k}^{n+1} - A_{j,k}^n}{\Delta t} + \frac{A_{j,k}^n - A_{j-1,k}^n}{\Delta a} + \gamma_a \frac{A_{j,k}^n - A_{j,k-1}^n}{\Delta m} + d_a(a) A_{j,k}^n,\\
&= (A_t + (\Delta t)K_1) + (A_a - (\Delta a)K_2) + \gamma_a(A_m - (\Delta m)K_3)\\
&\quad + d_a(a) A_{j,k}^n.
\end{aligned}$$

If we subtract the PDE from the difference scheme at the point $(n\Delta t, j\Delta a, m_0 + k\Delta)$, we obtain

$$P_\Delta A - PA = (\Delta t)K_1 - (\Delta a)K_2 - \gamma_a(\Delta m)K_3.$$

Then if we take Δt, Δa, and Δm to zero, we see that the difference of the PDE and the difference scheme also goes to zero. This shows the adult PDE is consistent.

Juvenile equations. To find the consistency of the juvenile difference scheme, we input the Taylor expansion into the scheme as

$$\begin{aligned} Q_\Delta J &= \frac{J^{n+1}_{j,k} - J^n_{j,k}}{\Delta t} + \frac{J^n_{j,k} - J^n_{j-1,k}}{\Delta a} + g(t,m)\frac{J^n_{j,k} - J^n_{j,k-1}}{\Delta m} \\ &\quad + \left(d_j(t) + w(a,m) + \frac{\partial g}{\partial m}\right) J^n_{j,k}, \\ &= (J_t + (\Delta t)K_4) + (J_a - (\Delta a)K_5) + g(t,m)(J_m - (\Delta m)K_6) \\ &\quad + \left(d_j(t) + w(a,m) + \frac{\partial g}{\partial m}\right) J^n_{j,k}. \end{aligned}$$

We subtract the PDE from the difference scheme at the point $(n\Delta t, j\Delta a, m_0 + k\Delta)$ and get

$$Q_\Delta J - QJ = (\Delta t)K_4 - (\Delta a)K_5 - g(t,m)(\Delta m)K_6,$$

which will go to zero when all step sizes go to zero, i.e. $\Delta t, \Delta a, \Delta m \to 0$. This shows that the difference scheme is indeed consistent. □

6.3.2 *Stability of difference method: von Neumann analysis*

In order to ensure that the initial value problem is stable, we consider von Neumann analysis on the base model. The von Neumann analysis with a system of PDEs requires finding an amplification matrix.

Theorem 6.2. *The numerical scheme given in Eqs.* (6.1)–(6.6) *is stable.*

Proof. Consider our PDEs for juvenile and adults where all parameters are non-negative. The amplification matrix is diagonal, since the adult and juvenile PDEs only interact in the boundary and not in the partial differential equations. Let G represent the amplification matrix. Then

$$\|G^n\|_2 = \left\| \begin{matrix} g_1^n & 0 \\ 0 & g_2^n \end{matrix} \right\|_2 = \max\{|g_1^n|, |g_2^n|\}.$$

This shows that if we bound both individual amplification factors, we bound the amplification matrix, and the initial value scheme is stable.

We use the quick method of von Neumann analysis for each equation. We do this by substituting $A^n_{j,k} = g^n \exp(i(j\Delta a\xi_1 + k\Delta m\xi_2)) = g^n \exp(i(j\theta_1 + k\theta_2))$ where g is the amplification factor and ξ_1 and ξ_2 are the angles from the transformation into the frequency domain. We plug this into the difference scheme, and then rearrange to solve for the amplification factor.

$$\frac{A^{n+1}_{j,k} - A^n_{j,k}}{\Delta t} + \frac{A^n_{j,k} - A^n_{j-1,k}}{\Delta a} + \gamma_a \frac{A^n_{j,k} - A^n_{j,k-1}}{\Delta m} + \delta_a A^n_{j,k} = 0,$$

$$\frac{g^{n+1}\mathrm{e}^{i(j\theta_1+k\theta_2)} - g^n\mathrm{e}^{i(j\theta_1+k\theta_2)}}{\Delta t} + \frac{g^n\mathrm{e}^{i(j\theta_1+k\theta_2)} - g^n\mathrm{e}^{i((j-1)\theta_1+k\theta_2)}}{\Delta a}$$

$$+ \gamma_a \frac{g^n\mathrm{e}^{i(j\theta_1+k\theta_2)} - g^n\mathrm{e}^{i(j\theta_1+(k-1)\theta_2)}}{\Delta m} + \delta_a g^n \mathrm{e}^{i(j\theta_1+k\theta_2)} = 0,$$

$$g^n\mathrm{e}^{i(j\theta_1+k\theta_2)} \left(\frac{g-1}{\Delta t} + \frac{1-\mathrm{e}^{-i\theta_1}}{\Delta a} + \gamma_a \frac{1-\mathrm{e}^{-i\theta_2}}{\Delta m} + \delta_a \right) = 0.$$

Now we solve for the amplification factor g.

$$g = 1 - \left(\frac{\Delta t}{\Delta a}(1 - \mathrm{e}^{-\mathrm{i}\theta_1}) + \gamma_a \frac{\Delta t}{\Delta m}(1 - \mathrm{e}^{-\mathrm{i}\theta_2}) + \Delta t \delta_a \right).$$

We let $\lambda_a = \frac{\Delta t}{\Delta a}$ and $\lambda_m = \frac{\Delta t}{\Delta m}$, such that

$$\begin{aligned} g &= 1 - \Big(\lambda_a(1 - \mathrm{e}^{-\mathrm{i}\theta_1}) + \gamma_a\lambda_m(1 - \mathrm{e}^{-\mathrm{i}\theta_2}) + \Delta t\delta_a \Big), \\ &= \Big(1 - \lambda_a - \gamma_a\lambda_m - \Delta t\delta_a + \lambda_a \cos(\theta_1) + \gamma_a\lambda_m \cos(\theta_2) \Big) \\ &\quad -\mathrm{i}\Big(\lambda_a \sin(\theta_1) + \gamma_a\lambda_m \sin(\theta_2) \Big). \end{aligned}$$

Now we need to bound $|g| \leq 1 + c\Delta t$ where $c > 0$ is a constant. We first note that

$$\begin{aligned}
|g| &= \Big|\Big(1 - \lambda_a - \gamma_a\lambda_m - \Delta t\delta_a + \lambda_a\cos(\theta_1) + \gamma_a\lambda_m\cos(\theta_2)\Big) \\
&\quad -\mathrm{i}\Big(\lambda_a\sin(\theta_1) + \gamma_a\lambda_m\sin(\theta_2)\Big)\Big|, \\
&\leq \Big|\Big(1 - \lambda_a - \gamma_a\lambda_m + \lambda_a\cos(\theta_1) + \gamma_a\lambda_m\cos(\theta_2)\Big) \\
&\quad -\mathrm{i}\Big(\lambda_a\sin(\theta_1) + \gamma_a\lambda_m\sin(\theta_2)\Big)\Big| + \Delta t\delta_a.
\end{aligned}$$

We label the left piece in the absolute value $\hat{g}$, so that $g = \hat{g} - \Delta t\delta_a$. If $|\hat{g}| < 1$, then $|g| < 1 + c\Delta t$ with $c = \delta_a$. We want to bound $|\hat{g}|^2 < 1$, which if true, implies that $|\hat{g}| < 1$. To do so, we consider

$$\begin{aligned}
|\hat{g}|^2 &= \Big(1 - \lambda_a - \gamma_a\lambda_m + \lambda_a\cos(\theta_1) + \gamma_a\lambda_m\cos(\theta_2)\Big)^2 \\
&\quad +\Big(\lambda_a\sin(\theta_1) + \gamma_a\lambda_m\sin(\theta_2)\Big)^2.
\end{aligned}$$

In order to bound this for all θ_1 and θ_2, we take the derivative with respect to both and find critical points. Let $q(\theta_1, \theta_2) = |\hat{g}|^2$. We only need to consider one period, so we will look for maxima in $[-\pi, \pi]$. Consider the derivatives

$$\begin{aligned}
\frac{dq}{d\theta_1} &= -2\lambda_a\sin(\theta_1)\Big(1 - \lambda_a - \gamma_a\lambda_m + \lambda_a\cos(\theta_1) + \gamma_a\lambda_m\cos(\theta_2)\Big) \\
&\quad +2\lambda_a\cos(\theta_1)\Big(\lambda_a\sin(\theta_1) + \gamma_a\lambda_m\sin(\theta_2)\Big), \\
&= -2\lambda_a\sin(\theta_1)(1 - \lambda_a - \gamma_a\lambda_m) - 2\gamma_a\lambda_m\lambda_a\sin(\theta_1)\cos(\theta_2) \\
&\quad + 2\gamma_a\lambda_m\lambda_a\cos(\theta_1)\sin(\theta_2), \\
&= -2\lambda_a\sin(\theta_1)(1 - \lambda_a - \gamma_a\lambda_m) + 2\gamma_a\lambda_m\lambda_a\sin(\theta_2 - \theta_1), \\
\frac{dq}{d\theta_2} &= -2\gamma_a\lambda_m\sin(\theta_2)\Big(1 - \lambda_a - \gamma_a\lambda_m + \lambda_a\cos(\theta_1) + \gamma_a\lambda_m\cos(\theta_2)\Big) \\
&\quad +2\gamma_a\lambda_m\cos(\theta_2)\Big(\lambda_a\sin(\theta_1) + \gamma_a\lambda_m\sin(\theta_2)\Big),
\end{aligned}$$

$$= -2\gamma_a\lambda_m \sin(\theta_2)(1 - \lambda_a - \gamma_a\lambda_m) - 2\gamma_a\lambda_m\lambda_a \sin(\theta_2)\cos(\theta_1)$$
$$+ 2\gamma_a\lambda_m\lambda_a \cos(\theta_2)\sin(\theta_1),$$
$$= -2\gamma_a\lambda_m \sin(\theta_2)(1 - \lambda_a - \gamma_a\lambda_m) + 2\gamma_a\lambda_m\lambda_a \sin(\theta_1 - \theta_2).$$

There are critical points when $\theta_1, \theta_2 \in \{-\pi, 0, \pi\}$. We plug these values directly into $|\hat{g}|^2$ to find the largest value in the following:

$$\begin{aligned} q(0, \pm\pi) &= (1 - \lambda_a - \gamma_a\lambda_m + \lambda_a - \gamma_a\lambda_m)^2 = (1 - 2\gamma_a\lambda_m)^2, \\ q(0, 0) &= (1 - \lambda_a - \gamma_a\lambda_m + \lambda_a + \gamma_a\lambda_m)^2 = 1, \\ q(\pm\pi, 0) &= (1 - \lambda_a - \gamma_a\lambda_m - \lambda_a + \gamma_a\lambda_m)^2 = (1 - \lambda_a)^2, \\ q(\pm\pi, \pm\pi) &= (1 - \lambda_a - \gamma_a\lambda_m - \lambda_a - \gamma_a\lambda_m)^2 = (1 - 2\lambda_a - 2\gamma_a\lambda_m)^2. \end{aligned}$$

Then for $q \leq 1$, we need that $\lambda_a + \gamma_a\lambda_m \leq 1$ for stability of the initial value problem for the adult equation, as all parameter values are assumed to be non-negative.

Now we repeat this for the juvenile equation in the base case. We again use the same notation and let g be the amplification factor and ξ_1 and ξ_2 be the angles from the transformation into the frequency domain. We replace $J_{j,k}^n = g^n \exp(i(j\Delta a\xi_1 + k\Delta m\xi_2)) = g^n \exp(i(j\theta_1 + k\theta_2))$ where

$$\frac{J_{j,k}^{n+1} - J_{j,k}^n}{\Delta t} + \frac{J_{j,k}^n - J_{j-1,k}^n}{\Delta a} + g_j \frac{J_{j,k}^n - J_{j,k-1}^n}{\Delta m}$$
$$+ \left(\delta_j + w(a, m)\right) J_{j,k}^n = 0,$$
$$\times \frac{g^{n+1}\exp(i(j\theta_1 + k\theta_2)) - g^n \exp(i(j\theta_1 + k\theta_2))}{\Delta t}$$
$$+ \frac{g^n \exp(i(j\theta_1 + k\theta_2)) - g^n \exp(i((j-1)\theta_1 + k\theta_2))}{\Delta a}$$
$$+ g_j \frac{g^n \exp(i(j\theta_1 + k\theta_2)) - g^n \exp(i(j\theta_1 + (k-1)\theta_2))}{\Delta m}$$
$$+ \left(\delta_j + w(a, m)\right) g^n \exp(i(j\theta_1 + k\theta_2)) = 0,$$

$$\times g^n \exp\Big(i(j\theta_1 + k\theta_2)\Big)$$
$$\times\left(\frac{g-1}{\Delta t} + \frac{1-\mathrm{e}^{-i\theta_1}}{\Delta a} + g_j\frac{1-\mathrm{e}^{-i\theta_2}}{\Delta m} + (\delta_j + w(a,m))\right) = 0.$$

Now we solve for the amplification factor g, and replace $\lambda_a = \dfrac{\Delta t}{\Delta a}$ and $\lambda_m = \dfrac{\Delta t}{\Delta m}$ such that

$$g = 1 - \Delta t\Bigg(\frac{1-\mathrm{e}^{-\mathrm{i}\theta_1}}{\Delta a} + g_j\frac{1-\mathrm{e}^{-\mathrm{i}\theta_2}}{\Delta m} + \big(\delta_j + w(a,m)\big)\Bigg),$$
$$= 1 - \Bigg(\lambda_a(1-\mathrm{e}^{-\mathrm{i}\theta_1}) + g_j\lambda_m(1-\mathrm{e}^{-\mathrm{i}\theta_2}) + \Delta t\big(\delta_j + w(a,m)\big)\Bigg).$$

Recall that

$$w(a,m) = \Big(1 - \exp\big(-\alpha_1 a^{n_1}\big)\Big)\Big(1 - \exp\big(-\alpha_2 m^{n_2}\big)\Big).$$

As we know that $w(a,m) \le 1$, we can say the following:

$$\left|1 - \Bigg(\lambda_a(1-\mathrm{e}^{-i\theta_1}) + g_j\lambda_m(1-\mathrm{e}^{-\mathrm{i}\theta_2}) + \Delta t\big(\delta_j + w(a,m)\big)\Bigg)\right|$$
$$\le \left|1 - \Bigg(\lambda_\mathrm{a}\Big(1-\mathrm{e}^{-\mathrm{i}\theta_1}\Big) + g_j\lambda_m\Big(1-\mathrm{e}^{-\mathrm{i}\theta_2}\Big)\Bigg)\right| + \Delta t(\delta_j + 1).$$

Let $\hat{g} = 1 - \Bigg(\lambda_a\Big(1-\mathrm{e}^{-\mathrm{i}\theta_1}\Big) + g_j\lambda_m\Big(1-\mathrm{e}^{-\mathrm{i}\theta_2}\Big)\Bigg)$. If $|\hat{g}|^2 \le 1$, then $g \le 1 + c\Delta t$ where $c = \delta_j + 1$. Note that $\hat{g}$ is the exact same as what we had previously. Thus, we already bounded this by one, so this implies that the initial value problem for the juvenile difference scheme is stable as long as $\lambda_a + \gamma_a\lambda_m \le 1$. Since the amplification matrix is diagonal, we can say that the difference scheme for the initial value problem is stable.[74] □

Theorem 6.3. *The forward time-backward space difference scheme for our initial value problem found in Eqs.* (6.1)–(6.6) *is convergent when* $\lambda_a + \gamma_a\lambda_m \le 1$.

Proof. From Theorem 6.1, the difference scheme is consistent. From Theorem 6.2, the difference scheme is stable if $\lambda_a + \gamma_a \lambda_m \leq 1$. Thus, from the Lax–Richtmyer equivalence theorem, the forward time–backward space difference scheme for our initial value problem is convergent when $\lambda_a + \gamma_a \lambda_m \leq 1$. □

6.4 Numerical Results

We first determined an optimal parameter set by fitting the model assuming decaying resource, $r_1(t)$, in the growth function. We did so to compare to data where the experimental setup began with a fixed amount of food which the larvae use over time. Once we determined these parameters, we assessed their sensitivity. Then we compared the incorporation of resources into the growth and death functions using both decay of resources and cyclic resources. Finally, we assessed the impact of resource cycle length.

6.4.1 *Optimal parameter sets*

Our model replicated the mass and development time distributions observed in the data. In the high density setting (Figure 6.1, top row), the data had more variation in development time as compared to low density (bottom row). The model was capable of reproducing both scenarios. The total development time, including an assumed three days to hatch, in the high density setting was observed to be approximately 9–24 days (top left panel). The model with parameters fit for the high density setting showed a development time from 6 to more than 26 days. In the low density setting, the data were highly clustered with development time primarily between 8 and 13 days. The model with parameters fit to the low density setting including a forced three-day hatching period predicts development time primarily between 6 and 16 days with values as low as 5 days. The model results covered the regions of development time and age where the data were observed, but were unable to capture strong clustering and were more dispersed. This dispersed pattern was in part the product of the use of the numerical difference scheme.

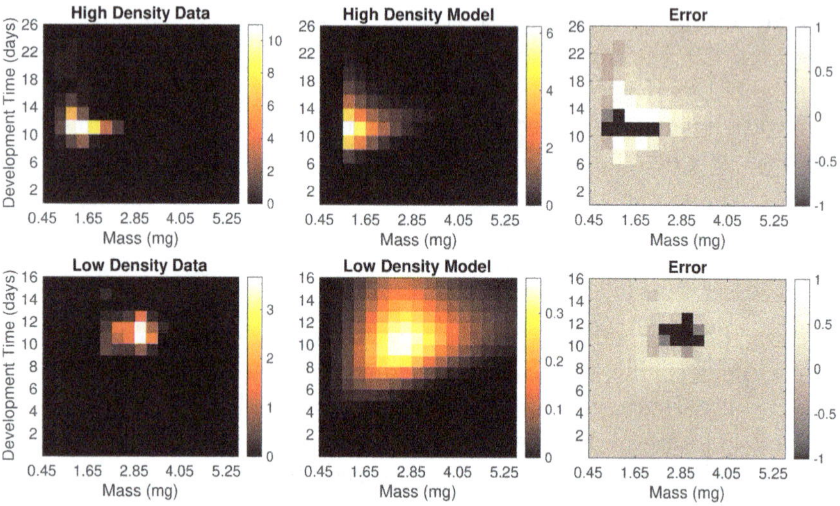

Fig. 6.1. Comparison of data and model fit for high and low density settings separately. The top row shows high density and bottom row low density. The left column displays the data as discretized by the histogram function in MATLAB by mass (mg) and development time (days). The middle column shows the best fit model output discretized in the same way as the data. The final column shows the absolute difference between the first two columns. Note that the scales are different for each panel, so colors are not comparable across panels.

Next, we determined a single parameter set using the combined low and high density data by fitting the model using a weighted average with the combined error accounting for two times the error compared to the low density data plus the error compared to the high density data (weighted to produce roughly equal contributions) of the data from the low and high density settings. A single parameter set was important so that as resources change, the system can transition smoothly between low and high density settings. The optimal model output with the parameter set from the combined error from both datasets did not show the same patterns as results from the parameter set determined using only the low density data. While the model results for development time fitted for the low density data were much more spread out, the model output for the parameters fitted to the combined data were more clustered in the low density setting (Figure 6.2). With parameters fit to the data sets separately or combined, the outputs covered the same regions as the data (Figures 6.1 and 6.2).

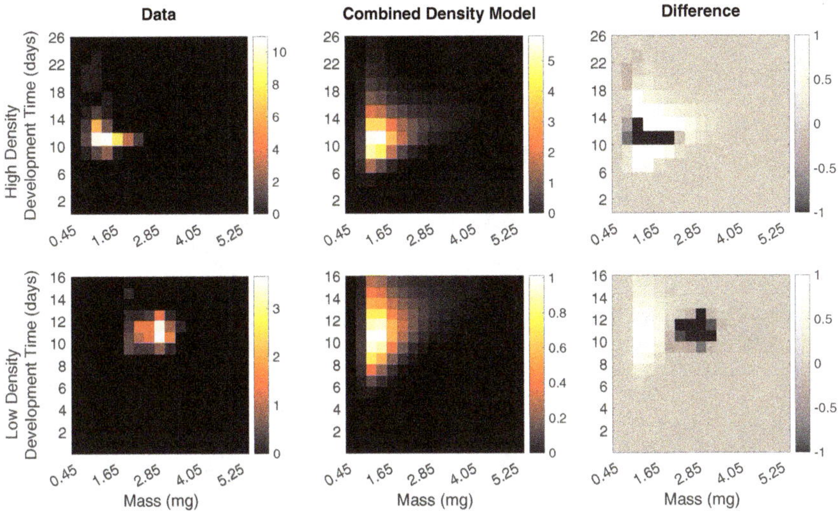

Fig. 6.2. Comparison of data and model fit for high and low density settings together. The top row shows high density and bottom row low density. The left column displays the data as discretized by the histogram function in MATLAB by mass (mg) and development time (days). The middle column shows the best fit model output discretized in the same way as the data with different initial conditions for high and low density settings. The final column shows the absolute difference between the first two columns. Note that the scales are different for each panel, so colors are not comparable across panels.

6.4.2 *Fitted parameter sensitivity*

The parameters with the most sensitivity were the juvenile growth parameter, γ_j, and the exponent for age in the maturation function, n_1. This was true for both the results fit to the high and low density settings separately as well as fit to the combined data (Figures 6.3 and 6.4). In the high density setting, variation in the exponent for age in the maturation function, n_1, resulted in the largest changes in error (Figure 6.3, top). The error increased by more than 40% when n_1 was increased or decreased by 20%. When the juvenile growth parameter, γ_j, was decreased or increased by 20%, a change in error was observed similar to what was found when increasing n_1. The results for the low density setting were most influenced by the juvenile growth parameter, γ_j, and by decreasing the exponent for age in the maturation function, n_1 (Figure 6.3, bottom). Additionally, the low density results were affected by a decrease of 20% in the

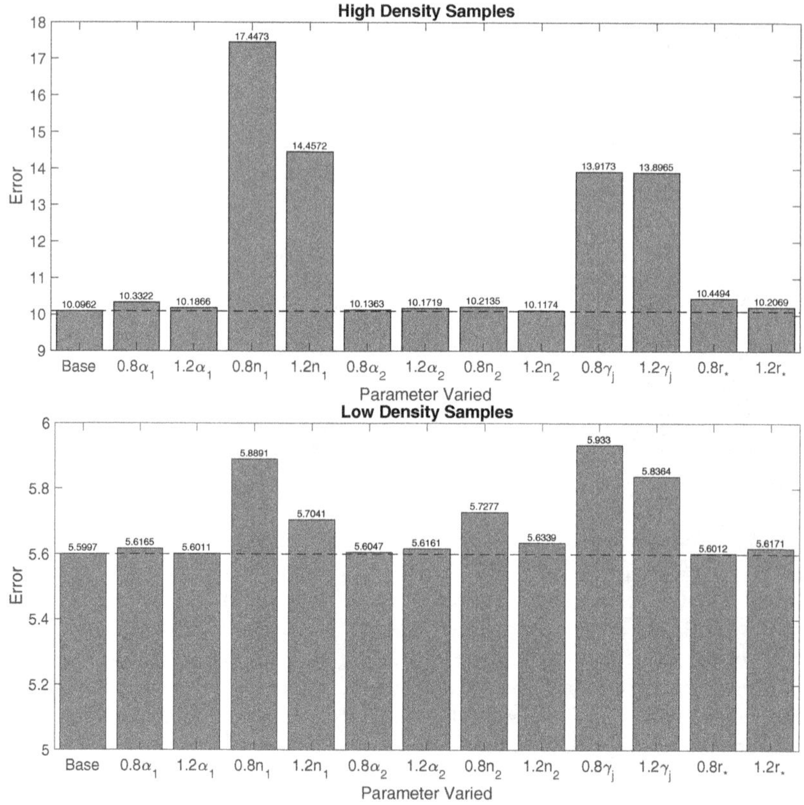

Fig. 6.3. Changes to error with univariate sensitivity for low and high density settings. The parameter variation is shown on the x-axis. Parameters were decreased (0.8) or increased (1.2) by 20% of the standard value determined from the original fits. The top row is the error in comparison to the high density data; the bottom row is the error in comparison to the low density data.

exponent for mass in the maturation function, n_2, but not when n_2 was increased. Correspondingly, the error in the combined low and high density data was affected the most by decreasing the juvenile growth parameter, γ_j, and varying the exponent for age in the maturation function, n_1 (Figure 6.2). Results when varying the other parameters showed very little change in the error.

6.4.3 *Decaying resources altering growth and death*

With decaying resources, the mass distribution was affected strongly by including resource in the growth function, but there was no

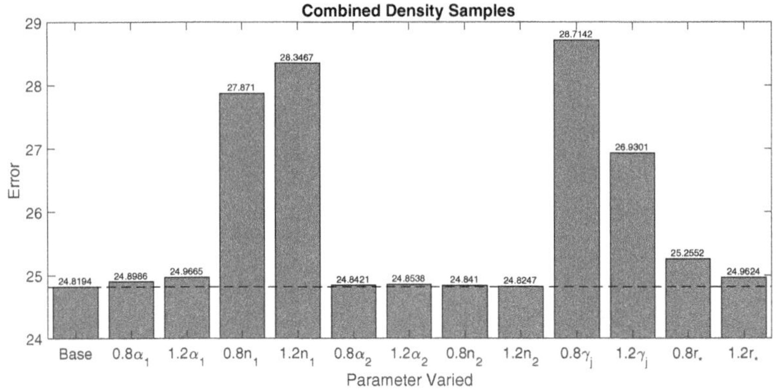

Fig. 6.4. Changes to error with uni-variate sensitivity for the combined data set. The parameter variation is shown on the x-axis. Parameters were decreased (0.8) or increased (1.2) by 20% of the standard value determined from the original fits.

effect when resource was incorporated into the death function (Figure 6.5). The initial mass distribution had a peak around 1.1 mg, a median around 1.6 mg, and a long right tail. The initial distribution was determined under the assumption that resources were plentiful ($r(t) = 1$), which would represent ideal conditions for larvae and produce larger adults. When modeling resource-dependent growth, as resources decay the median shifts lower and the proportion skewed to higher masses decreases. Initially, the median mass dropped before settling to a low value. Over time, the distribution became more symmetric. Eventually, the model predicted unrealistically small masses for adults; however, the density had dropped below one by this time, indicating the loss of the population in a true biological setting.

Resource-dependent growth altered the age distribution (Figure 6.6, left column). The median of the age distribution increased as the resource decayed. This was primarily because the initial cohort of adults was aging, while the rate individuals emerged occurred more slowly. With the fastest decay rate ($\rho = 0.20$), the initial cohort aged until death, and the second cohort was significantly smaller. Once this initial cohort died out, there were virtually no adults, and the ones that did exist were very small (Figure 6.5). Thus, although the age distribution in this setting shifted, it only represented very few total adults. Compared to resource-dependent

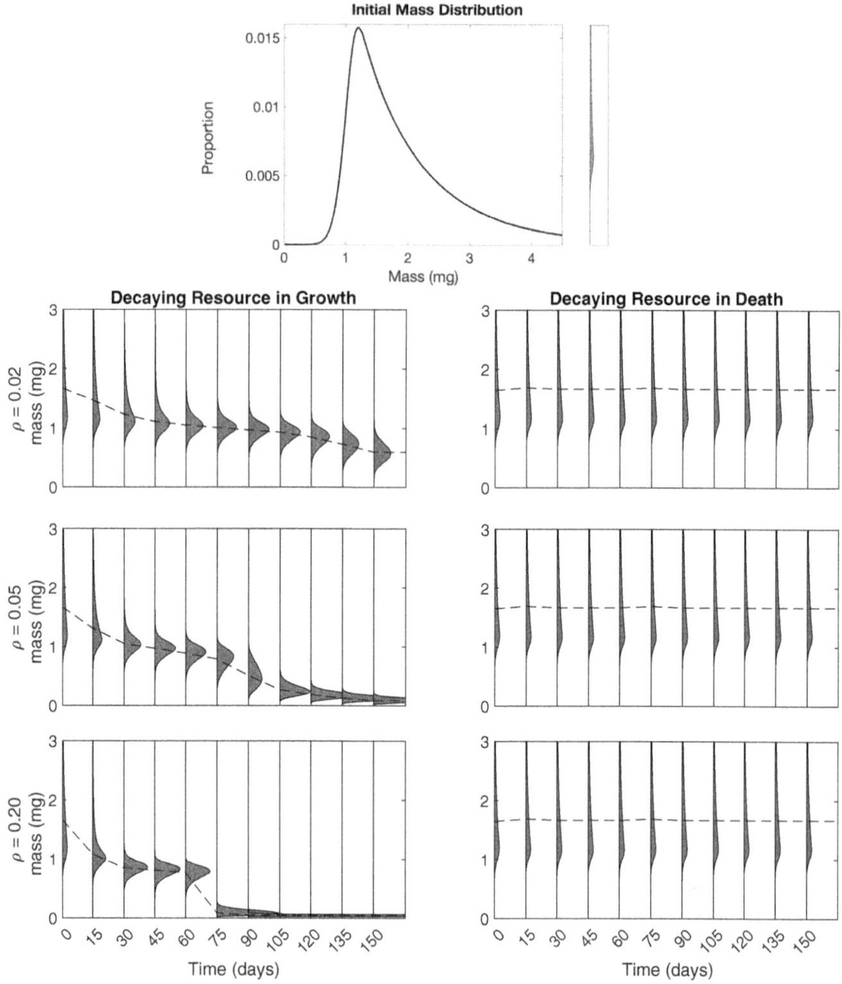

Fig. 6.5. Mass distribution of female adult mosquitoes with decaying resources. The top panel shows the initial mass (mg) distribution with the corresponding vertical representation. The colored areas show mass distribution at specific time points in days (x-axis), summed over all age groups. The y-axis shows the mass (mg) and the distributions indicate relative proportion at a particular mass. The left column shows inclusion of resource in the growth function for juveniles and the right shows inclusion in the death function of juveniles. Each row shows a different resource decay rate, ρ: 0.02 (top), 0.05 (middle), and 0.20 (bottom). The dashed line indicates the median of the distributions. The distributions are scaled to the same possible maximum, to be comparable across panels. The distribution at time 0 is identical for all panels.

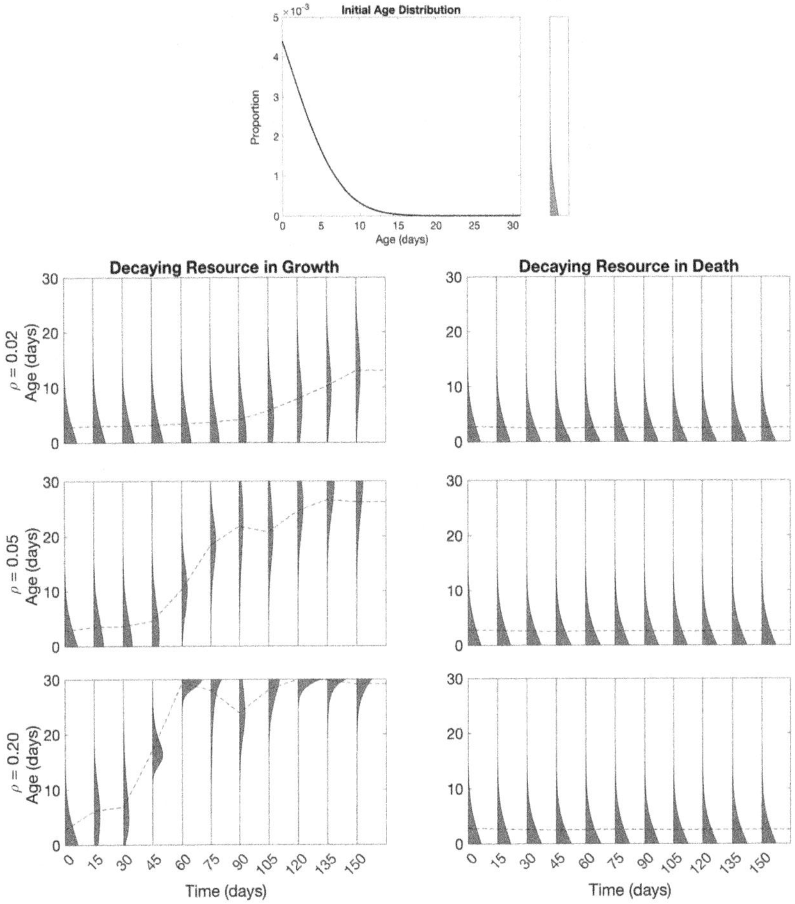

Fig. 6.6. Age distribution of female adult mosquitoes with decaying resources. The top panel shows the initial age distribution in days with the corresponding vertical representation. The colored areas show age distribution at specific time points in days (x-axis), summed over all mass groups. The y-axis shows the age and the distribution indicates relative proportion at a particular age. The left column shows inclusion of resource in the growth function for juveniles and the right shows inclusion in the death function of juveniles. Each row shows a different resource decay rate, ρ: 0.02 (top), 0.05 (middle), and 0.20 (bottom). The dashed line indicates the median of the distributions. The distributions are scaled to the same possible maximum, to be comparable across panels. The distribution at time 0 is identical for all panels.

growth, the age distribution did not change when the death function was resource-dependent (Figure 6.6, right column).

6.4.4 *Cyclic resources altering growth and death*

The mass distribution shifted with cycles of the resources (Figure 6.7). The median of the adult mass distributions oscillated over time (dashed line). This occurred for both incorporation of resource into the growth and death functions, but was more evident

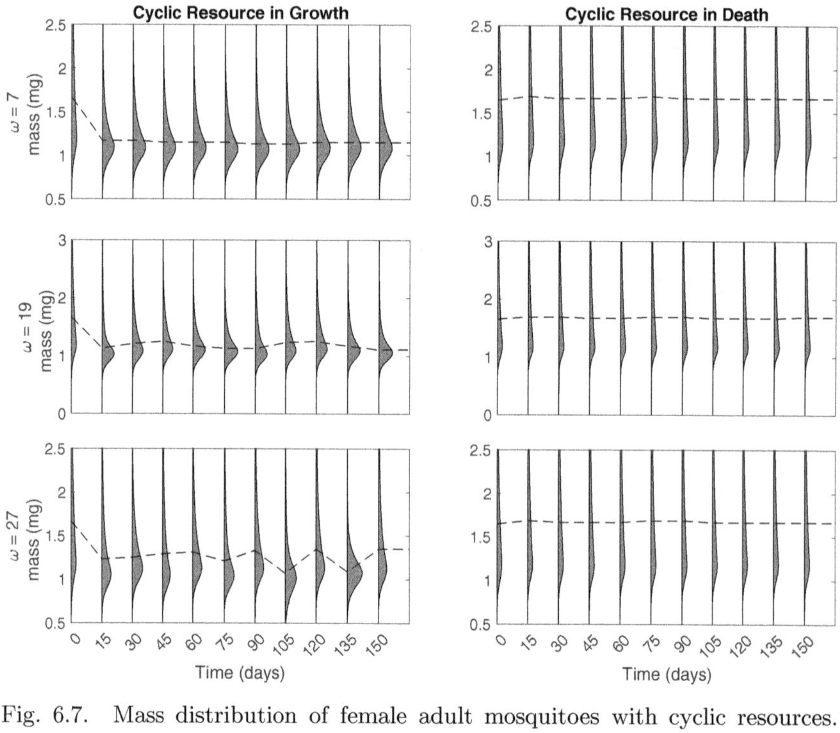

Fig. 6.7. Mass distribution of female adult mosquitoes with cyclic resources. The colored areas show mass distribution at specific time points in days (x-axis), summed over all age groups. The y-axis shows the mass (mg) and the distributions indicate relative proportion at a particular mass. The left column shows inclusion of resource in the growth function for juveniles and the right shows inclusion in the death function of juveniles. Each row shows a different resource cycling period, ω: 7 (top), 19 (middle), and 27 (bottom). The dashed line indicates the median of the distributions. The distributions are scaled to the same possible maximum, to be comparable across panels. The distribution at time 0 is identical for all panels.

with a resource-dependent growth function. As the period increased (lower rows), the oscillations in the median of the mass distributions became larger. For resource-dependent growth (left column), the initial distribution always had the highest median value as it was determined with ideal resource availability. Thus, the median value fell before beginning to oscillate.

The age distribution with resource dependence followed the cyclic nature of the resource function (Figure 6.8). Similar to changes in the mass distributions, this was more evident for resource-dependent growth but also occurred for resource-dependent death, where the

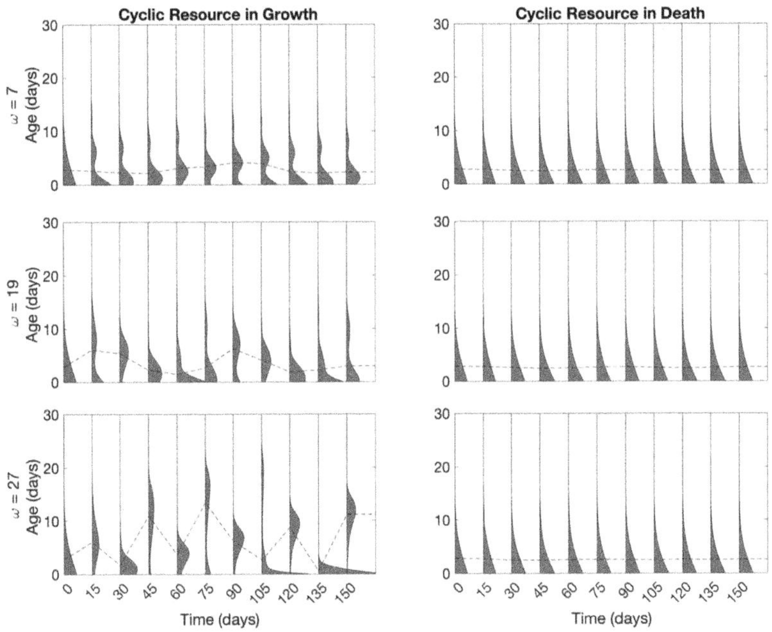

Fig. 6.8. Age distribution of female adult mosquitoes with cyclic resources. The colored areas show age distribution at specific time points in days (x-axis), summed over all mass groups. The y-axis shows the age in days and the distributions indicate relative proportion at a particular age. The left column shows inclusion of resource in the growth function for juveniles and the right shows inclusion in the death function of juveniles. Each row shows a different resource cycling period, ω: 7 (top), 19 (middle), and 27 (bottom). The dashed line indicates the median of the distributions. The distributions are scaled to the same possible maximum, to be comparable across panels. The distribution at time 0 is identical for all panels.

age distribution was only marginally affected. The initial age distribution was heavily weighted towards younger individuals. Through the course of the simulations, bursts of individuals emerged at various time points, which was related to fluctuations in resources. As the resource cycling period lengthened, there were more dramatic shifts in the age distribution. When $\omega = 7$, the median age never rose above 5 days, indicating a population of mostly young individuals. With a long period ($\omega = 27$), more than 50% of the mosquitoes were older than 10 days at multiple points. In such cases, individuals born during times of high resources thrived for long periods, while those born during times with poor resources were delayed in growth and died before emergence.

6.4.5 *Effect of cycle length with cyclic resources*

Longer cycles of resources produced larger mosquitoes (Figure 6.9). When the resources fluctuated over a period of less than 13 days, very few time points had mosquitoes with mass greater than 2.5 mg. In the data, the majority of the mosquitoes developed between 5 and 15 days (Figure 6.2 including 3 days for eggs), meaning that almost all mosquitoes would have experienced at least one full cycle of resources. With resources that had a period between 13 and 23 days, the median mass increased, demonstrating it was more common to have larger individuals. With cycles of 23 days or greater, more than 5% of the population had mass above 2.5 mg in more than 50% of the time points. This indicated that some juveniles developed entirely in a rich resource setting and others entirely in a poor resource setting. Additionally, mosquitoes that were greater than 1.5 mg constituted more than a third of the population for more 50% of the time points for resource cycles greater than 25 days.

The quartiles of the proportion of the population above a particular mass cut-off (either 1.5 or 2.5 mg) spread as the cycle lengthened. Along with this spreading, the median value for the proportion of large mosquitoes increased with the cycle length. The maximum proportion of the population greater than 1.5 mg was consistently found to be 0.59 over all cycle numbers (Figure 6.9, bottom row). Considering mosquitoes above 2.5 mg, the maximum proportion was consistently found to be 0.22, again across all cycle numbers (middle row).

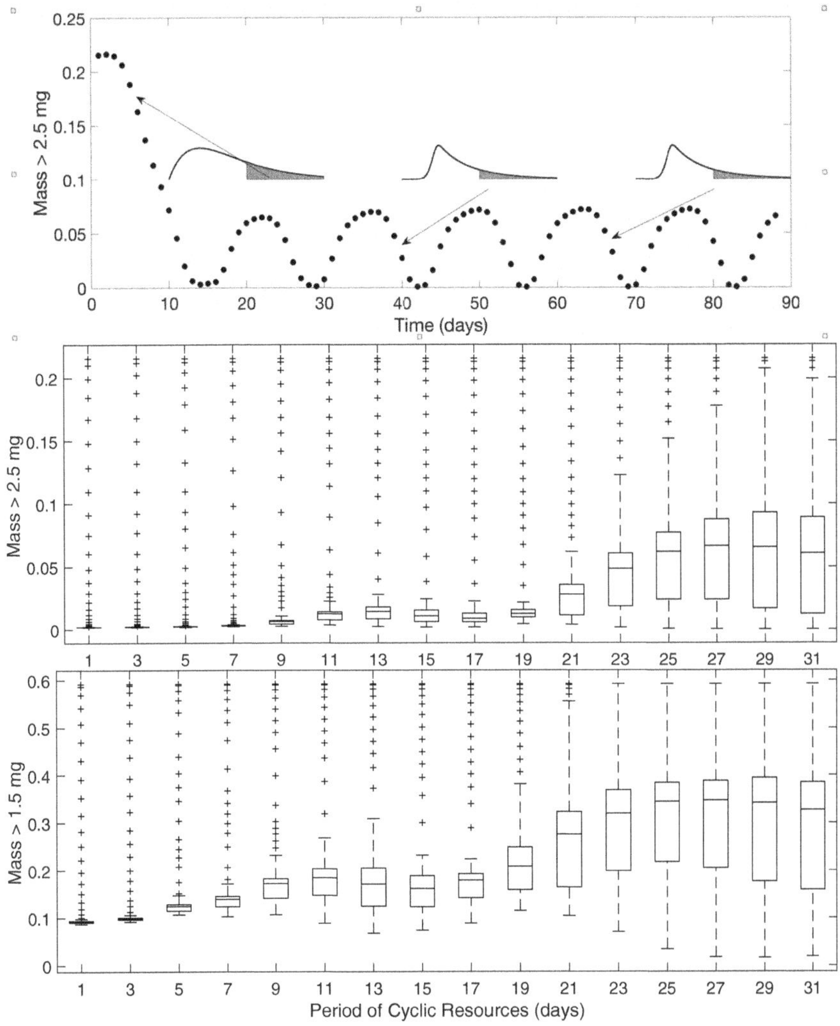

Fig. 6.9. Proportion of larger mosquitoes with cyclic resources in the growth function. The top row is a schematic representation of how the y-axis values for the box plots are determined. The proportion of the population above a mass cut-off is plotted through time. Each box plot incorporates all the 176 times from these temporal plots for a given resource cycle period (days). The x-axis shows the period of the cyclic resource, and the y-axis is the distribution of the proportion of individuals that are larger than 2.5 mg (middle row), and 1.5 mg (bottom row), as described above. Note that the y-axis scale for mass greater than 2.5 mg is half that for mass greater than 1.5 mg.

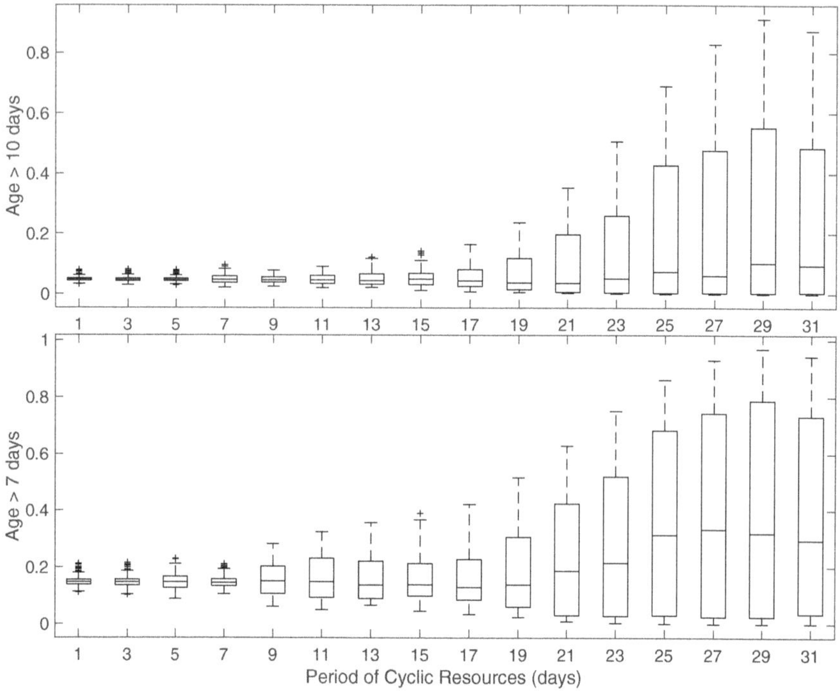

Fig. 6.10. Proportion of older mosquitoes with cyclic resources in the growth function. The x-axis shows the period of the cyclic resource (days), and the y-axis is the distribution of the proportion of individuals that are older than 10 days (top row) and 7 days (bottom row). Contributions to the box plot are as described in Figure 6.9.

Older mosquitoes were more common when the resource cycles were long (Figure 6.10). As with mass, the range of the proportion of older mosquitoes increased as the resource cycle lengthened. In contrast to mass, the median of the proportion of older individuals did not change significantly. With short cycles, the median proportion for mosquitoes older than 7 and 10 days was 0.15 and 0.05, respectively. This was approximately the same for cycles of length 1 through 21. With resource cycles greater in length, there was a slightly greater proportion of older individuals with approximately 0.06–0.10 of the population being older than 10 days. We see an increase in the median of those older than 7 and then a slight decrease with the maximum median of 0.33 of the population being older than 7 days with a cycle of 27 days. With longer cycles, there was more fluctuation of the proportion of older individuals through time, as demonstrated by the larger interquartile range (Figure 6.10).

6.5 Discussion

The age and size of adult mosquitoes impacts their ability to transmit disease. Mosquitoes are unable to spread diseases until they themselves become infected with a pathogen and this pathogen successfully survives its incubation period.[36] Thus, a population of mainly young mosquitoes will transmit fewer pathogens than an older population,[37] and disease will be less likely to spread. Additionally, large mosquitoes are often more fit.[34] Not only are they more able to successfully find a blood meal, they may seek hosts earlier, as they do not need to build a reserve of energy. In this chapter, we examined how resource dependence alters the age and mass distribution of adult mosquitoes.

When larvae lack nutrition, emerging adult females have higher mortality and lower fecundity.[26, 29–35] Using a PDE model, we showed that resource dependence in the juvenile stage creates non-intuitive changes in the adult population. In particular, the inclusion of resource dependence in the juvenile growth function led to both smaller and larger mosquitoes. We found that larger mosquitoes dominated the population when resource levels were high throughout the majority of the juvenile lifespan (Figure 6.9). In comparison, when resource levels fluctuated on a short time scale, restricting the resources available during the larval stage, the population was more often composed of smaller mosquitoes. These changes to the population could impact a control method that targets larval sites. Importantly, it is possible that reductions in larval population size may unintentionally boost the resources for a group of larvae, resulting in larger mosquitoes. In our modeling work, we observed such a phenomena when incorporating longer cycles of resources.

Previous research showed that the size and age of juveniles is an important indicator of when individuals emerge as adults.[20, 45] Our study of the growth of juveniles showed impacts not only on the mass of adults, but also the age distribution (Figures 6.5–6.8). We found that changes in the growth function, through resource dependence, strongly affected the mass and age distributions of adults. With decaying resources, the mass distribution shifted to small individuals and the age distribution exhibited large increases with oscillations. With cyclic resources, the mass and age distributions oscillated with fluctuations in the resources, but with less variation than seen with decaying resources. In contrast to changes observed when resource levels impacted growth, resource-dependent mortality had little effect on the age distribution.

This model does not include several important complexities known to impact mosquito dynamics. Larval density dependence is excluded to avoid the analytical difficulties of a complicated, nonlinear PDE. As density has previously been shown to be important in larval development,[22,53,75] we use time-dependent resources as a proxy for density dependence. This is not perfect as resource levels should be a function of the number of individuals; in particular, the level of resources available per larvae is important. For example, larger larvae populations decrease resource levels faster. Another limitation of the model is the lack of temperature dependence. It has been observed that temperature strongly influences the life history traits of mosquitoes.[76,77] While vector traits are influenced by temperature, we hypothesize that the inclusion of temperature would only accentuate the changes in age and mass already observed in our model. We expect that temperature extremes would have a significant impact on dynamics as they alter mosquito life history traits as well as resource levels. A third limitation is the consideration of only a single mosquito species, *Ae. aegypti.* Multiple mosquito species are often found in the same geographic regions, and even within the same local larval habitats.[78] In such a context, it will be important to consider competition between species during the larval stage, and impacts on age and mass at emergence, which will likely have effects on overall population dynamics and disease transmission. As multiple different *Aedes* species are known to coexist in areas with potentially different competitive interactions, a detailed examination is warranted.

Effectively reducing mosquito populations is expected to reduce disease spread, but mosquitoes are notoriously difficult to control due to their complicated life cycle involving multiple stages growing in different environments. Many control methods focus on reducing mosquito population size, e.g. Refs. 11 and 79, but vector control methods may have unintended outcomes. For example, Sanchez *et al.*

found that given selective vector control measures, vastly different vector densities were possible.[80] Such studies focus on simpler formulations of the mosquito life cycle, often with minimal complexity to the larval dynamics. Our model demonstrates that larval growth dynamics are important for determining features of the resulting adult population, such as age and mass distributions. Moreover, these distributions are affected by changes in resources over time in complex, nonlinear ways. Thus, larval growth should be considered carefully when estimating the effectiveness of mosquito control methods.

Acknowledgments

We thank C. Vinauger, K. Chandrasegaran, and O. F. Prosper for helpful conversations. LC and MW received support from Virginia Tech via a Center for Emerging, Zoonotic, and Arthropod-borne Pathogens grant. LC and MW were supported by NSF Grant #1853495. The funders had no role in study design, data collection and analysis, decision to publish, or preparation of the manuscript.

References

1. Mbacham, W. F., Ayong, L., Guewo-Fokeng, M., and Makoge, V. (2019). Current situation of malaria in Africa, in *Malaria Control and Elimination*, pp. 29–44. Springer, New York, NY.
2. Stanaway, J. D., Shepard, D. S., Undurraga, E. A., Halasa, Y. A., Coffeng, L. E., Brady, O. J., Hay, S. I., Bedi, N., Bensenor, I. M., Castañeda-Orjuela, C. A., *et al.* (2016). The global burden of dengue: An analysis from the global burden of disease study 2013, *Lancet Infect. Dis.* **16**, 6, 712–723.
3. Hung, T. M., Wills, B., Clapham, H. E., Yacoub, S., and Turner, H. C. (2019). The uncertainty surrounding the burden of post-acute consequences of dengue infection, *Trends Parasitol.* **35**, 9, 673–676.
4. World Health Organization Vector-borne diseases. https://www.who.int/news-room/fact-sheets/detail/vector-borne-diseases. Accessed 1 December 2021.
5. Shepard, D. S., Undurraga, E. A., Halasa, Y. A., and Stanaway, J. D. (2016). The global economic burden of dengue: A systematic analysis, *Lancet Infect. Dis.* **16**, 8, 935–941.

6. Göertz, G. P., Vogels, C. B., Geertsema, C., Koenraadt, C. J., and Pijlman, G. P. (2017). Mosquito co-infection with Zika and chikungunya virus allows simultaneous transmission without affecting vector competence of *Aedes aegypti*, *PLoS Neglect. Trop. Dis.* **11**, 6, e0005654.
7. McKenzie, B. A., Wilson, A. E., and Zohdy, S. (2019). *Aedes albopictus* is a competent vector of zika virus: A meta-analysis, *PLOS One* **14**, 5, e0216794.
8. Mweya, C. N., Kimera, S. I., Kija, J. B., and Mboera, L. E. (2013). Predicting distribution of *Aedes aegypti* and *Culex pipiens* complex, potential vectors of rift valley fever virus in relation to disease epidemics in East Africa, *Infect. Ecol. Epidemiol.* **3**, 1, 21748.
9. Napp, S., Petrić, D., and Busquets, N. (2018). West Nile virus and other mosquito-borne viruses present in Eastern Europe, *Pathogens Global Health* **112**, 5, 233–248.
10. Vega-Rúa, A., Marconcini, M., Madec, Y., Manni, M., Carraretto, D., Gomulski, L. M., Gasperi, G., Failloux, A.-B., and Malacrida, A. R. (2020) Vector competence of *Aedes albopictus* populations for chikungunya virus is shaped by their demographic history, *Commun. Biol.* **3**, 1, 1–13.
11. World Health Organization and UNICEF (2017). Global vector control response 2017-2030. https://apps.who.int/iris/bitstream/handle/10665/259205/9789241512978-eng.pdf. Accessed 29 October 2021.
12. Chung, H.-N., Rodriguez, S. D., Gonzales, K. K., Vulcan, J., Cordova, J. J., Mitra, S., Adams, C. G., Moses-Gonzales, N., Tam, N., Cluck, J. W., *et al.* (2018). Toward implementation of mosquito sterile insect technique: The effect of storage conditions on survival of male *Aedes aegypti* mosquitoes (Diptera: *Culicidae*) during transport, *J. Insect Sci.* **18**, 6, 2.
13. Ritchie, S. A., van den Hurk, A. F., Smout, M. J., Staunton, K. M., and Hoffmann, A. A. (2018). Mission accomplished? We need a guide to the 'post release' world of *Wolbachia* for *Aedes*-borne disease control, *Trends Parasitol.* **34**, 3, 217–226.
14. Chakradhar, S. (2015). Buzzkill: Regulatory uncertainty plagues rollout of genetically modified mosquitoes, *Nat. Med.* **21**, 416–419.
15. Noden, B. H., O'Neal, P. A., Fader, J. E., and Juliano, S. A. (2016). Impact of inter-and intra-specific competition among larvae on larval, adult, and life-table traits of *Aedes aegypti* and *Aedes albopictus* females, *Ecol. Entomol.* **41**, 2, 192–200.
16. Telang, A., Qayum, A., Parker, A., Sacchetta, B., and Byrnes, G. (2012). Larval nutritional stress affects vector immune traits in adult

yellow fever mosquito *Aedes aegypti* (*Stegomyia aegypti*), *Med. Vet. Entomol.* **26**, 3, 271–281.
17. Souza, R. S., Virginio, F., Riback, T. I. S., Suesdek, L., Barufi, J. B., and Genta, F. A. (2019). Microorganism-based larval diets affect mosquito development, size and nutritional reserves in the yellow fever mosquito *Aedes aegypti* (Diptera: Culicidae), *Front. Physiol.* **10**, 152.
18. Merritt, R. W., Dadd, R., and Walker, E. D. (1992). Feeding behavior, natural food, and nutritional relationships of larval mosquitoes, *Annu. Rev. Entomol.* **37**, 1, 349–374.
19. Yee, D. A., Kesavaraju, B., and Juliano, S. A. (2004). Larval feeding behavior of three co-occurring species of container mosquitoes, *J. Vector Ecol.* **29**, 2, 315.
20. Padmanabha, H., Correa, F., Legros, M., Nijhout, H. F., Lord, C., and Lounibos, L. P. An eco-physiological model of the impact of temperature on *Aedes aegypti* life history traits, *J. Insect Physiol.* **58**, 12, 1597–1608.
21. Telang, A., Frame, L., and Brown, M. R. (2007). Larval feeding duration affects ecdysteroid levels and nutritional reserves regulating pupal commitment in the yellow fever mosquito *Aedes aegypti* (Diptera: *Culicidae*), *J. Exp. Biol.* **210**, 5, 854–864.
22. Couret, J., Dotson, E., and M. Q. (2014). Benedict Temperature, larval diet, and density effects on development rate and survival of *Aedes aegypti* (Diptera: *Culicidae*), *PLoS One* **9**, 2, e87468.
23. Delatte, H., Gimonneau, G., Triboire, A., and Fontenille, D. (2009). Influence of temperature on immature development, survival, longevity, fecundity, and gonotrophic cycles of *Aedes albopictus*, vector of chikungunya and dengue in the Indian Ocean, *J. Med. Entomol.* **46**(1), 33–41.
24. Foster, W. A. (1995). Mosquito sugar feeding and reproductive energetics, *Annu. Rev. Entomol.* **40**, 1, 443–474.
25. Judson, C. L. (1967). Feeding and oviposition behavior in the mosquito *Aedes aegypti* (L.). I. Preliminary studies of physiological control mechanisms, *Biol. Bull.* **133**, 2, 369–377.
26. Farjana, T. and Tuno, N. (2013). Multiple blood feeding and host-seeking behavior in *Aedes aegypti* and *Aedes albopictus* (Diptera: *Culicidae*), *J. Med. Entomol.* **50**, 4, 838–846.
27. Maciel-de Freitas, R., Codeco, C., and Lourenço-de Oliveira, R. (2007). Body size-associated survival and dispersal rates of *Aedes aegypti* in Rio de Janeiro, *Med. Vet. Entomol.* **21**, 3, 284–292.
28. Nasci, R. S. (1991). Influence of larval and adult nutrition on biting persistence in *Aedes aegypti* (Diptera: *Culicidae*), *J. Med. Entomol.* **28**, 4, 522–526.

29. Briegel, H. Metabolic relationship between female body size, reserves, and fecundity of *Aedes aegypti*, *J. Insect Physiol.* **36**, 3, 165–172.
30. Chandrasegaran, K., Kandregula, S. R., Quader, S., and Juliano, S. A. (2018). Context-dependent interactive effects of non-lethal predation on larvae impact adult longevity and body composition, *PLoS One* **13**, 2, e0192104.
31. Jeffrey Gutiérrez, E. H., Walker, K. R., Ernst, K. C., Riehle, M. A., and Davidowitz, G. (2020). Size as a proxy for survival in *Aedes aegypti* (Diptera: *Culicidae*) mosquitoes, *J. Med. Entomol.* **57**, 1228–1238.
32. Reiskind, M. and Lounibos, L. (2009). Effects of intraspecific larval competition on adult longevity in the mosquitoes *Aedes aegypti* and *Aedes albopictus*, *Med. Vet. Entomol.* **23**, 1, 62–68.
33. Rocha-Santos, C., Dutra, A. C. V., P. L., Fróes Santos, R., Cupolillo, C. D. L. S., de Melo Rodovalho, C., Bellinato, D. F., dos Santos Dias, L., Jablonka, W., Lima, J. B. P., Silva Neto, M. A. C., *et al.* (2021). Effect of larval food availability on adult *Aedes aegypti* (Diptera: *Culicidae*) fitness and susceptibility to Zika infection, *J. Med. Entomol.* **58**, 2, 535–547.
34. Steinwascher, K. (2018). Competition among *Aedes aegypti* larvae, *PLoS One* **13**, 11, e0202455.
35. Yan, J., Kibech, R., and Stone, C. M. (2021). Differential effects of larval and adult nutrition on female survival, fecundity, and size of the yellow fever mosquito, *Aedes aegypti*, *Front. Zool.* **18**, 1, 1–9.
36. Goindin, D., Delannay, C., Ramdini, C., Gustave, J., and Fouque, F. Parity and longevity of *Aedes aegypti* according to temperatures in controlled conditions and consequences on dengue transmission risks, *PLOS One.* **10**(8), e0135489 (2015).
37. Mayton, E. H., Tramonte, A. R., Wearing, H. J., and Christofferson, R. C. (2020). Age-structured vectorial capacity reveals timing, not magnitude of within-mosquito dynamics is critical for arbovirus fitness assessment, *Parasites Vectors.* **13**, 1, 1–13.
38. Styer, L. M., Minnick, S. L., Sun, A. K., and Scott, T. W. (2007). Mortality and reproductive dynamics of *Aedes aegypti* (Diptera: *Culicidae*) fed human blood, *Vector-borne Zoonotic Dis.* **7**, 1, 86–98.
39. Watts, D. M., Burke, D. S., Harrison, B. A., Whitmire, R. E., and Nisalak, A. (1986). Effect of temperature on the vector efficiency of *Aedes aegypti* for dengue 2 virus. Technical report, Army Medical Research Institute of Infectious Diseases, Fort Detrick, MD.
40. Harrington, L. C., Edman, J. D., and Scott, T. W. (2001). Why do female *Aedes aegypti* (Diptera: *Culicidae*) feed preferentially and frequently on human blood? *J. Med. Entomol.* **38**, 3, 411–422.

41. Harrington, L. C., Françoisevermeylen, N., Jones, J. J., Kitthawee, S., Sithiprasasna, R., Edman, J. D., and Scott, T. W. (2014). Age-dependent survival of the dengue vector *Aedes aegypti* (Diptera: Culicidae) demonstrated by simultaneous release–recapture of different age cohorts, *J. Med. Entomol.* **45**, 2, 307–313.
42. Bong, L.-J., Tu, W.-C., and Neoh, K.-B. (2021). Interpopulation variations in life history traits and reproductive tactics in *Aedes aegypti*: A test on populations 50 km apart, *Acta Trop.* **213**, 105750.
43. Arrivillaga, J. and Barrera, R. (2004). Food as a limiting factor for *Aedes aegypti* in water-storage containers, *J. Vector Ecol.* **29**, 11–20.
44. Padmanabha, H., Lord, C., and Lounibos, L. (2011). Temperature induces trade-offs between development and starvation resistance in *Aedes aegypti* (L.) larvae, *Med. Vet. Entomol.* **25**, 4, 445–453.
45. Gilpin, M. E. and GAH, M. (1979). Systems analysis of the yellow fever mosquito *Aedes aegypti*, *Forts. Zool.* **25**, 355–388.
46. Focks, D. A., Haile, D., Daniels, E., and Mount, G. A. (1993). Dynamic life table model for *Aedes aegypti* (Diptera: Culicidae): analysis of the literature and model development, *J. Med. Entomol.* **30**, 6, 1003–1017.
47. Hopp, M. J. and Foley, J. A. (2001). Global-scale relationships between climate and the dengue fever vector, *Aedes aegypti*, *Clim. Change* **48**, 2, 441–463.
48. Magori, K., Legros, M., Puente, M. E., Focks, D. A., Scott, T. W., Lloyd, A. L., and Gould, F. (2009). Skeeter Buster: A stochastic, spatially explicit modeling tool for studying *Aedes aegypti* population replacement and population suppression strategies, *PLoS Neglect. Trop. Dis.* **3**, 9, e508.
49. Aznar, V. R., De Majo, M. S., Fischer, S., Francisco, D., Natiello, M. A., and Solari, H. G. (2015). A model for the development of *Aedes* (Stegomyia) *aegypti* as a function of the available food, *J. Theor. Biol.* **365**, 311–324.
50. Walker, M., Chandrasegaran, K., Vinauger, C., Robert, M. A., and Childs, L. M. (2021). Modeling the effects of *Aedes aegypti*'s larval environment on adult body mass at emergence, *PLoS Comput. Biol.* **17**, 11, e1009102.
51. Lunde, T. M., Korecha, D., Loha, E., Sorteberg, A., and Lindtjørn, B. (2013). A dynamic model of some malaria-transmitting anopheline mosquitoes of the Afrotropical region. I. Model description and sensitivity analysis, *Malaria J.* **12**, 1, 1–29.
52. Lunde, T. M., Bayoh, M. N., and Lindtjørn, B. (2013). How malaria models relate temperature to malaria transmission, *Parasites Vectors* **6**, 1, 1–10.

53. Walker, M., Robert, M., and Childs, L. M. (2020). The importance of density dependence in juvenile mosquito development and survival: A model-based investigation, *Ecol. Model.* **440**, 109357.
54. Amponsah, S. K., Asiedu, B., Avornyo, S. Y., Setufe, S. B., Afranewaa, N. A. B., and Failler, P. (2021). Growth, mortality and exploitation rates of African moonfish (*Selene dorsalis, Gill* 1863) encountered in the coast of Ghana (West Africa), *Eur. J. Biol. Biotechnol* **2**, 2, 6–10 (2021).
55. de Sousa, J. E. R., Façanha, D. A. E., Bermejo, L. A., Ferreira, J., Paiva, R. D. M., Nunes, S. F., and de Souza, M. d. S. M. (2021). Evaluation of non-linear models for growth curve in Brazilian tropical goats, *Trop. Anim. Health Prod.* **53**, 2, 1–6.
56. de Sousa, V. C., Biagiotti, D., Sarmento, J. L. R., Sena, L. S., Barroso, P. A., Barjud, S. F. L., Almeida, M. K. d. S., and Santos, N. P. d. S. (2021). Nonlinear mixed models for characterization of growth trajectory of New Zealand rabbits raised in tropical climate, *Asian-Australasian J. Anim. Sci.* **35**, 648–658.
57. Gilligan-Lunda, E. K., Stich, D. S., Mills, K. E., Bailey, M. M., and Zydlewski, J. D. (2021). Climate change may cause shifts in growth and instantaneous natural mortality of American Shad throughout their native range, *Trans. Am. Fish. Soc.* **150**, 407–421.
58. Jakes-Cota, U., Chavéz-Arellano, R., Sepulveda, C., Aalbers, S., and Ortega-García, S. (2021). Estimating age and growth of roosterfish (*Nematistius pectoralis*) from otoliths, *Fis. Res.* **240**, 105958.
59. Ramirez, M. D., Popovska, T., and Babcock, E. A. (2021). Global synthesis of sea turtle von Bertalanffy growth parameters through Bayesian hierarchical modeling, *Mar. Ecol. Prog. Ser.* **657**, 191–207.
60. Then, A. Y., Hoenig, J. M., Hall, N. G., Hewitt, D. A., and Jardim, H. E. (2015). Evaluating the predictive performance of empirical estimators of natural mortality rate using information on over 200 fish species, *ICES J. Mar. Sci.* **72**, 1, 82–92.
61. Villa-Diharce, E. R., Cisneros-Mata, M. A., Rodríguez-Félix, D., Ramírez-Félix, E. A., and Rodríguez-Domínguez, G. (2021). Molting and individual growth models of *Callinectes bellicosus*, *Fish. Res.* **239**, 105897.
62. Von Bertalanffy, L. (1957). Quantitative laws in metabolism and growth, *Rev. Biol.* **32**, 3, 217–231.
63. Dumas, A., France, J., and Bureau, D. (2010). Modelling growth and body composition in fish nutrition: Where have we been and where are we going? *Aquacult. Res.* **41**, 2, 161–181.
64. Marshall, D. J. and White, C. R. (2019). Have we outgrown the existing models of growth? *Trends Ecol. Evol.* **34**, 2, 102–111.

65. Day, T. and Taylor, P. D. (1997). Von Bertalanffy's growth equation should not be used to model age and size at maturity, *Am. Nat.* **149**, 2, 381–393.
66. Kooijman, S. (2001). Quantitative aspects of metabolic organization: a discussion of concepts, *Philos. Trans. Roy. Soc. Lond. Ser. B Biol. Sci.* **356**, 1407, 331–349.
67. Liu, X., Sha, Z., Wang, C., Li, D., and Bureau, D. P. (2018). A web-based combined nutritional model to precisely predict growth, feed requirement and waste output of gibel carp (*Carassius auratus gibelio*) in aquaculture operations, *Aquaculture* **492**, 335–348.
68. Quince, C., Abrams, P. A., Shuter, B. J., and Lester, N. P. (2008). Biphasic growth in fish I: Theoretical foundations, *J. Theor. Biol.* **254**, 2, 197–206.
69. Sinko, J. W. and Streifer, W. (1967). A new model for age-size structure of a population, *Ecology* **48**, 6, 910–918.
70. Sinko, J. W. and Streifer, W. (1969). Applying models incorporating age-size structure of a population to daphnia, *Ecology* **50**, 4, 608–615.
71. Chandrasegaran, K., Walker, M., Vinauger, C., Robert, M., and Childs, L. (2021). Effects of *Aedes aegypti*'s larval density on adult body mass at emergence. Mendeley Data.
72. Bellan, S. E. (2010). The importance of age dependent mortality and the extrinsic incubation period in models of mosquito-borne disease transmission and control, *PLoS One* **5**, 4, e10165.
73. Rueda, L., Patel, K., Axtell, R., and Stinner, R. (1990). Temperature-dependent development and survival rates of *Culex quinquefasciatus* and *Aedes aegypti* (Diptera: *Culicidae*), *J. Med. Entomol.* **275**, 892–898.
74. Strikwerda, J. C. (2004). *Finite Difference Schemes and Partial Differential Equations*, SIAM, Philadelphia, PA.
75. Hancock, P. A., White, V. L., Callahan, A. G., Godfray, C. H., Hoffmann, A. A., and Ritchie, S. A. (2016). Density-dependent population dynamics in *Aedes aegypti* slow the spread of wMel *Wolbachia*, *J. Appl. Ecol.* **53**, 3, 785–793.
76. El Moustaid, F. and Johnson, L. R. (2019). Modeling temperature effects on population density of the dengue mosquito *Aedes aegypti*, *Insects* **10**, 11, 393.
77. Mordecai, E. A., Cohen, J. M., Evans, M. V., Gudapati, P., Johnson, L. R., Lippi, C. A., Miazgowicz, K., Murdock, C. C., Rohr, J. R., Ryan, S. J., *et al.* (2017). Detecting the impact of temperature on transmission of zika, dengue, and chikungunya using mechanistic models, *PLoS Neglect. Trop. Dis.* **11**, 4, e0005568.

78. Braks, M. A., Honório, N. A., Lourenço-De-Oliveira, R., Juliano, S. A., and Lounibos, L. P. (2003). Convergent habitat segregation of *Aedes aegypti* and *Aedes albopictus* (Diptera: *Culicidae*) in southeastern Brazil and Florida, *J. Med. Entomol.* **40**, 6, 785–794.
79. Rose, R. I. (2001). Pesticides and public health: integrated methods of mosquito management. *Emerg. Infect. Dis.* **7**, 1, 17.
80. Sanchez, F., Engman, M., Harrington, L., and Castillo-Chavez, C. (2006). Models for dengue transmission and control, *Contemp. Math.* **410**, 311–326.

https://doi.org/10.1142/9789811263033_0007

Chapter 7

Novel Hybrid Continuous-Discrete-Time Epidemic Models*

Maia Martcheva[†,¶], **Abdul-Aziz Yakubu**[‡], **and Necibe Tuncer**[§]

[†]*Department of Mathematics, University of Florida, Gainesville, FL, USA*
[‡]*Department of Mathematics, Howard University, Washington DC, USA*
[§]*Department of Mathematical Sciences, Florida Atlantic University, Boca Raton, FL, USA*
[¶]*maia@ufl.edu*

Continuous-time (ODE and PDE) population models are typically used for describing the population dynamics of continuously reproducing populations, such as humans. However, discrete-time (difference equations DE) population models are used for describing the population dynamics of discretely reproducing populations, such as most insect and fish populations. In this chapter, we introduce a novel hybrid continuous-discrete-time epidemic model for describing infectious disease dynamics, where the host is a continuously reproducing species, while the disease vector is a discretely reproducing species. The hybrid continuous-discrete-time epidemic model, for example, can be used to model malaria or Lyme disease infections in human and mosquito or tick populations. In this chapter, we develop analytical methods for analyzing such hybrid models. The methods introduced are then used to compute the hybrid system's reproduction number. The disease-free equilibrium is locally

*This chapter is dedicated to the memory of Abdul-Aziz Yakubu.

asymptotically stable when the reproduction number is less than one and unstable when the reproduction number is more than one. The system is simulated using numerical techniques developed for such hybrid models.

Keywords: discrete models, continuous-discrete time models, reproduction number, two species models

7.1 Introduction

Mathematical modeling is an essential component in linking mathematics to its application disciplines, such as biology.[1] Development of new types of models is of paramount importance for improving the realism of mathematical models. In this chapter, we introduce novel "continuous-discrete-time" models of infectious diseases. Continuous-discrete-time models are not entirely new to the modeling world. In Ref. 2 reviews discrete-continuous-time models in energy and transportation, where these models have been more popular. However, the authors are unaware of hybrid discrete-continuous-time models being used in the context of biology. Most mathematical models in population biology are discrete or continuous. Discrete time models are appropriate for organisms with discrete growth events such as an annual plant or an animal population with breeding seasons[3] (e.g. tigers). Modeling with discrete-time models in population biology has a long history, and many of the popular introductory population and epidemic models have discrete analogues.[4–7] Continuous time models are appropriate for organisms with continuous growth in a non-seasonal environment such as humans (births happen at all times). Continuous growth populations are modeled typically with ODEs or other continuous time differential equation models. Examples of modeling with differential equation models in the infectious disease context abound.[7–9] Theoretically, discrete vs continuous model should be chosen based on the reproductive practices of the population; however, this choice is often made based on the preference of the modelers. An interesting question occurs: what if we model a disease epidemic with two interacting populations which have distinct breeding patterns — one is discrete, and the other is continuous? Thus far this scenario would have been modeled either with a discrete-time or with a continuous-time model. The goal of this chapter is to offer a more realistic approach: hybrid continuous-discrete-time models of two species epidemics. The models we introduce in this chapter are novel, in the sense that such hybrid models have not been previously used in infectious disease context.

Although the models we introduce here are SIR models for a general disease, and our goal is to develop the mathematical methods for those novel models, it is useful to give an example of a disease that might be better represented with such models. The example we want to discuss here is Lyme disease. Lyme disease is the most common vector-borne disease in the United States.[10,11] It is caused by the bacterium *Borrelia burgdorferi* and, in rare instances, *Borrelia mayonii*.[10] Lyme disease has a complex transmission cycle, but it is typically transmitted by blacklegged ticks to various animals, including humans. Blacklegged ticks (*Ixodes scapularis* and *Ixodes pacificus*) have a two-year life cycle. Eggs are laid in the spring, larvae hatch in the summer, and nymphs emerge in the second spring. By the fall, the nymphs mature into adult ticks, and the adult ticks lay eggs the following spring. Transmission of Lyme disease to animals and humans is highest in the spring and summer. Lyme disease has previously been modeled, primarily using ODEs.[12–14] Our novel models will use discrete-time equations for the ticks coupled with a continuous ODE for the humans.

In the next section, we set the stage by presenting singe-host models without vitals including both continuous and discrete SIR models. We take the opportunity to compute the within-species reproduction numbers in both the continuous and discrete cases. Our ultimate goal is to combine these single-species models. In Section 7.3, we introduce a two-species continuous-discrete-time model without vitals. In addition, we develop the mathematical techniques for studying such continuous-discrete-time models analytically. We derive the reproduction number both from first principles and from the next generation approach for discrete-time models whose application is not completely justified but gives the same result. In Section 7.4, we introduce and study a hybrid continuous-discrete-time model of two-species with vitals. We develop the mathematical techniques to obtain and study the stability of the disease-free equilibrium and obtain the endemic equilibrium. In Section 7.5, we present some simulation results with the two hybrid models: the model without vitals and the model with vitals. Finally, in Section 7.6, we summarize our results and draw conclusions.

7.2 Single Host Models without Vitals: Continuous and Discrete-Time

We begin addressing our goal of introducing a mixed continuous-discrete-time model by first presenting a continuous single-species model and a discrete single-species model. In this section, we disregard the vital dynamics, assuming the dynamics occurs on a short-time scale. We start by presenting a simple continuous SIR model without vital dynamics, which was first considered by Kermack and McKendrick in 1927.[15]

7.2.1 *SIR continuous-time model without vital dynamics (Human)*

To introduce the model, let S, I, and R denote susceptible, infected, and recovered individuals, respectively. We assume infected are also infectious. The infectious rate of humans is denoted by α_H, while the recovery rate is denoted by γ_H. For each $t \geq 0$, the model has the following form:

$$\text{SIR model with no vitals:} \quad \begin{cases} \dfrac{\mathrm{d}S}{\mathrm{d}t} = -\alpha_H S I, \\[2mm] \dfrac{\mathrm{d}I}{\mathrm{d}t} = \alpha_H S I - \gamma_H I, \\[2mm] \dfrac{\mathrm{d}R}{\mathrm{d}t} = \gamma_H I. \end{cases} \tag{7.1}$$

The model is equipped with the initial conditions:

$$(S, I, R) = (S(0), I(0), R(0)) \in \mathbb{R}^3_+.$$

The total population at time t is

$$N(t) = S(t) + I(t) + R(t).$$

Adding all three equations of model (7.1), we obtain

$$\frac{\mathrm{d}N}{\mathrm{d}t} = 0.$$

Hence, in model (7.1), the total population is constant and $N(t) = N(0)$ for all $t \geq 0$.

From the first equation of model (7.1), we see that $\frac{\mathrm{d}S}{\mathrm{d}t} \leq 0$ and the number of susceptibles is always declining. When $S(0) > 0$ and $I(0) > 0$, then $\frac{\mathrm{d}S}{\mathrm{d}t} < 0$. From the first and second equations of model (7.1), we obtain

$$\frac{\mathrm{d}S}{\mathrm{d}R} = -\frac{\alpha_H}{\gamma_H} S.$$

Hence, for all $t \geq 0$,

$$S(t) = S(0)\, e^{-\frac{\alpha_H}{\gamma_H} R} \geq S(0)\, e^{-\frac{\alpha_H}{\gamma_H} N} > 0,$$

whenever $S(0) > 0$.

From the second equation of model (7.1), we see that the prevalence first starts increasing if

$$\left.\frac{\mathrm{d}I}{\mathrm{d}t}\right|_{t=0} = (\alpha_H S(0) - \gamma_H)\, I(0) > 0.$$

Hence, the basic reproduction number for model (7.1) is

$$\mathcal{R}_{0H} = \frac{\alpha_H S(0)}{\gamma_H}.$$

That is, $\mathcal{R}_{0H} > 1$ implies an initial increase in the number of infectious individuals before a decline to zero occurs, while $\mathcal{R}_{0H} < 1$ implies an immediate decline to zero as $t \to \infty$.

From the third equation of model (7.1), we see that $\frac{\mathrm{d}R}{\mathrm{d}t} \geq 0$ and the number of recovered individuals is always increasing. When $I(0) > 0$, then $\frac{\mathrm{d}R}{\mathrm{d}t} > 0$. Since the number of recovered individuals is non-negative and bounded by the total population, we have

$$\lim_{t \to \infty} R(t) = R_\infty.$$

In the next section, we introduce a discrete-time model of a single species. We assume that the species has non-overlapping generations and can reasonably be approximated by a discrete model.

7.2.2 *SIR discrete-time model without vital dynamics (Animal)*

We denote again by S, I, R, the susceptible, infected (infectious), and recovered discretely reproducing species, respectively. The constant α_{HA} denotes the per capita infectious probability, and γ_A, the per capita probability of recovery. For each $t \in \{0, 1, 2, 3, \ldots\}$, the model takes the following form:[8, 16]

$$\text{SIR discrete model with no vitals:}$$
$$\begin{cases} S_{t+1} = S_t \mathrm{e}^{-\alpha_{HA} I_t} \\ I_{t+1} = S_t \left(1 - \mathrm{e}^{-\alpha_{HA} I_t}\right) + (1 - \gamma_A) I_t, \\ R_{t+1} = \gamma_A I_t + R_t \end{cases} \tag{7.2}$$

with initial conditions

$$(S, I, R) = (S_0, I_0, R_0) \in \mathbb{R}^3_+.$$

In this model, $\mathrm{e}^{-\alpha_{HA} I_t}$ denotes the probability that a susceptible individual remains susceptible in the next time period, while $1 - \mathrm{e}^{-\alpha_{HA} I_t}$ denotes the probability that a susceptible individual becomes infected in the present time period. In model (7.2), the total population at each time $t \in \{0, 1, 2, \ldots\}$ is

$$N_t = S_t + I_t + R_t.$$

Adding all three equations of model (7.2), we obtain

$$N_{t+1} = N_t.$$

That is, the total population is constant and $N_t = N_0$ at each time $t \in \{0, 1, 2, \ldots\}$.

From the first equation of model (7.2), we see that $S_{t+1} \leq S_t$ at each time $t \in \{0, 1, 2, \ldots\}$, and the number of susceptibles is always declining. When $I_0 > 0$, then $S_{t+1} < S_t$. Note that

$$P_0 = (S, I, R) = (S_0, 0, 0)$$

is a disease-free equilibrium of model (7.2). With linearization of the second equation of model (7.2) at P_0, we obtain that at each time $t \in \{0, 1, 2, \ldots\}$,

$$I_{t+1} \approx (\alpha_{HA} S_0 + (1 - \gamma_A)) I_t.$$

Hence, the basic reproduction number for model (7.2) is

$$\mathcal{R}_{0A} = \frac{\alpha_{HA} S_0}{\gamma_A}.$$

That is, as in model (7.1), $\mathcal{R}_{0A} > 1$ implies an initial increase in the number of infectious individuals before a decline to zero occurs while $\mathcal{R}_{0A} < 1$ implies an immediate decline to zero as $t \to \infty$.

7.3 Two-Host Hybrid Models without Vitals: Continuous-Discrete-Time

In this section, we introduce for the first time a mixed continuous-discrete-time model of two species. The model is designed to model the interaction of two species — one that reproduces continuously, and another that reproduces at discrete time intervals. These are two species (1) human and (2) animal. In the next section, we introduce the model. And in the following section, we perform analysis of the continuous-discrete-time model.

7.3.1 *The two-host hybrid model without vitals*

We assume that each species can infect the other, as well as that each species can infect itself. To introduce the continuous-discrete-time model of two species, we denote by S^H, I^H, and R^H the susceptible, infected, and recovered individuals, respectively. Furthermore, we denote by S^A, I^A, and R^A the susceptible, infected, and recovered animals, respectively. We note that the variables S^H, I^H, R^H are continuous functions of t, while S^A, I^A, and R^A are discrete-time variables which are defined at the integer values of t. We will interpret them as step functions with values specified at the integer values of t and remaining constant in the interval $[t, t+1)$. For each $t \geq 0$, we define the model as follows:

$$\begin{cases} \dfrac{\mathrm{d}S^H}{\mathrm{d}t} = -S^H\left(\alpha_H\theta_H I^H + \alpha_{HA}\left(1-\theta_H\right)I_t^A\right) \\ \dfrac{\mathrm{d}I^H}{\mathrm{d}t} = S^H\left(\alpha_H\theta_H I^H + \alpha_{HA}\left(1-\theta_H\right)I_t^A\right) - \gamma_H I^H \\ \dfrac{\mathrm{d}R^H}{\mathrm{d}t} = \gamma_H I^H \\ S_{t+1}^A = S_t^A\left(\theta_A \mathrm{e}^{-\alpha_A I_t^A} + \left(1-\theta_A\right)\mathrm{e}^{-\alpha_{AH} I^H}\right) \\ I_{t+1}^A = S_t^A\left(\theta_A\left(1-\mathrm{e}^{-\alpha_A I_t^A}\right) + \left(1-\theta_A\right)\left(1-\mathrm{e}^{-\alpha_{AH}I^H}\right)\right) \\ \qquad + \left(1-\gamma_A\right)I_t^A \\ R_{t+1}^A = \gamma_A I_t^A + R_t^A \end{cases} \tag{7.3}$$

with initial conditions

$$\left(S^H, I^H, R^H, S^A, I^A, R^A\right) = \left(S^H(0), I^H(0), R^H(0), S_0^A, I_0^A, R_0^A\right) \in \mathbb{R}_+^6.$$

The incidence terms for the humans have transmission coefficients α_H (transmission from humans to humans) and α_{HA} (transmission from animals to humans), while that of animals have transmission coefficients α_A (transmission from animals to animals) and α_{AH} (transmission from human to animal). Furthermore:

- $\theta_H \in (0,1)$ is the fraction of susceptible humans that interact with infectious humans, while $(1-\theta_H) \in (0,1)$ is the fraction of susceptible humans that interact with infectious animals. That is, the probability of multiple infection on an individual is small, and is ignored.
- $\theta_A \in (0,1)$ is the fraction of susceptible animals that interact with infectious animals, while $(1-\theta_H) \in (0,1)$ is the fraction of susceptible animals that interact with infectious humans. That is, the probability of multiple infection on an individual is small.
- α_A, α_{AH}, α_H, α_{HA}, $\gamma_H > 0$.

- $\gamma_A \in (0,1)$ is the fraction of exposed animals that progress to the infectious class while $(1-\gamma_A) \in (0,1)$ is the fraction of exposed animals that remain infectious.
- Denote by $N^H(0) = S^H(0) + I^H(0) + R^H(0)$ and $N_0^A = S_0^A + I_0^A + R_0^A$.

Because these hybrid continuous-discrete-time models are new, the mathematical tools for them have not been developed. Thus, we opt to analyze the model while developing the tools for that analysis.

7.3.2 *Analysis of the two-host hybrid model without vitals*

We anticipate that analysis will be possible if we "unify" the discrete and continuous parts of the model. The idea is to discretize the continuous-time ODE part and obtain a fully discrete model. We let $0 < h \leq 1$ be the step size, and assume that

$$h = \frac{1}{n}$$

for some positive integer n. That is, we are discretizing one time step. We use a backward difference to discretize the derivative and "linearize" the transition and the incidence terms around the previous step. The fully discrete hybrid model takes the form

$$\begin{cases} \dfrac{S^H(t+h) - S^H(t)}{h} = -S^H(t+h)\left(\alpha_H\theta_H I^H(t) \right. \\ \qquad\qquad \left. +\alpha_{HA}(1-\theta_H) I_t^A\right) \\ \dfrac{I^H(t+h) - I^H(t)}{h} = S^H(t+h)\left(\alpha_H\theta_H I^H(t) + \alpha_{HA}(1-\theta_H) I_t^A\right) \\ \qquad\qquad -\gamma_H I^H(t) \\ \dfrac{R^H(t+h) - R^H(t)}{h} = \gamma_H I^H(t) \\ S_{t+1}^A = S_t^A\left(\theta_A \mathrm{e}^{-\alpha_A I_t^A} + (1-\theta_A)\mathrm{e}^{-\alpha_{AH} I^H(t)}\right) \\ I_{t+1}^A = S_t^A\left(\theta_A\left(1-\mathrm{e}^{-\alpha_A I_t^A}\right)\right. \\ \qquad\qquad \left. +(1-\theta_A)\left(1-\mathrm{e}^{-\alpha_{AH} I^H(t)}\right)\right) \\ \qquad\qquad +(1-\gamma_A) I_t^A \\ R_{t+1}^A = \gamma_A I_t^A + R_t^A \end{cases} \tag{7.4}$$

We multiply by h and solve for $S^H(t+h)$, $I^H(t+h)$, and $R^H(t+h)$ as follows:

$$\begin{cases} S^H(t+h) = \dfrac{S^H(t)}{1+h\left(\alpha_H\theta_H I^H(t)+\alpha_{HA}(1-\theta_H)I_t^A\right)} \\ I^H(t+h) = \dfrac{hS^H(t)\left(\alpha_H\theta_H I^H(t)+\alpha_{HA}(1-\theta_H)I_t^A\right)}{1+h\left(\alpha_H\theta_H I^H(t)+\alpha_{HA}(1-\theta_H)I_t^A\right)} \\ \qquad +(1-h\gamma_H)I^H(t) \\ R^H(t+h) = R^H(t)+h\gamma_H I^H(t) \\ S_{t+1}^A = S_t^A\left(\theta_A e^{-\alpha_A I_t^A}+(1-\theta_A)e^{-\alpha_{AH}I^H(t)}\right) \\ I_{t+1}^A = S_t^A\left(\theta_A\left(1-e^{-\alpha_A I_t^A}\right)+(1-\theta_A)\left(1-e^{-\alpha_{AH}I^H(t)}\right)\right) \\ \qquad +(1-\gamma_A)I_t^A \\ R_{t+1}^A = \gamma_A I_t^A + R_t^A \end{cases} \tag{7.5}$$

Thus, if $h=\frac{1}{n}$, we need to make n steps with the discretization of the ODE before we make one step with the discrete model. As $n\to\infty$ and $h\to 0$, the following discrete-time model of $3(n+1)$ equations, which describes the progression of the hybrid model from time t to time $t+1$, converges to the hybrid model. The discretization of the continuous part of the above model only gives the values of S^H, I^H, and R^H at $t+h$, that is after one step after t. Given the values after j steps, the values in $j+1$ steps are obtained from the following system:

$$\begin{cases} S^H(t+(j+1)h) = \dfrac{S^H(t+jh)}{1+h\left(\alpha_H\theta_H I^H(t+jh)+\alpha_{HA}(1-\theta_H)I_t^A\right)} \\ I^H(t+(j+1)h) = \dfrac{\begin{array}{c} hS^H(t+jh)\left(\alpha_H\theta_H I^H(t+jh)\right. \\ \left.+\alpha_{HA}(1-\theta_H)I_t^A\right) \end{array}}{\begin{array}{c} 1+h\left(\alpha_H\theta_H I^H(t+jh)\right. \\ \left.+\alpha_{HA}(1-\theta_H)I_t^A\right) \end{array}} \\ \qquad +(1-h\gamma_H)I^H(t+jh) \\ R^H(t+(j+1)h) = R^H(t+jh)+h\gamma_H I^H(t+jh) \\ S_{t+1}^A = S_t^A\left(\theta_A e^{-\alpha_A I_t^A}+(1-\theta_A)e^{-\alpha_{AH}I^H(t)}\right) \\ I_{t+1}^A = S_t^A\left(\theta_A\left(1-e^{-\alpha_A I_t^A}\right)\right. \\ \qquad \left.+(1-\theta_A)\left(1-e^{-\alpha_{AH}I^H(t)}\right)\right)+(1-\gamma_A)I_t^A \\ R_{t+1}^A = \gamma_A I_t^A + R_t^A \end{cases} \tag{7.6}$$

for $j = 0, \ldots, n-1$. Now, we show that for sufficiently small step sizes the fully discrete model (7.6) is well-posed, and there is no population explosion.

Theorem 7.1. *In model* (7.6), *with step size*

$$h \in \left(0, \min\left\{\frac{1}{\gamma_H}, 1\right\}\right),$$

and non-negative initial conditions, for all $t \in \{1, 2, \ldots\}$ *and* $j \in \{0, 1, 2, \ldots, n-1\}$

$$\begin{aligned} &S^H(t+(j+1)h), I^H(t+(j+1)h), R^H(t+(j+1)h) \geq 0, \\ &S_t^A, I_t^A, R_t^A \geq 0 \end{aligned}$$

and there is no population explosion.

Proof. Since $(1 - h\gamma_H) > 0$, each equation of model (7.6) is a sum of non-negative terms whenever the initial population sizes are non-negative. Hence,

$$S^H(t+(j+1)h), I^H(t+(j+1)h), R^H(t+(j+1)h) \geq 0,$$

and

$$S_t^A, I_t^A, R_t^A \geq 0$$

for all $t \geq 0$ and $j \geq 0$.

By summing the first three equations of model (7.6), at each time $t \in \{0, 1, 2, \ldots\}$ and each $j \in \{0, 1, 2, \ldots, n-1\}$, the total human population

$$N^H(\cdot) = S^H(\cdot) + I^H(\cdot) + R^H(\cdot),$$

is governed by the equation

$$N^H(t+(j+1)h) = N^H(t+jh).$$

That is,

$$\frac{\mathrm{d}}{\mathrm{d}t} N^H(t+jh) = \lim_{h \to 0} \frac{N^H((t+jh)+h) - N^H(t+jh)}{h} = 0.$$

Thus, for sufficiently small h, as $t \to \infty$ the total population of humans remain fixed near its initial value, $N^H(0)$.

Similarly, by summing the last three equations of model (7.6), at each time $t \in \{0, 1, 2, \ldots\}$, the total population of animals,

$$N_t^A = S_t^A + I_t^A + R_t^A,$$

is governed by the equation

$$N_{t+1}^A = N_t^A,$$

and $N_t^A = N_0$ for all t. That is, the total population of the animal population also remains fixed at its initial values, and all orbits are bounded. □

Because model (7.6) is a fully discrete model, we can apply tools from discrete-time dynamical systems. However, those should be applied with caution, because the two subsystems of the model have distinct step sizes, as opposed to fully discrete models for which all variables progress with the same step size (usually equal to one). To determine the model equilibria and their stability, we begin by looking at a disease-free equilibrium (DFE) of model (7.6). The equilibria satisfy the system

$$\begin{cases} S^H = \dfrac{S^H}{1 + h\left(\alpha_H \theta_H I^H + \alpha_{HA}\left(1 - \theta_H\right) I^A\right)} \\ I^H = \dfrac{hS^H\left(\alpha_H \theta_H I^H + \alpha_{HA}\left(1 - \theta_H\right) I^A\right)}{1 + h\left(\alpha_H \theta_H I^H + \alpha_{HA}\left(1 - \theta_H\right) I^A\right)} + \left(1 - h\gamma_H\right) I^H \\ R^H = R^H + h\gamma_H I^H \\ S^A = S^A\left(\theta_A \mathrm{e}^{-\alpha_A I^A} + \left(1 - \theta_A\right) \mathrm{e}^{-\alpha_{AH} I^H}\right) \\ I^A = S^A\left(\theta_A\left(1 - \mathrm{e}^{-\alpha_A I^A}\right) + \left(1 - \theta_A\right)\left(1 - \mathrm{e}^{-\alpha_{AH} I^H}\right)\right) \\ \qquad + \left(1 - \gamma_A\right) I^A \\ R^A = \gamma_A I^A + R^A \end{cases} \tag{7.7}$$

We note that the $3(n+1)$ system collapses into a six-equation system since all the equations for the different values of j are actually the same. In model (7.6), $R^H(t) = N^H(0) - S^H(t) - I^H(t)$ and $R_t^A = N_0^A - S_t^A - I_t^A$. Consequently, $R^H(t)$ and R_t^A are determined once $S^H(t)$, $I^H(t)$, S_t, and I_t are known. Hence, we can drop the $R^H(t)$

and R_t^A equations. To determine the DFE we take $I_H = 0$ and $I_A = 0$, and the system for the DFE reduces to the following two equations:

$$\left.\begin{array}{l} S^H = S^H \\ S^A = S^A \end{array}\right\}. \tag{7.8}$$

Thus, the DFE consists of six-tuples $(*^H, 0, N^H(0) - *^H, *^A, 0, N_0^A - *^A)$ where $*^H$ and $*^A$, respectively, denote the potentially non-zero entries given by $S^H(0)$ and S_0^A. We will use the next generation approach to compute $\mathcal{R}_0$. Linearizing around the DFE, equations for the infectious classes in system (7.6) with $y^H(t) = I^H(t)$ and $y_t^A = I_t^A$, we have

$$\left\{\begin{aligned} y^H(t+(j+1)h) &= hS^H(0)\left(\theta_H \alpha_H y^H(t+jh)\right.\\ &\quad \left.+ (1-\theta_H)\,\alpha_{HA} y_t^A\right) + (1-h\gamma_H)\,y^H(t+jh) \\ &\quad j = 0,\ldots,n-1 \\ y_{t+1}^A &= S_0^A\left(\theta_A \alpha_A y_t^A + (1-\theta_A)\,\alpha_{AH} y^H(t)\right) \\ &\quad + (1-\gamma_A)\,y_t^A \end{aligned}\right. . \tag{7.9}$$

We note that the next generation approach for discrete systems as described in Ref. 17 does not apply here because it assumes that the step in all dependent variables is the same and equal to one. That is not the case here. As a consequence, in analyzing the fully discrete system (7.6) we use the following strategy: we first derive the stability based on first principles, then show that we could apply the next generation method in Ref. 17 and obtain the same result.

We look for solution of the form $y^H(t+jh) = y_0^H r^{t+jh}$ and $y^A = y_0^A r^t$. Substituting in the system (7.9) and canceling r^t, we obtain

$$\left\{\begin{aligned} y_0^H r^{(j+1)h} &= hS^H(0)\left(\theta_H \alpha_H y_0^H r^{jh} + (1-\theta_H)\,\alpha_{HA} y_0^A\right) \\ &\quad + (1-h\gamma_H)\,y_0^H r^{jh} \\ &\quad j = 0,\ldots,n-1 \\ y_0^A r &= S_0^A\left(\theta_A \alpha_A y_0^A + (1-\theta_A)\,\alpha_{AH} y_0^H\right) + (1-\gamma_A)\,y_0^A \end{aligned}\right. . \tag{7.10}$$

We think of $y^H(t+jh)$ as a different variable for all $j = 0,\ldots,n$. Taking the determinant, we have

$$\begin{vmatrix} hS^H(0)\theta_H\alpha_H + (1-h\gamma_H) - r^h & \dots & 0 & hS^H(0)\,(1-\theta_H)\,\alpha_{HA} \\ \vdots & \vdots & \dots & \vdots \\ 0 & \dots & hS^H(0)\theta_H\alpha_H + (1-h\gamma_h) - r^h & hS^H(0)\,(1-\theta_H)\,\alpha_{HA} \\ 0 & \dots & (1-\theta_A)\alpha_{AH}S_0^A & \theta_A\alpha_A S_0^A + (1-\gamma_A) - r \end{vmatrix}$$

The eigenvalues are

$$r = [hS^H(0)\theta_H\alpha_H + (1-h\gamma_H)]^{\frac{1}{h}}.$$

Let $\mathcal{R}_{0H} = \frac{S^H(0)\theta_H\alpha_H}{\gamma_H}$. If $\mathcal{R}_{0H} < 1$, then $r < 1$ and for h small enough $-1 < r < 1$. If $\mathcal{R}_{0H} > 1$, then $r > 1$ and the disease-free equilibrium is unstable. Assume $\mathcal{R}_{0H} < 1$. The remaining eigenvalues are solutions of

$$\begin{vmatrix} hS^H(0)\theta_H\alpha_H + (1-h\gamma_h) - r^h & hS^H(0)\,(1-\theta_H)\,\alpha_{HA} \\ (1-\theta_A)\alpha_{AH}S_0^A & \theta_A\alpha_A S_0^A + (1-\gamma_A) - r \end{vmatrix} = 0.$$

We write this as

$$\begin{vmatrix} a - r^h & b \\ c & d - r \end{vmatrix} = 0.$$

We define as $\mathcal{R}_0$ the principle root of

$$f(r) = (r^h - a)(r - d) - bc = 0.$$

The eigenvalues are the roots of $f(r) = 0$. Assume $\mathcal{R}_0 < 1$. Then

$$f(1) = (1-a)(1-d) - bc > 0.$$

For $r > 1$, we have $f(r) > f(1) > 0$. Hence, the principle root of $f(r) = 0$, $r_p < 1$. All other roots are smaller in absolute value. Hence, the disease-free equilibrium is stable. Assume now that $\mathcal{R}_0 > 1$. Then $f(1) < 0$. Since $\lim_{r\to\infty} r = \infty$, we have that $r_p > 1$. This concludes the analysis.

Next, we develop the next generation approach for hybrid models based on the next generation approach for discrete-time models, as described in Ref. 4. We define the fertility matrix,

$$F = \begin{pmatrix} hS^H(0)\theta_H\alpha_H & 0 & \dots & 0 & hS^H(0)\,(1-\theta_H)\,\alpha_{HA} \\ 0 & hS^H(0)\theta_H\alpha_H & \dots & 0 & hS^H(0)\,(1-\theta_H)\,\alpha_{HA} \\ \vdots & \vdots & \dots & \vdots & \vdots \\ 0 & 0 & \dots & hS^H(0)\theta_H\alpha_H & hS^H(0)\,(1-\theta_H)\,\alpha_{HA} \\ 0 & 0 & \dots & (1-\theta_A)\alpha_{AH}S_0^A & \theta_A\alpha_A S_0^A \end{pmatrix}$$

and the transitions matrix (see Ref. 4),

$$T = \begin{pmatrix} 1-h\gamma_H & 0 & \dots & 0 & 0 \\ 0 & 1-h\gamma_H & \dots & 0 & 0 \\ \vdots & \vdots & \dots & \vdots & \vdots \\ 0 & 0 & \dots & 1-h\gamma_H & 0 \\ 0 & 0 & \dots & 0 & 1-\gamma_A \end{pmatrix}.$$

Clearly, for $h < h^*$ the spectral radius of T, given by $\rho(T) < 1$, the next generation matrix $\mathcal{Q} = F(\mathbf{I} - T)^{-1}$ is given by

$$\mathcal{Q} = \begin{pmatrix} \frac{S^H(0)\theta_H\alpha_H}{\gamma_H} & 0 & \dots & 0 & \frac{S^H(0)(1-\theta_H)\alpha_{HA}}{\gamma_H} \\ 0 & \frac{S^H(0)\theta_H\alpha_H}{\gamma_H} & \dots & 0 & \frac{S^H(0)(1-\theta_H)\alpha_{HA}}{\gamma_H} \\ \vdots & \vdots & \dots & \vdots & \vdots \\ 0 & 0 & \dots & \frac{S^H(0)\theta_H\alpha_H}{\gamma_H} & \frac{S^H(0)(1-\theta_H)\alpha_{HA}}{\gamma_H} \\ 0 & 0 & \dots & \frac{(1-\theta_A)\alpha_{AH}S_0^A}{\gamma_A} & \frac{\theta_A\alpha_A S_0^A}{\gamma_A} \end{pmatrix}.$$

It is easy to see that the matrix $\mathcal{Q}$ has $n-1$ eigenvalues of $\mathcal{R}_{0H}$ and the eigenvalues of the matrix

$$\mathcal{Q}_H = \begin{pmatrix} \frac{S^H(0)\theta_H\alpha_H}{\gamma_H} & \frac{S^H(0)(1-\theta_H)\alpha_{HA}}{\gamma_H} \\ \frac{(1-\theta_A)\alpha_{AH}S_0^A}{\gamma_A} & \frac{\theta_A\alpha_A S_0^A}{\gamma_A} \end{pmatrix}.$$

We note that the diagonal elements of $\mathcal{Q}_H$ are $\mathcal{R}_{0H}$ and $\mathcal{R}_{0A}$, respectively. We denote the off-diagonal elements as $q_{12} = \mathcal{R}_{AH}$ and $q_{21} = \mathcal{R}_{HA}$. The characteristic equation of the 2×2 matrix is

$$\lambda^2 - (\mathcal{R}_{0H} + \mathcal{R}_{0A})\lambda + (\mathcal{R}_{0H}\mathcal{R}_{0A} - \mathcal{R}_{AH}\mathcal{R}_{HA}) = 0.$$

Hence, the maximal root, and the reproduction number of the hybrid model is given by

$$\mathcal{R}_0 = \frac{(\mathcal{R}_{0H} + \mathcal{R}_{0A}) + \sqrt{(\mathcal{R}_{0H} - \mathcal{R}_{0A})^2 + 4\mathcal{R}_{AH}\mathcal{R}_{HA}}}{2}.$$

It is not hard to see that $\mathcal{R}_0 > \mathcal{R}_{0H}$ and it therefore is the spectral radius of $\mathcal{Q}$ as well. One important observation is that $\mathcal{R}_0$ does not depend on h, and can be derived by assuming $h = 1$ (although in this case it may not be true that $\rho(T) < 1$). A few remarks are in order:

- We opted for a mixed numerical approximation which still requires restriction on h but allows us to use the next generation approach for discrete equations even if its application is not rigorous in this case. Alternatively, we can use implicit numerical method, and there will be no restriction on h for positivity, but it is not clear how and whether the next generation approach will work.
- It may be worthwhile to formalize the next generation approach for hybrid models in general.

7.4 SIR Hybrid Continuous-Discrete-Time Model with Vital Dynamics

In this section, we derive a hybrid model with vital dynamics. Such a model will have not only a disease-free equilibrium but also an endemic equilibrium. This allows for a wider array of techniques being developed for hybrid models. First, we derive the continuous-time model with vital dynamics.

7.4.1 *SIR continuous-time model with vital dynamics (Human)*

With the notation of the previous sections, we introduce recruitment rate in the humans and a natural mortality rate. For each $t \geq 0$, the model with vital dynamics takes the form

$$\begin{cases} \dfrac{\mathrm{d}S}{\mathrm{d}t} = \Lambda_H - \alpha_H SI - \mu_H S, \\ \dfrac{\mathrm{d}I}{\mathrm{d}t} = \alpha_H SI - \gamma_H I - \mu_H I, \\ \dfrac{\mathrm{d}R}{\mathrm{d}t} = \gamma_H I - \mu_H R, \end{cases} \tag{7.11}$$

with initial conditions

$$(S, I, R) = (S(0), I(0), R(0)) \in \mathbb{R}^3_+,$$

where μ_H is the positive mortality rate and Λ_H is the positive recruitment rate. The total population at time t is

$$N(t) = S(t) + I(t) + R(t).$$

Adding all three equations of model (7.11), we obtain

$$\frac{\mathrm{d}N}{\mathrm{d}t} = \Lambda_H - \mu_H N$$

and

$$N(t) = N(0)\,\mathrm{e}^{-\mu_H t} + \frac{\Lambda_H}{\mu_H}\left(1 - \mathrm{e}^{-\mu_H t}\right).$$

Hence, in model (7.11), the total population is asymptotically constant and

$$N(t) \to \frac{\Lambda_H}{\mu_H} \text{ as } t \to \infty.$$

The basic reproduction number for model (7.11) is[8]

$$\mathcal{R}_{0H} = \frac{\Lambda_H \alpha_H}{\mu_H(\gamma_H + \mu_H)}.$$

That is, $\mathcal{R}_{0H} > 1$ implies that the number of infectious individuals reaches to a non-zero value, while $\mathcal{R}_{0H} < 1$ implies the numbers of infectious individuals decline to zero as $t \to \infty$.

7.4.2 SIR discrete-time model with vital dynamics (Animal)

With positive recruitment rate and probability of dying, for each $t \in \{0, 1, 2, 3, \ldots\}$, the model takes the form

$$\begin{cases} S_{t+1} = \Lambda_A + (1-\mu_A)\, S_t e^{-\alpha_A I_t} \\ I_{t+1} = (1-\mu_A)\left(S_t\left(1-e^{-\alpha_A I_t}\right) + (1-\gamma_A)\, I_t\right) \\ R_{t+1} = (1-\mu_A)\left(\gamma_A I_t + R_t\right) \end{cases} \tag{7.12}$$

with initial conditions

$$(S, I, R) = (S_0, I_0, R_0) \in \mathbb{R}^3_+,$$

where the fraction that die per unit interval is $\mu_A \in (0,1)$ and Λ_A is a positive constant denoting the recruitment rate per unit interval. In model (7.12), the total population at each time $t \in \{0, 1, 2, \ldots\}$ is

$$N_t = S_t + I_t + R_t.$$

Adding all three equations of model (7.12), we obtain

$$N_{t+1} = \Lambda_A + (1-\mu_A)\, N_t,$$

and

$$\lim_{t\to\infty} N_t = N_* = \frac{\Lambda_A}{\mu_A}.$$

That is, the total population is asymptotically constant. When there is no disease and $I_t = R_t = 0$, model (7.12) reduces to the following equation:

$$S_{t+1} = \Lambda_A + (1-\mu_A)\, S_t.$$

The disease-free equilibrium (DFE) of model (7.12) is

$$P_0 = (S, I, R) = (N_*, 0, 0).$$

By the next generation matrix (NGM) method, the basic reproduction number of the model (7.12) is derived as

$$\mathcal{R}_{0A} = \frac{\alpha_A (1-\mu_A)\, N_*}{1-(1-\mu_A)(1-\gamma_A)}.$$

That is, $\mathcal{R}_{0A} > 1$ implies that the number of infectious individuals reaches to a non-zero number, while $\mathcal{R}_{0A} < 1$ implies the number of infectious individuals declines to zero as $t \to \infty$.

7.4.3 *Two-host hybrid models with vitals: Continuous and discrete*

We introduce a hybrid model and include recruitment for humans and animals, and natural death rate for humans and a probability of survival for animals. For each $t \geq 0$, the hybrid model with vitals takes the form

$$\begin{cases} \dfrac{\mathrm{d}S^H}{\mathrm{d}t} = \Lambda_H - S^H\left(\alpha_H\theta_H I^H + \alpha_{HA}\left(1-\theta_H\right)I_t^A\right) - \mu_H S^H \\ \dfrac{\mathrm{d}I^H}{\mathrm{d}t} = S^H\left(\alpha_H\theta_H I^H + \alpha_{HA}\left(1-\theta_H\right)I_t^A\right) - \gamma_H I^H - \mu_H I^H \\ \dfrac{\mathrm{d}R^H}{\mathrm{d}t} = \gamma_H I^H - \mu_H R^H \\ S_{t+1}^A = \Lambda_A + \left(1-\mu_A\right)S_t^A\left(\theta_A \mathrm{e}^{-\alpha_A I_t^A} + \left(1-\theta_A\right)\mathrm{e}^{-\alpha_{AH} I^H}\right) \\ I_{t+1}^A = \left(1-\mu_A\right)\left(S_t^A\left(\theta_A\left(1-\mathrm{e}^{-\alpha_A I_t^A}\right)\right.\right. \\ \qquad\qquad \left.\left. + \left(1-\theta_A\right)\left(1-\mathrm{e}^{-\alpha_{AH} I^H}\right)\right) + \left(1-\gamma_A\right)I_t^A\right) \\ R_{t+1}^A = \left(1-\mu_A\right)\left(\gamma_A I_t^A + R_t^A\right) \end{cases} \tag{7.13}$$

with initial conditions

$$\left(S^H, I^H, R^H, S^A, I^A, R^A\right) = \left(S^H(0), I^H(0), R^H(0), S_0^A, I_0^A, R_0^A\right) \in \mathbb{R}_+^6.$$

We begin by discretizing the continuous component. We use a similar approach to discretize by using a semi-implicit scheme. Let h be the step size. For the continuous component, we have

$$\begin{cases} S^H(t+(j+1)h) - S^H(t+jh) \\ = \Lambda_H h - S^H(t+(j+1)h)\left(\alpha_H \theta_H I^H(t+jh)\right. \\ \quad \left. + \alpha_{HA}\left(1-\theta_H\right) I_t^A\right) - \mu_H h S^H(t+jh) \\ I^H(t+(j+1)h) - I^H(t+jh) \\ = S^H(t+(j+1)h)\left(\alpha_H \theta_H I^H(t+jh) + \alpha_{HA}\left(1-\theta_H\right) I_t^A\right) \\ \quad -(\gamma_H+\mu_H) h I^H(t+jh) \\ R^H(t+(j+1)h) - R^H(t+jh) \\ = \gamma_H I^H(t+jh) - \mu_H h R^H(t+jh) \end{cases} \tag{7.14}$$

Here, the equation for S^H is discretized by implicit method, but the equation for I^H is discretized by implicit method and then "linearized" by evaluating I^H at the previous time level. We take this approach with the goal to preserve the positivity without serious restrictions on the step h and to have some symmetry with the discrete part of the model. The fully discrete model becomes

$$\begin{cases} S^H(t+(j+1)h) = \dfrac{S^H(t+jh)(1-\mu_H h) + \Lambda_H h}{1 + \left(\alpha_H \theta_H I^H(t+jh) + \alpha_{HA}\left(1-\theta_H\right) I_t^A\right) h} \\ I^H(t+(j+1)h) = S^H(t+(j+1)h) h \left(\alpha_H \theta_H I^H(t+jh)\right. \\ \qquad \left. + \alpha_{HA}\left(1-\theta_H\right) I_t^A\right) \\ \qquad + (1-(\gamma_H+\mu_H) h) I^H(t+jh) \\ R^H(t+(j+1)h) = \gamma_H I^H(t+(j+1)h) + (1-\mu_H h) R^H(t+jh) \\ S_{t+1}^A = \Lambda_A + \left(1-\mu_A\right) S_t^A\left(\theta_A \mathrm{e}^{-\alpha_A I_t^A} + \left(1-\theta_A\right) \mathrm{e}^{-\alpha_{AH} I^H(t)}\right) \\ I_{t+1}^A = \left(1-\mu_A\right)\left(S_t^A\left(\theta_A\left(1-\mathrm{e}^{-\alpha_A I_t^A}\right)\right.\right. \\ \qquad \left.\left. + \left(1-\theta_A\right)\left(1-\mathrm{e}^{-\alpha_{AH} I^H(t)}\right)\right) + \left(1-\gamma_A\right) I_t^A\right) \\ R_{t+1}^A = \left(1-\mu_A\right)\left(\gamma_A I_t^A + R_t^A\right). \end{cases} \tag{7.15}$$

7.4.4 *Analysis of the hybrid model with vitals: Disease-free equilibrium and stability*

Analysis of the hybrid model with vitals shares similar features to the analysis of the model without vitals. However, in this case we can expect, besides the disease-free equilibrium, also endemic equilibria. We look for equilibria. Those satisfy the system

$$\begin{cases} S^H &= \dfrac{S^H(1-\mu_H h)+\Lambda_H h}{1+(\alpha_H\theta_H I^H+\alpha_{HA}(1-\theta_H)I^A)h} \\ I^H &= I^H+S^H h\left(\alpha_H\theta_H I^H(t+jh)+\alpha_{HA}(1-\theta_H)I^A\right) \\ &\quad -(\gamma_H+\mu_H)hI^H \\ R^H &= R^H+\gamma_H I^H-\mu_H h I^H \\ S^A &= \Lambda_A+(1-\mu_A)S^A\left(\theta_A \mathrm{e}^{-\alpha_A I^A}+(1-\theta_A)\mathrm{e}^{-\alpha_{AH}I^H}\right) \\ I^A &= (1-\mu_A)\left(S^A\left(\theta_A\left(1-\mathrm{e}^{-\alpha_A I^A}\right)\right.\right. \\ &\quad \left.\left.+(1-\theta_A)\left(1-\mathrm{e}^{-\alpha_{AH}I^H}\right)\right)+(1-\gamma_A)I^A\right) \\ R^A &= (1-\mu_A)\left(\gamma_A I^A+R^A\right) \end{cases} \tag{7.16}$$

To find the disease-free equilibrium, we set $I^H=0$. From the second equation, we see that $I^A=0$. From the third equation, we see that $R^H=0$ and from the last equation we get $R^A=0$. With these values the system for the equilibria simplifies to

$$\left.\begin{array}{r} S^H = S^H(1-\mu_H h)+\Lambda_H h \\ S^A = \Lambda_A+(1-\mu_A)S^A \end{array}\right\}. \tag{7.17}$$

This system gives us the following disease-free equilibrium:

$$\mathcal{E}^0 = \left(\frac{\Lambda_H}{\mu_H},0,0\frac{\Lambda_A}{\mu_A},0,0\right) = (S_0^H,0,0,S_0^A,0,0)$$

We linearize around the disease-free equilibrium. We set $S^H=S_0^H+x^H$, $I^H=y^H$, $R^H=z^H$, $S^A=S_0^A+x^A$, $I^A=y^A$, $R^A=z^A$. We obtain the following linearized system of equations:

$$\begin{cases} x^H(t+(j+1)h) = [x^H(t+jh)(1-\mu_H h)-S_0^H \\ \qquad\qquad (\alpha_H\theta_H y^H(t+jh)+\alpha_{HA}(1-\theta_H)y_t^A)h] \\ y^H(t+(j+1)h) = [y^h(t+jh)(1-(\gamma_H+\mu_H)h) \\ \qquad\qquad +S_0^H(\alpha_H\theta_H y^H(t+jh)+\alpha_{HA}(1-\theta_H)y_t^A)h] \\ \qquad x_{t+1}^A = -(1-\mu_a)[S_0^A(\theta_A\alpha_A y_t^a \\ \qquad\qquad +(1-\theta_a)\alpha_{AH}y^H(t))-x_t^A] \\ \qquad y_{t+1}^A = (1-\mu_A)[S_0^A(\theta_A\alpha_A y_t^A+(1-\theta_A)\alpha_{AH}y^H(t)) \\ \qquad\qquad +(1-\gamma_A)y_t^A] \end{cases}, \tag{7.18}$$

where we have used the equalities

$$S_0^H(1-\mu_H h)+\Lambda_H h = S_0^H$$

and

$$\Lambda_A+(1-\mu_a)S_0^A = S_0^A.$$

The equations for the y^H and y^A separate, so we consider them separately.

$$\begin{cases} y^H(t+(j+1)h) = [y^H(t+jh)(1-(\gamma_H+\mu_H)h) \\ \qquad\qquad + S_0^H\left(\alpha_H\theta_H y^H(t+jh)+\alpha_{HA}\left(1-\theta_H\right)y_t^A\right)h] \\ \qquad y_{t+1}^A = (1-\mu_A)[S_0^A(\theta_A\alpha_A y_t^A+(1-\theta_A)\alpha_{AH}y^H(t)) \\ \qquad\qquad + (1-\gamma_A)y_t^A] \end{cases} \tag{7.19}$$

We look for solutions of the form $y^H(t+jh)=y_j^H r^t$ and $y_t^A = y^A r^t$. Substituting these tentative solutions in the system above, we obtain

$$\begin{cases} y_j^H r^h = \left[y_j^H(1-(\gamma_H+\mu_H)h)\right. \\ \qquad\qquad \left. + S_0^H\left(\alpha_H\theta_H y_j^H+\alpha_{HA}\left(1-\theta_H\right)y^A\right)h\right] \\ y^A r = (1-\mu_A)\left[S_0^A(\theta_A\alpha_A y^A+(1-\theta_A)\alpha_{AH}y_0^H)+(1-\gamma_A)y^A\right] \end{cases} . \tag{7.20}$$

We note that this is not quite an eigenvalue problem. We consider the following characteristic equation for r:

$$\begin{vmatrix} A-r^h & 0 & \dots & 0 & [S_0^H\alpha_{HA}(1-\theta_H)h] \\ & & \vdots & & \\ 0 & 0 & \dots & A-r^h & [S_0^H\alpha_{HA}(1-\theta_H)h] \\ 0 & 0 & \dots & (1-\mu_A)S_0^A(1-\theta_A)\alpha_{AH} & (1-\mu_a)[S_0^A\theta_A\alpha_A+(1-\gamma_A)]-r \end{vmatrix} = 0, \tag{7.21}$$

where $A=[1-(\gamma_H+\mu_H)h+S_0^H\alpha_H\theta_H h]$. This characteristic equation has a repeated root

$$r=[1-(\gamma_H+\mu_H)h+S_0^H\alpha_H\theta_H h]^{\frac{1}{h}}.$$

Denote by

$$\mathcal{R}_{0H} = \frac{S_0^H\alpha_H\theta_H}{\mu_H+\gamma_H}$$

the reproduction number of the humans. If $\mathcal{R}_{0H} < 1$, $-1 < r < 1$ for h small enough. If $\mathcal{R}_{0H} > 1$, $r > 1$ and the disease-free equilibrium is unstable. If $\mathcal{R}_{0H} < 1$ is smaller than one, the stability of the disease-free equilibrium is controlled by the roots of the characteristic equation

$$\begin{vmatrix} [1-(\gamma_H+\mu_H)h + S_0^H\alpha_H\theta_H h] - r^h & [S_0^H\alpha_{HA}(1-\theta_H)h] \\ (1-\mu_A)S_0^A(1-\theta_A)\alpha_{AH} & (1-\mu_a)[S_0^A\theta_A\alpha_A \\ & +(1-\gamma_A)] - r \end{vmatrix} = 0. \tag{7.22}$$

We denote the entries of the determinant by a, b, c, d, respectively. Multiplying the first row by r^{1-h}, we have that r also satisfies the characteristic equation

$$\begin{vmatrix} ar^{1-h} - r & br^{1-h} \\ c & d-r \end{vmatrix} = 0. \tag{7.23}$$

We denote by A_r the matrix such that the characteristic equation is $|A_r - rI| = 0$. Thus, we consider the following cases:

(1) $0 < r < 1$. Then $r^{1-h} > r$

$$r = \rho(A_r) < \rho(A) = r_0.$$

We rewrite the above characteristic equation as

$$\begin{vmatrix} ar^{1-h} - r & br^{1-h} \\ c & d-r \end{vmatrix} = 0.$$

We note that $a, b, c, d > 0$. Then we have

$$\begin{vmatrix} ar - r & br \\ c & d-r \end{vmatrix} < 0.$$

Hence,

$$\begin{vmatrix} a-1 & b \\ c & d-r \end{vmatrix} < 0.$$

Since $r_0 > r$, we have

$$\begin{vmatrix} a-1 & b \\ c & d-r_0 \end{vmatrix} < \begin{vmatrix} a-1 & b \\ c & d-r \end{vmatrix} < 0.$$

Hence,

$$\begin{vmatrix} a-1 & b \\ c & d-r_0 \end{vmatrix} < \begin{vmatrix} a-r_0 & b \\ c & d-r_0 \end{vmatrix}$$

which implies that $r_0 < 1$.

(2) Assume $r > 1$. As before we can see that $r_0 < r$. Then $r^{1-h} < r$. Reversing the inequalities in the argument above, we have that $r_0 > 1$.

The above result means that we can apply the next generation approach[4] to compute $\mathcal{R}_0$.

$$F = \begin{pmatrix} hS_0^H\theta_H\alpha_H & 0 & \dots & 0 & hS_0^H(1-\theta_H)\alpha_{HA} \\ 0 & hS_0^H\theta_H\alpha_H & \dots & 0 & hS_0^H(1-\theta_H)\alpha_{HA} \\ \vdots & \vdots & \dots & \vdots & \vdots \\ 0 & 0 & \dots & hS_0^H\theta_H\alpha_H & hS_0^H(1-\theta_H)\alpha_{HA} \\ 0 & 0 & \dots & (1-\mu_A)(1-\theta_A)\alpha_{AH}S_0^A & (1-\mu_A)\theta_A\alpha_A S_0^A \end{pmatrix}$$

and the transitions matrix (see Ref. 4)

$$T = \begin{pmatrix} 1-h(\gamma_H+\mu_H) & 0 & \dots & 0 & 0 \\ 0 & 1-h(\gamma_H+\mu_H) & \dots & 0 & 0 \\ \vdots & \vdots & \dots & \vdots & \vdots \\ 0 & 0 & \dots & 1-h(\gamma_H+\mu_H) & 0 \\ 0 & 0 & \dots & 0 & (1-\mu_A)(1-\gamma_A) \end{pmatrix}.$$

Clearly, for $h < h^*$ the spectral radius of T, given by $\rho(T) < 1$, the next generation matrix $\mathcal{Q} = F(\mathbf{I}-T)^{-1}$ is given by

$$\mathcal{Q} = \begin{pmatrix} \frac{S_0^H\theta_H\alpha_H}{\gamma_H+\mu_H} & 0 & \dots & 0 & \frac{S^H(0)(1-\theta_H)\alpha_{HA}}{\gamma_H+\mu_H} \\ 0 & \frac{S_0^H\theta_H\alpha_H}{\gamma_H+\mu_H} & \dots & 0 & \frac{S_0^H(1-\theta_H)\alpha_{HA}}{\gamma_H+\mu_H} \\ \vdots & \vdots & \dots & \vdots & \vdots \\ 0 & 0 & \dots & \frac{S_0^H\theta_H\alpha_H}{\gamma_H+\mu_H} & \frac{S_0^H(1-\theta_H)\alpha_{HA}}{\gamma_H+\mu_H} \\ 0 & 0 & \dots & (1-\mu_A)\frac{(1-\theta_A)\alpha_{AH}S_0^A}{1-(1-\mu_A)(1-\gamma_A)} & (1-\mu_A)\frac{\theta_A\alpha_A S_0^A}{1-(1-\mu_A)(1-\gamma_A)} \end{pmatrix}.$$

It is easy to see that the matrix $\mathcal{Q}$ has $n-1$ eigenvalues of

$$\mathcal{R}_{0H} = \frac{S_0^H \theta_H \alpha_H}{\gamma_H + \mu_H}$$

and the eigenvalues of the matrix

$$\mathcal{Q}_H = \begin{pmatrix} \frac{S_0^H \theta_H \alpha_H}{\gamma_H + \mu_H} & \frac{S_0^H (1-\theta_H)\alpha_{HA}}{\gamma_H + \mu_H} \\ \frac{(1-\mu_A)(1-\theta_A)\alpha_{AH} S_0^A}{1-(1-\mu_A)(1-\gamma_A)} & \frac{(1-\mu_A)\theta_A \alpha_A S_0^A}{1-(1-\mu_A)(1-\gamma_A)} \end{pmatrix}.$$

We note that the diagonal elements of $\mathcal{Q}_H$ are $\mathcal{R}_{0H}$ and $\mathcal{R}_{0A}$, respectively. We denote the off-diagonal elements as $q_{12} = \mathcal{R}_{AH}$ and $q_{21} = \mathcal{R}_{HA}$. The characteristic equation of the 2×2 matrix is

$$\lambda^2 - (\mathcal{R}_{0H} + \mathcal{R}_{0A})\lambda + (\mathcal{R}_{0H}\mathcal{R}_{0A} - \mathcal{R}_{AH}\mathcal{R}_{HA}) = 0.$$

Hence, the maximal root, and the reproduction number of the hybrid model is given by

$$\mathcal{R}_0 = \frac{(\mathcal{R}_{0H} + \mathcal{R}_{0A}) + \sqrt{(\mathcal{R}_{0H} - \mathcal{R}_{0A})^2 + 4\mathcal{R}_{AH}\mathcal{R}_{HA}}}{2}.$$

It is not hard to see that $\mathcal{R}_0 > \mathcal{R}_{0H}$ and it therefore is the spectral radius of $\mathcal{Q}$ as well. One important observation is that $\mathcal{R}_0$ does not depend on h, and can be derived by assuming $h = 1$ (although in this case it may not be true that $\rho(T) < 1$).

7.4.5 *Endemic equilibrium of the system with vitals*

Theorem 7.2. *If $\mathcal{R}_0 > 1$, then the system with vitals always has at least one endemic equilibrium.*

Proof. As with the other model, an endemic equilibrium is a time-independent solution of the system given as solutions of the system (7.16). Looking for endemic equilibrium, we need $I^H \neq 0$ and $I^A \neq 0$. We denote by

$$\begin{aligned} D &= \alpha_H \theta_H I^H + \alpha_{HA}(1-\theta_H) I^A \\ Q &= \theta_A \mathrm{e}^{-\alpha_A I^A} + (1-\theta_A)\mathrm{e}^{-\alpha_{HA} I^H}. \end{aligned} \tag{7.24}$$

We note that while $Q < 1$, D can take on any positive value. We have that

$$S^H = \frac{\Lambda_H}{D+\mu_H} \qquad I^H = \frac{\Lambda_H D}{(D+\mu_H)(\mu_H+\gamma_H)}.$$

The discrete equations take the form

$$S^A = \Lambda_A + (1-\mu_A) S^A Q.$$

Hence,

$$S^A = \frac{\Lambda_A}{1-(1-\mu_A)Q}.$$

We also have

$$I^A = \frac{(1-\mu_A)S^A(1-Q)}{\Gamma_A} = \frac{(1-\mu_A)(1-Q)\Lambda_A}{(1-(1-\mu_A)Q)\Gamma_A},$$

where

$$\Gamma_A = 1-(1-\mu_A)(1-\gamma_A).$$

We note that $0 < \Gamma_A < 1$. The system in (D, Q) becomes

$$\begin{aligned} D &= \alpha_H \theta_H \tfrac{\Lambda_H D}{(D+\mu_H)(\mu_H+\gamma_H)} + \alpha_{HA}(1-\theta_H)\tfrac{(1-\mu_A)(1-Q)\Lambda_A}{(1-(1-\mu_A)Q)\Gamma_A} \\ Q &= \theta_A \mathrm{e}^{-\alpha_A \frac{(1-\mu_A)(1-Q)\Lambda_A}{(1-(1-\mu_A)Q)\Gamma_A}} + (1-\theta_A)\mathrm{e}^{-\alpha_{AH}\frac{\Lambda_H D}{(D+\mu_H)(\mu_H+\gamma_H)}} \end{aligned} \tag{7.25}$$

For this system, the disease-free equilibrium is the point $(0, 1)$, which is also a solution. We want to show that the system has at least one positive solution, that is a solution where $D > 0$ and $0 < Q < 1$. First we will translate the disease-free equilibrium to $(0, 0)$. We set

$$y = \frac{1-Q}{D} \longrightarrow Q = 1 - yD.$$

With this new notation, after dividing by D the first equation in (7.25) becomes

$$1 = \alpha_H \theta_H \frac{\Lambda_H}{(D+\mu_H)(\mu_H+\gamma_H)} + \alpha_{HA}(1-\theta_H)\frac{(1-\mu_A)y\Lambda_A}{(1-(1-\mu_A)(1-yD))\Gamma_A}.$$

This is equivalent to a linear equation in y and can be solved in terms of D. This defines a function $y = \mathcal{F}(D)$. The function $\mathcal{F}(D)$ can be explicitly computed. Thus, we have

$$y = \mathcal{F}(D) = \frac{(1 - \frac{\alpha_H\theta_H\Lambda_H}{(D+\mu_H)(\mu_H+\gamma_H)})\mu_A}{\alpha_{HA}(1-\theta_H)\frac{(1-\mu_A)\Lambda_A}{\Gamma_A} - (1 - \frac{\alpha_H\theta_H\Lambda_H}{(D+\mu_H)(\mu_H+\gamma_H)})D}.$$

y is not always a positive function of D, and we need it to be positive. We will address that later. Now, from the second equation in (7.25), we have

$$1-D\mathcal{F}(D) = \theta_A \mathrm{e}^{-\alpha_A \frac{(1-\mu_A)D\mathcal{F}(D)\Lambda_A}{(1-(1-\mu_A)(1-D\mathcal{F}(D))\Gamma_A}} + (1-\theta_A)\mathrm{e}^{-\alpha_{AH}\frac{\Lambda_H D}{(D+\mu_H)(\mu_H+\gamma_H)}}$$

Rearranging terms we have

$$D\mathcal{F}(D) = \theta_A\left(1 - \mathrm{e}^{-\alpha_A \frac{(1-\mu_A)D\mathcal{F}(D)\Lambda_A}{(1-(1-\mu_A)(1-D\mathcal{F}(D))\Gamma_A}}\right) + (1-\theta_A)\left(1 - \mathrm{e}^{-\alpha_{AH}\frac{\Lambda_H D}{(D+\mu_H)(\mu_H+\gamma_H)}}\right)$$

Dividing by D, we have

$$\mathcal{F}(D) = \frac{1}{D}\theta_A\left(1 - \mathrm{e}^{-\alpha_A \frac{(1-\mu_A)D\mathcal{F}(D)\Lambda_A}{(1-(1-\mu_A)(1-D\mathcal{F}(D))\Gamma_A}}\right) + \frac{1}{D}(1-\theta_A)\left(1 - e^{-\alpha_{AH}\frac{\Lambda_H D}{(D+\mu_H)(\mu_H+\gamma_H)}}\right) \quad (7.26)$$

This is an equation for D. Let us denote the function on the right-hand side of this equation as $\mathcal{G}(D)$. Clearly, $\mathcal{G}(D)$ is always a positive function. We have to consider two cases.

Case 1: $\mathcal{R}_{0H} < 1$. In this case, the numerator of y is always positive for any D nonnegative, including for $D = 0$. The denominator of y is positive at $D = 0$, but it becomes zero for some $\bar{D}$. Thus, for

$0 < D < \bar{D}$, y is always positive. The functions $\mathcal{F}(D)$ and $\mathcal{G}(D)$ are positive and continuous on that interval of D. We compare the left-hand side with the right-hand side of (7.26). For $D = 0$, $\mathcal{F}(0)$ is given by

$$\mathcal{F}(0) = \frac{(1 - \frac{\alpha_H \theta_H \Lambda_H}{\mu_H(\mu_H+\gamma_H)})\mu_A}{\alpha_{HA}(1-\theta_H)\frac{(1-\mu_A)\Lambda_A}{\Gamma_A}}.$$

Using "L. Hopital's rule, we can find the value of $\mathcal{G}(0)$. We have

$$\lim_{D\to 0} \mathcal{G}(D) = \theta_A \alpha_A \frac{(1-\mu_A)\mathcal{F}(0)\Lambda_A}{\mu_A \Gamma_A} + (1-\theta_A)\alpha_{AH} \frac{\Lambda_H}{\mu_H(\mu_H+\gamma_H)}$$

Since, we have assumed that $\mathcal{R}_0 > 1$, we have

$$\frac{\mathcal{R}_{0H} + \mathcal{R}_{0A} + \sqrt{(\mathcal{R}_{0H} - \mathcal{R}_{0A})^2 + 4\mathcal{R}_{AH}\mathcal{R}_{HA}}}{2} > 1.$$

Rewriting this inequality, we get

$$\mathcal{R}_{AH}\mathcal{R}_{HA} > 1 - \mathcal{R}_{0H} - \mathcal{R}_{0A} + \mathcal{R}_{0H}\mathcal{R}_{0A}.$$

On the other hand, the inequality $\mathcal{F}(0) < \mathcal{G}(0)$ is equivalent to the above inequality. Thus, $\mathcal{R}_0 > 1$ implies $\mathcal{F}(0) < \mathcal{G}(0)$. On the other hand, $\bar{D}$ makes the denominator of $\mathcal{F}(D)$ become zero, while the numerator is not. Thus, we have

$$\lim_{D\to \bar{D}^-} \mathcal{F}(D) = \infty$$

while $\mathcal{G}(\bar{D})$ is positive and finite. Thus, $\mathcal{F}(\bar{D}) > \mathcal{G}(\bar{D})$. Hence, there must be a point D^* in the interval $(0, \bar{D})$ such that $\mathcal{F}(D^*) = \mathcal{G}(D^*)$. This completes the proof in this case.

Case 2: $\mathcal{R}_{0H} > 1$. In this case, the numerator of y is negative when $D = 0$, but at the same time, the denominator is positive. So near $D = 0$ we do not have a viable solution. Since the numerator of y is an increasing function of D, at some point $\bar{D}_1$ it becomes positive. At that point $\bar{D}_1$ the denominator is still positive since the part that is subtracted there has just turned positive. At some other larger $\bar{D}_2$, the denominator becomes negative, while the numerator of y is still positive. Thus, for $\bar{D}_1 < D < \bar{D}_2$, y is positive. We also have that

$\mathcal{F}(\bar{D}_1) = 0$, while $\mathcal{G}(\bar{D}_1)$ is positive and finite. Thus, $\mathcal{F}(\bar{D}_1) < \mathcal{G}(\bar{D}_1)$. Since the denominator of y becomes zero at $\bar{D}_2$, we have

$$\lim_{D \to \bar{D}_2^-} \mathcal{F}(D) = \infty$$

while $\mathcal{G}(\bar{D}_2)$ is finite and positive. Thus, we have $\mathcal{F}(\bar{D}_2) > \mathcal{G}(\bar{D}_2)$. The two functions are continuous, thus, there is a point D^* in the interval $(\bar{D}_1, \bar{D}_2)$ such that $\mathcal{F}(D^*) = \mathcal{G}(D^*)$. This completes the proof of Case 2. We note that in Case 2 we do not need $\mathcal{R}_0 > 1$. However, since $\mathcal{R}_{0H} > 1$, that implies that $\mathcal{R}_0 > 1$. □

7.5 Numerical Simulations of the Two-Host Hybrid Model with and without Vitals

In this section, we present the simulations of the hybrid models (7.13), (7.3), with and without vitals, respectively, using the MATLAB programming language. Figure 7.1 shows the results of the simulation of the hybrid model with no vitals when the basic reproduction number is higher than one for three different step sizes. Since $\mathcal{R}_0 > 1$, there is an initial increase in the number of infected individuals before it decays to zero eventually (see Figure 7.1(b)). Three distinct step sizes, $h = 0.1$, $h = 0.25$, and $h = 0.5$, are used while approximating the solutions of the hybrid model (7.3). However, since the step size in the discrete part of the hybrid model (7.3) is fixed at one, this step size is only utilized for discretizing the continuous portion of the hybrid model.

In Figures 7.2 and 7.3, we present the dynamics of the hybrid model with vitals when the reproduction number is greater than one, and less than one respectively. When the basic reproduction number $\mathcal{R}_0$ is 4.4, the system reaches an endemic equilibrium, with infected classes remaining endemic in both populations (see Figure 7.2). Simulations for the hybrid model with vitals are carried out with three different step sizes, $h = 0.1$, $h = 0.5$, and $h = 1$. As can be seen from Figures 7.2 and 7.3, the step size greatly influences the approximate value of hybrid model state variables. The step size is only used to discretize the ODE portion of the hybrid model, not the discrete portion since in that case the step size is fixed at 1. But, the step size used for discretizing the ODE part of the model has a considerable

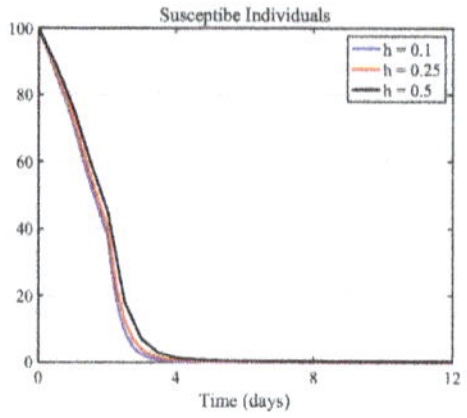

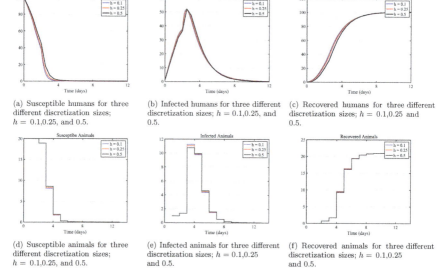

(a) Susceptible humans for three different discretization sizes; $h =$ 0.1,0.25, and 0.5.

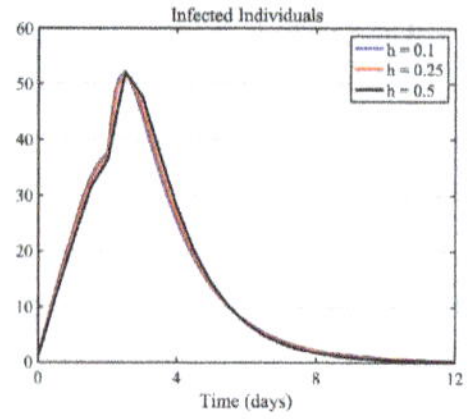

(b) Infected humans for three different discretization sizes; $h = $ 0.1,0.25, and 0.5.

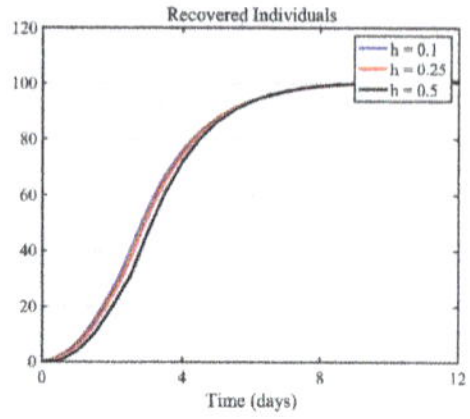

(c) Recovered humans for three different discretization sizes; $h =$ 0.1,0.25 and 0.5.

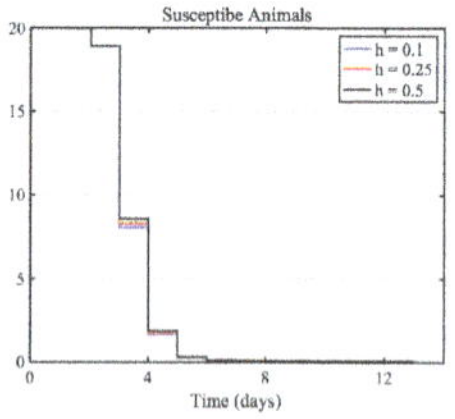

(d) Susceptible animals for three different discretization sizes; $h =$ 0.1,0.25, and 0.5.

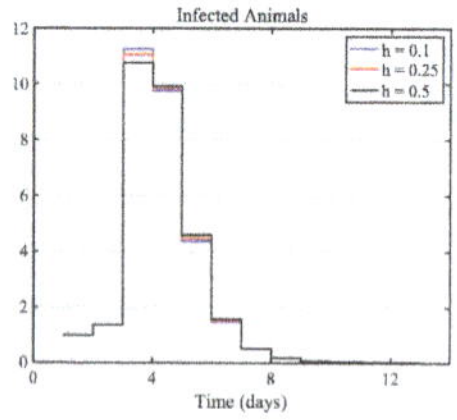

(e) Infected animals for three different discretization sizes; $h =$ 0.1,0.25 and 0.5.

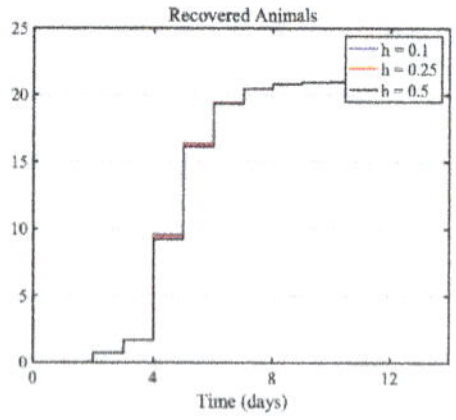

(f) Recovered animals for three different discretization sizes; $h =$ 0.1,0.25 and 0.5.

Fig. 7.1. Simulation of the hybrid model (7.3) with no vitals. The parameter values used for simulations are $\alpha_H = 0.05$, $\gamma_H = 0.6$, $\theta_H = 0.2$, and $\alpha_{HA} = 0.3$, $\alpha_A = 0.2$, $\gamma_A = 0.7$, $\theta_A = 0.1$, and $\alpha_{AH} = 0.04$. Initial values are $S^H(0) = 100, I^H(0) = 1, R^H(0) = 0$, $S_0^A = 20, I_0^H = 1, R_0^H = 0$, $\mathcal{R}_0 = 7.6$.

impact on the discrete portion of the hybrid model (see Figures 7.2 and 7.3). That is why we do not recommend taking the step size equal to one in order to discretize both ODE and discrete models with the same step $h = 1$.

7.6 Discussion

In this chapter, we introduce the novel continuous-discrete-time model of two-species epidemics. The main thrust for these models is that the species involved have distinct breeding patterns. One of the species, called “animal”, has breeding seasons, while the other one, called “human”, breeds continuously. Thus, the first species has to be modeled by discrete-time equations, while the second, by continuous. We combine the two modeling tools into one, devising for the first time infectious diseases continuous-discrete-time epidemic models. Clearly, the discrete part is defined only at discrete points, while the continuous is defined for all t. To tie these two types, we

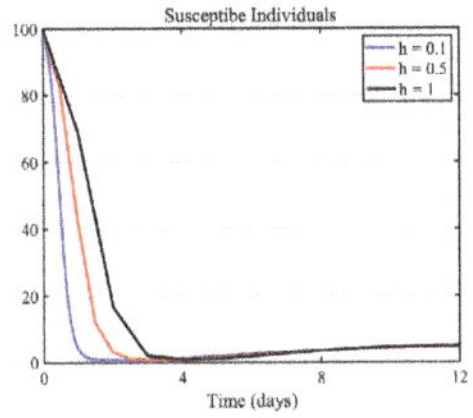

(a) Susceptible humans for three different discretization sizes; $h =$ 0.1,0.5, and 1.

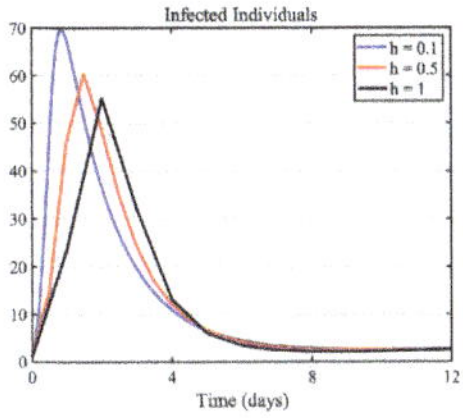

(b) Infected humans for three different discretization sizes; $h =$ 0.1,0.5, and 1.

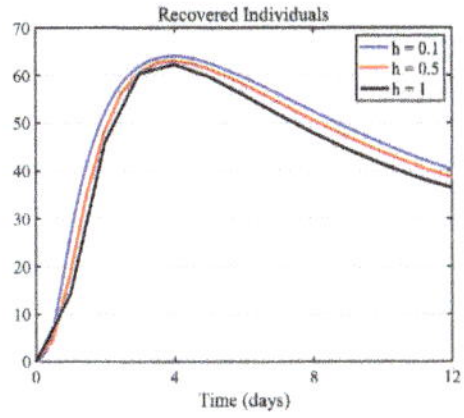

(c) Recovered humans for three different discretization sizes; $h=$ 0.1,0.5, and 1.

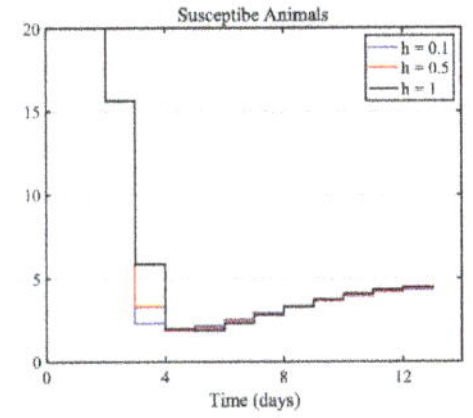

(d) Susceptible animals for three different discretization sizes; $h =$ 0.1,0.5, and 1.

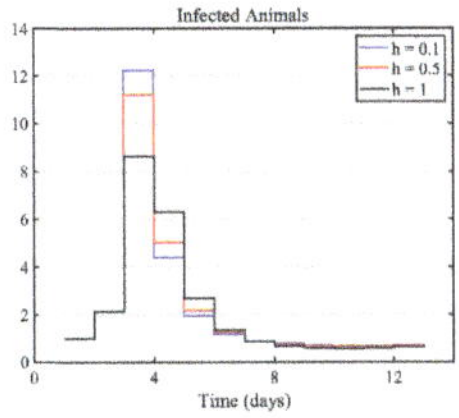

(e) Infected animals for three different discretization sizes; $h =$ 0.1,0.5, and 1.

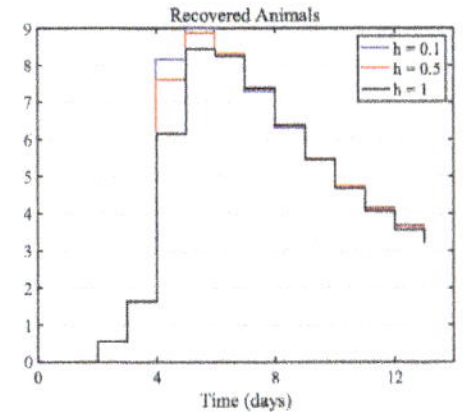

(f) Recovered animals for three different discretization sizes; $h =$ 0.1,0.5, and 1.

Fig. 7.2. Simulation of the hybrid model (7.13) with vitals. The parameter values used for simulations are $\lambda_H = 2.5, \mu_H = 0.1, \alpha_H = 0.5, \gamma_H = 0.6, \theta_H = 0.2$, and $\alpha_{HA} = 0.3$, $\lambda_A = 1.5, \mu_A = 0.2$, $\alpha_A = 1.8$, $\gamma_A = 0.7$, $\theta_A = 0.1$, and $\alpha_{AH} = 0.04$. Initial values are $S^H(0) = 100, I^H(0) = 1, R^H(0) = 0$, $S_0^A = 20, I_0^H = 1, R_0^H = 0$, $\mathcal{R}_0 = 4.4$.

imagine that the discrete variables are defined as constants for all $t \in [t_n, t_{n+1})$. Since these models are neither discrete nor continuous, all currently existing tools for analyzing non hybrid models do not apply. To make consistent the two intertwined types of equations, we discretize the continuous part with step $h << 1$. After the discretization, we have a fully discrete-time model, however, unlike the models that are generated as discrete models, this fully discrete model has two distinct steps — one and h, which again makes application of currently existing tools for discrete equations in many cases not justified. We choose a discretization scheme that guarantees positivity without serious restrictions on h.

We introduce two types of models — an SIR–SIR model without vitals and an SIR–SIR model with vitals. We develop the mathematical tools to derive the reproduction number for the model without vitals. In the model without vitals, all infectious classes converge to zero, thus $\mathcal{R}_0$ controls whether an outbreak occurs or not. We develop the mathematical tools to determine the stability of the disease-free

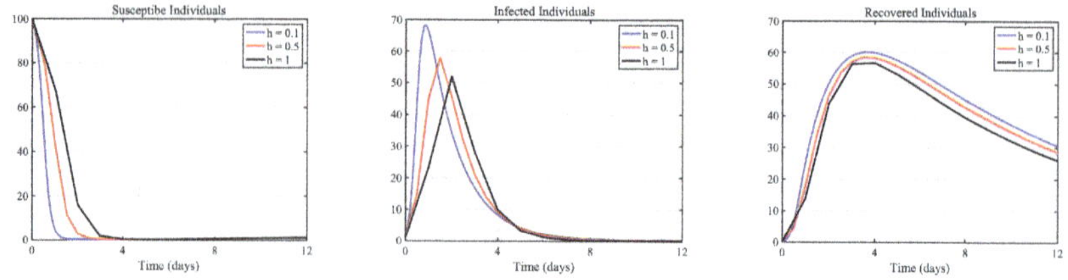

(a) Susceptible humans for three different discretization sizes; $h = 0.1$,0.5, and 1.

(b) Infected humans for three different discretization sizes; $h = 0.1$,0.5, and 1.

(c) Recovered humans for three different discretization sizes; $h =$ 0.1,0.5, and 1.

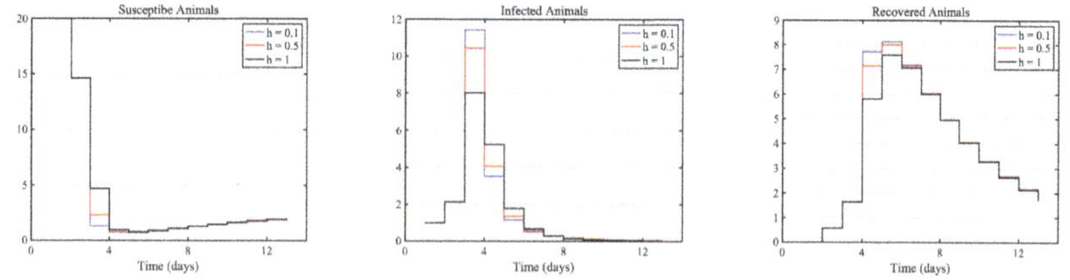

(d) Susceptible animals for three different discretization sizes; $h =$ 0.1,0.5, and 1.

(e) Infected animals for three different discretization sizes; $h = 0.1$,0.5, and 1.

(f) Recovered animals for three different discretization sizes; $h =$ 0.1,0.5, and 1.

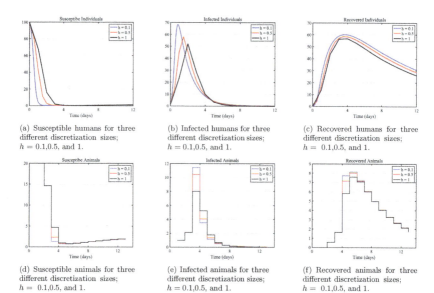

Fig. 7.3. Simulation of the hybrid model (7.13) with vitals. The parameter values used for simulations are $\lambda_H = 0.25, \mu_H = 0.1$, $\alpha_H = 0.5$, $\gamma_H = 0.6$, $\theta_H = 0.2$, and $\alpha_{HA} = 0.3$, $\lambda_A = 0.5, \mu_A = 0.2$, $\alpha_A = 1.8$, $\gamma_A = 0.7$, $\theta_A = 0.1$, and $\alpha_{AH} = 0.04$. Initial values are $S^H(0) = 100, I^H(0) = 1, R^H(0) = 0$, $S_0^A = 20, I_0^H = 1, R_0^H = 0$. $\mathcal{R}_0 = 0.7$

equilibrium. Next, we consider the SIR–SIR model with vital dynamics. This model has more possible outcomes. We find that it has a disease-free equilibrium and, if $\mathcal{R}_0 > 1$, we prove that it also has at least one endemic equilibrium. We use first principles to determine the stability of the disease-free equilibrium based on the value of the reproduction number. We find that if $\mathcal{R}_0 < 1$, the disease-free equilibrium is locally asymptotically stable, if $\mathcal{R}_0 > 1$, the disease-free equilibrium is unstable. We also apply in both cases the next generation approach for discrete models, and we show that the derived reproduction number is exactly the same. In particular, the reproduction number does not depend on the discretization step h — a result which should be expected, since h is not a part of the original model. Regardless, we believe that in the future, the next generation approach should be extended for hybrid continuous-discrete-time models.

References

1. Mathematical model. https://en.wikipedia.org/wiki/Mathematical_model.
2. Derakhshan, A., Khademi, A., Khademi, S., Yusof, N. M. and Lee, M. H. (2015). Review of discrete-continuous models in energy and transportation, *Procedia CIRP* **26**, pp. 281–286.
3. Understanding discrete vs. continuous growth. https://betterexplained.com/articles/understanding-discrete-vs-continuous-growth/.
4. Allen, L. J. S. and van den Driessche, P. (2008). The basic reproduction number in some discrete-time epidemic models, *J. Diff. Equations* **14**, 10–11, pp. 1127–1147.
5. Lefevre, C. and Picard, P. (1989). On the formulation of discrete-time epidemic models, *Mathe. Biosci.* **95**, 1, pp. 27–35.
6. Brauer, F., Feng Z. and Castillo-Chavez, C. Discrete epidemic models, *Math. Biosci. Eng.* **7**, 1, pp. 1–15.
7. Murray, J. (2002). *Mathematical Biology*, Springer, New York.
8. Martcheva, M. (2015). *An Introduction to Mathematical Epidemiology*, Texts in Applied Mathematics 61, Springer, New York.
9. Nowak, M. and May, R. (2001). *Virus Dynamics: Mathematical Principles of Immunology and Virology*, Oxford University Press, USA.
10. Centers for Disease Control and Prevention. Lyme disease, https://www.cdc.gov/lyme/index.html.
11. Lyme Disease. https://en.wikipedia.org/wiki/Lyme_disease.
12. Lou, Y. and Wu, J. (2017). Modeling Lyme disease transmission, *Infect. Dis. Model.* **2**, 2, pp. 229–243.
13. Porco, T. C. (1999). A mathematical model of the ecology of Lyme disease, *IMA J. Math. Appl. Med. Biol.* **16**, 3, pp. 261–296.
14. Nguyen, A., Mahaffy, J. and Vaidya, N. K. (2019). Modeling transmission dynamics of lyme disease: Multiple vectors, seasonality, and vector mobility, *Infect. Dis. Model* **4**, pp. 28–43.
15. Kermack, W. and McKendrick, A. (1927). A contribution to the mathematical theory of epidemics, *Proc. Roy. Soc. Lond. A* **115**, pp. 700–721.
16. Castillo-Chavez, C. and Yakubu, A. A. (2001). Discrete time S-I-S models with complex dynamics. In *Proceedings of the Third World Congress of Nonlinear Analysits, Part 7* (Catania 2000), Vol. 47, pp. 4753–4762.
17. Allen, L. J. S. (1994). Some discrete-time SI, SIR, and SIS epidemic models, *Math. Biosci.* **124**, pp. 83–105.

https://doi.org/10.1142/9789811263033_0008

Chapter 8

Extending Analytical Solutions to Age–Mass Models of a Population

Lauren M. Childs[*,‡] **and Olivia F. Prosper**[†,§]

[*]*Department of Mathematics, Virginia Tech, Blacksburg, Virginia, USA; Center for Emerging Zoonotic and Arthropod-Borne Pathogens, Virginia Tech, Blacksburg, Virginia, USA.*

[†]*Department of Mathematics, University of Tennessee, Knoxville, TN, USA*

[‡]*lchilds@vt.edu*

[§]*oprosper@utk.edu*

There has been a long history of population models accounting for variations in physiological characteristics of individuals in the mathematical literature. Classically represented as the McKendrick partial differential equation (PDE) in time and age, there are multiple extensions that include, for example, mass and size. These extended equations rapidly become intractable analytically, without further simplifying assumptions, due to interconnections between the characteristics and fecundity. Here, we consider population dynamics using a three-dimensional PDE incorporating time, age, and an additional characteristic, which we consider to be mass, but without renewal. Such a scenario could represent pathogen development in a host prior to transmission. In advancement of previous work, our growth function remains a function of all three variables. Under conditions of separability, we obtain analytical results for the population dynamics. We confirm and extend these results with numerical simulations of our three-dimensional PDE. Our results can be used to understand age and characteristic, e.g., mass or size-dependent growth of populations.

Keywords: McKendrick, Malthus, von Foerster, Oldfield, Nisbet and Gurney, Integral Projection Model (IPM), kernel, Boundary value problem, Sinko and Streifer, Age structured (age-structured), Age-dependent, Mass-dependent, Density-dependent

8.1 Introduction

Population growth has a rich history of mathematical description and analysis, beginning several centuries ago. In 1798, Malthus showed that populations would grow or shrink geometrically, dependent on the relative contributions of the birth and death rates.[1] While this may be an accurate description over short time periods, limitations in resources led to expectation in limits in growth over time. Thus, various restrictions to the growth rate at large population size were introduced, producing the classic logistic equation, e.g., in Ref. 2. Numerous extensions to the basic logistic equation have appeared.

Metz and Diekmann[3] argued the need for models that bridge the gap between simulation-based models that can incorporate high degrees of complexity in individual characteristics and simpler ordinary differential equation models that track population numbers by ignoring these individual-level characteristics. In particular, they emphasized the importance of developing continuous, infinite-dimensional systems that incorporate physiological characteristics like size.[3] In addition to providing a tool-kit to construct and analyze these systems, the text provides a rich set of examples that include age-structure, size-structure, metapopulation-structure, and mixtures of the three. Many of the examples are extensions of predator–prey models, cell population models, and epidemiological models. These fall into well-studied differential equation approaches for describing continuous-time, infinite-dimensional systems such as partial differential equations (PDE) and integral differential equations. A more recent approach that is discrete in time but allows for the incorporation of continuous traits is the integral projection model (IPM). In this work, we focus on extending the PDE formulation beyond time and age.

8.1.1 *PDE framework*

General formulations were introduced multiple times.[4–11] In particular, we consider the linear partial differential equation in time and age introduced by McKendrick[6] in 1926. This equation was independently rediscovered by Scherbaum and Rasch in 1957[10] and von Foerster in 1959[8] to describe populations of cells,[8,10] as discussed in Refs. 3 and 11. Much work has extended this framework to consider a variety of additional traits, most notably mass and size.[9,12–16] A limitation of many of these models is the need for strong simplifying assumptions to produce an analytically tractable model. Some models reduce the number of independent variables to two, while others simplify the growth function to be a function of a single variable or choose singular boundary conditions.

Bell and Anderson developed a model of a population of cells with growth, division, and death as a function of time, age, and cell volume.[9] Transitional functions for all three are assumed to depend on age and cell volume. They find that when cell growth is proportional to cell volume, the results are intrinsically linked to the initial conditions as formulated, although this can be altered through the incorporation of an independent dispersive mechanism. For their analytical work, they focused on populations in the exponential growth phase or in steady state. Furthermore, they employ some simplifying assumptions such as that division leads to two cells of exactly equal volume. This is later relaxed in computer simulations.

Oldfield expanded the original PDE of cell growth to examine two additional arbitrary cell properties.[14] He found an analytical solution using piecewise continuous functions as coefficients in his PDE and with a number of cell-specific assumptions, such as the splitting into two identical daughter cells at the end of a cell's life span.[14,15] Interestingly, he restricts both additional properties to be identical between newborn and parent cells; thus, seemingly restricting the use of mass.[15] Furthermore, he mathematically restricts the coefficients of the partial derivative terms to be functions of time and the argument of the derivative only.

Sinko developed an extension of the McKendrick equation to incorporate size in addition to age.[15–17] Sinko applied this model to populations of *Dugesia tigrina* and *Daphnia pulex*.[17] Due to the complexity that the derived boundary condition depends on the density function itself, simplifications are required for analytical solutions. Sinko assumed that the growth function is dependent on age and time, but independent of mass, allowing for an integral solution.[17]

Nisbet and Gurney expanded the analytical results of the McKendrick framework with a different simplifying assumption.[13,18]

Beginning from a description using age, mass, and time for a single species model, they restricted their rate functions to only depend on mass and time. This simplification allowed them to reformulate the PDE structure in terms of two variables: mass and time. From this, they determined integro-differential equations describing the population change.[12]

8.1.2 *IPM framework*

The Integral Projection Model (IPM) is an alternative to the PDE framework that allows for continuous variation in other categories, such as size. An extension to classic matrix models, IPMs use a discrete time step and are able to capture key features such as the stable distribution, asymptotic growth rates, and sensitivities.[19] Compared to matrix models, however, IPMs do not require arbitrary breaks in continuous classification categories to form discrete groups. This is important, because the choice of discretization can artificially skew results, in particular, sensitivities.[19] IPMs have been widely applied in population ecology.[20,21] Their reach has expanded following the development of a user-friendly *R* package.[22]

Introduced by Easterling,[23] IPMs use information on the state of individuals at a given time point to determine the state of individuals (themselves and any offspring) at a future time. This relationship is codified through a function, called the kernel, which is often broken apart into separate functions to describe aspects of population growth including fecundity, survival, and growth.[19,22] Mathematically, this kernel maps the state distribution at time t to the distribution at time $t + 1$, and uses a joint likelihood, with the ability to combine information from multiple datasets.[19,24] Components entering into the kernel are typically parameterized using regression models from a variety of datasets.[22,25] In particular, data on individuals is used to produce models of various rates, which contribute to the formation of the joint probability distribution.[22] This allows for predictions of population dynamics and the observation of emergent phenomena. Standard analyses such as final state distribution, passage time to particular life events, and elasticity follow from the mathematics of matrix population models.[19,22]

IPMs rely on a variety of datasets in order to develop underlying models for the rates. In most cases, these involve measurements on the same individuals across multiple time steps to understand changes in properties such as growth. As such many of the examples involve plant populations.[19,22,25] Information not requiring the same individual, such as population counts, can also be incorporated.[24,26] In general, IPMs can be generated from individual or population level data or a combination of both.[20,26] The challenge remains what structure is best to retain the ability for population level predictions while accurately capturing individual level processes.

8.1.3 *Extending the PDE framework*

The aim of this chapter is to determine situations where a more complicated growth function, incorporating time, age, and mass, is possible in the PDE framework. To do so, we consider a population that does not renew, such as pathogen development within a host. This allows us to track the growth of the population without the complication of density-dependent birth. Under this simplification, we find analytical solutions to a more diverse set of growth functions.

In Section 8.2, we introduce our PDE in time, age, and mass; and in Section 8.3, we provide an analytical solution underlying simplifying conditions. The necessary simplifying conditions on growth are discussed in more detail in Section 8.3.1. In Section 8.4, we introduce and analyze dimensionally reduced versions of our PDE and compare to previous work. We provide exemplary cases of our full solution implemented numerically in Section 8.5 and discuss our findings in Section 8.7.

8.2 The General PDE Population Model

We develop a population model, which tracks the size of the population, $f(t, a, m)$ of age a and size m at time t. Here, age refers to chronological age, which progresses at the same rate as the system time, rather than physiological age, which could accumulate in a nonlinear manner. For biological consistency we assume $t \geq 0$, $a \geq 0$, and $m \geq m_0$. In other words, we only consider nonnegative time, individuals of nonnegative age, and individuals with a minimum mass, m_0, that can only be obtained at birth, $a = 0$. Following Refs. 12 and 16, we formalize this as follows:

$$\frac{\partial f(t,a,m)}{\partial t} + \frac{\partial f(t,a,m)}{\partial a} + \frac{\partial}{\partial m}\Big(g(t,a,m)f(t,a,m)\Big)$$
$$= -d(t,a,m)f(t,a,m), \tag{8.1}$$

with $g(t,a,m)$ the growth function and $d(t,a,m)$ the loss function.

For the boundary condition, we consider a distribution of individuals at age $a = 0$ with different masses. We assume all individuals have a minimum mass of m_0 and that minimum is only obtained at birth, $a = 0$. Furthermore, our distribution $\psi(t,m)$ may vary with time, but we assume independence from the current population size. This is a deviation from other models, which assume that the boundary condition depends on the current population size, e.g., Ref. 12. Thus, our boundary conditions can be represented by

$$f(t,a,m_0) = \begin{cases} \psi(t,m_0), & a = 0, t \geq 0, \\ 0, & a > 0, t \geq 0, \end{cases}$$
$$f(t,0,m) = \psi(t,m), \qquad m \geq m_0, t \geq 0.$$

Our initial condition at $t = 0$, assuming no individuals are present, is expressed by

$$f(0,a,m) = 0.$$

We reformulate our boundary value problem to an initial value problem by considering some time $t_0 > 0$, such that all individuals have entered the population. Thus, there is a distribution of individuals in age and mass, denoted by $\Phi(a,m)$, at time t_0. Since all individuals appeared by $t = t_0$, there are no individuals of age $a = 0$ or mass $m = m_0$. Thus, we can revise the boundary conditions for $t > t_0$ to be

$$f(t,0,m) = 0,$$
$$f(t,a,m_0) = 0$$

and the initial condition to be

$$f(t_0,a,m) = \begin{cases} \Phi(a,m), & m > m_0, \quad a > 0 \\ 0, & m = m_0, \quad a = 0. \end{cases}$$

We begin with the most general assumption that growth and loss are functions of age a, mass m, and system time t. While we do not expect these to explicitly be functions of system time t, we consider so in Section 8.4.2 using system time as a proxy for age for a population in which most individuals are formed in a narrow window of time.

8.2.1 *Generalized PDE system*

Consider our PDE (8.1) in terms of the general form

$$A(t,a,m)f_t + B(t,a,m)f_a + C(t,a,m)f_m = D(t,a,m,f), \qquad (8.2)$$

where

$$\begin{aligned} A(t,a,m) &= A, \\ \beta(t,a,m) &= B, \\ C(t,a,m) &= g(t,a,m), \\ D(t,a,m,f) &= -\left(d(t,a,m) + \frac{\partial g(t,a,m)}{\partial m}\right) f(t,a,m). \end{aligned}$$

We use initial and boundary conditions at $t_0 > 0$, such that all individuals have appeared. Thus, the boundary conditions are

$$\begin{aligned} f(t,0,m) &= 0, \\ f(t,a,m_0) &= 0, \end{aligned}$$

and the initial conditions are

$$f(t_0,a,m) = \begin{cases} \Phi(a,m), & m > m_0, \quad a > 0 \\ 0, & m = m_0, \quad a = 0. \end{cases}$$

8.3 Solving the Full Model

To begin, we consider the case $A = B = 1$. Thus, our general equation becomes

$$\begin{aligned} &f_t(t,a,m) + f_a(t,a,m) + g(t,a,m)f_m(t,a,m) \\ &\quad = -\left(d(t,a,m) + \frac{\partial g(t,a,m)}{\partial m}\right) f(t,a,m). \end{aligned}$$

To simplify the notation, we denote

$$\delta(t,a,m) = \left(d(t,a,m) + \frac{\partial g(t,a,m)}{\partial m}\right).$$

This formulation is similar to extensions of the McKendrick model made by Sinko and Streifer[15, 16] and Nisbet and Gurney.[12] In both cases, simplifying assumptions were necessary on $C = g(t,a,m)$, the coefficient to the derivative of mass, to analytically simplify and solve the equations. In Ref. 15, they assume the growth function is independent of mass, i.e., $g(t,a,m) = g(t,a)$, and in Ref. 12, that growth and death are independent of age, i.e., $g(t,a,m) = g(t,m)$ and $\delta(t,a,m) = \delta(t,m)$. Similarly, we will need to impose restrictions on our growth function, discussed in detail in what follows.

In the full three-dimensional case, our system is equivalent to the following system of ordinary differential equations:

$$\mathrm{d}t = \mathrm{d}a = \frac{\mathrm{d}m}{g(t,a,m)} = \frac{\mathrm{d}f}{-\delta(t,a,m)f}.$$

Because we assume that all individuals enter the system (for example, through birth) prior to time $t = 0$, we always satisfy the condition $a > t$, and all solutions presented assume this condition.

From our system, we find

$$\int_{t0}^{t} \mathrm{d}\hat{t} = \int_{a_0}^{a} \mathrm{d}\hat{a},$$

which can be rewritten as $a = t - t_0 + a_0$.

Assuming that our function for growth is separable, such that $g(t,a,m) = g_{\mathrm{TA}}(t,a)g_{\mathrm{M}}(m)$, and using our known relationship between a and t, we have

$$\int_{t_0}^{t} g_{\mathrm{TA}}(\hat{t}, \hat{t} - t_0 + a_0)\mathrm{d}\hat{t} = \int_{m_0}^{m} \frac{\mathrm{d}\hat{m}}{g_{\mathrm{M}}(\hat{m})}. \tag{8.3}$$

Assuming $\dfrac{1}{g_{\mathrm{M}}}$ has an invertible antiderivative, which we will denote as G_{M}, we can solve Eq. (8.3) to obtain

$$m_0 = G_{\mathrm{M}}^{-1}\left[G_{\mathrm{M}}(m) - \int_{t_0}^{t} g_{\mathrm{TA}}(\hat{t}, \hat{t} - t_0 + a_0)\mathrm{d}\hat{t}\right].$$

Using our expressions for a_0 and m_0, we find that

$$\ln(f) - \ln\left(\Phi(a_0, m_0)\right) = -\int_{t_0}^{t} \delta\Bigg(\hat{t}, \hat{t} - t_0 + a_0, G_{\mathrm{M}}^{-1}\Bigg[G_{\mathrm{M}}(m_0) + \int_{t_0}^{\hat{t}} g_{\mathrm{TA}}\left(\tilde{t}, \tilde{t} - t_0 + a_0\right)\mathrm{d}\tilde{t}\Bigg]\Bigg)\mathrm{d}\hat{t}. \quad (8.4)$$

Using the substitutions

$$-t_0 + a_0 = a - t$$

and

$$m_0 = G_{\mathrm{M}}^{-1}\left[G_{\mathrm{M}}(m) - \int_{t_0}^{t} g_{\mathrm{TA}}\left(\hat{t}, \hat{t} - t_0 + a_0\right)\mathrm{d}\hat{t}\right],$$

we can rewrite Eq. (8.4) as

$$f(t, a, m) = \Phi\Bigg(a - t + t_0,\ G_{\mathrm{M}}^{-1}\Bigg[G_{\mathrm{M}}(m) - \int_{t_0}^{t} g_{\mathrm{TA}}\left(\hat{t},\ \hat{t} + a - t\right)\mathrm{d}\hat{t}\Bigg]\Bigg)$$
$$\times \exp\Bigg\{-\int_{t_0}^{t} \delta\Bigg(\hat{t},\ \hat{t} + a - t,\ G_{\mathrm{M}}^{-1}\Bigg[G_{\mathrm{M}}(m) - \int_{\hat{t}}^{t} g_{\mathrm{TA}}\left(\tilde{t},\ \tilde{t} + a - t\right)\mathrm{d}\tilde{t}\Bigg]\Bigg)\mathrm{d}\hat{t}\Bigg\}.$$

A detailed derivation of $f(t, a, m)$ can be found in the Appendix.

Sinko and Streifer derive a simpler version of this solution in the case where $g(t, a, m)$ is only a function of time and age.[15] This would be equivalent to a choice of $g(t, a, m) = g_{\mathrm{TA}}(t, a)$ in our notation. With their simplification, $g_{\mathrm{M}}(m) = 1$; thus, $1/g_{\mathrm{M}}$ always has an invertible antiderivative, and Eq. (8.3) is straightforward to determine for m. Under this simplification, our solution collapses to their solution, which can handle more complicated boundary conditions. Similarly, Oldfield maintains full dimensionality in the model but restricts coefficients of the partial derivative terms to only depend on the corresponding variable.[14] With this simplification on the coefficients, the model is solvable generally, even with the inclusion of additional traits.

8.3.1 *Sufficient conditions*

Our analytical solution is guaranteed to be valid if the mass-dependent growth function g_M is chosen such that $1/g_M$ has an antiderivative that is invertible on $\mathbb{R}$, and whose inverse is defined on all of $\mathbb{R}$. We note that this is a sufficient condition for our solution to be valid; however, in Section 8.5.2, we demonstrate how we can relax this condition in a particular example by choosing appropriate bounds on the masses considered.

8.3.2 *Special case solutions*

For particular choices of our growth function, where we have simple analytical forms for our antiderivative, we can simplify our solution for $f(t, a, m)$. For example, if the growth function is independent of mass, i.e., $g_{\mathrm{M}}(m) = c$ (where c is a constant), then we have $G_{\mathrm{M}}(m) = \dfrac{m}{c}$ and $G_{\mathrm{M}}^{-1}(m) = cm$ and our full solution becomes

$$\begin{aligned} f(t,a,m) = \Phi\left(a - t + t_0, m - c\int_{t_0}^{t} g_{\mathrm{TA}}(\hat{t},\ \hat{t} + a - t)\mathrm{d}\hat{t}\right) \\ \times \exp\left\{ - \int_{t_0}^{t} \delta\left(\hat{t},\ \hat{t} + a - t, m \right.\right. \\ \left.\left. - c\int_{\tilde{t}}^{t} g_{\mathrm{TA}}(\tilde{t},\ \tilde{t} + a - t)\mathrm{d}\tilde{t}\right)\mathrm{d}\hat{t}\right\}. \end{aligned}$$

We examine additional solutions numerically in Section 8.5.

8.4 Simplified Models

8.4.1 *Independence from system time*

Under the case $A = 0$ and $B = 1$, the PDE (8.2) does not depend on system time, t. As time is an inherent component of these biological systems, no authors, to our knowledge, have made this type of simplification. For example, our function for growth must be interpreted as "growth per unit age." We find this reduction to be less intuitive than other simplifications.

To continue with this simplification, it requires the assumptions that all functions are independent of time, i.e., $g(t, a, m) = g(a, m)$, and $d(t, a, m) = d(a, m)$. Furthermore, our boundary condition becomes $\psi(t, m) = \psi(m)$. Our PDE becomes

$$f_a(a, m) + g(a, m) f_m(a, m) = -\left(d(a, m) + \frac{\partial g(a, m)}{\partial m}\right) f(a, m).$$

We employ the boundary conditions

$$f(0, m) = \psi(m), \quad m \geq m_0,$$
$$f(a, m_0) = \begin{cases} \psi(m_0), & a = 0, \\ 0, & a > 0. \end{cases}$$

This assumes that individuals have a distribution of masses at age zero, and there are no individuals of age $a > 0$ with minimum mass m_0.

In order to obtain an explicit solution, we impose an additional constraint on our assumption: the separability of $g(a, m) = g_A(a) g_M(m)$. Here $g_A(a)$ is the component of $g(a, m)$ not dependent on m and $g_M(m)$ is the component not dependent on a.

Our PDE is equivalent to the following system of ODEs:

$$\mathrm{d}a = \frac{\mathrm{d}m}{g_A(a) g_M(m)} = \frac{\mathrm{d}f}{-\delta(a, m) f}.$$

Solving this system, we obtain

$$f(a, m) = \psi\left(G_M^{-1}\left[G_M(m) - \int_{a_0}^{a} g_A(\hat{a})\mathrm{d}\hat{a}\right]\right)$$
$$\times \exp\left\{-\int_{a_0}^{a} \delta\left(\hat{a}, G_M^{-1}\left[G_M(m) - \int_{\hat{a}}^{a} g_A(\tilde{a})\mathrm{d}\tilde{a}\right]\right)\mathrm{d}\hat{a}\right\}.$$

8.4.2 *Independence from age*

Under the case $B = 0$ and $A = 1$, the PDE (8.2) does not depend on age, a. This requires the assumptions on the functions that $g(t, a, m) = g(t, m)$ and $d(t, a, m) = d(t, m)$. In addition, we assume that all individuals have entered the population by the time $t = t_0$ such that the initial condition simplifies to $\Phi(a, m) = \Phi(m)$.

Our PDE becomes

$$f_t(t,m) + g(t,m)f_m(t,m) = -\left(d(t,m) + \frac{\partial g(t,m)}{\partial m}\right) f(t,m).$$

Our boundary and initial conditions are given by

$$f(t, m_0) = 0,$$

$$f(t_0, m) = \begin{cases} \Phi(m), & m > m_0, \\ 0, & m < m_0, \end{cases}$$

such that all individuals are present at time $t = t_0$, and each individual's mass is greater than m_0. Again, in order to obtain an explicit solution, we impose an additional constraint on our assumption: the separability of $g(t,m) = g_{\mathrm{T}}(t)g_{\mathrm{M}}(m)$.

Similar to our previous example, the above PDE is equivalent to the following system of ODEs:

$$\mathrm{d}t = \frac{\mathrm{d}m}{g_{\mathrm{T}}(t)g_{\mathrm{M}}(m)} = \frac{\mathrm{d}f}{-\delta(t,m)f}.$$

Solving this system, we obtain

$$f(t,m) = \Phi\left(G_{\mathrm{M}}^{-1}\left[G_{\mathrm{M}}(m) - \int_{t_0}^{t} g_{\mathrm{T}}(\hat{t})\mathrm{d}\hat{t}\right]\right)$$
$$\times \exp\left\{-\int_{t_0}^{t} \delta\left(\hat{t}, G_{\mathrm{M}}^{-1}\left[G_{\mathrm{M}}(m) - \int_{\hat{t}}^{t} g_{\mathrm{T}}(\tilde{t})\mathrm{d}\tilde{t}\right]\right)\mathrm{d}\hat{t}\right\}.$$

The independence from age may be a reasonable assumption if all individuals in the population are present by a certain time. In such a case, as age refers to chronological age, which progresses at the same rate as system time, then the system time could be used as a proxy for age, facilitating the removal of the third variable.

Nisbet and Gurney mention that the mathematical complexity is greatly reduced with the assumption that growth, birth, and death are independent of age.[12] Thus, they reformulate their equation without dependence on age. Their resulting solution has an additional intricacy compared to ours, because their boundary condition depends on the population size. As a result, they end up with a delay in their renewal equation. Furthermore, they draw upon the assumption that all individuals are born with identical mass, thus reducing their boundary condition to be non-zero at a single point.[12] The appearance of such delays is common in age and stage structured population models.[27]

8.4.3 *Independence from mass*

Under the case $A = 1$, $B = 1$, and $C = 0$, the PDE (8.2) does not depend on mass, m. This requires the assumption $d(t, a, m) = d(t, a)$ and $\phi(a, m) = \phi(a)$. We do not discuss this further as it reduces to the McKendrick equation,[6] which has been extensively studied elsewhere, e.g., Refs. 6, 8, 10, 11 and 3.

8.5 Example Solutions Implemented Numerically

Our analytical solution can be used to compare the effects of different types of mass and age-dependent growth rate functions $g(t, a, m) = g_{TA}(t, a)g_{\mathrm{M}}(m)$ on population density $f(t, a, m)$. See Appendix 8.A.2 for details on the numerical implementation of our analytical solution. In the absence of external environmental factors, like changes in temperature or resources, for which time can be used as a proxy, growth is more likely to depend on mass and age than on time. Consequently, we omit the explicit dependence on time t in our numerical experiments, such that $g(t, a, m) = g_{TA}(a)g_{\mathrm{M}}(m)$. Exploring the effects of external forces that affect growth, either directly or through resource competition, is an interesting subject for future work.

8.5.1 *Specific choices of g_M and g_{TA}*

Here, we consider three broad categories for g_M: (1) increasing for all m, (2) increasing to a positive constant as $m \to \infty$, and (3) decreasing to zero as $m \to \infty$. Tables 8.1 and 8.2 summarize the combinations of functional forms considered, all of which satisfy the sufficient conditions for the analytical solution to be valid. In Section 8.5.2 that follows, we illustrate an example that relaxes these conditions, but remains valid over a particular domain.

We consider analytical solutions with various growth and decay functions that incorporate mass, as summarized in Tables 8.1 and 8.2. Results relating to these forms are found in Sections 8.5.3–8.5.8 and Figures 8.1–8.8.

8.5.2 *Violation of sufficient conditions*

The sufficient condition that G_M, the antiderivative of $1/g_M$, be defined and invertible on all $\mathbb{R}$ with an inverse defined on all of $\mathbb{R}$ is

Table 8.1. Forms considered for loss function, $\delta(t, a, m)$.

$\delta(t, a, m)$	Figure
0	8.1, 8.2, 8.3, 8.4, 8.5, 8.6 first row
$\frac{a}{2}$	8.6 second row
$\frac{m}{2}$	8.6 third row, 8.7
$\frac{am}{4}$	8.6 fourth row, 8.7

Note: Growth functions are found in Table 8.2.

Table 8.2. Forms considered for growth function, $g(t, a, m) = g_{\mathrm{M}}(m)g_{\mathrm{TA}}(t, a)$.

	$g_{\mathrm{M}}(m)$					$g_{\mathrm{TA}}(t, a)$	
Figures	c	cm	$\frac{m^2}{m^2+c^2}$	$c - cm$	$1 - \frac{m^2}{m^2+c^2}$	c	ca
8.1	a–e					a–e	
8.2		a,c,e	b,d,f			a–d	
8.3			a–d			a–d	
8.4	b,d,f	a,c,e				a,c,e	b,d,f
8.5				a,d	a,c,d,f	a,b,d,e	c,f
8.6	a,b	c				a–c	
8.7	8.7	8.7				8.7	

Note: Here, g_{TA} and g_{M} are the time/age-dependent and mass-dependent portions of the growth function, respectively. The numbers followed by letters indicate which subfigures used the corresponding functional forms for g_{M} and g_{TA}. Loss functions are found in Table 8.1.

rather restrictive. Here, we demonstrate that this condition can be relaxed, provided that care is taken to guarantee that G_M^{-1} is defined at $\sigma(t, m) = G_{\mathrm{M}}(m) - \int_{t_0}^{t} g_{\mathrm{TA}}(\hat{t}, \hat{t} - t_0 + a_0)d\hat{t}$ for all times (t) and masses (m).

As an example, we consider

$$g_{\mathrm{M}}(m) = \frac{1}{2}\left(1 - \frac{m}{2}\right)$$

for $m < 2$, with $g_{TA}(t, a) = 1$, and $\delta(t, a, m) = 0$. This choice for g_M yields

$$g_{\mathrm{M}}(m) = -4\log(2-m) \quad \text{and} \quad G_M^{-1}(m) = 2 - e^{-\frac{m}{4}}.$$

With this choice for g_M, although G_M is only defined for $m < 2$, its inverse is defined for all real numbers, and therefore, is well defined at $\sigma(t, m)$ for all pairs (t, m). Thus, as long as we restrict to considering $m < 2$, the presented solution is mathematically valid.

8.5.3 *The effect of mass-dependent growth*

Different growth functions, as expected, can lead to very different population distributions in mass and age across time. Using numerical simulations of our analytical solution, we can visualize how population density progresses over time depending on the functional forms for growth. Some mass-dependent growth functions lead to populations that have roughly uniformly large or uniformly small mass, while others lead to populations consisting of individuals with a broad range of masses. Furthermore, we demonstrate instances where there is a bifurcation in mass across the population, with many individuals remaining small while others grow large, and with fewer intermediate masses.

Figure 8.1 illustrates a baseline case where growth is constant: $g(t, a, m) = g_{\mathrm{M}}(m)g_{\mathrm{TA}}(a) = 1$. As expected, similar to a standard wave equation, when the growth rate is constantly equal to one, the population distribution simply shifts to larger masses linearly with time, as seen in Figure 8.1(a) when viewing with age and in Figure 8.1(b) when viewing with mass. The initial density in age (leftmost peak in Figure 8.1(c)) increases through time, as does density in mass in Figure 8.1(d). The main difference between viewing through the lens of mass or age is at what value of t the density leaves the bounds of the simulation ($m_{\max} = 5$ and $a_{\max} = 8$). The total density remains constant until it reaches the bounds of the numerical simulation at $m = 5$ and $a = 8$, which leads to a decline between $t = 4$ and $t = 5$ in Figure 8.1(e).

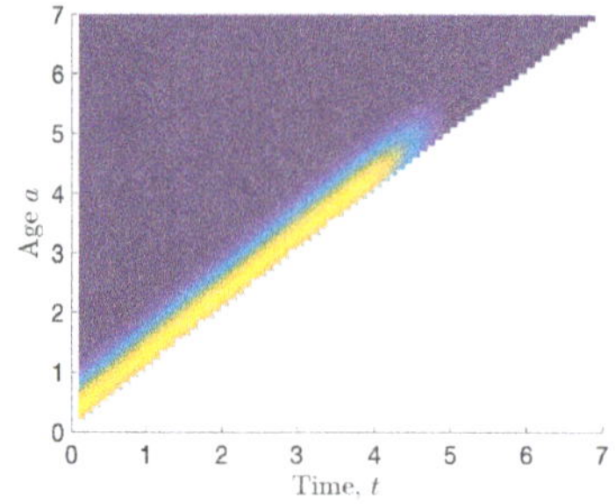

(a) Age density

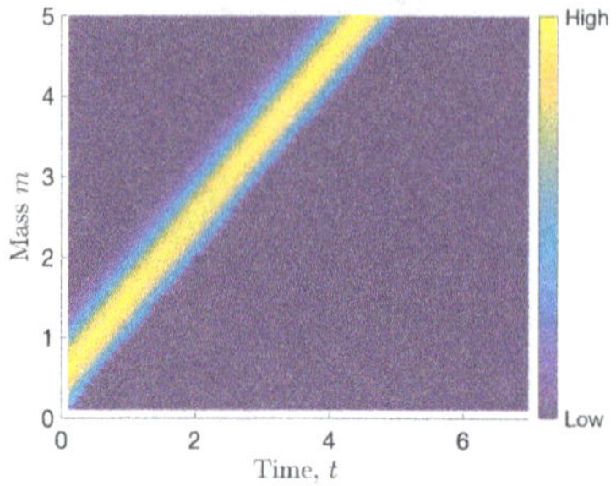

(b) Mass density

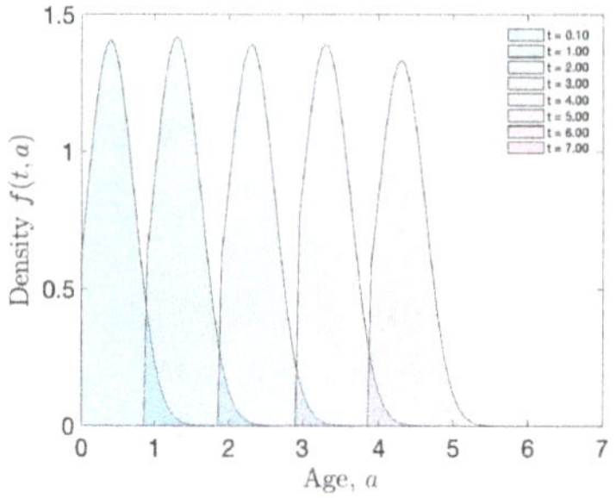

(c) Snapshots in time of age density

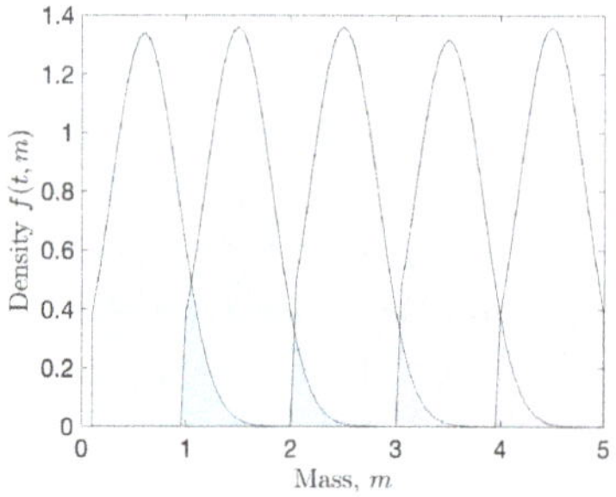

(d) Snapshots in time of mass density

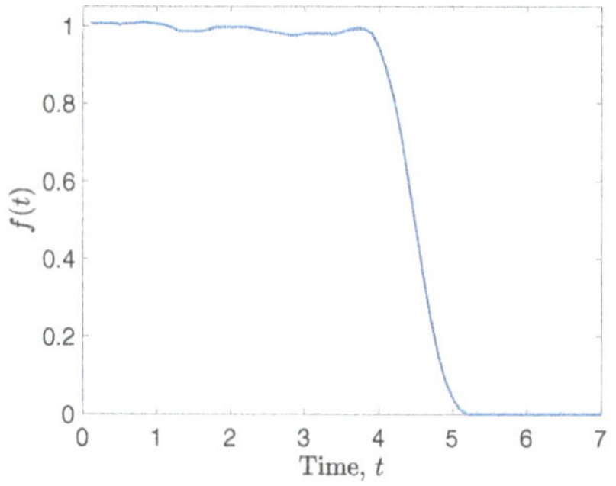

(e) Total density over time

Fig. 8.1. Density plots when growth is a constant function, with no dependence on mass or age: $g(t, a, m) = g_{\mathrm{M}}(m)g_{\mathrm{TA}}(a) = 1$. In (a) and (b), density scales from low (blue) to high (yellow). White areas are where the solution is undefined. In (c) and (d), snapshots of the distributions at particular time points are shown from $t_0 = 0.1$ (light blue) to $t = 7$ (dark pink). Once the density moves beyond the boundary of our numerical simulation, the peaks disappear. The raw values for density depend on the discretization step. In (e), the total density, across age and mass, should always equal one, unless the simulation exceeds the maximum value considered in one of the variables ($t_{\max} = 7.1$, $a_{\max} = 8$, $m_{\max} = 5$). Fluctuations are due to discretization step.

8.5.4 *Age and time-independent growth*

In Figure 8.2, we consider two examples where the growth rate is increasing, but where there is no time or age-dependent growth, that is $g_{TA}(t, a) = 1$. Figures 8.2(a)–8.2(b) demonstrate the mass-distribution of population density across time when growth is either linear in mass ($g_M(m) = m$) or increasing asymptotically in mass to one ($g_M(m) = \frac{m^2}{m^2+c^2}$, $c = 2$).

When growth is linear in mass, the change in mass is exponential, and the distribution across mass becomes more diffuse, but with the population skewed toward larger masses. When growth increases asymptotically to a constant of one, we observe very slow growth initially, with small individuals remaining small, and larger individuals growing more rapidly, eventually bifurcating away from the population that began at lower mass. This coincides with the observations from Figure 8.3, which illustrates the time-progression of the density f as a function of age and mass when the growth function is an asymptotically increasing function of mass. As time advances, the density becomes more diffuse and spreads out along the mass axis, with more of the density remaining concentrated at smaller masses. In fact, by time $t = 6$, we see a bifurcation from a population distribution with a single peak to one with a bi-modal distribution in mass, with a higher peak in density at very low mass, and a second peak at intermediate mass.

Figures 8.2(e) and 8.2(f) demonstrate, for linearly increasing g_M and asymptotically increasing g_M, respectively, the change in total population density over time. Figures 8.2(c) and 8.2(d) show the distribution of masses (of any age) at different times in the third row. Total density declines rapidly beginning around $t = 2$ and $t = 6$ under linear and asymptotic growth, respectively, because the highest mass individuals in the population have grown beyond the mass-bounds of the numerical simulation at those times. In the case of linear increasing growth in mass ($g_M(m) = m$), the mass-distribution plots in Figure 8.2(c) emphasize the diffusion and shift to larger masses described above. In the case of asymptotically increasing growth in mass, Figure 8.2(d) shows more clearly the population density skewed toward smaller masses with the peak density remaining fairly stationary, but becoming more diffuse over time, with a subtle shift to a bi-modal distribution of masses at later times.

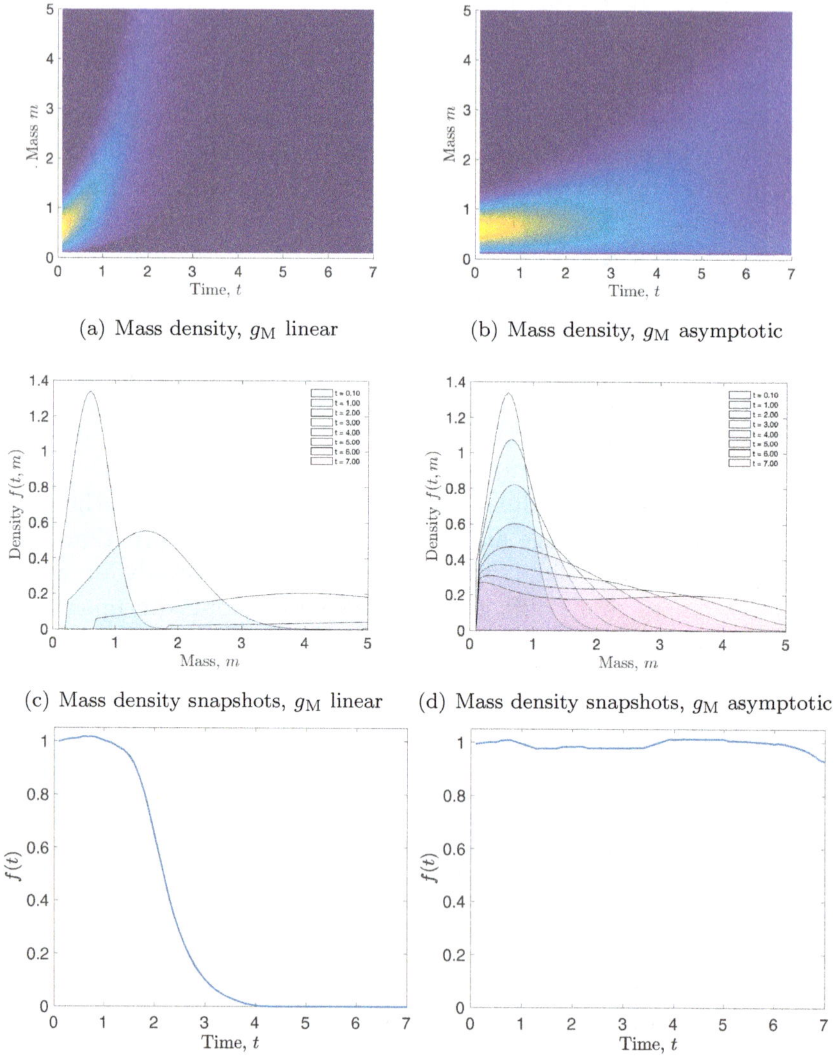

(a) Mass density, g_{M} linear

(b) Mass density, g_{M} asymptotic

(c) Mass density snapshots, g_{M} linear

(d) Mass density snapshots, g_{M} asymptotic

(e) Total density, g_{M} linear

(f) Total density, g_{M} asymptotic

Fig. 8.2. Monotonic increasing functions of mass with different asymptotic behavior. For (a),(c),(e), the growth function of mass is linear, $g_{\mathrm{M}}(m) = m$. For (b),(d),(f), the growth function of mass is asymptotic, $g_{\mathrm{M}}(m) = \frac{m^2}{m^2+c^2}$, $c = 2$. In all cases, the growth function of time and age is constant, $g_{\mathrm{TA}}(a) = 1$. See Figure 8.1 for a detailed description of the figure properties.

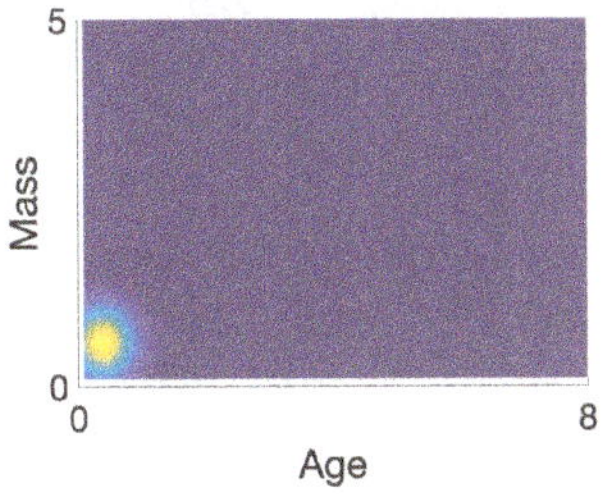

(a) $t = 0.1$

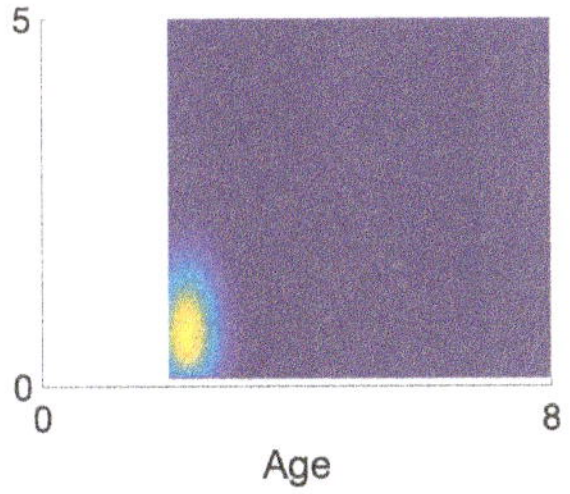

(b) $t = 2$

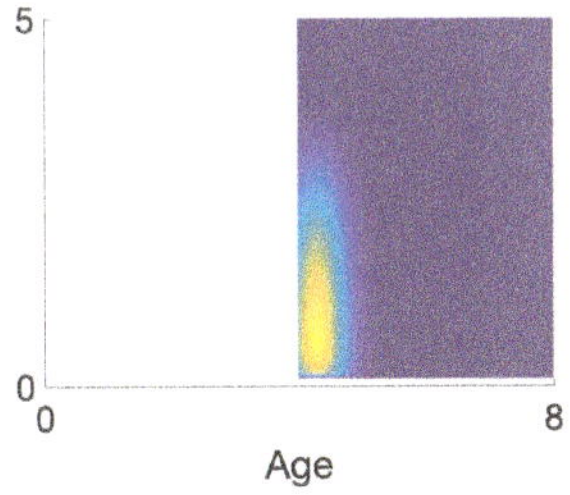

(c) $t = 4$

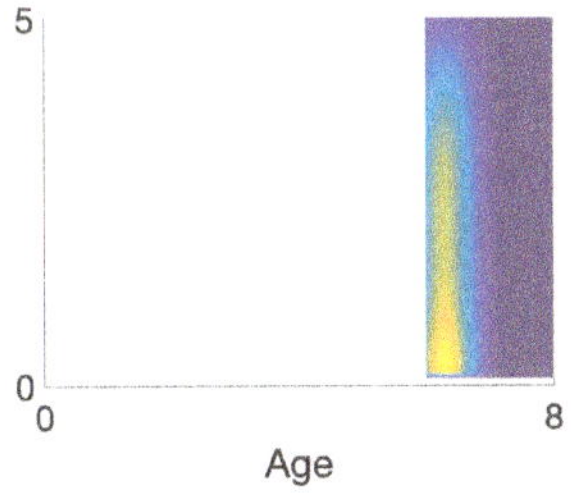

(d) $t = 6$

Fig. 8.3. Distribution of age and mass through time with $g_{\mathrm{M}}(m) = m^2/(m^2+c^2)$, $c = 2$, $g_{\mathrm{TA}}(a) = 1$. Density scales from low (blue) to high (yellow). White areas are where the solution is undefined. The subplots are scaled to the highest density for the individual subplot: colors are not comparable across subplots. See Figure 8.1 for a detailed description of the figure properties.

8.5.5 *Comparing increasing growth that is linear in mass vs linear in age*

Figure 8.4 compares changes in population density over time when growth is linearly increasing in age to the case where growth is linearly increasing in mass. The plots of total density over time are fairly similar; however, when growth linearly increases with mass, the disappearance of density from the simulation as it crosses $m = 5$ is more varied. Figures 8.4(a) and 8.4(c) show the change in mass-distribution over time and highlight clear differences in the population dynamics when we isolate the effect on mass. When growth is a linearly increasing function of age, as in Figures 8.4(b) and 8.4(d), the peak mass shifts at a similar rate, but the density diffuses more slowly than in the scenario where growth is a linearly increasing function of mass. Furthermore, when growth is linear in age, the distribution of masses remains fairly symmetric about this peak, whereas the density becomes more skewed toward higher masses when growth is linear in mass.

8.5.6 *Comparing decreasing functions of mass*

In Figure 8.5, we consider three examples where the growth rate is positive but decreasing as mass decreases. For Figures 8.5(b) and 8.5(c), we consider a functional form for mass-dependent growth, $g_M(m) = \frac{m^2}{m^2+c^2}$, $c = 1$, which satisfies all our assumptions, while in Figure 8.5(a), we choose a form, $g_M(m) = \frac{1}{2}\left(1 - \frac{m}{2}\right)$, that only satisfies our conditions if $m < 2$. Thus, we restrict our numerical simulation to where $m < 2$. The difference between Figures 8.5(b) and 8.5(c) is our choice of age-dependent growth with $g_{TA}(a) = 1/10$ and $g_{TA}(a) = a/10$, respectively.

In all cases, the density of mass "accumulates" over time near $m = 2$. This occurs because those with larger mass grow more slowly. This allows those of smaller mass to catch up. This is evident in the density by mass shown in Figures 8.5(e) and 8.5(f), which grow in peak height but whose width narrows significantly. Importantly, no mass at or above $m = 2$ is observed in Figure 8.5(d), as we have the restriction at $m = 2$, in order to satisfy our conditions. Figures 8.5(b) and 8.5(c) have identical mass-dependent growth functions but differ in age-dependent growth, with Figure 8.5(f)

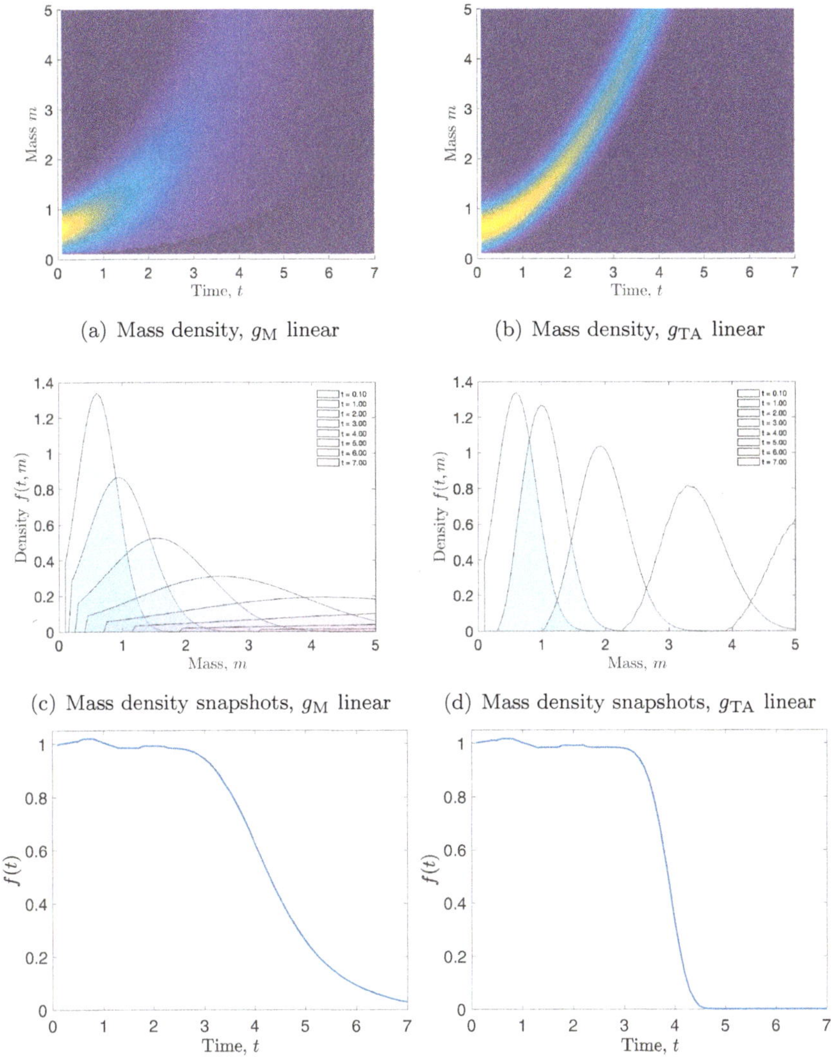

(a) Mass density, g_{M} linear

(b) Mass density, g_{TA} linear

(c) Mass density snapshots, g_{M} linear

(d) Mass density snapshots, g_{TA} linear

(e) Total density, g_{M} linear

(f) Total density, g_{TA} linear

Fig. 8.4. Linear increasing functions either of mass or age. For (a),(c),(e), growth is linear in mass and constant in time and age, $g_{\text{M}}(m) = m$, $g_{\text{TA}}(a) = 1/2$. For (b),(d),(f), growth is constant in mass and linear in age, $g_{\text{M}}(m) = 1/2$, $g_{\text{TA}}(a) = a$. See Figure 8.1 for a detailed description of the figure properties.

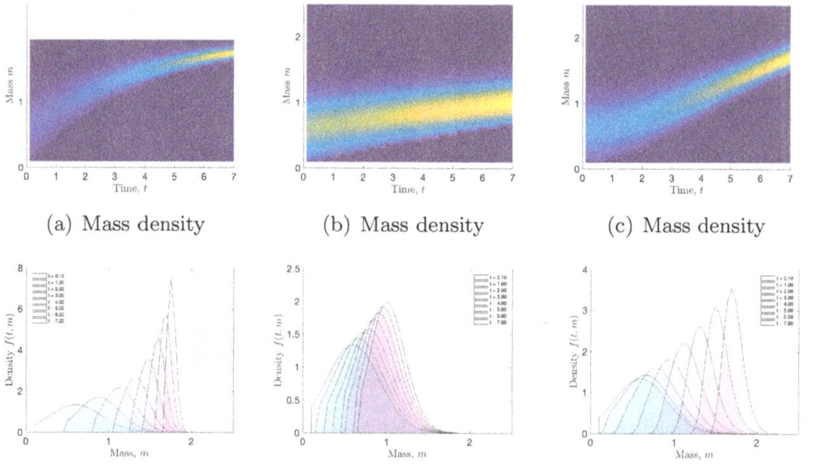

(a) Mass density (b) Mass density (c) Mass density

(d) Mass density snapshots (e) Mass density snapshots (f) Mass density snapshots

Fig. 8.5. Monotonic decreasing functions of mass with different asymptotic behavior. For (a)–(f), growth is decreasing with mass, but the functional forms differ. For (a),(d), $g_{\mathrm{M}}(m) = \frac{1}{2}\left(1 - \frac{m}{2}\right)$, $g_{\mathrm{TA}}(a) = \frac{1}{10}$; for (b),(e), $g_{\mathrm{M}}(m) = 1 - \frac{m^2}{m^2+c^2}$, $c = 1$, $g_{\mathrm{TA}}(a) = \frac{1}{10}$; for (c),(f), growth is linear with age, $g_{\mathrm{M}}(m) = 1 - \frac{m^2}{m^2+c^2}$, $c = 1$, $g_{\mathrm{TA}}(a) = \frac{a}{10}$. In (d), the gray-shaded area is where m is not defined. See Figure 8.1 for a detailed description of the figure properties.

incorporating dependence on age. This is why growth increases more sharply for mass below 2; however, as the mass-dependent growth component decreases to nearly zero, the mass accumulates around and mostly below two. This is in contrast to Figure 8.5(d), where mass is restricted to be below two, and to Figure 8.5(e), where growth stalls before reaching masses of two.

8.5.7 *The effect of age and mass-dependent decay functions*

Figures 8.6 and 8.7 illustrate how the dynamics change in the base case when the decay function, $\delta(t, a, m)$, is non-zero. In Figure 8.6, we consider three versions of growth: constant in mass with a constant of one (Figure 8.6(a)); constant in mass with a constant of one-half (Figure 8.6(b)); and linear in mass with $g_{\mathrm{M}}(m) = m/5$ (Figure 8.6(c)). All growth is constant in age and time, $g_{\mathrm{TA}}(a) = 1$. In each row of Figure 8.6, the density changes due to loss from the

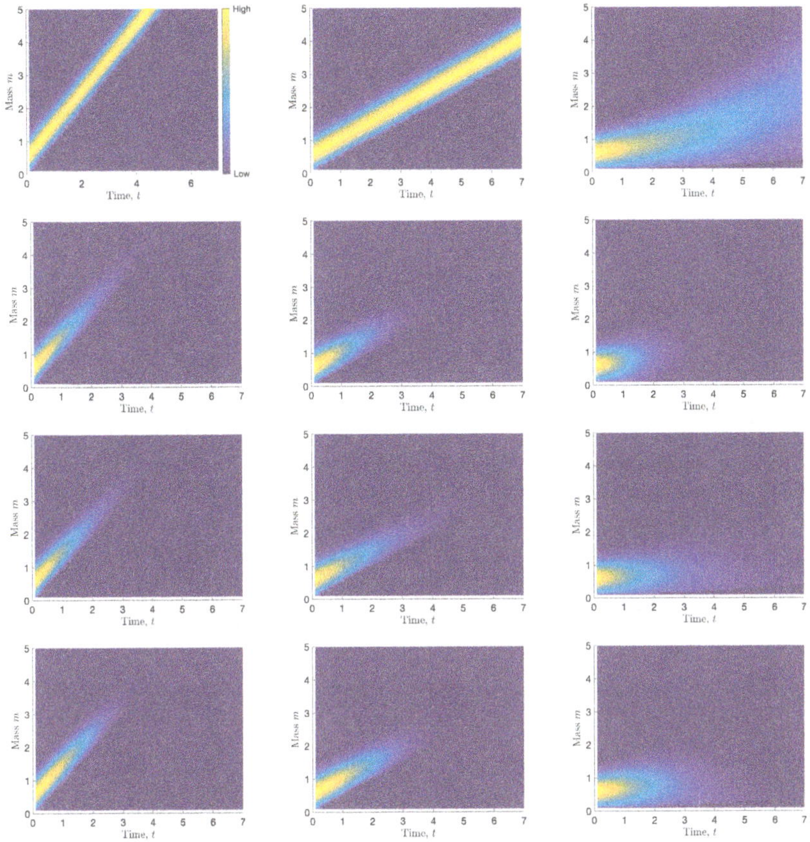

(a) g_{M} constant of 1 (b) g_{M} constant of 1/2 (c) g_{M} linear

Fig. 8.6. Mass density as the mortality function (δ) and growth function (g) vary. For (a), growth in mass is constant $g_{\mathrm{M}}(m) = 1$; for (b), growth in mass is constant, but smaller $g_{\mathrm{M}}(m) = 1/2$; for (c), growth in mass is linear, $g_{\mathrm{M}}(m) = m/5$. For row 1, $\delta(t, a, m) = 0$; for row 2, $\delta(t, a, m) = a/2$; for row 3, $\delta(t, a, m) = m/2$; for row 4, $\delta(t, a, m) = am/4$. For all subplots, growth is constant in age, $g_{\mathrm{TA}}(a) = 1$. Density scales from low (blue) to high (yellow). White areas are where the solution is undefined.

system, while the dynamics of growth remain the same in each column. The decrease in density, reflected in the earlier transition to solid blue in Figure 8.6 moving down a column, is different than that observed previously: whereas loss of density to the bounds of the numerical simulation can be remedied by increasing the bounds

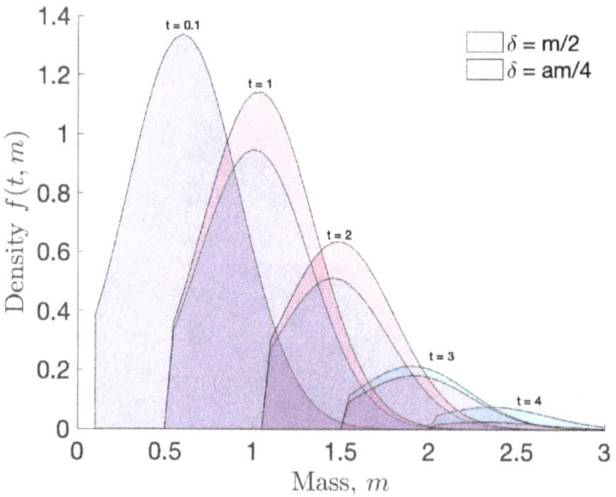

Fig. 8.7. Mass density snapshots with different mortality functions. These distributions of mass at various times reflect the mass density plots from Figure 8.6(b) where $g_{\mathrm{M}}(m) = 1/2$, $g_{\mathrm{TA}}(a) = 1$. In light blue, $\delta(t, a, m) = m/2$ (third row in Figure 8.6(b)) and in pink $\delta(t, a, m) = am/4$ (fourth row in Figure 8.6(b)). The purple is overlap of the two colors, as seen with identical distributions at $t = 0.1$, our initial condition.

considered, this decrease is removal of density from the system due to the loss term. In the base case, where $g(t, a, m) = 1$, increases in age and mass behave symmetrically. Thus, decreases in Figure 8.6(a) middle rows are nearly identical, except for the minor differences in the asymmetric initial condition.

When the loss function is a combination of age and mass, as in Figure 8.6 bottom row with $\delta(t, a, m) = am/4$, the loss occurs slower at small age and mass and even more quickly for large age and mass. This is seen more clearly in Figure 8.7, where the mass-density decreases more quickly for $\delta(t, a, m) = m/2$ at early time (green shading). In comparison, for $\delta(t, a, m) = am/4$, the decrease is slower at earlier times but more pronounced at later times (red shading).

8.5.8 *The effect of the initial condition distribution on growth*

Figure 8.8 shows linearly increasing mass-dependent growth starting from two initial distributions in age and mass: one with higher mean density in mass with increasing variation of mass (Figures 8.8(a),

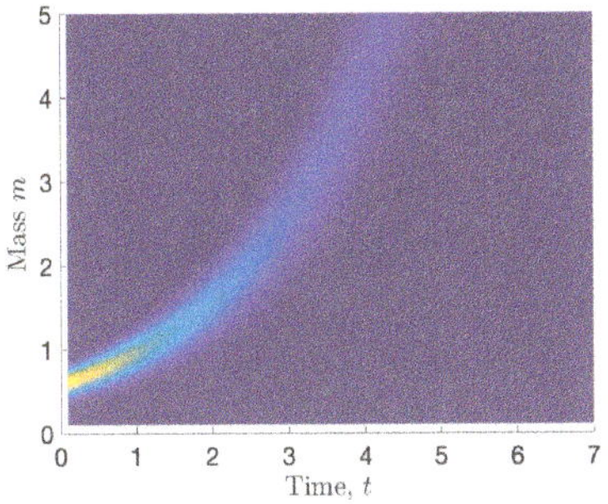

(a) IC larger in mass, $\sigma_m = 0.01$

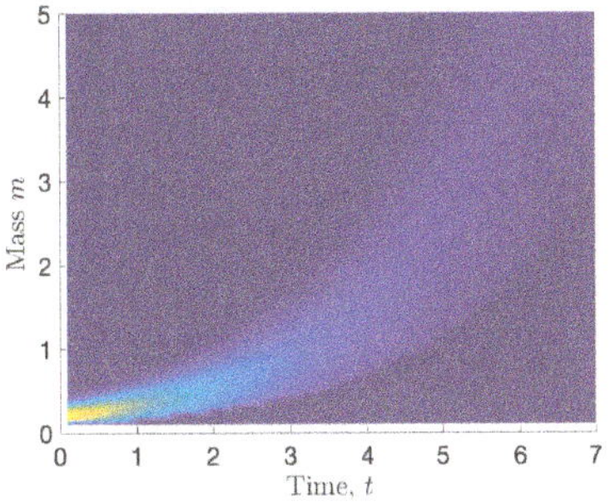

(b) IC larger in age, $\sigma_m = 0.01$

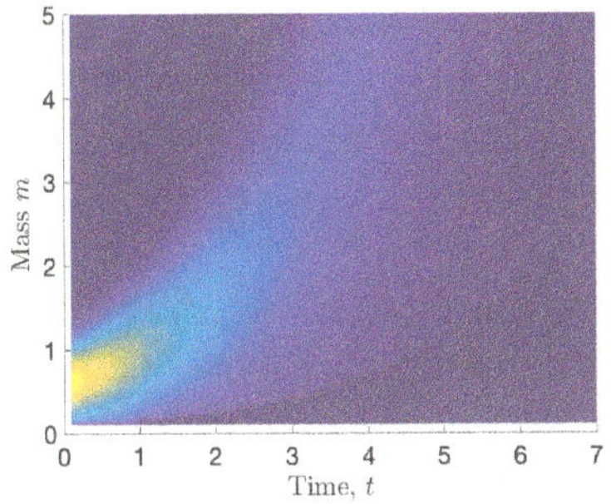

(c) IC larger in mass, $\sigma_m = 0.1$

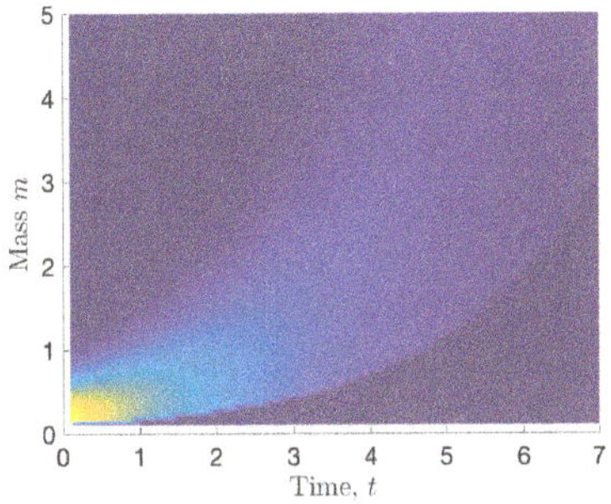

(d) IC larger in age, $\sigma_m = 0.1$

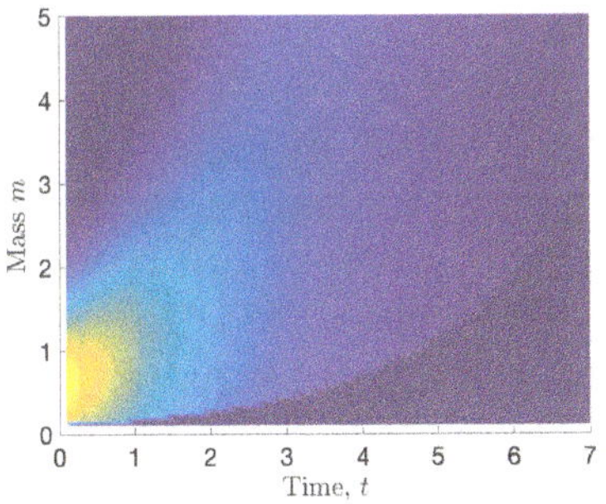

(e) IC larger in mass, $\sigma_m = 0.5$

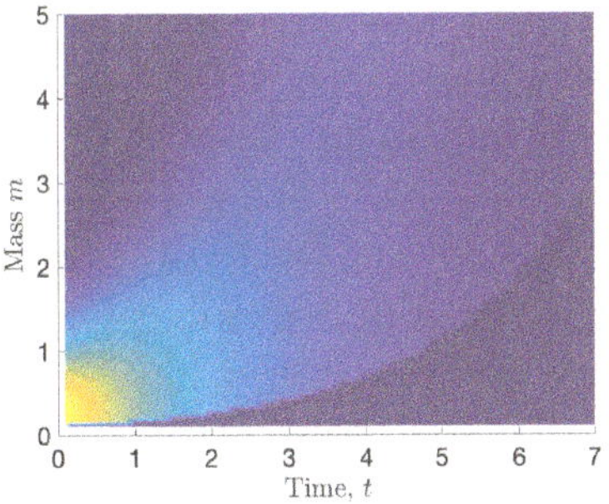

(f) IC larger in age, $\sigma_m = 0.5$

Fig. 8.8. Mass density with varying initial conditions with $g_{\mathrm{M}}(m) = \frac{m}{2}$, $g_{\mathrm{TA}}(a) = 1$. In (a), (c), (e), the mean initial mass is 0.6 and the mean initial age is 0.2. For (b), (d), (f), the mean initial mass is 0.2 and the mean initial age is 0.6. The diagonal of the variance matrix is smaller for mass, $\sigma_m = 0.01$, in (a)–(b); intermediate for mass, $\sigma_m = 0.1$, in (c)–(d); and larger for mass, $\sigma_m = 0.5$ in (e)–(f). In all cases, $\sigma_a = 0.1$, and the off-diagonal elements of the variance matrix are zero. Density scales from low (blue) to high (yellow). White areas are where the solution is undefined.

8.8(c), 8.8(e)) and the other with higher mean density in age with increasing variation of mass (Figures 8.8(b), 8.8(d), 8.8(f)).

We focus on different variances in the initial distribution of mass rather than in age because mass-dependent growth creates a feedback loop in the mass-distribution. The feedback is more nuanced in age. Once the initial variance in age is set, the age-distribution will remain the same across time, with the mean age shifting linearly with time. However, changing the variation in age, assuming the inclusion of age-dependent growth, can alter the progression of the mass-distribution through time. When the initial condition has higher mean density in age, there is less density at higher mass to grow more quickly. This is reflected in Figures 8.8(a), 8.8(c), 8.8(e) moving to higher masses at earlier times than in Figures 8.8(b), 8.8(d), 8.8(f). This is also observed in the total density dropping earlier and more rapidly, as the density moves above $m = 5$, the maximum in the simulation (not shown).

When the initial density has higher variation in mass, more spread in mass occurs over the course of the simulation (Figure 8.8, comparison of rows). This is because those with lower mass grow slowly and those with higher mass grow faster, facilitating their movement to higher mass and even faster growth. Interestingly, a higher initial density in mass with low variation in mass leads to less spread than a higher initial density in age with low variation in mass (Figures 8.8(a) and 8.8(b), respectively). This is the opposite pattern seen when variation in mass is larger ($\sigma_m = 0.1$ in Figures 8.8(c), 8.8(d) and $\sigma_m = 0.5$ in Figures 8.8(e), 8.8(f)). In these cases, the initial spread of mass across both low and high masses, which is greater when the initial mean density is greater in mass, leads to a wide range of growth and more spread. Importantly, these patterns are dependent on the choices of our initial conditions. In other words, which initial conditions lead to wider distributions of mass will be an interplay between the mean initial density, its variance, and the formulation of the growth function.

8.6 An Alternative to Partial Differential Equation Models

The integral projection model (IPM), as described in Section 8.1.2, is a well-studied alternate mathematical framework for including continuous age and size-structure. While IPMs have a long history in ecological modeling, few have leveraged this tool for modeling the dynamics of infectious disease.[21] Instead, PDEs have been the predominant approach for modeling within-host pathogen dynamics. However, these two approaches — PDEs and IPMs — are not interchangeable. Which approach to use will depend on the biological system under study, the type of data that is possible to collect for that system, and the type of model output required to address the question of interest.

IPMs, which are discrete in time, are implemented as data-driven models. The kernel that is used to project the population from one discrete time-step to the next is typically the result of a regression analysis using repeated measurements across time of a particular phenotype, such as size, for a set of individuals, along with data related to fecundity and survival of tracked individuals. While this type of data is easy to obtain for plant systems or other organisms that can be readily marked, tracked, and repeatedly measured, there are certain biological systems for which these repeated measurements across time are impossible to obtain. For example, individual parasites within a host cannot be repeatedly measured since their measurement requires the dissection of the host, or removal from the host. More recent techniques allow IPMs to be defined using population-level data, which may be more amenable to within-host systems.[24,26] In contrast to this data-driven approach, PDE models are built from a mechanistic understanding of the underlying process. These models can be calibrated using measurements of different individuals at different time points, which is more easily suited for biological systems in which repeated measurements across time are impossible.

The output of IPMs and PDEs is also fundamentally different. The kernel, described above, is used to project the distribution function $n(x,t)$ of individuals of state x (such as size) at time t using the equation $n(y,t+1) = \int_L^U K(y,x)n(x,t)\mathrm{d}x$, where $x \in [L,U]$ is the continuous range of possible values for the state x.[28] In contrast, the PDE models described here model the expected population density at time t, age a, and state (mass in our case) m. The distributions produced by the IPMs tend to stabilize quickly, whereas the PDEs show more variability across time.

A challenge for all mathematical models is the problem of identifiability — the ability to unambiguously estimate model parameters from data. Metcalf *et al.*[21] note that this is a major challenge for IPMs. On the other hand, some techniques, including the use of differential algebra, have been applied to age-structured PDE models[29] to determine model identifiability.

Finally, the two modeling approaches handle the time variable differently: IPMs assume time is a discrete random variable, whereas PDEs assume time is continuous. The reliability of the IPM output depends intimately on the coarseness of the available data used for the regression over the desired timescale. This may limit the applicability in terms of making projections for a specific population, however, IPMs may still be useful in hypothesis generation,[21] as is the PDE framework, and may be an important alternative for systems where individual data needed to parameterize or formulate other model structures are challenging to obtain.

8.7 Conclusion

This work contributes to the history of population models that build on the McKendrick framework. The formulation of our PDE (Eq. (8.1)) in time, age, and size extends this framework in line with previous derivations.[10,12,15] In order to maintain analytical tractability, previous work has made various simplifications including dimensional reductions,[12] simplifications of the growth function,[14,15] or assumptions on the distribution of individuals at the boundary.[12,14,30] Other authors choose to forgo analytical tractability and focus on more situations with more biological fidelity, but which necessitate numerical solutions, e.g., Ref. 31, or other model structures, e.g., Refs. 30 and 32.

Indeed, if age and size (or another feature of interest) depend on time, such as through changing environment or density dependence, we expect this system to be analytically intractable but that could be assessed numerically.

Here, we derive an analytical solution to a time-, age-, and mass-dependent population density model with growth dependent on age and mass with certain restrictions. One simplification is the omission of population renewal through births or immigration. While this may initially seem restrictive, there are many biological examples at the micro-scale for which this assumption is suitable, including pathogen population growth or cell growth *in vivo* or *in vitro*. This analytical solution could be fit to empirical data with different functional forms for growth and mortality rates to gain better understanding of the mechanisms underlying population growth in these micro-scale systems, and in particular, to inform the qualitative form of the functional response of growth and mortality rates based on mass and age. The assumption of a non-renewing population also has relevance to population dynamics at larger ecological scales, like the population dynamics of insects that emerge at discrete intervals. The model can be used to track population growth of a single generation of these insects, or can be applied iteratively, with new initial conditions at each discrete time-step.

Our numerical simulations of the analytical solution suggest that a population of similarly sized individuals may, for example, remain of similar size, become more widely distributed in mass, or may bifurcate into two mass groups, depending on the functional form of the growth rate. This relatively simple model may provide insight into what types of growth rate functions lead to different size-distribution patterns observed empirically.

Acknowledgments

We thank Eyvindur Ari Palsson, Michael Robert, and Melody Walker for helpful discussions. This work was supported by National Science Foundation Grants 1853495 and 1953838.

Appendix 8

8.A.1 Derivation of $f(t, a, m)$

In the full three-dimensional case, our system is equivalent to the following system of ordinary differential equations:

$$\mathrm{d}t = \mathrm{d}a = \frac{\mathrm{d}m}{g(t, a, m)} = \frac{\mathrm{d}f}{-\delta(t, a, m) f}. \tag{8.A.1}$$

From the first equality in Eq. (8.A.1),

$$\mathrm{d}t = \mathrm{d}a,$$

we integrate as follows:

$$\int_{t_0}^{t} \mathrm{d}\hat{t} = \int_{a_0}^{a} \mathrm{d}\hat{a},$$

where a_0 is the age of the individuals at time t_0 and t_0 is the time at which we first observe the system. This results in the following solutions:

$$a = t - t_0 + a_0, \tag{8.A.2}$$

$$t = a + t_0 - a_0, \tag{8.A.3}$$

$$a_0 = a - t + t_0, \tag{8.A.4}$$

$$-t_0 + a_0 = a - t. \tag{8.A.5}$$

Now, consider the first and third equality from Eq. (8.A.1) using our assumption that $g(t, a, m)$ is separable, i.e., $g(t, a, m) = g_{\mathrm{TA}}(t, a)g_{\mathrm{M}}(m)$, then

$$\mathrm{d}t = \frac{\mathrm{d}m}{g(t, a, m)} = \frac{\mathrm{d}m}{g_{\mathrm{TA}}(t, a)g_{\mathrm{M}}(m)}.$$

However, a is a function of t, as seen in Eq. (8.A.2), so this becomes

$$\mathrm{d}t = \frac{\mathrm{d}m}{g_{\mathrm{TA}}(t, a)g_{\mathrm{M}}(m)} = \frac{\mathrm{d}m}{g_{\mathrm{TA}}(t, t - t_0 + a_0)g_{\mathrm{M}}(m)}.$$

Thus, using separation of variables, we write the integral of this system as

$$\int_{t_0}^{t} g_{\mathrm{TA}}(\hat{t}, \hat{t} - t_0 + a_0)\mathrm{d}\hat{t} = \int_{m_0}^{m} \frac{\mathrm{d}\hat{m}}{g_{\mathrm{M}}(\hat{m})}.$$

Using our choice that G_{M} is the antiderivative of $\dfrac{1}{g_{\mathrm{M}}}$, then we have

$$\int_{t_0}^{t} g_{\mathrm{TA}}(\hat{t}, \hat{t} - t_0 + a_0)\mathrm{d}\hat{t} = G_{\mathrm{M}}(m) - G_{\mathrm{M}}(m_0).$$

We can rewrite this in terms of m_0 or m as

$$m_0 = G_{\mathrm{M}}^{-1}\left[G_{\mathrm{M}}(m) - \int_{t_0}^{t} g_{\mathrm{TA}}(\hat{t}, \hat{t} - t_0 + a_0)\mathrm{d}\hat{t}\right], \qquad (8.A.6)$$

$$m = G_{\mathrm{M}}^{-1}\left[G_{\mathrm{M}}(m_0) + \int_{t_0}^{t} g_{\mathrm{TA}}(\hat{t}, \hat{t} - t_0 + a_0)\mathrm{d}\hat{t}\right]. \qquad (8.A.7)$$

Finally, we must consider the first and fourth equality from Eq. (8.A.1)

$$\mathrm{d}t = \frac{\mathrm{d}f}{-\delta(t, a, m)f},$$

where we can write a and m in terms of t using Eqs. (8.A.2) and (8.A.7) to get

$$\mathrm{d}t = \frac{\mathrm{d}f}{-\delta\left(t, t - t_0 + a_0, G_{\mathrm{M}}^{-1}\left[G_{\mathrm{M}}(m_0) + \int_{t_0}^{t} g_{\mathrm{TA}}(\tilde{t}, \tilde{t} - t_0 + a_0)\mathrm{d}\tilde{t}\right]\right) f}.$$

Using separation of variables, we can integrate to get

$$-\int_{t_0}^{t} \delta\left(\hat{t}, \hat{t} - t_0 + a_0, G_{\mathrm{M}}^{-1}\left[G_{\mathrm{M}}(m_0) + \int_{t_0}^{\hat{t}} g_{\mathrm{TA}}(\tilde{t}, \tilde{t} - t_0 + a_0)\mathrm{d}\tilde{t}\right]\right) \mathrm{d}\hat{t} = \int_{\Phi(a_0, m_0)}^{f} \frac{\mathrm{d}\hat{f}}{\hat{f}}.$$

Thus, we can solve for f as follows:

$$\ln(f) - \ln\left(\Phi(a_0, m_0)\right) = -\int_{t_0}^{t} \delta\left(\hat{t}, \hat{t} - t_0 + a_0, G_{\mathrm{M}}^{-1}\left[G_{\mathrm{M}}(m_0) + \int_{t_0}^{\hat{t}} g_{\mathrm{TA}}(\tilde{t}, \tilde{t} - t_0 + a_0)\mathrm{d}\tilde{t}\right]\right) \mathrm{d}\hat{t}.$$

Substituting $a_0 = a - t + t_0$ (8.A.4), $m_0 = G_{\mathrm{M}}^{-1}\left[G_{\mathrm{M}}(m) - \int_{t_0}^{t} g_{\mathrm{TA}}(\hat{t}, \hat{t} - t_0 + a_0)\mathrm{d}\hat{t}\right]$ (8.A.6), and $-t_0 + a_0 = a - t$ (8.A.5), we obtain

$$\begin{aligned}\ln(f) - \ln\left(\Phi\left(a - t + t_0, G_{\mathrm{M}}^{-1}\left[G_{\mathrm{M}}(m) - \int_{t_0}^{t} g_{\mathrm{TA}}(\hat{t}, \hat{t} + a - t)\mathrm{d}\hat{t}\right]\right)\right) \\ = -\int_{t_0}^{t} \delta\left(\hat{t}, \hat{t} + a - t, G_{\mathrm{M}}^{-1}\left[G_{\mathrm{M}}(m) \right.\right. \\ \left.\left. - \int_{\hat{t}}^{t} g_{\mathrm{TA}}(\tilde{t}, \tilde{t} + a - t)\mathrm{d}\tilde{t}\right]\right)\mathrm{d}\hat{t}.\end{aligned}$$

Note that when we substitute m_0 into our function δ, the bounds of the integral for g_{TA} in the third variable shift. Isolating f gives our final solution

$$\begin{aligned}f(t, a, m) = \Phi\Bigg(a - t + t_0,\ G_{\mathrm{M}}^{-1}\Bigg[G_{\mathrm{M}}(m) \\ - \int_{t_0}^{t} g_{\mathrm{TA}}(\hat{t},\ \hat{t} + a - t)\mathrm{d}\hat{t}\Bigg]\Bigg) \\ \times \exp\Bigg\{-\int_{t_0}^{t} \delta\Bigg(\hat{t},\ \hat{t} + a - t,\ G_{\mathrm{M}}^{-1}\Bigg[G_{\mathrm{M}}(m) \\ - \int_{\hat{t}}^{t} g_{\mathrm{TA}}(\tilde{t},\ \tilde{t} + a - t)\mathrm{d}\tilde{t}\Bigg]\Bigg)\mathrm{d}\hat{t}\Bigg\}.\end{aligned}$$

8.A.2 Numerical Implementation

Simulations and plots were performed in MATLAB (2020b). Parameters used in the simulations are given in Table 8.A.1. All code is available at https://github.com/laurenchilds/PDE_time_age_mass.

8.A.2.1 *Numerical implementation of integration*

For our numerical implementation of our analytical solution, we discretize our domain. We approximate the integrals using the trapezoid method via the `trapz` function in MATLAB. Although we normalize

Table 8.A.1. Parameters for simulation.

Symbol	Description	Standard value
t_0	Initial time	0.1
t_{step}	Step size in time	0.1
$t_{\max}$	Final time	7.0
m_0	Minimum mass	0.1
m_{step}	Step size in mass	0.05
$m_{\max}$	Maximum mass considered	5.0
a_0	Minimum age	0.0
a_{step}	Step size in age	0.05
$a_{\max}$	Maximum age considered time	8.0
$\Psi(a, m)$	Initial distribution of mass and age	$\mathcal{N}(\mu_i, \sigma_{ij})$
μ_a	Mean initial age	0.4
μ_m	Mean initial mass	0.6
σ_{ij}	Initial covariate for $i = j$	0.1
σ_{ij}	Initial covariate for $i \neq j$	0.0

the total density over our domain at our initial time to be one, the density values (for a given age and mass) are determined by the choice of the discretization. Smaller step sizes lead to lower density values. Thus, one must be careful to compare situations where step sizes are consistent. All results presented here use step sizes as given in Table 8.A.1.

The total density of our solutions should integrate to one, especially early in simulations when no density has left the numerical domain. However, fluctuations are observed. This is due to "edge effects," where the crude discretization and integration method slightly over or under approximate the density.

References

1. Malthus, T. R. (1986). An essay on the principle of population (1798), in *The Works of Thomas Robert Malthus*, Vol. 1, Pickering & Chatto Publishers, London, pp. 1–139.
2. Verhulst, P.-F. (1838). Notice sur la loi que la population suit dans son accroissement, *Correspondence Mathematique et Physique (Ghent)* **10**, pp. 113–126.

3. Metz, J. A. and Diekmann, O. (2014). *The Dynamics of Physiologically Structured Populations*, Vol. 68, Springer.
4. Lotka, A. J. (1907). Relation between birth rates and death rates, *Science* **26**, 653, pp. 21–22 (1907).
5. Lotka, A. J. (1913). A natural population norm. I, *J. Washington Acad. Sci.* **3**, 9, pp. 241–248.
6. McKendrick, A. (1926). Mathematics applied to medical problems, *Proc. Edinb. Math. Soc.* Vol. 44, pp. 98–130.
7. Sharpe, F. R. and Lotka, A. J. L. (1911). A problem in age-distribution, *Lond. Edinb. Dublin Philos. Mag. J. Sci.* **21**, 124, pp. 435–438.
8. von Foerster, H. (1959). Some remarks on changing populations, in F. Stohlman (ed.), *The Kinetics of Cellular Proliferation*, Grune and Stratton, New York, pp. 382–407.
9. Bell, G. I. and Anderson, E. C. (1967). Cell growth and division: I. A mathematical model with applications to cell volume distributions in mammalian suspension cultures, *Biophys. J.* **7**, 4, pp. 329–351.
10. Scherbaum, O. and Rasch, G. (1957). Cell size distribution and single cell growth in *Tetrahymena pyriformis gl* 1, *Acta Pathol. Microb. Scand.* **41**, 2, pp. 161–182.
11. Inaba, H. (2017). *Age-Structured Population Dynamics in Demography and Epidemiology*, Singapore: Springer.
12. Nisbet, R. and Gurney, W. (1983). The systematic formulation of population models for insects with dynamically varying instar duration, *Theor. Popul. Biol.* **23**, 1, pp. 114–135.
13. Nisbet, R. M. and Gurney, W. (2003). *Modelling Fluctuating Populations: Reprint of First Edition (1982).* John Wiley & Sons, Caldwell, NJ.
14. Oldfield, D. A continuity equation for cell populations, *Bull. Math. Biophys.* **28**, 4, pp. 545–554 (1966).
15. Sinko, J. W. and Streifer, W. (1967). A new model for age-size structure of a population, *Ecology* **48**, 6, pp. 910–918.
16. Sinko, J. W. and Streifer, W. (1969). Applying models incorporating age-size structure of a population to daphnia, *Ecology* **50**, 4, 608–615.
17. Sinko, J. W. (1970). A new mathematical model for describing the age-size structure of a population of simple animals. University of Rochester, thesis, Rochester, NY.
18. Gurney, W., Nisbet, R., and Lawton, J. (1983). The systematic formulation of tractable single-species population models incorporating age structure, *J. Anim. Ecol.* pp. 479–495.
19. Easterling, M. R., Ellner, S. P., and Dixon, P. M. (2000). Size-specific sensitivity: applying a new structured population model, *Ecology* **81**, 3, pp. 694–708.

20. Ellner, S. P. (2012). Comments on: Inference for size demography from point pattern data using integral projection models, *J. Agri. Biol. Environ. Stat.* **17**, 4, pp. 682–689.
21. Metcalf, C. J. E., Graham, A. L., Martinez-Bakker, M. and Childs, D. Z. (2016). Opportunities and challenges of Integral Projection Models for modelling host–parasite dynamics, *J. Anim. Ecol.* **85**, 2, pp. 343–355.
22. Merow, C., Dahlgren, J. P., Metcalf, C. J. E., Childs, D. Z., Evans, M. E., Jongejans, E., Record, S., Rees, M., Salguero-Gómez, R., and McMahon, S. M. (2014). Advancing population ecology with integral projection models: A practical guide, *Methods Ecol. Evol.* **5**, 2, pp. 99–110.
23. Easterling, M. R. (1998). *The Integral Projection Model: Theory, Analysis and Application*, North Carolina State University, Raleigh, NC.
24. Schaub, M. (2020). Combining counts of unmarked individuals and demographic data using integrated population models, *Popul. Ecol. Pract.* pp. 215–236.
25. Rees, M., Childs, D. Z., and Ellner, S. P. (2014). Building integral projection models: A user's guide, *J. Anim. Ecol.* **83**, 3, pp. 528–545.
26. Ghosh, S., Gelfand, A. E., and Clark, J. S. (2012). Inference for size demography from point pattern data using integral projection models, *J. Agric. Biol. Environ. Stat.* **17**, 4, pp. 641–677.
27. Robertson, S. L., Henson, S. M., Robertson, T., and Cushing, J. M. (2018). A matter of maturity: To delay or not to delay? Continuous-time compartmental models of structured populations in the literature 2000–2016, *Nat. Resource Model.* **31**, 1, p. e12160.
28. Ellner, S. P. and Rees, M. (2006). Integral projection models for species with complex demography, *Am. Nat.* **167**, 3, 410–428.
29. Renardy, M., Kirschner, D., and Eisenberg, M. (2021). Structural identifiability analysis of PDEs: A case study in Continuous age-structured epidemic models, arXiv preprint arXiv:2102.06178.
30. Di Cola, G., Gilioli, G., and Baumgartner, J. (1998). Mathematical models for age-structured population dynamics, in J. Baumgartner, P. Brandmayr, and B. F. J. Manly (eds.), *Population and Community Ecology for Insect Management and Conservation*, pp. 45–61. CRC Press, London.
31. Rossini, L., Severini, M., Contarini, M., and Speranza, S. (2019). A novel modelling approach to describe an insect life cycle *vis-à-vis* plant protection: Description and application in the case study of tuta absoluta, *Ecol. Model* **409**, p. 108778.
32. Li, X.-Z., Yang, J., and Martcheva, M. (2020). *Age Structured Epidemic Modeling*, Vol. 52, Springer Nature, Cham, Switzerland.

https://doi.org/10.1142/9789811263033_0009

Chapter 9

Solving Singular Control Problems in Mathematical Biology Using PASA

Summer Atkins*,§, Mahya Aghaee†,¶, Maia Martcheva‡,||, and William Hager‡,**

**Department of Mathematics, Louisiana State University, Baton Rouge, LA, USA*

†Laboratory for Systems Medicine, College of Medicine, University of Florida Gainesville, Florida, USA

‡Department of Mathematics, University of Florida Gainesville, Florida, USA

§srnatkins@lsu.edu

¶mahya.aghaee@medicine.ufl.edu

||maia@ufl.edu

***hager@ufl.edu*

In this chapter, we demonstrate how to use a nonlinear polyhedral constrained optimization solver called the Polyhedral Active Set Algorithm (PASA) for solving a general singular control problem. We present a method for discretizing a general optimal control problem involving the use of the gradient of the Lagrangian for computing the gradient of the cost functional so that PASA can be applied. When a numerical solution contains artifacts that resemble "chattering," a phenomenon where the control oscillates wildly along the singular region, we recommend a method of regularizing the singular control problem by adding a term to the cost functional that measures a scalar multiple of the total variation of the control, where the scalar is viewed as a tuning parameter. We then demonstrate PASA's performance on three singular control problems that give rise to different applications of mathematical biology. We also provide some exposition on the heuristics that we use in determining an appropriate size for the tuning parameter.

Keywords: Singular control, Total variation, Bounded variation Regularization, Pontryagin's Minimum Principle, Switching function, Fishery problem, Plant problem, SIR problem

9.1 Introduction

Optimal control theory is a tool that is used in mathematical biology for observing how a dynamical system behaves when employing one or many variables that can be controlled outside of that system. Mathematical biologists apply optimal control theory to disease models of immunologic and epidemic types,[1–4] to management decisions in harvesting,[5,6] and to resource allocation models.[7,8] In practice, mathematical biologists tend to construct optimal control problems with quadratic dependence on the control. Problems of this structure are well-behaved in the sense that there are established methods of proving existence and uniqueness of an optimal control.[9–11] In addition, for numerically solving problems of this form, many employ the forward–backward sweep method, a numerical procedure presented in Lenhart and Workman's book[12] that involves combinations of the forward application and the backward application of a fourth-order Runge–Kutta method. We direct the reader to Refs. 13–16 which are excellent surveys of other numerical procedures, such as gradient methods, quasi-Newton methods, shooting methods, and collocation methods, that are used within the optimization community for solving optimal control problems.

Control problems in biology tend to depend quadratically with respect to the control due to the construction of the objective or cost functional, which is the functional that is being optimized with respect to the control variables. The construction of the objective functional is an essential component to optimal control theory because it measures our criteria for determining what control strategy is deemed "best." In mathematical biology, the costs are frequently nonlinear and depend on the states and the controls, and a cost term for a particular control may be the sum of a bilinear term in that control and one for the states and a quadratic term in the control. Frequently, the quadratic term has a lower coefficient.

With regard to the principle of parsimony, it is difficult to justify the use of a quadratic term for representing the cost of administering a control. A linear term would be a more realistic representation of the cost of applying a control. For control problems that are linear in the control, it is possible to obtain a solution that is piecewise constant where the constant values correspond to the bounds of the control. An optimal control of this structure, which is often called a "bang-bang" control, can be readily interpreted and implemented. These characteristics compel many (see Refs. 5,6,8 and 17–20), to use control problems with linear dependence on the control for biological models.

There are however evident setbacks to using optimal control problems in which the control appears linearly. For one thing, these problems are much more difficult to solve analytically due to the potential existence of a singular subarc. As demonstrated in Refs. 5,6 and 18, procedures for obtaining an explicit formula for the singular case involve taking one or many time derivatives of the switching function as a means to gain a system of equations. Often the first-order and second-order necessary conditions for optimality, which are respectively named Pontryagin's Maximum Principle[10] and the Generalized Legendre–Clebsch Condition/Kelley's Condition,[21–24] are checked. Naturally, methods for explicitly solving singular control problems increase in difficulty when multiple state variables and multiple control variables are involved.

Additionally, numerically solving singular control problems are inevitably problematic. If the parameters of an optimal control problem that is linear in the control are set to where all optimal control variables are bang-bang, then the forward–backward sweep method[12] can successfully run. However, if the presence of a singular control is a possibility, then the forward–backward sweep method is not advisable. In Foroozandeh *et al.*,[25] Foroozandeh, De Pinho, Shamsi test four numerical solvers, including the Imperial College London Optimal Control Software (ICLOCS)[26] and the Gauss Pseudospectral Optimization Software (GPOPS),[27] by solving a singular optimal control problem for Autonomous Underwater Vehicles (AUV). They find that three of these methods have difficulty in detecting the structure of the optimal control and in accurately computing the switching points without *a priori* information. Switching points are the corresponding points in time when an optimal control switches from singular to non-singular and vice versa. When solving for the AUV problem, both ICLOCS and GPOPS obtain a control that exhibits oscillations within the singular region, causing both methods to be unable to provide direct information about the switching points. Foroozandeh *et al.* conclude that only the mixed binary nonlinear programming method (MBNLP)[28] is successful in accurately approximating the optimal control to the AUV problem and its switching points.

One of the predominate issues associated with singular control problems is the concept of "chattering," which is also known as the "Fuller Phenomenon" (see Ref. 24). As mentioned in Zelikin and Borisov's book,[24] an optimal control is said to be *chattering* if the control oscillates infinitely many times between the bounds of the control over a finite region. It is thought that such an event occurs in singular control problems when the optimal control has a singular subarc that cannot be directly joined to a non-singular (bang) region without promoting oscillations. In MacDanell and Powers,[21] MacDanell and Powers present some necessary conditions for joining singular and nonsingular subarcs. Zelikin and Borisov[24] mention other theorems that can be used to verify when chattering is present. It is possible that the discretization of an optimal control problem causes a numerical method to generate numerical artifacts that resemble chattering even though the optimal control does not chatter. However, it is difficult to determine when a numerical solution is exhibiting many oscillations due to chattering or due to numerical artifacts.

Regardless of the situation, it is evident that a wildly oscillating optimal control is an unrealistic procedure to implement. A way to bypass this issue is to solve for a regularized version of the optimal control problem. In Ref. 29, Yang *et al.* present methods of regularizing optimal control problems by adding a penalty term to the cost functional of the original problem. The penalization terms suggested in Yang *et al.*[29] are: a weighted parameter times the L^2 norm of the control, a weighted parameter times the L^2 norm of the derivative of the control, and a weighted parameter times the L^2 norm of the second derivative of the control. In Ding and Lenhart,[6] Ding and Lenhart employ a penalty term to a harvesting optimal control problem to avoid a potential chattering result found in the control h. The penalty term that was applied for this problem consisted of a weighted parameter times $\|\nabla h\|_{L^2}^2$. The penalty term that is used in Ding and Lenhart[6] adds convexity properties to the control problem which can be beneficial for verifying existence and uniqueness of an optimal control. Additionally, $\|\nabla h\|_{L^2}^2$ can be discretized to where it can serve as a crude estimate for the total variation of the control. However, Lenhart and Ding[6] need to use variational inequalities to solve for their problem, and incorporating such a penalty term restricts their set of admissible controls into a functional space that requires its controls to be differentiable.

The penalty term that we believe to show the most promise has been recently suggested in Capognigro, Ghezzi, Piccoli, and Trelat's work.[30] In Ref. 30, recommend a method of regularizing chattering in optimal control problems by adding a penalty term that represents a tuning parameter times the total variation of the control. Total variation regularization or bounded variation regularization influences the numerical solution in a way that reduces the number of oscillations. In Caponigro *et al.*,[30] this penalty is applied to the Fuller problem, which is the classical example that introduced the concept of chattering, and they obtain a "quasi" optimal solution to the Fuller problem that does not chatter. Caponigro *et al.* also prove that the optimal value of the penalized problem converges to the associated optimal value of the original problem as the penalty weight parameter p tends to zero.

In this chapter, we demonstrate how to use a nonlinear polyhedral constrained optimization solver called PASA,[1] developed by Hager and Zhang,[31] to solve a general singular control problem that is being regularized by use of a total variation term.[30] We recommend PASA because it is user friendly to those who are not as acquainted with optimization techniques for optimal control problems, and it is freely accessible to use on MATLAB for Linux and Unix operating systems. According to Hager and Zhang,[31] PASA consists of two phases, with the first phase being the gradient projection algorithm and the second phase being any algorithm that is used for solving linearly constrained optimization problems. The gradient projection algorithm is an optimization solver commonly used for bounded constrained optimization problems. When applying gradient descent to a bounded constrained optimization problem, it is possible to obtain

[1]To access PASA software that can be used on MATLAB for Linux and Unix operating systems, download the SuiteOPT Version 1.0.0 software given on https://people.clas.ufl.edu/hager/software. For future reference, any updates to the software will be uploaded to this link, and to access older versions of the software, use the same link and then select "Software Archive."

an iterate that lies outside of the feasible set due to the negative direction of the gradient. Projected gradient method, takes this issue into consideration by adding additional steps involving projecting points outside the feasible set onto the feasible set. For more information on projected gradient methods we direct the reader to Refs. 31–36. Using PASA to solve optimal control problems involves converting an optimal control problem into a discretized optimization problem.

In this chapter, the discretization for the general singular control problem involves using explicit Euler's method for the state equations, and left-rectangular integral approximation for the original cost functional. Additionally, we need to ensure that the discretized and regularized cost functional is differentiable, which requires performing a decomposition of the absolute value terms associated with the total variation penalty. We use the gradient of the Lagrangian function of the discretized optimal control problem for computing the gradient to the discretized cost functional. Conveniently, the process of ensuring that the gradient of the Lagrangian is equal to the gradient of the discretized objective functional yields a discretization procedure for the adjoint equations.

Further, we demonstrate PASA's performance on three singular control problems that are being regularized via bounded variation. The order of these examples increases in difficulty based upon the number of state variables and the number of control variables. Additionally, each example gives rise to different applications of mathematical biology. Explicit formulas for the singular case and for the switching points are given for the first two examples, which allow us to compare PASA's numerical results with the exact solution. For each problem, we illustrate the discretization process and then present numerical results that were obtained when solving for both the unpenalized and penalized problem. We also provide some exposition on the heuristics that we use in determining an appropriate size for the tuning parameter ρ.

The first example is a fishery harvesting problem that was originally presented in Clark's[37] and later restated in Lenhart and Workman's.[12] The fishery problem consists of one state variable and one control variable, where the state variable represents the fish population and the control variable represents harvesting effort. In Lenhart and Workman,[12] an explicit formula for the singular case is obtained by using Pontryagin's Maximum Principle[10]; however, forward–backward sweep is unable to solve for the problem whenever parameters are set to ensure existence of a singular subarc.

The second example is from King and Roughgarden's work,[8] and the optimal control problem is a resource allocation model for studying an annual plant's allocation procedure in distributing photosynthate. The control problem consists of two state variables and one control variable. One variable measures the weight of the components of a plant that correspond to vegetative growth, while the other variable measures the weight of the components of a plant that correspond to reproductive growth. The control variable used in this problem represents the fraction of photosynthate being reserved for vegetative growth. King and Roughgarden use Pontryagin's Maximum Principle[10] and find conditions based upon the parameters of the problem for determining when the optimal control would be bang–bang or concatenations of singular and bang control. They verify that their explicit formula for the singular subarc satisfies the generalized Legendre–Clebsch Condition and the strengthened Legendre–Clebsch Condition.[21–24] Additionally, King and Roughgarden use one of MacDanell and Power's Junction theorems[21] to show the optimal control to the problem satisfies the necessary conditions for joining singular and non-singular subarcs. When using PASA to solve for this plant problem, we only need to use the regularization term for a degenerate case of the problem.

The final example, which is from Ledzewicz, Aghaee, and Schättler's,[17] is an optimal control problem where the three state variables correspond to an SIR model with demography. An SIR model is a compartmental model that is used for modeling the spread of an infectious disease in a population, where the population is divided into the following three classes: (1) S is the class of individuals who are susceptible to the disease; (2) I is the class of infected individuals who are assumed to be infectious; and (3) R is the class of individuals who have recovered from the disease and are considered immune to the disease. For books covering mathematical models for epidemiology, we recommend Brauer and Castillo-Chavez's[38] and Martcheva's.[39] The optimal control problem used in Ledzewicz *et al.*[17] consists of two control variables where one control represents vaccination while the other represents treatment. Ledzewicz *et al.* numerically solve this problem with parameters set to where a singular subarc is present in the optimal vaccination strategy while the optimal treatment strategy obtained appears to be bang-bang. When using PASA, we regularize only the vaccination control via bounded variation since the treatment control contains no oscillations. In the Appendix section, we provide the MATLAB code that was used for solving the last example to illustrate how to write up these optimal control problems.

9.2 Discretization of the Regularized Control Problem via Bounded Variation

The following problem is the optimal control problem of interest:

$$\begin{aligned} \min_{u\in\mathcal{A}} \quad & J(\boldsymbol{u}) = \textstyle\int_0^T g(\boldsymbol{x}(t), \boldsymbol{u}(t))\mathrm{d}t \\ \text{sub. to} \quad & \dot{x}_i(t) = f_i(\boldsymbol{x}(t), \boldsymbol{u}(t)) \quad \text{for all } i = 1, \dots, n, \\ & x_i(0) = x_{i,0}, \quad \text{for all } i = 1, \dots, n, \\ & \xi_j \le u_j(t) \le \omega_j \quad \text{for all } t \in [0, T] \quad \text{and} \quad \text{for all } j = 1, \dots, m, \end{aligned} \tag{9.1}$$

where functions $f_1, \dots, f_n$, and g are assumed to be continuously differentiable in all arguments. In addition, we assume that if any of the m control variables u_i appear in functions $f_1, \dots, f_n$ and g, then u_i appears linearly. The class of admissible controls is the following:

$$\mathcal{A} = \{\boldsymbol{u} \in (L^1(0,T))^m | \boldsymbol{\xi} \le \boldsymbol{u}(t) \le \boldsymbol{\omega} \quad \text{for all } t \in [0,T]\}.$$

We assume that the conditions for the Filippov–Cesari Existence Theorem[9] hold for problem (9.1). State vector $\boldsymbol{x} \in \mathbb{R}^n$ consists of n state variables that satisfy the state equations and the initial conditions given in problem (9.1).

The common procedure for solving problem (9.1) is employing Pontryagin's Maximum Principle[10] to generate the first-order necessary conditions for optimality. We first define the Hamiltonian function to problem (9.1), $H(\boldsymbol{x}, \boldsymbol{u}, \boldsymbol{\lambda})$, as

$$\begin{aligned} H(\boldsymbol{x}, \boldsymbol{u}, \boldsymbol{\lambda}) &= g(\boldsymbol{x}, \boldsymbol{u}) + \boldsymbol{\lambda}^T \boldsymbol{f}(\boldsymbol{x}, \boldsymbol{u}) \\ &= g(\boldsymbol{x}, \boldsymbol{u}) + \sum_{i=0}^{n} \lambda_i(t) f_i(\boldsymbol{x}, \boldsymbol{u}), \end{aligned}$$

where $\boldsymbol{\lambda} \in \mathbb{R}^n$ is the adjoint vector and the superscript T means transpose. We have from Pontryagin's Maximum Principle[10] that if $\boldsymbol{u}^*$ is the optimal control to problem (9.1) with corresponding trajectory $\boldsymbol{x}^*$, then there exists a non-zero adjoint vector $\boldsymbol{\lambda}^*$ that is a solution to the following adjoint system:

$$\dot{\lambda}_\ell(t) = -\frac{\partial H(\boldsymbol{x}, \boldsymbol{u}, \boldsymbol{\lambda})}{\partial x_\ell}, \quad \text{for all } \ell = 1, \ldots, n,$$

$$\lambda_\ell(T) = 0, \quad \text{for all } \ell = 1, \ldots, n,$$

and satisfies

$$H(\boldsymbol{u}^*, \boldsymbol{x}^*, \boldsymbol{\lambda}^*) = \min_{\boldsymbol{u} \in \mathcal{A}} H(\boldsymbol{u}, \boldsymbol{x}^*, \boldsymbol{\lambda}^*).$$

Using our definition of the Hamiltonian function, the adjoint equations are

$$\dot{\lambda}_\ell(t) = -\frac{\partial g(\boldsymbol{x}, \boldsymbol{u})}{\partial x_\ell} - \sum_{i=1}^{n} \lambda_i(t) \frac{\partial f_i(\boldsymbol{x}, \boldsymbol{u})}{\partial x_\ell} \quad \text{for all } \ell = 1, \ldots, n, \tag{9.2}$$

with transversality conditions being

$$\lambda_\ell(T) = 0, \quad \text{for all } \ell = 1, \ldots, n. \tag{9.3}$$

Based on assumptions of functions g and $f_1, \ldots, f_n$ in problem (9.1), the Hamiltonian is linear in the control.

For demonstration purposes, we assume that every component of control vector $\boldsymbol{u}$ in problem (9.1) needs to be regularized via bounded variation.[30] This means that when numerically solving problem (9.1) without this regularization term, we obtain oscillatory numerical artifacts or other unwarranted numerical artifacts in each control. To regularize problem (9.1) via bounded variation, we introduce a tuning vector $\boldsymbol{\rho}$ where $0 \leq \rho_j < 1$ for all $j = 1, \ldots, m$, and we present Conway's[40] definition of the total variation function V of a real or complex valued function u that is defined on the interval $[a, b]$. Let $\mathbb{K}$ be the field of complex numbers or the field of real numbers. Given $u : [a, b] \to \mathbb{K}$, the total variation of u on $[a, b]$ is defined to be

$$V(u) = \sup_{\mathcal{P}} \sum_{k=0}^{N_P - 1} |u(t_{k+1}) - u(t_k)|,$$

where $\mathcal{P} = \{P = \{t_0, t_1, \dots, t_{N_P}\} : a \leq t_0 \leq t_1 \leq \cdots \leq t_{N_p} \leq b\}$. Note that if we assume that function u is real-valued and piecewise constant on $[a, b]$, then the total variation of u on $[a, b]$ is the sum of the absolute value of the jumps in u.

The regularization of problem (9.1) via bounded variation is

$$\begin{aligned} \min_{\boldsymbol{u} \in \mathcal{A}} \quad & J_{\boldsymbol{\rho}}(\boldsymbol{u}) = \int_0^T g(\boldsymbol{x}(t)\boldsymbol{u}(t))\mathrm{d}t + \sum_{j=1}^{m} \rho_j V(u_j) \\ \text{sub.to} \quad & \dot{x}_i(t) = f_i(\boldsymbol{x}(t), \boldsymbol{u}(t)), \quad \text{for all } i = 1, \dots, n, \\ & x_i(0) = x_{i,0}, \quad \text{for all } i = 1, \dots, n, \\ & \xi_j \leq u_j(t) \leq \omega_j \quad \text{for all } t \in [0, T] \text{ and for all } j = 1, \dots, m, \end{aligned} \tag{9.4}$$

where $V(u_j)$ is the total variation of u_j on interval $[0, T]$ and $0 \leq \rho_j < 1$ is the bounded variation penalty parameter associated with control variable u_j for all $j = 1, \dots, m$. For this problem, we assume that all control variables need to be penalized. However, if (based upon observations of the numerical solutions for the unregularized problem (9.1)) we notice that one of the control variables u_ℓ exhibits no numerical artifacts or unusual oscillations, then we recommend solving problem (9.4) with the corresponding tuning parameter ρ_ℓ set to being zero. We can construct the Hamiltonian function that corresponds to problem (9.4), and the Hamiltonian gives the same adjoint equations (9.2) and transversality conditions (9.3).

For using PASA to numerically solve the regularized problem, we need to discretize problem (9.4) and discretize the adjoint equations (9.2). We present the method of discretizing the regularized problem only, but we emphasize that we can use PASA to solve the discretized unpenalized problem by numerically solving the discretized regularized problem when the vector $\boldsymbol{\rho}$ is set to being the zero vector. We begin by partitioning time interval $[0, T]$, by using $N+1$ equally spaced nodes. For all $i = 1, \dots, n$ and $k = 0, \dots, N$, we denote $x_{i,k} = x_i(t_k)$, and we emphasize that component $x_{i,0}$ is the initial value given in problem (9.4). So for each $i = 1, \dots, n$, we have that $\boldsymbol{x}_i \in \mathbb{R}^{N+1}$ with $x_{i,0}$ being the initial value given. We assume that for all $j = 1, \dots, m$, control variable u_j is constant over each mesh interval. For all $j = 1, \dots, m$ and $k = 0, \dots, N-2$, we denote $u_{j,k} = u_j(t)$ for all $t_k \leq t < t_{k+1}$, and for all $j = 1, \dots, m$, we denote

$u_{j,N-1} = u_j(t)$ for all $t_{N-1} \leq t \leq t_N = T$. So for each $j = 1, \ldots, m$ we have that $\boldsymbol{u}_j \in \mathbb{R}^N$. The assumption of each control variable u_j being constant on each mesh interval allows us to express the total variation of u_j on $[0, T]$ in terms of the particular partition of $[0, T]$ that is used for the discretization:

$$V(u_j) = \sum_{k=0}^{N-2} |u_{j,k+1} - u_{j,k}|.$$

For simplicity of discussion, we use left-rectangular integral approximation to discretize the integral used in objective functional J_ρ, and we use forward Euler's method to discretize the state equations given in problem (9.4). In general, we recommend using an explicit scheme for discretizing the state equations if the dynamics of the system have only initial conditions involved. We then have the following:

$$\begin{aligned}
\min \quad & J_\rho(\boldsymbol{u}_1, \ldots, \boldsymbol{u}_m) \\
& = \sum_{k=0}^{N-1} hg(\boldsymbol{x}_{\cdot,k}, \boldsymbol{u}_{\cdot,k}) + \sum_{j=1}^{m} \rho_j \sum_{k=0}^{N-2} |u_{j,k+1} - u_{j,k}| \\
& x_{i,k+1} = x_{i,k} + hf_i(\boldsymbol{x}_{\cdot,k}, \boldsymbol{u}_{\cdot,k}) \quad \text{for all } i = 1, \ldots, n \text{ and} \\
& \quad k = 0, \ldots, N-1, \\
& \quad \xi_j \leq u_{j,k} \leq \omega_j, \quad \text{for all } j = 1, \ldots, m \text{ and } k = 0, \ldots, N-1,
\end{aligned} \tag{9.5}$$

where $h = \frac{T}{N}$ is the mesh size, $\boldsymbol{x}_{\cdot,k} = [x_{1,k}, x_{2,k}, \ldots, x_{n,k}]$, and $\boldsymbol{u}_{\cdot,k} = [u_{1,k}, u_{2,k}, \ldots, u_{m,k}]$ for all $k = 0, \ldots, N-1$.

Since PASA consists of a phase that uses a projected gradient method, we need the objective function in problem (9.5) to be differentiable, which is not the case due to the absolute value terms that correspond to the discretization of the total variation function. We need to perform a decomposition of each absolute value term so that J_ρ can be differentiable. For each $j = 1, \ldots, m$, we introduce two $N-1$ dimensional vectors $\boldsymbol{\zeta}_j$ and $\boldsymbol{\iota}_j$ whose entries are non-negative, and every entry of $\boldsymbol{\zeta}_j$ and $\boldsymbol{\iota}_j$ is defined as

$$|u_{j,k+1} - u_{j,k}| = \zeta_{j,k} + \iota_{j,k} \quad \text{for all } k = 0, \ldots, N-2.$$

Another way of viewing $\boldsymbol{\zeta}_j$ and $\boldsymbol{\iota}_j$, is that each component $\zeta_{j,k}$ and $\iota_{j,k}$ will be defined based upon the following conditions:

$$\begin{aligned}
\text{Condition 1}: \text{If } & u_{j,k+1} - u_{j,k} > 0, \text{ then} \\
& \zeta_{j,k} = u_{j,k+1} - u_{j,k} \text{ and } \iota_{j,k} = 0, \\
\text{Condition 2}: \text{If } & u_{j,k+1} - u_{j,k} \leq 0, \text{ then} \\
& \zeta_{j,k} = 0 \quad \text{and} \quad \iota_{j,k} = -(u_{j,k+1} - u_{j,k}).
\end{aligned}$$

Employing this decomposition to problem (9.5) yields

$$\begin{aligned}
\min \quad & J_\rho(\boldsymbol{u}_1, \boldsymbol{\zeta}_1, \ldots, \boldsymbol{u}_m, \boldsymbol{\zeta}_m, \boldsymbol{\iota}_m) \\
& = \sum_{k=0}^{N-1} hg(\boldsymbol{x}_{\cdot,k}, \boldsymbol{u}_{\cdot,k}) + \sum_{j=1}^{m} \left(\rho_j \sum_{k=0}^{N-2} (\zeta_{j,k} + \iota_{j,k}) \right) \\
& x_{i,k+1} = x_{i,k} + hf_i(\boldsymbol{x}_{\cdot,k}, \boldsymbol{u}_{\cdot,k}), \quad \text{for all } i = 1, \ldots, n \text{ and} \\
& \qquad\qquad k = 0, \ldots, N-1, \\
& \xi_j \leq u_{j,k} \leq \omega_j, \quad \text{for all } j = 1, \ldots, m \text{ and } k = 0, \ldots, N-1, \\
& u_{j,k+1} - u_{j,k} = \zeta_{j,k} - \iota_{j,k}, \quad \text{for all } j = 1, \ldots, m \text{ and} \\
& \qquad\qquad k = 0, \ldots, N-2, \\
& 0 \leq \boldsymbol{\zeta}_j, \quad 0 \leq \boldsymbol{\iota}_j \quad \text{for all } j = 1, \ldots, m.
\end{aligned} \tag{9.6}$$

For the above problem, we are now minimizing J_ρ with respect to vectors $\boldsymbol{u}_1, \boldsymbol{\zeta}_1, \boldsymbol{\iota}_1, \ldots, \boldsymbol{u}_m, \boldsymbol{\zeta}_m$, and $\boldsymbol{\iota}_m$. The constraints associated with each $\boldsymbol{\zeta}_j$ and $\boldsymbol{\iota}_j$, are constraints that PASA can interpret. For all $j = 1, \ldots, m$, the equality constraints associated with each $\boldsymbol{\zeta}_j$ and $\boldsymbol{\iota}_j$ in problem (9.6) are linear and can be expressed as

$$\left[\boldsymbol{A}_j \middle| -\boldsymbol{I}_{N-1} \middle| \boldsymbol{I}_{N-1} \right] \begin{bmatrix} \boldsymbol{u}_j \\ \hline \boldsymbol{\zeta}_j \\ \hline \boldsymbol{\iota}_j \end{bmatrix} = \boldsymbol{0},$$

where $\boldsymbol{I}_{N-1}$ is the identity matrix of dimension $N-1$, $\boldsymbol{0}$ is an $N-1$ dimensional all zeros vector, and $\boldsymbol{A}_j$ is an $N-1$ by N matrix defined as

$$\boldsymbol{A_j} = \begin{bmatrix} -1 & 1 & 0 & \cdots & 0 \\ 0 & \ddots & \ddots & \ddots & \vdots \\ \vdots & \ddots & \ddots & \ddots & 0 \\ 0 & \cdots & 0 & -1 & 1 \end{bmatrix}. \tag{9.7}$$

Moreover, the equality constraints of the decomposition vectors given in problem (9.6) can be written as

$$\left[\boldsymbol{A_1}\middle|-\boldsymbol{I}_{N-1}\middle|\boldsymbol{I}_{N-1}\middle|\cdots\middle|\boldsymbol{A_j}\middle|-\boldsymbol{I}_{N-1}\middle|\boldsymbol{I}_{N-1}\middle|\cdots\middle|\boldsymbol{A_m}\middle|-\boldsymbol{I}_{N-1}\middle|\boldsymbol{I}_{N-1}\right]$$

$$\begin{bmatrix} \boldsymbol{u}_1 \\ \boldsymbol{\zeta}_1 \\ \boldsymbol{\iota}_1 \\ \vdots \\ \boldsymbol{u}_j \\ \boldsymbol{\zeta}_j \\ \boldsymbol{\iota}_j \\ \vdots \\ \boldsymbol{u}_m \\ \boldsymbol{\zeta}_m \\ \boldsymbol{\iota}_m \end{bmatrix} = \mathbf{0},$$

where $\boldsymbol{A_j}$ is defined as (9.7) for all $j = 1, \ldots, m$ and $\mathbf{0} \in \mathbb{R}^{N-1}$.

From problem (9.6), we wish to find the gradient of J_ρ with respect to $\boldsymbol{u}_1, \boldsymbol{\zeta}_1, \boldsymbol{\iota}_1, \ldots, \boldsymbol{u}_m, \boldsymbol{\zeta}_m$, and $\boldsymbol{\iota}_m$. We compute the gradient of the Lagrangian of problem (9.6) with respect to $\boldsymbol{u}_1, \boldsymbol{\zeta}_1, \boldsymbol{\iota}_1, \ldots, \boldsymbol{u}_m, \boldsymbol{\zeta}_m$, and $\boldsymbol{\iota}_m$ to find ∇J_ρ. This is necessary because based upon problem (9.6), state variables $\boldsymbol{x}_1, \ldots, \boldsymbol{x}_n$ can be viewed as functions of $\boldsymbol{u}_1, \ldots, \boldsymbol{u}_m$. So when computing the gradient of J_ρ with respect to $\boldsymbol{u}_1, \boldsymbol{\zeta}_1, \boldsymbol{\iota}_1, \ldots, \boldsymbol{u}_m, \boldsymbol{\zeta}_m$, and $\boldsymbol{\iota}_m$, we should consider that vectors $\boldsymbol{x}_1, \ldots, \boldsymbol{x}_m$ depend on the controls.

The discretized problem can be put in the following general form for which the technique of using the Lagrangian to compute the gradient of a cost functional can be used:

$$\begin{aligned} \min \quad & G(\boldsymbol{x}, \boldsymbol{u}) \\ & \boldsymbol{F}(\boldsymbol{x}, \boldsymbol{u}) = \mathbf{0}, \end{aligned} \tag{9.8}$$

where $\boldsymbol{u} \in \mathbb{R}^m$, $\boldsymbol{x} \in \mathbb{R}^n$, $G : \mathbb{R}^n \times \mathbb{R}^m \to \mathbb{R}$, and $\boldsymbol{F} : \mathbb{R}^n \times \mathbb{R}^m \to \mathbb{R}^n$ are differentiable; moreover, it is assumed in problem (9.8) that we can uniquely solve for $\boldsymbol{x}$ in term of $\boldsymbol{u}$. Based on these assumptions, we can rewrite problem (9.8) as

$$\begin{aligned} \min \quad & \mathcal{J}(\boldsymbol{u}) = G(\boldsymbol{x}(\boldsymbol{u}), \boldsymbol{u}) \\ & \boldsymbol{F}(\boldsymbol{x}(\boldsymbol{u}), \boldsymbol{u}) = \boldsymbol{0}, \end{aligned} \tag{9.9}$$

where $\boldsymbol{x} = \boldsymbol{x}(\boldsymbol{u})$ denotes the unique solution of $\boldsymbol{F}(\boldsymbol{x}, \boldsymbol{u}) = \boldsymbol{0}$ for a given $\boldsymbol{u} \in \mathbb{R}^m$. The following result can be deduced from the implicit function theorem and the chain rule (see Ref. [41, Remark 3.2], Ref. [42]):

Theorem 9.1. *If the Jacobian $\nabla_{\boldsymbol{x}} \boldsymbol{F}(\boldsymbol{x}, \boldsymbol{u})$ is invertible for each $\boldsymbol{u} \in \mathbb{R}^m$ and $\boldsymbol{x} = \boldsymbol{x}(\boldsymbol{u})$, then for each $\boldsymbol{u} \in \mathbb{R}^m$ the gradient of $\mathcal{J}$ in problem (9.9) is*

$$\nabla_{\boldsymbol{u}} \mathcal{J}(\boldsymbol{u}) = \nabla_{\boldsymbol{u}} \mathcal{L}(\boldsymbol{x}, \boldsymbol{u}, \boldsymbol{\lambda})_{|\boldsymbol{x}=\boldsymbol{x}(\boldsymbol{u})}, \tag{9.10}$$

where $\mathcal{L} : \mathbb{R}^n \times \mathbb{R}^m \times \mathbb{R}^n \to \mathbb{R}$ is the Lagrangian of problem (9.9),

$$\mathcal{L}(\boldsymbol{x}, \boldsymbol{u}, \boldsymbol{\lambda}) = G(\boldsymbol{x}, \boldsymbol{u}) + \boldsymbol{\lambda}^T \boldsymbol{F}(\boldsymbol{x}, \boldsymbol{u}),$$

and $\boldsymbol{\lambda}$ is chosen such that

$$\nabla_{\boldsymbol{x}} \mathcal{L}(\boldsymbol{x}, \boldsymbol{u}, \boldsymbol{\lambda}) = \nabla_{\boldsymbol{x}} G(\boldsymbol{x}, \boldsymbol{u}) + \boldsymbol{\lambda}^T \nabla_{\boldsymbol{x}} \boldsymbol{F}(\boldsymbol{x}, \boldsymbol{u}) = \boldsymbol{0}. \tag{9.11}$$

Before computing the Lagrangian to problem (9.6), we rewrite the state equations accordingly:

$$\begin{aligned} & -x_{i,k+1} + x_{i,k} + h f_i(\boldsymbol{x}_{\cdot,k}, \boldsymbol{u}_{\cdot,k}) = 0, \\ & \text{for all } i = 1, \ldots, n \quad \text{and} \quad k = 0, \ldots, N-1. \end{aligned}$$

The Lagrangian to problem (9.6) is then

$$\begin{aligned} & \mathcal{L}(\boldsymbol{u}_1, \boldsymbol{\zeta}_1, \ldots, \boldsymbol{u}_m, \boldsymbol{\zeta}_m, \boldsymbol{\iota}_m) \\ & \quad = \sum_{k=0}^{N-1} h g(\boldsymbol{x}_{\cdot,k}, \boldsymbol{u}_{\cdot,k}) + \sum_{j=1}^{m} \rho_j \sum_{k=0}^{N-2} (\zeta_{j,k} + \iota_{j,k}) \\ & \qquad + \sum_{i=1}^{n} \sum_{k=0}^{N-1} \lambda_{i,k} (-x_{i,k+1} + x_{i,k} + h f_i(\boldsymbol{x}_{\cdot,k}, \boldsymbol{u}_{\cdot,k})), \end{aligned} \tag{9.12}$$

where $\boldsymbol{\lambda}_1, \ldots, \boldsymbol{\lambda}_n \in \mathbb{R}^{N-1}$ are the Lagrange multiplier vectors. For all $j = 1, \ldots, m$ and for all $k = 0, \ldots, N-1$, we compute the partial derivative of $\mathcal{L}$ with respect to $u_{j,k}$ and obtain

$$\frac{\partial}{\partial u_{j,k}} \mathcal{L} = h \frac{\partial}{\partial u_{j,k}} g(\boldsymbol{x}_{\cdot,k}, \boldsymbol{u}_{\cdot,k}) + \sum_{i=1}^{n} \lambda_{i,k} \left(h \frac{\partial}{\partial u_{j,k}} f_i(\boldsymbol{x}_{\cdot,k}, \boldsymbol{u}_{\cdot,k}) \right).$$

For all $j = 1, \ldots, m$, we have that

$$\nabla_{\boldsymbol{u}_j} \mathcal{L} = h \begin{bmatrix} \dfrac{\partial g(\boldsymbol{x}_{\cdot,0}, \boldsymbol{u}_{\cdot,0})}{\partial u_{j,0}} + \sum_{i=1}^{n} \lambda_{i,0} \left(\frac{\partial f_i(\boldsymbol{x}_{\cdot,0}, \boldsymbol{u}_{\cdot,0})}{\partial u_{j,0}} \right) \\ \vdots \\ \dfrac{\partial g(\boldsymbol{x}_{\cdot,k}, \boldsymbol{u}_{\cdot,k})}{\partial u_{j,k}} + \sum_{i=1}^{n} \lambda_{i,k} \left(\frac{\partial f_i(\boldsymbol{x}_{\cdot,k}, \boldsymbol{u}_{\cdot,k})}{\partial u_{j,k}} \right) \\ \vdots \\ \dfrac{\partial g(\boldsymbol{x}_{\cdot,N-1}, \boldsymbol{u}_{\cdot,N-1})}{\partial u_{j,N-1}} + \sum_{i=1}^{n} \lambda_{i,N-1} \left(\frac{\partial f_i(\boldsymbol{x}_{\cdot,N-1}, \boldsymbol{u}_{\cdot,N-1})}{\partial u_{j,N-1}} \right) \end{bmatrix} \tag{9.13}$$

where $\nabla_{\boldsymbol{u}_j} \mathcal{L} \in \mathbb{R}^N$. For all $j = 1, \ldots, m$ and for all $k = 0, \ldots, N-2$, we compute the partial derivative of $\mathcal{L}$ with respect to $\zeta_{j,k}$ and the partial derivative of $\mathcal{L}$ with respect to $\iota_{j,k}$, as follows:

$$\frac{\partial}{\partial \zeta_{j,k}} \mathcal{L} = \rho_j,$$

$$\frac{\partial}{\partial \iota_{j,k}} \mathcal{L} = \rho_j.$$

We can then say for all $j = 1, \ldots, m$ that

$$\nabla_{\zeta_j} \mathcal{L} = \begin{bmatrix} \rho_j \\ \vdots \\ \rho_j \end{bmatrix}, \tag{9.14}$$

and

$$\boldsymbol{\nabla}_{\iota_j}\boldsymbol{\mathcal{L}} = \begin{bmatrix} \rho_j \\ \vdots \\ \rho_j \end{bmatrix} \tag{9.15}$$

where $\boldsymbol{\nabla}_{\zeta_j}\boldsymbol{\mathcal{L}} \in \mathbb{R}^{N-2}$ and $\boldsymbol{\nabla}_{\iota_j}\boldsymbol{\mathcal{L}} \in \mathbb{R}^{N-2}$. By Theorem 9.1, provided that Lagrange multiplier vectors $\boldsymbol{\lambda}_1, \ldots, \boldsymbol{\lambda}_n$ satisfy theorem condition (9.11), we have that

$$\boldsymbol{\nabla}\boldsymbol{J} = \boldsymbol{\nabla}\boldsymbol{\mathcal{L}} = \left[\begin{array}{c} \boldsymbol{\nabla}_{\boldsymbol{u}_1}\boldsymbol{\mathcal{L}} \\ \boldsymbol{\nabla}_{\zeta_1}\boldsymbol{\mathcal{L}} \\ \boldsymbol{\nabla}_{\iota_1}\boldsymbol{\mathcal{L}} \\ \hline \vdots \\ \hline \boldsymbol{\nabla}_{\boldsymbol{u}_k}\boldsymbol{\mathcal{L}} \\ \boldsymbol{\nabla}_{\zeta_k}\boldsymbol{\mathcal{L}} \\ \boldsymbol{\nabla}_{\iota_k}\boldsymbol{\mathcal{L}} \\ \hline \vdots \\ \hline \boldsymbol{\nabla}_{\boldsymbol{u}_m}\boldsymbol{\mathcal{L}} \\ \boldsymbol{\nabla}_{\zeta_m}\boldsymbol{\mathcal{L}} \\ \boldsymbol{\nabla}_{\iota_m}\boldsymbol{\mathcal{L}} \end{array}\right] \tag{9.16}$$

where entries of $\boldsymbol{\nabla}\boldsymbol{\mathcal{L}}$ are defined in equations (9.13)–(9.15). Conveniently, our method of finding vectors $\boldsymbol{\lambda}_1, \ldots, \boldsymbol{\lambda}_n$ that satisfy condition (9.11) produces a discretization of the adjoint equations (9.2). For all $i = 1, \ldots, n$ and for $k = 1, \ldots, N-1$, we compute the partial derivative of $\mathcal{L}$ with respect to $x_{i,k}$, as follows:

$$\frac{\partial \mathcal{L}}{\partial x_{i,k}} = h\frac{\partial g(\boldsymbol{x}_{\cdot,k}, \boldsymbol{u}_{\cdot,k})}{\partial x_{i,k}} - \lambda_{i,k-1} + \lambda_{i,k} + \sum_{\ell=1}^{n} \lambda_{\ell,k}\left(h\frac{\partial f_\ell(\boldsymbol{x}_{\cdot,k}, \boldsymbol{u}_{\cdot,k})}{\partial x_{i,k}}\right).$$

Additionally, we have that

$$\frac{\partial \mathcal{L}}{\partial x_{i,N}} = -\lambda_{i,N-1}.$$

We did not take the partial derivative of $\mathcal{L}$ with respect to $x_{i,0}$ since $x_{i,0}$ is a known value for all $i = 1, \dots, n$.

To satisfy theorem condition (9.11), we set $\frac{\partial}{\partial x_{i,k}}\mathcal{L}$ equal to zero for all $i = 1, \dots, n$ and $k = 1, \dots, N$, and solve for $\lambda_{i,k-1}$. We obtain the following for all $i = 1, \dots, n$:

$$\lambda_{i,k-1} = h\frac{\partial g(\boldsymbol{x}_{\cdot,k}, \boldsymbol{u}_{\cdot,k})}{\partial x_{i,k}} + \lambda_{i,k} + \sum_{\ell=1}^{n} \lambda_{\ell,k}\left(h\frac{\partial f_\ell(\boldsymbol{x}_{\cdot,k}, \boldsymbol{u}_{\cdot,k})}{\partial x_{i,k}}\right) \tag{9.17}$$

$$\text{for } k = 1, \dots, N-1, \quad \text{and}$$

$$\lambda_{i,N-1} = 0. \tag{9.18}$$

The above equations not only allow us to use the gradient of the Lagrangian of problem (9.5) to find the gradient of J_ρ, but also give us a discretization for adjoint equations (9.2). Additionally, for all $i = 1, \dots, n$, equation (9.18) is analogous to the transversality condition given in (9.3).

9.3 Example 1: The Fishery Problem

In this section, we focus on a basic resource model on harvesting, which was first presented in Clark's.[37] We present the fishery problem as stated in Lenhart and Workman [12, Example 17.4], where a logistic growth function is utilized within the fishery model.

$$\begin{aligned}
\max_{u\in\mathcal{A}} \quad & J(u) = \textstyle\int_0^T (pqx(t) - c)u(t)\,\mathrm{d}t \\
\text{sub.to} \quad & x'(t) = x(t)(1 - x(t)) - qu(t)x(t), \\
& x(0) = x_0 > 0, \\
& \qquad 0 \le u(t) \le M
\end{aligned} \tag{9.19}$$

In problem (9.19), $u(t)$ is the control variable that measures the effort put into harvesting fish at time t, and $x(t)$ measures the total population of fish at time t. We are assuming that there is a maximum harvesting rate, M. Parameter q represents the "catchability" of a fish and parameter c represents the cost of harvesting one unit of fish. Parameter p is the selling price for one unit of harvested fish.

The objective functional J is constructed to represent the total profit, revenue less cost, of harvesting fish over time interval $[0, T]$. The set of admissible controls, $\mathcal{A}$, for problem (9.19) is

$$\mathcal{A} = \{u \in L^1(0,T): \quad \text{for all } t \in [0,T],\ u(t) \in A,\ A = [0, M]\}.$$

Existence of an optimal control for problem (9.19) follows from the Filippov–Cesari Existence Theorem.[9]

According to Lenhart and Workman,[12] the standard forward–backward sweep method does not converge if parameters in problem (9.19) are set to where the optimal control u^* contains a singular region. In this section, we present an explicit formula for the singular case, which is obtained via Pontryagin's Maximum Principle,[10] and we show that the singular case satisfies the generalized Legendre–Clebsch Condition.[21–24] Additionally, we present a set of assumptions on the parameters to the problem in order to obtain an optimal harvesting strategy that begins singular and switches to the maximum harvesting rate. For this scenario, an explicit formula for the switching point is obtained. With parameters set to meet those particular assumptions, we use the explicit solution to problem (9.19) to test PASA's accuracy in solving for the regularized problem for varying values of the tuning parameter. We also discuss how to discretize for both the fishery problem and the regularized version of the problem in which a bounded variation regularization term is applied. And finally, we present some empirical evidence for convergence between the numerical solution obtained by PASA for the regularized fishery problem and the exact solution to the fishery problem.

9.3.1 *Explicitly solving the fishery problem*

Lenhart and Workman demonstrate a method of using Pontryagin's Maximum Principle[10, 12] and properties of the switching function to solve for the singular case to problem (9.19). Lenhart and Workman also discuss conditions for existence of a singular case. Since we use a numerical solver that is used for solving minimization problems, we provide an analytical solution to the minimization problem that is equivalent to problem (9.19). The equivalent minimization problem is obtained by negating the objective functional $J(u)$ in problem (9.19):

$$\begin{aligned} \min_{u} \quad & J(u) = \textstyle\int_0^T -(pqx(t) - c)u(t)\mathrm{d}t \\ \text{sub. to} \quad & x'(t) = x(t)(1 - x(t)) - qu(t)x(t), \\ & x(0) = x_0 > 0, \\ & \quad 0 \le u(t) \le M. \end{aligned} \tag{9.20}$$

Our procedure for solving problem (9.20) is analogous to what is used in Lenhart and Workman.[12] We use Pontryagin's Maximum Principle[10] to solve problem (9.20). The Hamiltonian for the above problem is

$$H(x, u, \lambda) = (c - pqx)u + \lambda(x - x^2 - qux), \tag{9.21}$$

where λ is the adjoint variable. By taking the partial derivative of the Hamiltonian with respect to state variable x, we obtain the adjoint equation associated with adjoint variable λ, which is

$$\lambda'(t) = -\frac{\partial H}{\partial x} = pqu - \lambda + 2\lambda x + q\lambda u \tag{9.22}$$

with the transversality condition being

$$\lambda(T) = 0. \tag{9.23}$$

We also use the Hamiltonian given in (9.21) to compute the switching function corresponding to problem (9.20)

$$\psi(t) = \frac{\partial H}{\partial u} = c - pqx(t) - q\lambda(t)x(t). \tag{9.24}$$

Based on Pontryagin's Maximum Principle,[10] if there exists an optimal pair (u^*, x^*) for problem (9.20), then there exists λ^*, satisfying adjoint equation (9.22) and terminal condition (9.23), where $H(x^*, u^*, \lambda^*) \le H(x^*, u, \lambda^*)$ for all admissible controls u. Additionally, if u^* is the optimal control, then u^* must have the following form:

$$u^*(t) = \begin{cases} 0 & \text{whenever } \psi(t) > 0, \\ \text{singular} & \text{whenever } \psi(t) = 0, \\ M & \text{whenever } \psi(t) < 0. \end{cases} \tag{9.25}$$

To solve for the singular case we suppose $\psi(t) \equiv 0$ on some subinterval $I \subset [0,T]$. Assuming that parameter $c > 0$, then we have from the switching function given in (9.24) that both x and $pq + q\lambda$ are nonzero on interval I. By setting the ψ equal to zero and solving for λ, we have that

$$\lambda^*(t) = \frac{c - pqx(t)}{qx(t)} \tag{9.26}$$

on the interval I. We differentiate equation (9.26) and use the state equation given in problem (9.20) to obtain the following:

$$\begin{aligned} \lambda'(t) &= -\frac{c}{qx^2}(x(1-x) - qux) \\ &= -\frac{c}{qx} + \frac{c}{q} + \frac{cu}{x}. \end{aligned} \tag{9.27}$$

We rewrite equation (9.22) by using expression (9.26), and we get

$$\lambda'(t) = -\frac{c}{qx} + p + \frac{2c}{q} - 2px + \frac{cu}{x}. \tag{9.28}$$

Equating expressions (9.27) and (9.28) gives a solution for x on the singular interval which is

$$x^*(t) = \frac{c + pq}{2pq}.$$

Since x^* is constant on I, we have that $x'(t) = 0$ on I. We use the singular solution for x^* and set the state equation found in problem (9.20) equal to zero to solve for u^*. On the interval I, we have that

$$\begin{aligned} 0 = x'(t) &= x^*(1 - x^*) - qux^* \\ u^*(t) &= \frac{1 - x^*}{q} \\ u^*(t) &= \frac{pq - c}{2pq^2}. \end{aligned}$$

Additionally, x^* being constant on the singular region and expression (9.26) would imply that λ^* is also constant on the singular region with constant value being as follows:

$$\lambda^*(t) = \frac{p(c - pq)}{c + pq}.$$

By substituting in the constant solutions for u^*, x^*, and λ^* into adjoint equation (9.22), we have that the right-hand side of the adjoint equation is zero, as desired. To conclude, we find that if a singular region, I, exists, then u^*, x^*, and λ^* are all constant on I, where

$$u^* = \frac{pq - c}{2pq^2}, \quad x^* = \frac{c + pq}{2pq}, \quad \lambda^* = \frac{p(c - pq)}{c + pq}. \tag{9.29}$$

We wish to show that the singular case solution satisfies the second-order necessary condition of optimality, which is referred to as the generalized Legendre–Clebsch Condition [21,22] or Kelley's condition.[23,24,43] Before showing that the singular cases given in (9.29) satisfy the Legendre–Clebsch Condition and/or Kelley's condition, we present some parameter assumptions on problem (9.20).

Assumption 9.1. Parameters $p, c, M, q > 0$ are set to satisfy $0 < pq - c < 2pq^2M$.

Assumption 9.2. Initial value x_0 is set to equal $\frac{c+pq}{2pq}$, which is the constant value that is associated to the state solution corresponding to singular u^*.

Note that Assumption 9.1 implies that the singular case solution u^* satisfies the bounded constraints that are assumed on the control. Assumption 9.2 dynamically forces the problem to yield an optimal control that begins singular. The generalized Legendre–Clebsch Condition involves finding what is called the *order* of a singular arc, which is defined as being the integer q such that $(\frac{d^{2q}}{dt^{2q}} \frac{\partial H}{\partial u})$ is the lowest order total derivative of the partial derivative of the Hamiltonian with respect to u, in which control u appears explicitly. We use the state equations given in problem (9.20) and the adjoint equations given in (9.22) to find the first and second time derivative of the switching function (9.24):

$$\frac{\mathrm{d}}{\mathrm{d}t}\psi = pq(x^2 - x) - q\lambda x^2,$$

$$\frac{\mathrm{d}^2}{\mathrm{d}t^2}\psi = pq(2x - 1)(x - x^2) - q\lambda x^2 + [pq^2x + q^2\lambda x^2 - 3pq^2x^2]u. \tag{9.30}$$

From above, we have that the order of the singular arc is $q = 1$. We need to show that if u^* is an optimal singular control on some interval of order q, then it is necessary that

$$(-1)^q \frac{\partial}{\partial u}\left[\frac{\mathrm{d}^2}{\mathrm{d}t^2}\left(\frac{\partial H}{\partial u}\right)\right]\Bigg|_{x=x^*,\lambda=\lambda^*} \geq 0. \tag{9.31}$$

We take the partial derivative of (9.30) with respect to u and evaluate at the corresponding singular case solutions for x and λ given in (9.29):

$$\frac{\partial}{\partial u}\left[\frac{\mathrm{d}^2}{\mathrm{d}t^2}\left(\frac{\partial H}{\partial u}\right)\right]\Bigg|_{x=\frac{c+pq}{2pq},\lambda=\frac{p(c-pq)}{2pq^2}}$$
$$= \frac{q(c+pq)}{2} - \frac{3(c+pq)^2}{4p} - \frac{(pq-c)(c+pq)^2}{8p^2q^2}.$$

Note that the term $\frac{(pq-c)(c+pq)^2}{8p^2q^2}$ is positive by Assumption 9.1. Using algebra, we combine the first two terms in the above equation to obtain

$$\frac{\partial}{\partial u}\left[\frac{\mathrm{d}^2}{\mathrm{d}t^2}\left(\frac{\partial H}{\partial u}\right)\right]\Bigg|_{x=\frac{c+pq}{2pq},\lambda=\frac{p(c-pq)}{2pq^2}}$$
$$= -\frac{3c^2}{4p} - cq - \frac{pq^2}{4} - \frac{(pq-c)(c+pq)^2}{8p^2q^2} < 0.$$

By multiplying the above inequality by $(-1)^q$ where $q = 1$, we then have the second-order necessary condition of optimality (9.31) being satisfied.

Note that only Assumption 9.1 is needed in proving that the singular case satisfies the Legendre–Clebsch condition; however, we can use both assumptions to obtain a control that begins singular and switches to the maximum harvesting rate, where an explicit formula for the switching point can be obtained. This particular scenario for problem (9.20) makes it an excellent candidate problem to use for testing PASA's accuracy in solving the regularized variant of this problem. We first verify Assumptions 9.1 and 9.2 imply that u^* is not singular on the entire time interval $[0, T]$. By using the transversality condition, $\lambda(T) = 0$, we recognize that the singular case solution for λ given in (9.26) is 0 if and only if parameters are set to either satisfy $p = 0$ or $c - pq = 0$, and Assumption 9.1 ensures that both cases are not possible. Let $0 < t^* < T$ be the time when u^* switches from being singular to non-singular, and let $I = [0, t^*)$ be the interval corresponding to when the optimal control u^* is singular.

By looking at the objective functional for problem (9.20), intuition tells us that $u^* \neq 0$ on $[t^*, T]$, and we can prove this by using proof by contradiction. Assume that

$$u^*(t) = \begin{cases} \dfrac{pq - c}{2pq^2} & 0 \leq t < t^* \\ 0 & t^* \leq t \leq T \end{cases}$$

is the optimal control to problem (9.20). Consider the following admissible control $\hat{v}$ where $\hat{v}$ is singular over the entire interval, i.e.,

$$\hat{v}(t) = \frac{pq - c}{2pq^2} \quad \text{for all } t \in [0, T].$$

By assumption of u^* being optimal for problem (9.20), we have $J(\hat{v}) \geq J(u^*)$. Since $\hat{v} = u^*$ on the interval $[0, t^*]$ and $u^* \equiv 0$ on interval $[t^*, T]$, we obtain the following:

$$J(\hat{v}) = J(u^*) - \int_{t^*}^{T} ((pqx_{\hat{v}}(t) - c)\hat{v}(t))\mathrm{d}t,$$

where $x_{\hat{v}}(t)$ is the corresponding solution to the state equation given in problem (9.20). A contradiction is obtained if we can show that $pqx_{\hat{v}} - c > 0$ for all $t \in [t^*, T]$ because the following implies $J(\hat{v}) \leq J(u^*)$. Since $\hat{v}$ is singular on the interval $[t^*, T]$, we have that $x_{\hat{v}}$ is the corresponding singular case solution given in (9.29). We then have that

$$\begin{aligned} pqx_{\hat{v}} - c &= pq\left(\frac{c + pq}{2pq}\right) - c \\ &= \frac{1}{2}(pq - c) \\ &> 0, \end{aligned}$$

where the above inequality holds by Assumption 9.1. Therefore, we have our contradiction.

It then follows by Pontryagin's Maximum Principle that the optimal harvesting policy for problem (9.20) with parameters set to satisfy Assumptions 9.1 and 9.2 is a control that begins singular and switches once to the maximal harvesting effort, meaning $u^* \equiv M$ on the interval $I = [t^*, T]$. We can find an explicit expression for t^*. Assume that $u^*(t) = M$ for all $t \in [t^*, T]$, then the adjoint equation (9.22) becomes

$$\lambda'(t) = pqM - \lambda + 2\lambda x + q\lambda M. \tag{9.32}$$

Now, λ is continuous at t^*. Hence,

$$\lambda(t^*) = \lim_{t \to t^{*-}} \lambda(t) = \frac{p(c - pq)}{c + pc},$$

which is the solution for equation (9.22) when u is singular. We can then solve for equation (9.32) to where λ must satisfy the terminal condition, $\lambda(T) = 0$, and the condition that

$$\lambda(t^*) = \frac{p(c - pq)}{c + pc}.$$

This condition allows us to obtain an explicit solution for t^*. When solving differential equation (9.32), we use a standard method that is used in solving linear differential equations. We rewrite (9.32) as

$$\lambda'(t) + \lambda(t)(\alpha - 2x(t)) = pqM, \tag{9.33}$$

where $\alpha = 1 - qM$. We let $v(t)$ be the appropriate integrating factor, which is defined as

$$v(t) = \mathrm{e}^{\int_{t^*}^{t} (\alpha - 2x(\tau))\mathrm{d}\tau}. \tag{9.34}$$

Multiplying $v(t)$ to both sides of Eqs. (9.33), integrating over the interval (t^*, t), and rearranging terms yields

$$\lambda(t) = \frac{1}{v(t)} \left[v(t^*)\lambda(t^*) + pqM \int_{t^*}^{t} v(\tau)\mathrm{d}\tau \right]. \tag{9.35}$$

Now in order to evaluate $v(t)$ and $\int_{t^*}^{t} v(\tau)\mathrm{d}\tau$, we need an explicit solution for $x(t)$ over the interval $[t^*, T]$.

Given that $u^*(t) = M$ for all $t \in [t^*, T]$, the state equation given in problem (9.20) over the specified interval becomes

$$\frac{\mathrm{d}x}{\mathrm{d}t} = x(1-x) - qMx, \tag{9.36}$$

which is separable. To solve the above equation, we separate variables x and t

$$\frac{\mathrm{d}x}{x(\alpha - x)} = \mathrm{d}t,$$

where $\alpha = 1 - qM$. We perform a partial fraction decomposition on the left-hand side of the above equation, integrate both sides, and exponentiate to obtain the following:

$$\frac{x}{|\alpha - x|} = K\mathrm{e}^{\alpha t}, \tag{9.37}$$

where K is some constant. We have that state variable x is continuous at switching point t^*. By continuity and Assumption 9.2 we have

$$x^*(t^*) = \lim_{t \to t^{*-}} x^*(t) = x_0 = \frac{c + pq}{2pq},$$

which is the state solution value associated with the singular case. We can use the above equality to solve for constant value K found in Eq. (9.37). Evaluating Eq. (9.37) at $t = t^*$ yields the following:

$$K = \frac{x_0 \mathrm{e}^{-\alpha t^*}}{|\alpha - x_0|} = \frac{(c + pq)\mathrm{e}^{-\alpha t^*}}{\gamma}, \tag{9.38}$$

where $\gamma = |2\alpha pq - c - pq|$. We obtain an explicit solution of Eq. (9.36), by rewriting Eq. (9.37) as

$$\frac{x}{\alpha - x} = \hat{K}\mathrm{e}^{\alpha t} \quad \text{where } \hat{K} = \begin{cases} K & \alpha - x \geq 0, \\ -K & \alpha - x < 0 \end{cases}.$$

Solving for the above equation yields

$$x(t) = \frac{\alpha \hat{K} \mathrm{e}^{\alpha t}}{1 + \hat{K}\mathrm{e}^{\alpha t}} \quad \text{for } t \in [t^*, T]. \tag{9.39}$$

Note that the value of $\hat{K}$ seems to depend on the sign of $\alpha - x(t)$. However, we can use continuity of x and the structure of the solutions for x on $[t^*, T]$ to show that the value of $\hat{K}$ only depends on the sign of $\alpha - x_0$. But first, we prove the following:

Proposition 9.1. *Assumptions* 9.1 *and* 9.2 *imply* $\alpha - x_0 < 0$ *where* $\alpha = 1 - qM$.

Proof. By Assumption 9.1, parameters $c, p, q, M > 0$ are chosen to satisfy $0 < pq - c$ and $pq - c < 2pq^2M$. Dividing both sides of the second inequality by $2pq$ yields

$$\frac{pq - c}{2pq} < qM.$$

Negating the inequality and adding one to both sides yields

$$\alpha < 1 + \frac{c - pq}{2pq} = \frac{c + pq}{2pq},$$

and the right-hand side of the inequality is x_0, by Assumption 9.2. Therefore, we have $\alpha - x_0 < 0$. □

Recall that by Assumption 9.2 and Eqs. (9.29) and (9.39), we have the following solution for the state variable:

$$x(t) = \begin{cases} x_0 = \dfrac{c + pq}{2pq} & 0 \le t \le t^* \\ \dfrac{\alpha \hat{K} e^{\alpha t}}{1 + \hat{K} e^{\alpha t}} & t^* \le t \le T \end{cases}, \tag{9.40}$$

where the sign of $\hat{K}$ is determined by the sign $\alpha - x(t)$. Since x is continuous on the entire time interval and since x is constant on the singular region, we have that $x(t^*) = x_0$. Hence, at $t = t^*$, $\hat{K}$ is determined by $\alpha - x(t^*) = \alpha - x_0$. Differentiating $x(t)$ along the non-singular region yields

$$x'(t) = \frac{\alpha^2 \hat{K} e^{\alpha t}}{(1 + \hat{K} e^{\alpha t})^2},$$

which is either strictly positive or strictly negative based upon the sign of $\hat{K}$. Since $\hat{K}$ is negative at $t = t^*$, there is some open interval containing t^* such that the function x is a non-increasing function. Note also

$$\lim_{t\to\infty} \frac{\alpha \hat{K} \mathrm{e}^{\alpha t}}{1 + \hat{K} \mathrm{e}^{\alpha t}} = \alpha,$$

so this function has a horizontal asymptote being $x_{\text{hor}} = \alpha$ on the tx-plane. This implies that $x(t)$ given in (9.40) remains above the horizontal asymptote on the interval $[t^*, T]$ even though the function is non-increasing on (t^*, T). In conclusion, the sign of $\alpha - x_0$ determines the structure of $x(t)$ on the non-singular region. Using Proposition 9.1 and Eq. (9.40) we then conclude $x(t)$ is of the following form:

$$x(t) = \begin{cases} x_0 = \dfrac{c + pq}{2pq} & 0 \leq t \leq t^* \\[2ex] \dfrac{\alpha K \mathrm{e}^{\alpha t}}{-1 + K \mathrm{e}^{\alpha t}} & t^* \leq t \leq T \end{cases},$$

where K is defined on Eq. (9.38). We now use the explicit solution for variable $x(t)$ to evaluate $v(t)$ given in (9.34) as follows:

$$v(t) = \exp\left(\alpha(t - t^*) - 2\int_{t^*}^{t} \frac{\alpha K \mathrm{e}^{\alpha \tau}}{-1 + K \mathrm{e}^{\alpha \tau}} \mathrm{d}\tau\right).$$

We use a u-substitution to evaluate the integral term in v. After applying some logarithm rules, we obtain the following:

$$v(t) = \left(\frac{-1 + K \mathrm{e}^{\alpha t^*}}{-1 + K \mathrm{e}^{\alpha t}}\right)^2 \mathrm{e}^{\alpha(t - t^*)}. \tag{9.41}$$

To evaluate $\int_{t^*}^{t} v(\tau)\mathrm{d}\tau$, we need to use a u-substitution method. Let $\sigma(\tau) = -1 + K\mathrm{e}^{\alpha\tau}$, then $\int_{t^*}^{t} v(\tau)\mathrm{d}\tau$ with v given in (9.41) becomes

$$\begin{aligned} \int_{t^*}^{t} v(\tau)\mathrm{d}\tau &= \int_{t^*}^{t} \left(\frac{-1 + K \mathrm{e}^{\alpha t^*}}{-1 + K \mathrm{e}^{\alpha \tau}}\right)^2 \mathrm{e}^{\alpha(\tau - t^*)} \mathrm{d}\tau \\ &= \int_{\sigma(t^*)}^{\sigma(t)} \frac{(-1 + K \mathrm{e}^{\alpha t^*})^2 \mathrm{e}^{-\alpha t^*}}{\alpha K \sigma^2} \mathrm{d}\sigma \\ &= \frac{(-1 + K \mathrm{e}^{\alpha t^*})^2}{\alpha K} \mathrm{e}^{-\alpha t^*} \left[\frac{1}{-1 + K \mathrm{e}^{\alpha t^*}} - \frac{1}{-1 + K \mathrm{e}^{\alpha t}}\right] \end{aligned}$$

$$= \frac{e^{-\alpha t^*}(-1 + Ke^{\alpha t^*})}{\alpha K}\left[1 - \frac{-1 + Ke^{\alpha t^*}}{-1 + Ke^{\alpha t}}\right]$$

$$= \frac{e^{-\alpha t^*}(-1 + Ke^{\alpha t^*})}{\alpha(-1 + Ke^{\alpha t})}\left[e^{\alpha t} - e^{\alpha t^*}\right]$$

$$\int_{t^*}^{t} v(\tau)\mathrm{d}\tau = \frac{(-1 + Ke^{\alpha t^*})}{\alpha(-1 + Ke^{\alpha t})}\left[e^{\alpha(t-t^*)} - 1\right].$$

Now we use equation (9.35) to evaluate $\lambda(t)$ as follows:

$$\lambda(t) = \left(\frac{-1 + Ke^{\alpha t}}{-1 + Ke^{\alpha t^*}}\right)^2 e^{\alpha(t^*-t)} f(t). \tag{9.42}$$

where

$$f(t) = \frac{p(c - pq)}{c + pq} + \frac{pqM}{\alpha}\left(\frac{-1 + Ke^{\alpha t^*}}{-1 + Ke^{\alpha t}}\right)\left(e^{\alpha(t-t^*)} - 1\right)$$

To find t^*, we set $\lambda(T) = 0$ and solve for t^*. Note that the term $-1 + Ke^{\alpha t}$ used in Eq. (9.42) must be non-zero for all $t \in [t^*, T]$, otherwise the state variable solution given in Eq. (9.39) is not defined for all $t \in [t^*, T]$. Using $-1+Ke^{\alpha t}$ for all $t \in [t^*, T]$ with $t^* < T$ allows us to conclude that $\lambda(T) = 0$ if and only if $f(T) = 0$. Consequently, for finding t^* we set $f(T) = 0$ and solve for t^*.

$$0 = f(T)$$

$$0 = \frac{c - pq}{c + pq} + \frac{qM}{\alpha}\left(\frac{-1 + Ke^{\alpha t^*}}{-1 + Ke^{\alpha T}}\right)\left(e^{\alpha(T-t^*)} - 1\right)$$

We multiply both sides of the above equation by $\alpha(c+pq)(-1+Ke^{\alpha T})$ to obtain

$$0 = \alpha(c-pq)(-1+Ke^{\alpha T}) + qM(c+pq)(-1+Ke^{\alpha t^*})\left(e^{\alpha(T-t^*)} - 1\right).$$

Substituting the value for K given in (9.38) into the above equation and multiplying everything by γ yields

$$0 = \alpha(c - pq)\left(-\gamma + (c + pq)e^{\alpha(T-t^*)}\right)$$
$$+ qM(c + pq)[-\gamma + (c + pq)]\left(e^{\alpha(T-t^*)} - 1\right).$$

We rearrange terms from the above equation to isolate expression $e^{\alpha(T-t^*)}$

$$e^{\alpha(T-t^*)} = \frac{\gamma\alpha(c-pq) + qM(c+pq)(-\gamma+(c+pq))}{(c+pq)[\alpha(c-pq)+qM(-\gamma+(c+pq))]}.$$

We take the natural logarithm of both sides of the above equation and rearrange terms to find that

$$t^* = T - \frac{1}{\alpha}\ln\left[\frac{\gamma\alpha(c-pq) + qM(c+pq)(-\gamma+(c+pq))}{(c+pq)[\alpha(c-pq)+qM(-\gamma+(c+pq))]}\right],$$

where $\alpha = 1 - qM$ and $\gamma = |pq - c - 2pq^2M|$. We simplify t^* more by substituting in $\alpha = 1 - qM$ into the above expression:

$$t^* = T - \frac{1}{1-qM}\ln\left[\frac{-\gamma(pq-c) - 2\gamma cqM + qM(c+pq)^2}{(c+pq)[(c-pq)+2pq^2M - \gamma qM]}\right]. \quad (9.43)$$

To summarize, Assumptions 9.1 and 9.2 imply that the optimal control u^* to problem (9.20) must begin singular and switch once to the non-singular case where $u^* \equiv M$ on $[t^*, T]$ with t^* given in (9.43). Additionally, the solutions for u^*, x^*, and λ^* are the following:

$$u^*(t) = \begin{cases} \dfrac{pq-c}{2pq^2} & 0 \le t < t^*, \\ M & t^* \le t \le T, \end{cases} \quad (9.44)$$

$$x^*(t) = \begin{cases} \dfrac{c+pq}{2pq} & 0 \le t \le t^*, \\ \dfrac{\alpha K e^{\alpha t}}{-1+Ke^{\alpha t}} & t^* \le t \le T, \end{cases} \quad (9.45)$$

and

$$\lambda^*(t) = \begin{cases} \dfrac{p(c-pq)}{c+pq} & 0 \le t \le t^*, \\ \left(\dfrac{-1+Ke^{\alpha t}}{-1+Ke^{\alpha t^*}}\right)^2 e^{\alpha(t^*-t)} f(t) & t^* \le t \le T, \end{cases} \quad (9.46)$$

where

$$f(t) = \frac{p(c-pq)}{c+pq} + \frac{pqM}{\alpha}\left(\frac{-1+Ke^{\alpha t^*}}{-1+Ke^{\alpha t}}\right)\left(e^{\alpha(t-t^*)} - 1\right),$$

$$\alpha = 1 - qM,$$

$$K = \frac{x_0 \mathrm{e}^{-\alpha t^*}}{|\alpha - x_0|} = \frac{(c + pq)\mathrm{e}^{-\alpha t^*}}{\gamma},$$

and

$$\gamma = |2\alpha pq - c - pq|.$$

9.3.2 *Discretization of the fishery problem*

For numerically solving problem (9.20), we first discretize and then optimize. We use the polyhedral active set algorithm (PASA), which was developed by Hager and Zhang,[31] to find an optimal solution to the discretized problem. Additionally, we need to discretize the adjoint equation associated with problem (9.20), which is

$$\lambda'(t) = pqu - \lambda + 2\lambda x + q\lambda u, \tag{9.47}$$

with the transversality condition being

$$\lambda(T) = 0.$$

For discretizing problem (9.20), we assume that control u is constant over each mesh interval. We partition time interval $[0, T]$, by using $N + 1$ equally spaced nodes, $0 = t_0 < t_1 < \cdots < t_N = T$. For all $k = 0, 1, \ldots, N$, we assume that the $x_k = x(t_k)$. For the control, we denote $u_k = u(t)$ for all $t_k \leq t < t_{k+1}$ when $k = 0, \ldots, N - 2$ and $u_{N-1} = u(t)$ for all $t_{N-1} \leq t \leq t_N$. So we have $\boldsymbol{x} \in \mathbb{R}^{N+1}$ while $\boldsymbol{u} \in \mathbb{R}^N$. We use a left-rectangular integral approximation for objective function J in (9.20), and we use forward Euler's method to approximate the state equation in (9.20). The discretization of problem (9.20) is then

$$\begin{aligned} \min \quad J(\boldsymbol{u}) &= \sum_{k=0}^{N-1} h(c - pqx_k)u_k \\ x_{k+1} &= x_k + h(1 - x_k - qu_k)x_k \quad \text{for all } 0 \leq k \leq N-1, \\ & x_0 > 0, \\ & 0 \leq u_k \leq M \qquad \text{for all } 0 \leq k \leq N-1, \end{aligned} \tag{9.48}$$

where $h = T/N$ is the mesh size and the first component of state vector, x_0, is set to being the initial condition associated with the state equation given in problem (9.20).

Since PASA uses the gradient projection algorithm for one of its phases, we need to compute the gradient of the cost functional for problem (9.48). We use Theorem 9.1 to find $\nabla_{\boldsymbol{u}} J$, which requires finding the Lagrangian to problem (9.48) and its gradient. Additionally, we need to construct a Lagrange multiplier vector $\boldsymbol{\lambda}$ that satisfies Eq. (9.11). Consequently, the Lagrange multiplier vector that satisfies Eq. (9.11) produces the numerical scheme that is used for discretizing adjoint equation (9.47) and produces the transversality condition (9.23). To compute the Lagrangian to problem (9.48), we first need to arrange the discretized state equations accordingly

$$-x_{k+1} + x_k + h(x_k - x_k^2 - qu_k x_k) = 0 \quad \text{for all } k = 0, 1, \ldots, N-1. \tag{9.49}$$

The Lagrangian to problem (9.48) is

$$\mathcal{L}(\boldsymbol{x}, \boldsymbol{u}, \boldsymbol{\lambda}) = \sum_{k=0}^{N-1} (h(c - pqx_k)u_k) + \sum_{k=0}^{N-1} \lambda_k(-x_{k+1} + x_k + h(x_k - x_k^2 - qu_k x_k)),$$

where $\boldsymbol{\lambda} \in \mathbb{R}^{N-1}$ is the Lagrange multiplier vector. Note that we need not worry about the inequality constraints associated with the bounds of the control when computing the Lagrangian to problem (9.48) because these bounds are not being entered into the cost functional J. By taking the partial derivative of $\mathcal{L}$ with respect to u_k, we obtain

$$\frac{\partial \mathcal{L}}{\partial u_k} = h(c - pqx_k - q\lambda_k x_k) \quad \text{for } k = 0, \ldots, N-1.$$

By Theorem 9.1, we have that

$$\nabla_{\boldsymbol{u}} J = \nabla_{\boldsymbol{u}} \mathcal{L} = \begin{bmatrix} h(c - pqx_0 - q\lambda_0 x_0) \\ \vdots \\ h(c - pqx_k - q\lambda_k x_k) \\ \vdots \\ h(c - pqx_{N-1} - q\lambda_{N-1} x_{N-1}) \end{bmatrix}, \tag{9.50}$$

provided that Eq. (9.11) is satisfied.

To satisfy Eq. (9.11), we take the partial derivative of $\mathcal{L}$ with respect to x_k for all $k = 1, \ldots, N$, and note that we do not take the partial derivative of $\mathcal{L}$ with respect to x_0 because x_0 is a known value. Taking the partial derivative of $\mathcal{L}$ with respect to the state vector components yield the following expressions:

$$\frac{\partial \mathcal{L}}{\partial x_k} = h(-pqu_k + \lambda_k(1 - 2x_k - qu_k)) \\ + \lambda_k - \lambda_{k-1} \quad \text{for } k = 1, \ldots, N-1, \tag{9.51}$$

$$\frac{\partial \mathcal{L}}{x_N} = -\lambda_{N-1} \tag{9.52}$$

To align with Eq. (9.11), we set expressions (9.51) and (9.52) equal to zero and solve for λ_{k-1} for all $k = 1, \ldots, N$:

$$\lambda_{k-1} = \lambda_k + h(-pqu_k + \lambda_k(1 - 2x_k - qu_k)) \\ \text{for } k = 1, \ldots, N-1, \tag{9.53}$$

$$\lambda_{N-1} = 0. \tag{9.54}$$

Expressions (9.53) and (9.54) serve as the discretization for the costate equation (9.47) and the transversality condition (9.23).

We use PASA to solve for the regularized version of problem (9.20) where the penalty applied to the problem is a bounded variation penalty, as suggested in Capognigro *et al.*,.[30] The regularized version of problem (9.20) is as follows:

$$\begin{aligned} \min \quad J_\rho(u) &= \int_0^T (c - pqx)udx + \rho V(u) \\ x'(t) &= (1 - x)x - qux, \; x(0) = x_0 > 0, \\ 0 &\leq u(t) \leq M, \end{aligned} \tag{9.55}$$

where $0 \leq \rho < 1$ is a tuning parameter and $V(u)$ measures the total variation of u which is

$$V(u) = \sup_{\mathcal{P}} \sum_{i=0}^{n_P - 1} |u(t_{i+1}) - u(t_i)| \tag{9.56}$$

where $\mathcal{P} = \{P = \{t_0, t_1, \ldots, t_{n_P}\} : \text{P is a partition of } [0, T]\}$. Note that if we numerically solve problem (9.55) with $\rho = 0$, then the problem is not being regularized via bounded variation. We use the same procedure as before to discretize problem (9.55). Assuming control u to be constant over each mesh interval allows us to express the total variation of u as being the sum of the absolute value of the jumps of u. For a sufficiently small mesh size h, we have

$$V(u) = \sum_{k=0}^{N-1} |u(t_{k+1}) - u(t_k)|.$$

The discretized version of problem (9.55) is

$$\begin{aligned}
\min \quad J_\rho(\boldsymbol{u}) &= \sum_{k=0}^{N-1} (h(c - pqx_k)u_k) + \rho \sum_{k=0}^{N-1} |u_{k+1} - u_k| \\
x_{k+1} &= x_k + h(1 - x_k - qu_k)x_k \quad \text{for all } 0 \le k \le N-1, \\
& x_0 > 0, \\
& 0 \le u_k \le M \quad \text{for all } 0 \le k \le N-1.
\end{aligned} \tag{9.57}$$

Because PASA involves a gradient scheme, we should be concerned about the absolute value terms that are used in problem (9.57)'s cost function. We suggest a decomposition of each absolute value term in J_ρ to ensure that J_ρ is differentiable. We introduce two $N-1$ vectors $\boldsymbol{\zeta}$ and $\boldsymbol{\iota}$ whose entries are non-negative. Each entry of $\boldsymbol{\zeta}$ and $\boldsymbol{\iota}$ is defined as

$$|u_{k+1} - u_k| = \zeta_k + \iota_k, \quad \text{for all } k = 0, \dots, N-2.$$

An equivalent way of expressing the above equation is to assign values to the components of $\boldsymbol{\zeta}$ and $\boldsymbol{\iota}$ based upon the following conditions:

Condition 1: If $u_{k+1} - u_k > 0$, then $\zeta_k = u_{k+1} - u_k$ and $\iota_k = 0$;

Condition 2: If $u_{k+1} - u_k \le 0$, then $\zeta_k = 0$ and $\iota_k = -(u_{k+1} - u_k)$.

This decomposition converts problem (9.57) into the following:

$$\begin{aligned}
\min \quad J_\rho(\boldsymbol{u}, \boldsymbol{\zeta}, \boldsymbol{\iota}) &= \sum_{k=0}^{N-1} (h(c - pqx_k)u_k) + \rho \sum_{k=0}^{N-2} (\zeta_k + \iota_k) \\
x_{k+1} &= x_k + h(1 - x_k - qu_k)x_k \quad \text{for all } k = 0, \dots, N-1, \\
& x_0 > 0, \\
& 0 \le u_k \le M \quad \text{for all } k = 0, \dots, N-1, \\
& u_{k+1} - u_k = \zeta_k - \iota_k \quad \text{for all } k = 0, \dots, N-2, \\
& \zeta_k \ge 0 \text{ and } \iota_k \ge 0 \quad \text{for all } 1, \dots, N-1.
\end{aligned} \tag{9.58}$$

For problem (9.58), we are minimizing the regularized objective function with respect to vectors $\boldsymbol{u}, \boldsymbol{\zeta}$, and $\boldsymbol{\iota}$. The constraints associated with $\boldsymbol{\zeta}$ and $\boldsymbol{\iota}$, are constraints that PASA can interpret. The equality constraints associated with $\boldsymbol{\zeta}$ and $\boldsymbol{\iota}$ can be written accordingly as

$$\left[\boldsymbol{A} \middle| -\boldsymbol{I}_{N-1} \middle| \boldsymbol{I}_{N-1}\right] \begin{bmatrix} \boldsymbol{u} \\ \hline \boldsymbol{\zeta} \\ \hline \boldsymbol{\iota} \end{bmatrix} = \mathbf{0}, \tag{9.59}$$

where $\boldsymbol{I}_{N-1}$ is the identity matrix with dimension $N-1$, $\mathbf{0}$ is the $N-1$ dimensional all zeros vector, and $\boldsymbol{A}$ is the $N-1 \times N$ dimensional sparse matrix given in (9.7). Since problem (9.58) is optimizing J_ρ with respect to $\boldsymbol{u}$, $\boldsymbol{\zeta}$, and $\boldsymbol{\iota}$, the Lagrangian of problem (9.58) is

$$\begin{aligned}\mathcal{L}(\boldsymbol{x}, \boldsymbol{u}, \boldsymbol{\zeta}, \boldsymbol{\iota}, \boldsymbol{\lambda}) = &\sum_{k=0}^{N-1} (h(c - pqx_k)u_k) + \rho \sum_{k=0}^{N-2} (\zeta_k + \iota_k) \\ &+ \sum_{k=0}^{N-1} \lambda_k(-x_{k+1} + x_k + h(x_k - x_k^2 - qu_k x_k)).\end{aligned}$$

If we use Theorem (9.1) to find the gradient of J_ρ, we have the following:

$$\nabla_{\boldsymbol{u},\boldsymbol{\zeta},\boldsymbol{\iota}} J_\rho = \nabla_{\boldsymbol{u},\boldsymbol{\zeta},\boldsymbol{\iota}} \mathcal{L} = \begin{bmatrix} \nabla_{\boldsymbol{u}} J \\ \hline \rho \\ \vdots \\ \rho \\ \hline \rho \\ \vdots \\ \rho \end{bmatrix}, \tag{9.60}$$

where $\nabla_{\boldsymbol{u}} J$ is defined in Eq. (9.50), provided that the theorem's condition (9.11) is satisfied. As before, we satisfy condition (9.11) by taking the partial derivative of $\mathcal{L}$ with respect to each x_k for $k = 1, \ldots, N$ and set each partial derivative equal to 0 and solve for the components of $\boldsymbol{\lambda}$. Performing this procedure yields Eqs. (9.53) and (9.54), and they serve as our discretization procedure of the adjoint equation (9.22) and our generalization of the transversality condition (9.23).

9.3.2.1 *Summary of the discretization*

We convert the regularized harvesting problem (9.55) into problem (9.58) by performing the following steps.

(1) *Discretization of the state equation*: We use forward Euler's method to discretize the state equation which is given in Eq. (9.49).
(2) *Discretization of the objective functional and the decomposition of absolute value terms*: We use a left-rectangular integral approximation for discretizing the integral shown in (9.55). We also assume u as being piecewise constant over each mesh interval, which allows us to convert the total variation term given in (9.56) into the finite series of absolute value terms that is used in problem (9.57). Because we use a gradient scheme for solving the discretized problem, we need to decompose the absolute value terms by introducing two $N-1$ vectors $\boldsymbol{\zeta}$ and $\boldsymbol{\iota}$ that satisfy the constraints that are included in problem (9.58).
(3) *Finding the gradient of the regularized objective functional*: By Theorem 9.1, we use the gradient of the Lagrangian to problem (9.58) to find $\nabla_{[\boldsymbol{u},\boldsymbol{\zeta},\boldsymbol{\iota}]} J_\rho$. The formula for $\nabla_{[\boldsymbol{u},\boldsymbol{\zeta},\boldsymbol{\iota}]} J_\rho$ is given in Eq. (9.60), where as $\nabla_{\boldsymbol{u}} J$ is given in equation (9.50).
(4) *Discretization of the adjoint variable*: In order to apply Theorem 9.1 for computing the gradient of the penalized objective function, adjoint vector $\boldsymbol{\lambda}$ needs to satisfy condition (9.11). We set each partial derivative of the Lagrangian $\mathcal{L}$ with respect to x_k equal to 0 and solve for λ_{k-1}. This results in discretizing adjoint equation (9.22) and the transversality condition $\lambda(T) = 0$. The discretized equations are given in Eqs. (9.53) and (9.54).

9.3.3 *Numerical results of the fishery problem*

We wish to numerically solve problem (9.20) with the following parameters defined in Table 9.1. Note that these parameter values satisfy Assumptions 9.1 and 9.2, so the optimal harvesting strategy is of the form given in (9.44). Based on Table 9.1 and Eq. (9.43), $t^* \approx 9.5392$, which is between values 0 and $T = 10$. Additionally, from these parameter values the singular control solution is $u^* = 0.1875$ which is between the bounds 0 and $M = 1$, as desired. After using Table 9.1 to evaluate t^* from (9.43), u^* from (9.44), x^* from (9.45),

Table 9.1. Parameter settings for the fishery problem.

Parameter	Description	Value
T	Terminal time	10
p	Selling price of one unit of fish	2
q	"Catchability" of the fish	2
c	Cost of harvesting one unit of fish	1
M	Maximum harvest effort	1
x_0	Initial population size of fish	$\frac{c+pq}{2pq} = 0.625$

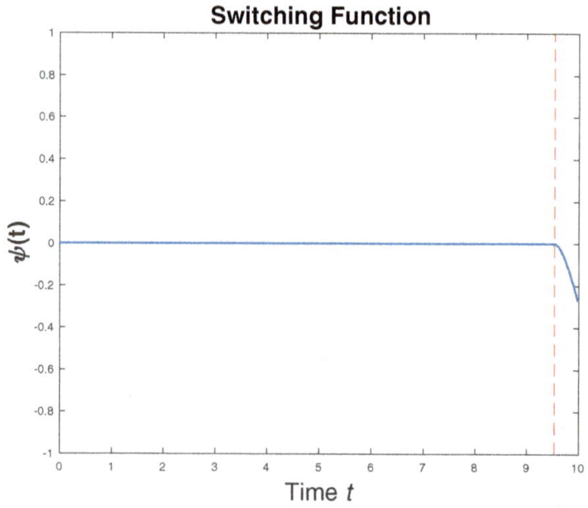

Fig. 9.1. Plot of the Switching Function. The red dotted line is the vertical line $t = t^* \approx 9.5392$.

and λ^* from (9.46), we substitute these solutions into the switching function (9.24) to see if u^* satisfies (9.25). In Figure 9.1, we plot the switching function ψ that we obtained. Regard that the switching function is zero when u^* is singular and becomes negative at the interval $[t^*, T]$, which is when $u^* = M$. So we have that the harvesting policy u^* for (9.44) satisfies (9.25). This is equivalent to saying that u^* satisfies Pontryagin's Maximum Principle,[10] which is the first-order necessary condition for optimality.

We use PASA to solve problem (9.20) with parameter settings given in Table 9.1. Our initial guess for control u is $u(t) = 0$ for all $t \in [0, T]$, and the stopping tolerance is set to 10^{-10}. We partition

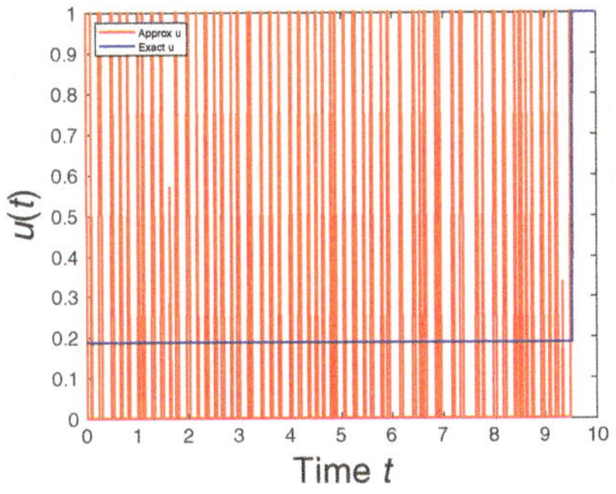

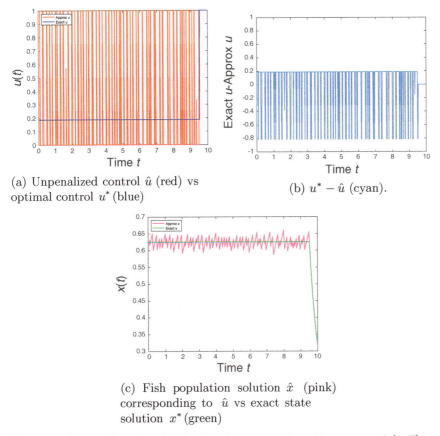

(a) Unpenalized control $\hat{u}$ (red) vs optimal control u^* (blue)

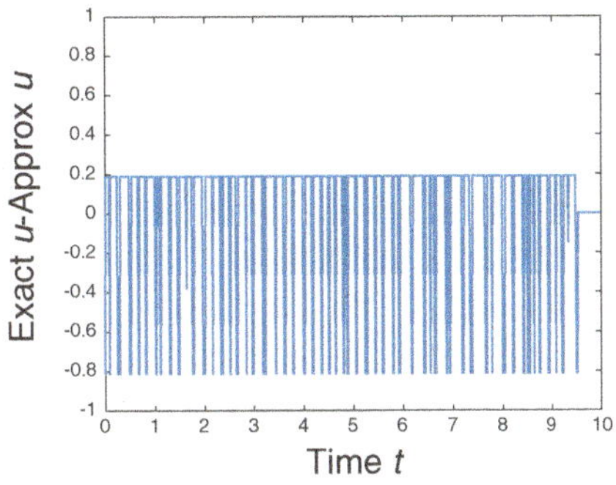

(b) $u^* - \hat{u}$ (cyan).

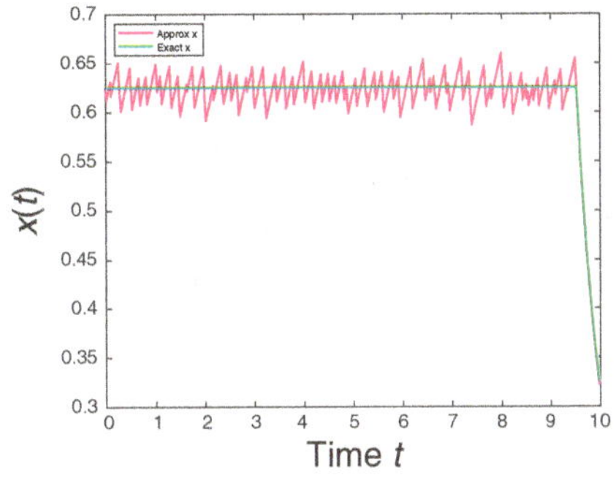

(c) Fish population solution $\hat{x}$ (pink) corresponding to $\hat{u}$ vs exact state solution x^* (green)

Fig. 9.2. Unpenalized Results for the fishery problem: Time interval $[0, T]$ is partitioned to have $N = 750$ mesh intervals and PASA stopping tolerance is tol$=10^{-10}$.

$[0, T]$ to where there are $N = 750$ mesh intervals, and we use the discretization method that is presented in Section 9.3.2. We first observe PASA's approximation of the optimal control for problem (9.20) when no bounded variation regularization is being applied, which we denote as being $\hat{u}$. In Figure 9.2(a), a plot of $\hat{u}$ is shown in red and is compared to the exact solution, u^*, which is shown in blue. In Figure 9.2(a), $\hat{u}$ has oscillations that resemble chattering on the singular region. Additionally, we reran experiments with a tighter stopping tolerance and a finer partition of $[0, T]$, and PASA still obtained an oscillatory solution. It is likely that the discretization of problem (9.20) causes PASA to generate the oscillatory artifacts. We computed a left-rectangular integral approximation of the total profit of the harvested fish over time interval $[0, T]$, (i.e., the cost functional for the equivalent maximization problem (9.19)) when employing harvesting policy, $\hat{u}$, and optimal harvesting policy, u^*, and found that the $-J(\hat{u}) \approx 3.074$ and $-J(u^*) \approx 3.0575$ where J is the cost functional used in (9.20).

The major concern associated with the approximate solution obtained in Figure 9.2(a) is that it is an unrealistic harvesting strategy to use. We wish to regularize problem (9.20) by adding a bounded variation term that reduces the number of oscillations. Before getting into the results, we first discuss our methods for determining which penalty parameter values give the best solution. We have some advantages for choosing an appropriate tuning parameter value due to knowing the analytic solution. However, for problems in which we do not know what u^* looks like we recommend looking at the plots of PASA's approximation of the regularized control u_ρ. Firstly, we would rule out a penalty parameter value if the plot of u_ρ contained any unusual jumps and/or oscillations. If such a solution occurred, we suggest that the penalty parameter value is too small and is generating a solution that is similar to the unpenalized solution. Secondly, we would rule out a ρ value if the corresponding penalized solution u_ρ does not closely align with Pontryagin's Maximum Principle.[10] In this example, Pontryagin's Maximum Principle implies that u^* is a piecewise constant function where $u^*(t) \in \{0, \frac{pq-c}{2pq^2}, M\}$ for all $t \in [0, T]$. So if the penalized solution u_ρ ever takes on values that were not $0, M$, or $\frac{pq-c}{2pq^2}$, then we find value ρ to be suspect. Thirdly, one might find it useful to look at the plots of the switching function that are associated with the penalized solution u_ρ. This third suggestion might be the most useful if one were penalizing an optimal control problem where the singular case solution cannot be found explicitly. The sign of the switching function can help one verify whether or not u_ρ aligns with Pontryagin's Maximum Principle.

Since we have the advantage of comparing the regularized solution u_ρ with the exact solution u^*, we can determine which penalty parameter value is appropriate by comparing the plots between u^* and u_ρ as well as the plots of $u^* - u_\rho$. Additionally, we look at the instance in time when the penalized solution u_ρ switched from singular to non-singular to see if it is close to when the switching point should occur, i.e., when $t^* \approx 9.5392$.

When calculating the associated switching point for u_ρ, we look at the first instance when $u_\rho = M$. Due to the discretization of problem (9.55), each approximated switching point is some mesh point value of $[0, T]$. We also use the L_1 norm difference between u^* and u_ρ for determining which penalty parameter value gives the closest approximation to u^*. We observe the L_∞ norm difference between u^* and u_ρ as well; however, this computation may not be helpful since approximated switching points are restricted to mesh points of the partitioned time interval.

Using parameter settings given in Table 9.1, we use PASA to solve problem (9.55) for varying values of penalty parameter $\rho \in \{10^{-9}, 10^{-8}, 10^{-7}, 10^{-6}, 10^{-5}, 10^{-4}, 10^{-3}, 10^{-2}, 10^{-1}\}$. Our initial guess for problem (9.55) is $u(t) = 0$ for all $t \in [0, T]$. We partition $[0, T]$ to where there are $N = 750$ mesh intervals, and we use the discretization method described in Section 9.3.2.1. Additionally, the stopping tolerance is set to 10^{-10}. Results corresponding to the penalized solutions that PASA obtained are found in Figures 9.3 and 9.4 and Table 9.2. Note that from looking at the plots of the penalized controls of Figure 9.2 alone, we find that $\rho = 10^{-2}$ is the most appropriate penalty parameter value. We believe that even if we do not know what the explicit formula for the singular case was for problem (9.55), the plots of the regularized controls should lead us to choose $\rho = 10^{-2}$ as being the appropriate tuning parameter because no oscillations are occurring and the end behavior of the penalized control matches the non-oscillating region from the unpenalized control. This yields some evidence that we could potentially use PASA for solving penalized control problems without *a priori* information of the optimal control.

We continue to explain how we ruled out all cases except for $\rho = 10^{-2}$. In Figure 9.3(b), we rule out tuning parameter value $\rho = 10^{-8}$, since its corresponding solution u_ρ is oscillating in the same manner as the unpenalized solution, whose plot is given in Figure 9.3(a). We do not provide plots of penalized solution u_ρ that corresponded to when $\rho = 10^{-9}, 10^{-7}, 10^{-6}$ because their plots possessed many oscillations similar to Figure 9.3(b). Additionally, one could look at

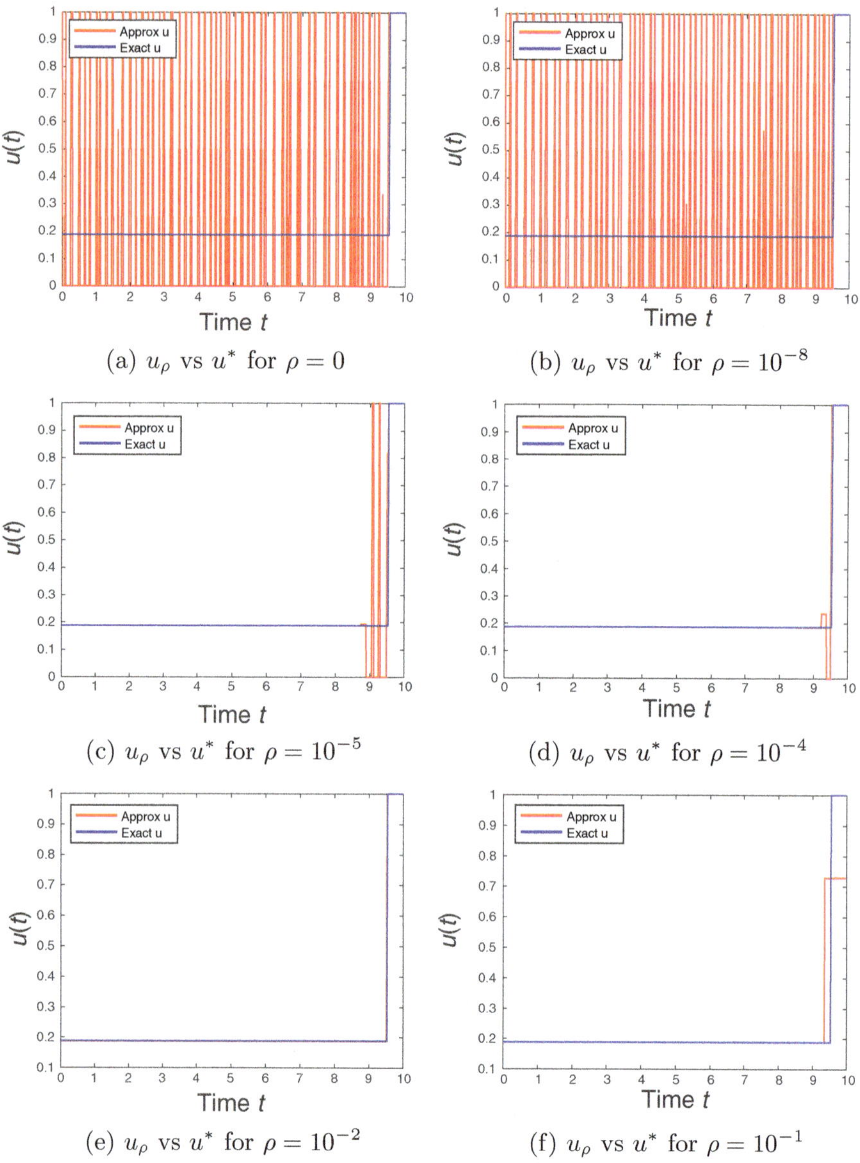

(a) u_ρ vs u^* for $\rho = 0$

(b) u_ρ vs u^* for $\rho = 10^{-8}$

(c) u_ρ vs u^* for $\rho = 10^{-5}$

(d) u_ρ vs u^* for $\rho = 10^{-4}$

(e) u_ρ vs u^* for $\rho = 10^{-2}$

(f) u_ρ vs u^* for $\rho = 10^{-1}$

Fig. 9.3. Plots of the regularized control u_ρ (red) vs the optimal control u^* (blue) for varying values of the tuning parameter ρ. Time interval $[0, T]$ is partitioned to have $N = 750$ mesh intervals. PASA stopping tolerance is tol=10^{-10}.

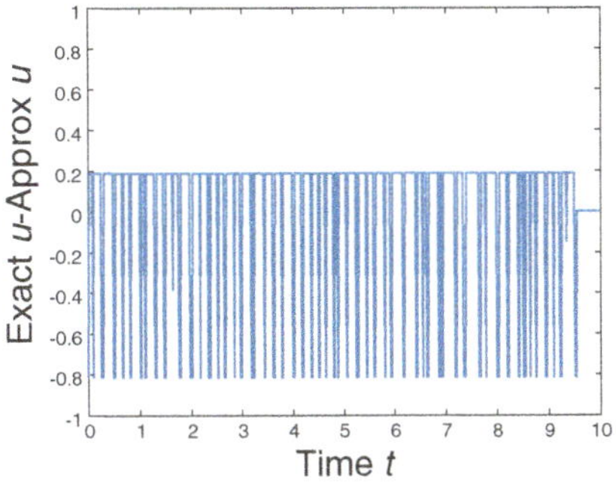

(a) $u^* - u_\rho$ for $\rho = 0$

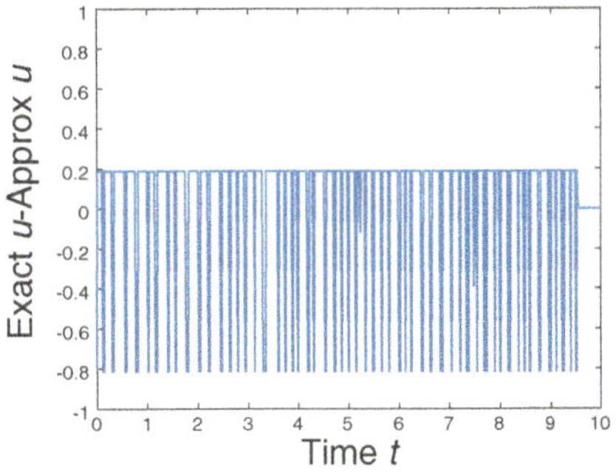

(b) $u^* - u_\rho$ for $\rho = 10^{-8}$

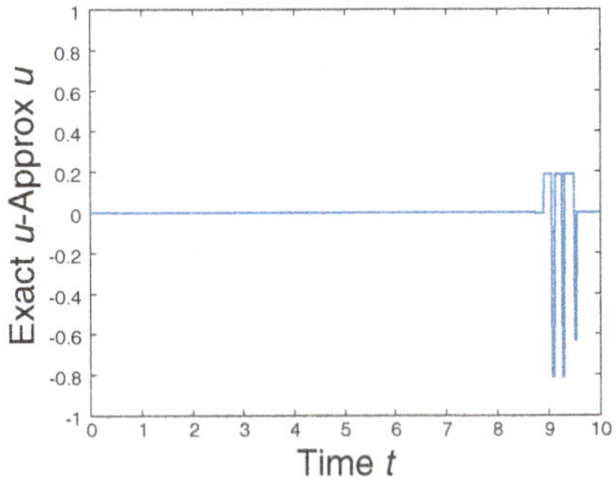

(c) $u^* - u_\rho$ for $\rho = 10^{-5}$

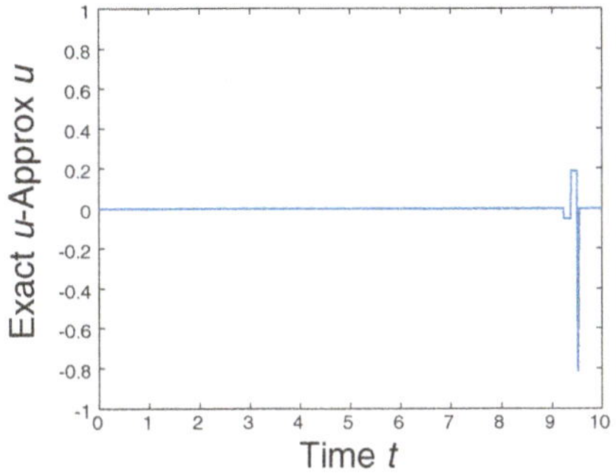

(d) $u^* - u_\rho$ for $\rho = 10^{-4}$

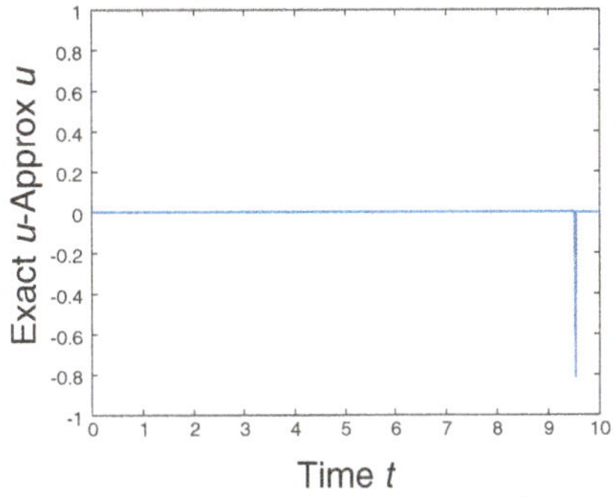

(e) $u^* - u_\rho$ for $\rho = 10^{-2}$

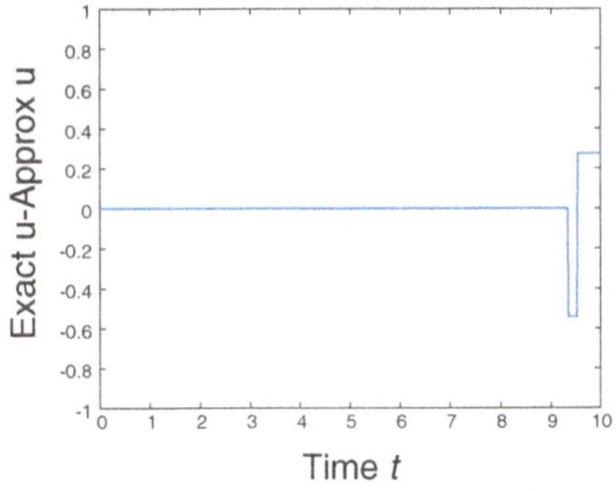

(f) $u^* - u_\rho$ for $\rho = 10^{-1}$

Fig. 9.4. Plots of the optimal control minus the regularized control $u^* - u_\rho$ (cyan) for varying values of the tuning parameter ρ. Time interval $[0, T]$ is partitioned to have $N = 750$ mesh intervals. PASA stopping tolerance is tol$=10^{-10}$.

Table 9.2. Varying tuning parameter table for the fishery problem.

Parameters	ρ	$\|u^* - u_\rho\|_{L^1}$	$\|u^* - u_\rho\|_\infty$	Switch	Runtime
$N = 750$	0	2.89548286	0.8125	0^*	8.82
$\text{tol} = 10^{-10}$	10^{-9}	2.89262350	0.8125	0^*	14.44
$T = 10$	10^{-8}	2.89708395	0.8125	0^*	22.11
$M = 1$	10^{-7}	2.89799306	0.8125	0^*	28.45
$p = 2$	10^{-6}	2.88774180	0.8125	0.04^*	34.88
$q = 2$	10^{-5}	0.19705717	0.8125	9.0667	6.58
$c = 1$	10^{-4}	0.05427589	0.8125	9.5200	1.25
$x_0 = 0.625$	10^{-3}	0.01628685	0.8125	**9.5333**	0.68
$h = 0.0133$	10^{-2}	**0.01282480**	0.8125	**9.5333**	**0.38**
	10^{-1}	0.22780044	0.5402	9.9867	0.43

Note: The total variation of our exact solution is $v(u^*) = 0.8125$, so if our approximated solution u_ρ switched from the singular case to m at a time different from u^*, then $\|u^* - u_\rho\|_\infty = v(u^*)$. the switch column is indicating the first instance in time t when $u_\rho(t) = m$. the $*$-values on the switch column indicate that the numerical solution is oscillating. The Runtime column indicates the time (in seconds) it takes for PASA to solve the problem.

the $\|u^* - u_\rho\|_{L^1}$ column in Table 9.2 to see that the penalized solution for u_ρ for $\rho \in \{10^{-9}, 10^{-8}, 10^{-7}, 10^{-6}\}$ are negligible in comparison to the unpenalized solution. In Figures 9.3(c) and 9.4(c), we start to see some improvements in the singular region of u_ρ when $\rho = 10^{-5}$. However, due to the oscillations appearing in time interval $(8.5, 9.5)$, we dismiss $\rho = 10^{-5}$. We observe in Figure 9.3(d) that increasing the penalty parameter ρ to 10^{-4} significantly reduces the number of oscillations along the singular region. However, around time interval $[9.24, 9.37]$, $u_\rho(t) \approx 0.2366$, which is a value different from the bounds of the control and the singular case solution. We rule out penalty $\rho = 10^{-4}$ for this reason. We do not provide a figure of the solution we obtained when penalty parameter value $\rho = 10^{-3}$, but we dismiss this penalty parameter value due to similar reasoning that was used when $\rho = 10^{-4}$. We also rule out tuning parameter $\rho = 10^{-1}$ because as illustrated in Figure 9.3(f), u_ρ is constant around $[9.36, 10]$ with constant value being approximately 0.7277. For this case, we say that $\rho = 10^{-1}$ is over-penalizing the control along the non-singular region. Based on Table 9.2 and Figures 9.3(e) and 9.4(e), we find that penalty parameter $\rho = 10^{-2}$ is the most appropriate penalty parameter. For $\rho = 10^{-2}$, we have that u_ρ switches to the non-singular case at the

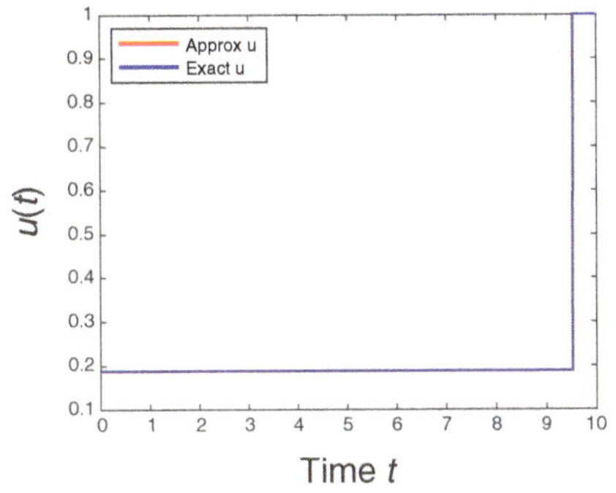

(a) u_ρ (red) vs u^* (blue) with $\rho = 10^{-2}$

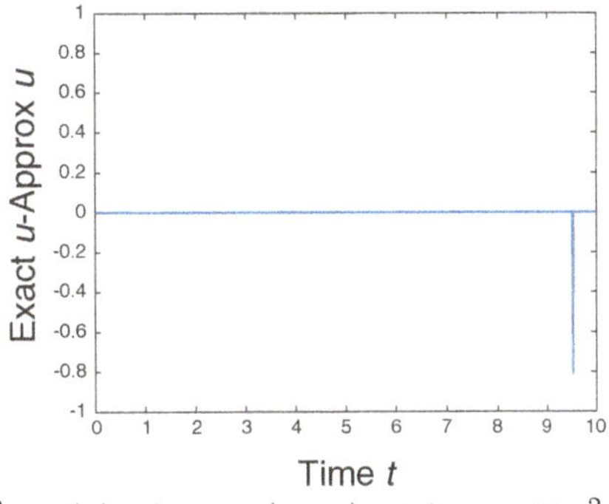

(b) $u^* - u_\rho$ (cyan) with $\rho = 10^{-2}$

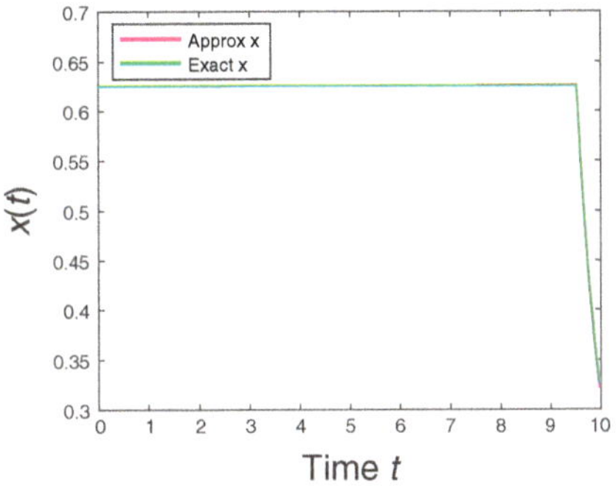

(c) Approximate State x_ρ (pink) vs Exact State x^* (green) with $\rho = 10^{-2}$

Fig. 9.5. Results for when $\rho = 10^{-2}$: Time interval $[0, T]$ is partitioned to have $N = 750$ mesh intervals and PASA stopping tolerance is tol=10^{-10}.

node that is closest to the actual switching point. In addition, u_ρ have the lowest L^1 norm error in Table 9.2 when the penalty parameter is set to $\rho = 10^{-2}$. In Figure 9.5(c), we have a plot of the state solution corresponding to u_ρ when $\rho = 10^{-2}$. Observe that the plot of the approximated fish population corresponding to u_ρ lines up almost perfectly with the true solution.

After finding an appropriate tuning parameter value for problem (9.55), we then study the numerical convergence rate between PASA's solution to the discretized regularized problem with tuning parameter $\rho = 10^{-2}$ and the exact solution to problem (9.20). We observe the L^1 norm error between the exact solution u^* and the regularized solution u_h for varying mesh sizes $h \in \{0.2, 0.1, 0.05, 0.025, 0.0125, 0.00625, 0.003125\}$. In Table 9.3, we have

Table 9.3. Convergence analysis between the penalized solution and the exact solution to fishery problem.

Parameters	h	err_h	$\frac{\text{err}_h}{\text{err}_{h/2}}$	$\log_2\left(\frac{\text{err}_h}{\text{err}_{h/2}}\right)$
tol $= 10^{-10}$	0.2	0.28091381		
$\rho = 10^{-2}$	0.1	0.12111065	2.31948057	1.21380176
	0.05	0.00089762	134.92448804	7.07600840
	0.025	0.02736103	0.03280644	−4.92987718
	0.0125	0.01089531	2.51126802	1.32841601
	0.00625	0.00527063	2.06717301	1.04765914
	0.003125	0.00165509	3.18450288	1.67106818

Note: Here $err_h = \|u_h - u^*\|_{L^1}$, where u_h is the penalized solution for Fishery problem when the mesh size is h.

err_h representing the $\|u^* - u_h\|_{L^1}$. Note that for almost all mesh size values the L^1 norm error between u_h and u^* is relatively close to the mesh size value. However, in the case when $h = 0.05$, we have that the L^1 norm error is significantly less than the mesh size. This is because the mesh size influences the discretization of time interval $[0, T]$ to where one mesh point is exceedingly close to the true switching point value. We look at the last column of Table 9.3 to determine what the rate of convergence between the penalized solution and the exact solution is. All but two entries in the last column of Table 9.3 are values that are slightly greater than one, and the entries that are not close to one involved using the L^1 norm error for when $h = 0.05$. Based on the values of the last column of Table 9.3, we find the convergence rate as being slightly better than a linear rate. Additionally, form Table 9.3, we take the natural logarithm of values found in columns 2 and 3, and use least squares method to indicate if there is a linear relationship between $\ln(h)$ and $\ln(\text{err}_h)$. Since we view data point $(\ln(0.05), \ln(0.00089762))$ as being an outlier, we also perform least squares between $\ln(h)$ and $\ln(\text{err}_h)$ without the outlier point. In Figure 9.6, the blue line does a better job with fitting to the data than the red line. In Table 9.4, we see that the goodness of fit associated with the blue line is very strong. We look at the slope associated with the blue line to determine the rate of convergence. Since the slope of the blue line is approximately 1.200738, we have further indication that the rate of convergence between the penalized solution and the exact solution is better than linear.

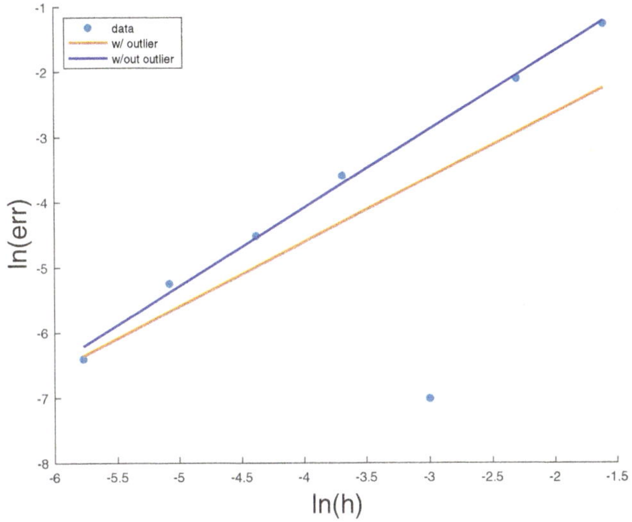

Fig. 9.6. Linear fit of data points $(\ln(h), \ln(\text{err}_h))$ from Table 9.3. Red line is the linear fit of the data points including outlier point $(\ln(0.05), \ln(0.00089762))$. Blue line is the linear fit of data points excluding the outlier point.

Table 9.4. Linear fit table.

$\ln(\text{err}_h) = m(\ln(h)) + b$	Slope m	y-intercept b	r^2
With outlier	0.988064	−0.6645189	0.4826212
Without outlier	1.200738	0.7100375	0.9959617

9.4 Example 2: The Plant Problem

In this section, we implement a biological optimal control problem from David King and Jonathan Roughgarden[8] that can be solved analytically. In Ref. 8, King and Roughgarden use optimal control theory to study the allocation strategies that annual plants possess when distributing photosynthate to components of the plant that pertain to *vegetative growth* and to components that pertain to *reproductive growth*. The control variable $u(t)$ involved represents the fraction of photosynthate being reserved for vegetative growth. The remaining photosynthate, $1 - u(t)$, is assigned to aid in the reproductive growth processes. The annual plants that are being modeled are assumed to live in an environment where consecutive seasons are identical. Since $u(t)$ represents a fraction, the control variable is assumed to be bounded in the obvious way:

$$0 \leq u(t) \leq 1 \quad \text{for all } t \in [0, T], \tag{9.61}$$

where T represents the maximum season length. The optimal control problem has two state variables where one state x_1 represents the weight of the vegetative part of a plant, while the second state x_2 represents the weight of the reproductive part of a plant. The construction of the state equations is a continuous version of a discrete time model of reproduction for annual plants that was developed by Cohen[44] and is given by

$$x_1'(t) = u(t)x_1, \qquad x_1(0) > 0, \tag{9.62}$$

$$x_2'(t) = (1 - u(t))x_2, \quad x_2(0) \geq 0. \tag{9.63}$$

For the construction of an objective functional, King and Roughgarden[8] aim to find the strategy that maximizes total reproduction and assert that such a strategy is "expected in organisms if natural selection maximizes total reproduction." Their construction of a cost functional that measures the total reproduction produced by a plant is based on Cohen's discrete model in[45] which uses the expectation of the logarithm of seed yield to express the long-term rate of population increases observed in annual plants. The objective functional is chosen to have natural logarithm of x_2, which helps this problem to have singular arcs. The optimal control problem is then

$$\max_{u \in \mathcal{A}} J(u) = \int_0^T \ln(x_2(t))dt \tag{9.64}$$

subject to the state equations (9.62)–(9.63) and control bounds (9.61). The set of the admissible controls is defined $\mathcal{A} = \{u \in L^1(0,T) : \text{for all } t \in [0,T],\ u(t) \in A,\ A = [0,1]\}$. Existence of an optimal control follows from the Filippov–Cesari Existence Theorem.[9] An explicit solution is obtained for problem (9.64) that satisfies Pontryagin's Maximum Principle,[10] the Generalized Legendre Clebsch Condition or Kelley's Condition,[22–24,43] the Strengthened Generalized Legendre Clebsch Condition, and McDanell and Powers's Junction Theorem.[21] The structure of the optimal allocation strategy u^* depends on the maximum season length T, the initial weight of the vegetative part of the plant $x_1(0)$, and the initial weight of the reproductive part of the plant $x_2(0)$.

In this section, we provide a summary of the solution to the equivalent minimization problem as follows:

$$\begin{aligned} \min \quad & -J(u) = -\int_0^T \ln(x_2(t))dt \\ \text{s.t.} \quad & \dot{x}_1(t) = u(t)x_1, \\ & \dot{x}_2(t) = (1-u(t))x_1, \\ & x_1(0) = x_{1,0} > 0, \quad x_2(0) = x_{2,0} \geq 0, \\ & 0 \leq u(t) \leq 1. \end{aligned} \tag{9.65}$$

Additionally we demonstrate how to discretize the penalized version of problem (9.65)

$$\begin{aligned} \min \quad & J_\rho(u) = -\int_0^T \ln(x_2(t))dt + \rho V(u) \\ \text{s.t.} \quad & \dot{x}_1(t) = u(t)x_1, \\ & \dot{x}_2(t) = (1-u(t))x_1, \\ & x_1(0) = x_{1,0} > 0, \quad x_2(0) = x_{2,0} \geq 0, \\ & 0 \leq u(t) \leq 1, \end{aligned} \tag{9.66}$$

where $0 \leq \rho < 1$ is the bounded variation tuning parameter and $V(u)$ measures the total variation of control u, as defined in Eq. (9.56). We then use PASA to numerically solve problems (9.65) and (9.66) when parameters $x_{1,0}$, $x_{2,0}$, and T are set in such a way where u^* contains a singular subarc.

9.4.1 *Solving for the singular case to the plant problem*

Before we present a summary of the solution to problem (9.65), we discuss some terminology that is used in King and Roughgarden's work[8] to describe the events when $u^*(t) = 0$ and when $u^*(t) = 1$. We say that u^* is *purely reproductive* at time t when $u^*(t) = 0$. We say that u^* is *purely vegetative* at time t when $u^*(t) = 1$. Our method for obtaining a solution to the minimization problem (9.65) parallels with what is used in King and Roughgarden.[8] Our method for solving problem (9.65) involves Pontryagin's Maximum Principle.[10] We compute the Hamiltonian to problem (9.65) as

$$H = -\ln x_2 + x_1(\lambda_1 - \lambda_2)u + x_1\lambda_2.$$

We take the partial derivatives of the Hamiltonian with respect to x_1 and x_2 to construct differential equations for costate variables λ_1 and λ_2 as follows:

$$\lambda_1'(t) = -\frac{\partial H}{\partial x_1} = (\lambda_2 - \lambda_1)u - \lambda_2, \qquad (9.67)$$

$$\lambda_2'(t) = -\frac{\partial H}{\partial x_2} = \frac{1}{x_2}, \qquad (9.68)$$

and use the transversality conditions to obtain boundary conditions for the costate equations

$$\lambda_1(T) = \lambda_2(T) = 0. \qquad (9.69)$$

We construct the switching function $\psi(t)$ by taking the partial derivative of the Hamiltonian with respect to control u as:

$$\psi(t) = x_1(\lambda_1 - \lambda_2) \qquad (9.70)$$

Since $x_1(t) > 0$ for all t, the sign of ψ depends on the sign of $\lambda_1 - \lambda_2$. By Pontryagin's Maximum Principle, the optimal allocation strategy $u^*(t)$ is as follows:

$$u^*(t) = \begin{cases} 0 & \text{whenever } \lambda_1(t) > \lambda_2(t), \\ \text{singular} & \text{whenever } \lambda_1(t) = \lambda_2(t), \\ 1 & \text{whenever } \lambda_1(t) < \lambda_2(t). \end{cases} \qquad (9.71)$$

To solve for the singular case, we assume that there is an interval $I \subset [0, T]$ where $\lambda_1(t) = \lambda_1(t)$ for all $t \in I$. This implies that $\lambda_1'(t) \equiv \lambda_2'(t)$ on I. From adjoint equations (9.67)–(9.68), we have that the following holds on I:

$$(\lambda_2(t) - \lambda_1(t))u(t) - \lambda_2(t) = \frac{1}{x_2(t)},$$

which yields the following on interval I:

$$\lambda_2(t) = -\frac{1}{x_2(t)}. \tag{9.72}$$

It follows from Eqs. (9.68) and (9.72) that

$$\dot{\lambda}_2 = -\lambda_2$$

for all $t \in I$. Solving for the above differential equation yields

$$\lambda_2(t) = A\mathrm{e}^{-t},$$

where $A = -\frac{1}{x_2(t_1)}\mathrm{e}^{t_1}$ and t_1 is the time when the singular subarc begins.

We can construct a differential equation for $x_2(t)$ on interval I by differentiating Eq. (9.72) with respect to time and rearranging the terms as follows:

$$\dot{x}_2 = x_2. \tag{9.73}$$

Equations (9.63) and (9.73) yield

$$(1 - u(t))x_1 = x_2,$$

which implies that

$$\frac{x_2(t)}{x_1(t)} = 1 - u(t) \tag{9.74}$$

on interval I. We differentiate (9.74) to get

$$-\dot{u}(t) = \frac{x_1\dot{x}_2 - \dot{x}_1 x_2}{x_1^2}.$$

We replace $\dot{x}_1$ and $\dot{x}_2$ in the above equation with state Eq. (9.62) and equation (9.73), respectively, and obtain the following:

$$-\dot{u}(t) = \frac{x_2}{x_1}(1 - u(t)). \tag{9.75}$$

Substituting (9.74) into (9.75) yields

$$\frac{\dot{u}(t)}{(1 - u(t))^2} = -1.$$

We solve for the separable equation above and obtain the following solution:

$$u(t) = 1 - \frac{1}{C - t}, \tag{9.76}$$

on interval I, where C is some constant. Observe that from Eq. (9.76), (9.74) becomes

$$\frac{x_2(t)}{x_1(t)} = \frac{1}{C - t}, \tag{9.77}$$

on interval I.

In order to find constant term C we need to find the time, t_2, when the singular subarc ends. As mentioned in,[8] we can be certain that u^* is not singular at time T because the singular case solution for adjoint variable $\lambda_2(t)$ given in equation (9.72) would not satisfy the transversality condition (9.69). Now from the adjoint equations (9.67) and (9.68) and the transversality conditions (9.69), we have that $\dot{\lambda}_1(t) \to 0$ and $\dot{\lambda}_2(t) \to \frac{1}{x_2(T)}$ as $t \to T$. The derivatives of λ_1 and λ_2 near T and the transversality conditions imply that $\lambda_1(t) > \lambda_2(t)$ for some interval, namely $(t_2, T]$, and so from (9.71), we have that $u^*(t) \equiv 0$ on $(t_2, T]$. So, an optimal control that possesses a singular region switches to being purely reproductive on t_2 and remains purely reproductive on the interval $(t_2, T]$.

In order to find t_2, we must first solve for the state equations and the associated adjoint equations on time interval $[t_2, T]$. From the state equations (9.62)–(9.63), $u(t) \equiv 0$ on $[t_2, T]$ implies the following:

$$\begin{aligned} x_1(t) &= x_1(t_2), \\ x_2(t) &= x_1(t_2)(t - t_2) + x_2(t_2), \end{aligned} \tag{9.78}$$

where

$$x_1(t_2) = (C - t_2)x_2(t_2), \quad \text{and} \quad x_2(t_2) = x_2(t_1)e^{t_2 - t_1}.$$

Differentiating $x_2(t)$ given in Eq. (9.78) yields

$$\dot{x}_2(t) = x_1(t) = x_1(t_2). \tag{9.79}$$

We use the differential found from the above separable equation and the transversality condition (9.69) to solve for the adjoint equation (9.68).

$$\lambda_2(T) - \lambda_2(t) = \int_{x_2(t)}^{x_2(T)} \frac{1}{x_2} \frac{dx_2}{x_1(t_2)}$$

$$\lambda_2(t) = \frac{1}{x_1(t_2)} \ln\left(\frac{x_2(t)}{x_2(T)}\right) \text{ for } t \in [t_2, T]. \tag{9.80}$$

We use Eq. (9.80) and adjoint equation (9.67) to obtain the following differential equation for λ_1:

$$\dot{\lambda}_1(t) = -\frac{1}{x_1(t_2)} \ln\left(\frac{x_2(t)}{x_2(T)}\right). \tag{9.81}$$

Again, we use the differential obtained from separable equation (9.79) and the terminal condition for λ_1 (9.69) to solve the above differential equation as follows:

$$\lambda_1(T) - \lambda_1(t) = -\frac{1}{(x_1(t_2))^2} \int_{x_2(t)}^{x_2(T)} (\ln x_2 - \ln x_2(T))dx_2$$

$$-\lambda_1(t) = -\frac{1}{(x_1(t_2))^2} \left[x_2(t) - x_2(T) + x_2(t) \ln\left(\frac{x_2(T)}{x_2(t)}\right)\right].$$

By negating the above equality and applying logarithmic rules, we obtain a solution for $\lambda_1(t)$:

$$\lambda_1(t) = -\frac{1}{(x_1(t_2))^2} \left[x_2(t) \ln\left(\frac{x_2(t)}{x_2(T)}\right) + x_2(T) - x_2(t)\right]$$
$$\text{for } t \in [t_2, T]. \tag{9.82}$$

After finding the solutions for the state and adjoint equations on interval $(t_2, T]$, we are ready to solve for t_2. Since $\lambda_1(t)$ and $\lambda_2(t)$ are continuous on $[0, T]$ and $\lambda_1(t) \equiv \lambda_2(t)$ on the singular case, we have $\lambda_1(t_2) = \lambda_2(t_2)$ We equate Eqs. (9.82) and (9.80) at $t = t_2$ and solve for t_2 as follows:

$$-\frac{1}{(x_1(t_2))^2} \left[x_2(t_2) \ln\left(\frac{x_2(t_2)}{x_2(T)}\right) + x_2(T) - x_2(t_2)\right]$$
$$= -\frac{1}{x_1(t_2)} \ln\left(\frac{x_2(t_2)}{x_2(T)}\right).$$

We multiply the above equation by $x_1(t_2)$ and rearrange the terms to obtain the following equation:

$$x_2(T) - x_2(t_2) = (x_1(t_2) + x_2(t_2)) \ln\left(\frac{x_2(T)}{x_2(t_2)}\right). \tag{9.83}$$

We use (9.78) to evaluate $x_2(T)$, which yields $x_2(T) = x_1(t_2)$ $(T - t_2) + x_2(t_2)$, and apply it to Eq. (9.83) as follows:

$$T - t_2 = \left[1 + \frac{x_2(t_2)}{x_1(t_2)}\right]\left[\ln\left(\frac{x_1(t_2)}{x_2(t_2)}(T - t_2) + 1\right)\right]. \tag{9.84}$$

We also have $\dot{\lambda}_1(t_2) = \dot{\lambda}_2(t_2)$, so using (9.72), (9.81), and $x_2(T) = x_1(t_2)(T - t_2) + x_2(t_2)$, we find that

$$\frac{x_1(t_2)}{x_2(t_2)} = \ln\left(\frac{x_1(t_2)}{x_2(t_2)}(T - t_2) + 1\right). \tag{9.85}$$

We rewrite Eqs. (9.84) and (9.85) into terms of y and z, where $y = T - t_2$ and $z = \frac{x_2(t_2)}{x_1(t_2)}$, to obtain a nonlinear system of equations that can be solved numerically through a nonlinear solver. Consequently, we have the following:

$$y = T - t_2 \approx 2.79328213, \quad \text{and} \tag{9.86}$$

$$z = \frac{x_2(t_2)}{x_1(t_2)} \approx 0.55763674. \tag{9.87}$$

Using $x_2(T) = x_1(t_2)(T - t_2) + x_2(t_2)$ and $x_1(t) = x_1(t_2)$ for $t \in [t_2, T]$ yields

$$\frac{x_2(T)}{x_1(T)} = (T - t_2) + \frac{x_2(t_2)}{x_1(t_2)}.$$

From Eqs. (9.86) and (9.87), we then have

$$\frac{x_2(T)}{x_1(T)} \approx 3.35091887. \tag{9.88}$$

Additionally, from Eqs. (9.84) and (9.85) we have

$$T - t_2 = \left[1 + \frac{x_2(t_2)}{x_1(t_2)}\right]\frac{x_1(t_2)}{x_2(t_2)} = 1 + \frac{x_1(t_2)}{x_2(t_2)},$$

which implies that

$$\frac{x_2(t_2)}{x_1(t_2)} = \frac{1}{(T-1)-t_2}.$$

Finally, we can evaluate Eq. (9.77) at $t = t_2$ and use the above equation to solve for constant C, and we have that $C = T - 1$.

9.4.1.1 *Summary of singular case solution to the plant problem*

In Section 9.4.1, we find that if the optimal control u^* to problem (9.65) contains a singular subarc where u^* becomes singular at time t_1, then u^* is the following on the interval $[t_1, T]$:

$$u^*(t) = \begin{cases} 1 - \dfrac{1}{T-1-t} & \text{when } t \in [t_1, t_2), \\ 0, & \text{when } t \in [t_2, T], \end{cases} \tag{9.89}$$

where $t_2 \approx T - 2.79328213$. The corresponding state solutions x_1 and x_2 are as follows:

$$x_1(t) = \begin{cases} (T-1-t)x_2(t) & \text{when } t \in [t_1, t_2], \\ x_1(t_2) & \text{when } t \in [t_2, T], \end{cases} \tag{9.90}$$

$$x_2(t) = \begin{cases} x_2(t_1)\mathrm{e}^{t-t_1} & \text{when } t \in [t_1, t_2], \\ x_1(t_2)(t-t_2) + x_2(t_2) & \text{when } t \in [t_2, T], \end{cases} \tag{9.91}$$

where $x_1(t_2) = (T-1-t_2)x_2(t_2)$ and $x_2(t_2) = x_2(t_1)\mathrm{e}^{t_2-t_1}$. The corresponding solution to the adjoint variables λ_1 and λ_2 are as follows:

$$\lambda_1(t) = \begin{cases} \lambda_2(t) & \text{when } t \in [t_1, t_2], \\ -\dfrac{1}{(x_1(t_2))^2}\left[x_2(t)\ln\left(\dfrac{x_2(t)}{x_2(T)}\right)\right. & \\ \qquad \left. + x_2(T) - x_2(t)\right] & \text{when } t \in [t_2, T], \end{cases} \tag{9.92}$$

$$\lambda_2(t) = \begin{cases} -\dfrac{1}{x_2(t_1)}\mathrm{e}^{t_1-t} & \text{when } t \in [t_1, t_2], \\ \dfrac{1}{x_1(t_2)}\ln\left(\dfrac{x_2(t)}{x_2(T)}\right) & \text{when } t \in [t_2, T]. \end{cases} \tag{9.93}$$

Additionally, King and Roughgarden's work[8] uses the generalized Legendre–Clebsch Condition,[21–23] also referred to as Kelley's condition,[24,43] to show that the second order necessary condition for optimality is satisfied. The generalized Legendre–Clebsch Condition involves finding what is called the *order* of a singular arc, which is defined as the integer q being such that $\left(\frac{\mathrm{d}^{2q}}{\mathrm{d}t^{2q}}\frac{\partial H}{\partial u}\right)$ is the lowest order total derivative of the partial derivative of the Hamiltonian with respect to u, in which control u appears explicitly. By using Eqs. (9.62), (9.63), (9.67), (9.68), and (9.70), we have

$$\frac{\mathrm{d}}{\mathrm{d}t}\frac{\partial H}{\partial u} = \frac{\mathrm{d}}{\mathrm{d}t}\psi = -x_1\lambda_2 - \frac{x_1}{x_2},$$

$$\frac{\mathrm{d}^2}{\mathrm{d}t^2}\frac{\partial H}{\partial u} = \frac{\mathrm{d}^2}{\mathrm{d}t^2}\psi = \left(-x_1\lambda_2 - \frac{x_1}{x_2} - \frac{x_1^2}{x_2^2}\right)u - \frac{x_1}{x_2} + \frac{x_1^2}{x_2^2}.$$

So for problem (9.65), the order of singular subarc u^* is 1. By General Legendre Clebsch Condition, if u^* is an optimal singular control on some interval $[t_1, t_2)$ of order q, then it is necessary that

$$(-1)^q \frac{\partial}{\partial u}\left[\frac{\mathrm{d}^{2q}}{\mathrm{d}t^{2q}}\left(\frac{\partial H}{\partial u}\right)\right] \geq 0$$

if the extremum is a minimum, and the inequality is reversed if the extremum is a maximum.[21] Additionally, if the above inequality is strict, then the strengthened Legendre–Clebsch Condition holds. On singular region $[t_1, t_2)$, we have that

$$(-1)^1 \frac{\partial}{\partial u}\left[\frac{\mathrm{d}^2}{\mathrm{d}t^2}\left(\frac{\partial H}{\partial u}\right)\right] = \frac{x_1^2}{x_2^2} > 0,$$

so the singular arc is of order 1 and satisfies both the generalized Legendre–Clebsch Condition and the strengthened Legendre–Clebsch Condition.

What we have yet to discuss is conditions for problem (9.65) that determine when the optimal control u^* is bang-bang or concatenations of bang and singular controls. In King and Roughgarden,[8] King and Roughgarden construct a series of conditions for determining the structure of the optimal control solution u^* to problem (9.65). Most of the conditions are based upon the value of terminal time T and the ratio of the reproductive to vegetative weight at time $t = 0$. We provide a summary of these conditions (see Ref. 8 for details on how these conditions are constructed):

(1) The optimal allocation strategy cannot be purely reproductive on the entire time interval ($u^*(t)$ cannot be zero over the entire time interval $[0, T]$) unless $T \leq 3.35091887$ and $\frac{x_2(0)}{x_1(0)} \leq 3.35091887 - T$.

(2) If $T > 3.35091887$ and $0 \leq \frac{x_2(0)}{x_1(0)} < 0.55763674e^{T-2.79328213}$, the optimal allocation, u^*, contains a singular subarc.

   (a) If $\frac{x_2(0)}{x_1(0)} = \frac{1}{T-1}$, then the optimal allocation strategy u^* begins with a singular subarc and switches to being purely reproductive at time $t_2 = T - 2.79328213$.

   (b) If $0 \leq \frac{x_2(0)}{x_1(0)} < \frac{1}{T-1}$, then the optimal control u^* contributes to reproductive growth before and after the singular subarc occurs.

   (c) If $\frac{1}{T-1} < \frac{x_2(0)}{x_1(0)} < 0.55763674e^{T-2.79328213}$, then u^* begins contributing to purely vegetative growth before the singular subarc occurs. At time $t_2 = T - 2.79328213$, u^* switches from being singular to being purely reproductive.

(3) If $\frac{x_2(0)}{x_1(0)} \geq 0.55763674e^{T-2.79328213}$, then the optimal control must be bang-bang, where u^* begins with purely vegetative growth and switches once to being purely reproductive.

The value $\frac{1}{T-1}$ corresponds to evaluating the right-hand side of equation (9.77) at $t = 0$. Additionally, some of the numerical values shown in the above conditions correspond to Eqs. (9.86), (9.87), and (9.88).

We are only interested in finding the exact solutions to problem (9.65) in the cases where the optimal control contains a singular subarc, i.e., Cases 2a, 2b, and 2c. For Case 2a, we use Eqs. (9.89)–(9.93) with $t_1 = 0$ to construct the exact solution which is as follows:

Case 2a: If $T > 3.35091887$, $0 \leq \frac{x_2(0)}{x_1(0)} < 0.55763674e^{T-2.79328213}$, and $\frac{x_2(0)}{x_1(0)} = \frac{1}{T-1}$, then

$$u^*(t) = \begin{cases} 1 - \dfrac{1}{T - t - 1} & 0 \leq t \leq t_2, \\ 0 & t_2 < t \leq T, \end{cases} \tag{9.94}$$

$$x_1(t) = \begin{cases} x_2(t)(T-t-1) & 0 \le t \le t_2, \\ x_1(t_2) & t_2 \le t \le T, \end{cases}$$

$$x_2(t) = \begin{cases} x_2(0)e^t & 0 \le t \le t_2, \\ x_1(t_2)(t-t_2) + x_2(t_2) & t_2 \le t \le T, \end{cases}$$

$$\lambda_1(t) = \begin{cases} \lambda_2(t) - \dfrac{1}{(x_1(t_2))^2} & 0 \le t \le t_2, \\ \quad \times \left[x_2(t) \ln\left(\dfrac{x_2(t)}{x_2(T)}\right) + x_2(T) - x_2(t) \right] & t_2 \le t \le T, \end{cases}$$

$$\lambda_2(t) = \begin{cases} -\dfrac{1}{x_2(0)} \mathrm{e}^{-t} & 0 \le t \le t_2, \\ \dfrac{1}{x_1(t_2)} \ln\left(\dfrac{x_2(t)}{x_2(T)}\right) & t_2 \le t \le T, \end{cases}$$

where $t_2 = T - 2.79328213$, $x_1(t_2) = (T - 1 - t_2)x_2(t_2)$, and $x_2(t_2) = x_2(0)\mathrm{e}^{t_2}$.

For Case 2b, we solve the state equations (9.62) and (9.63) on the interval $[0, t_1]$ with u set to being 0, and then we use continuity of the state variables to get an explicit solution for t_1. Then we solve the adjoint equations (9.67) and (9.68) on $[0, t_1]$ where we use Eqs. (9.92)–(9.93) and continuity of λ_1 and λ_2 at t_1 to gain the following boundary conditions: $\lambda_1(t_1) = \lambda_2(t_1) = -\frac{1}{x_2(t_1)}$.

The exact solution for Case 2b is as follows:

Case 2b: If $T > 3.35091887$, $0 \le \frac{x_2(0)}{x_1(0)} < 0.55763674\mathrm{e}^{T-2.79328213}$, and $\frac{x_2(0)}{x_1(0)} < \frac{1}{T-1}$, then

$$u^*(t) = \begin{cases} 0 & 0 \le t < t_1, \\ 1 - \dfrac{1}{T-t-1} & t_1 \le t \le t_2, \\ 0 & t_2 < t \le T, \end{cases}$$

$$x_1(t) = \begin{cases} x_1(0) & 0 \le t \le t_1, \\ (T - t - 1)x_2(t) & t_1 \le t \le t_2, \\ x_1(t_2) & t_2 \le t \le T, \end{cases}$$

$$x_2(t) = \begin{cases} x_1(0)t + x_2(0) & 0 \le t \le t_1, \\ x_2(t_1)\mathrm{e}^{t-t_1} & t_1 \le t \le t_2, \\ x_1(t_2)(t - t_2) + x_2(t_2) & t_2 \le t \le T, \end{cases}$$

$$\lambda_1(t) = \begin{cases} \dfrac{x_2(t)}{(x_1(0))^2} \ln\left(\dfrac{x_2(t_1)}{x_2(t)}\right) + \dfrac{x_2(t) - x_2(t_1)}{(x_1(0))^2} \\ \quad + \dfrac{x_2(t) - x_2(t_1)}{x_2(t_1)x_1(0)} - \dfrac{1}{x_2(t_1)} & 0 \le t \le t_1, \\ \lambda_2(t) - \dfrac{x_2(t)}{(x_1(t_2))^2} & t_1 \le t \le t_2, \\ \quad \times \left(\ln\left(\dfrac{x_2(t)}{x_2(T)}\right) - 1\right) - \dfrac{x_2(T)}{(x_1(t_2))^2} & t_2 \le t \le T, \end{cases}$$

$$\lambda_2(t) = \begin{cases} \dfrac{1}{x_1(0)} \ln\left(\dfrac{x_2(t)}{x_2(t_1)}\right) - \dfrac{1}{x_2(t_1)} & 0 \le t \le t_1, \\ -\dfrac{1}{x_2(t_1)}\mathrm{e}^{t_1 - t} & t_1 \le t \le t_2, \\ \dfrac{1}{x_1(t_2)} \ln\left(\dfrac{x_2(t)}{x_2(T)}\right) & t_2 \le t \le T, \end{cases}$$

where $t_2 \approx T - 2.7932821329007607$, $x_1(t_2) = x_2(t_2)(T - t_2 - 1)$, and $x_2(t_1) = x_1(0)t_1 + x_2(0)$. Also, switch t_1 is the solution to the following equation:

$$\frac{x_2(t_1)}{x_1(t_1)} = \frac{1}{T - t_1 - 1}.$$

The solution to the above equation is the following expression:

$$t_{1,\pm} = \frac{1}{2}\left[T - 1 - \frac{x_2(0)}{x_1(0)} \pm \sqrt{T^2 + 2T\left(\frac{x_2(0)}{x_1(0)} - 1\right) - 3 + \left(\frac{x_2(0)}{x_1(0)}\right)^2}\right],$$

and we choose t_1 to be the solution that is between the values 0 and t_2.

For Case 2c, we solve the state equations (9.62) and (9.63) on the interval $[0, t_1]$ with u set to being 1, and then use continuity of the state variables to obtain an equation that can be used to solve for t_1. Then we solve the adjoint equations (9.67) and (9.68) on $[0, t_1]$, where we used Eqs. (9.92)–(9.93) and continuity of λ_1 and λ_2 at t_1 to gain the following boundary conditions: $\lambda_1(t_1) = \lambda_2(t_1) = -\frac{1}{x_2(t_1)}$.

The exact solution for Case 2c is as follows:

Case 2c: If $T > 3.35091887$ and $\frac{1}{T-1} < \frac{x_2(0)}{x_1(0)} \leq 0.55763673\mathrm{e}^{T-2.79328213}$, then

$$u^*(t) = \begin{cases} 1 & 0 \leq t < t_1, \\ 1 - \dfrac{1}{T-t-1} & t_1 \leq t \leq t_2, \\ 0 & t_2 < t \leq T, \end{cases}$$

$$x_1(t) = \begin{cases} x_1(0)\mathrm{e}^t & 0 \leq t \leq t_1, \\ x_2(t)(T-t-1) & t_1 \leq t \leq t_2, \\ x_1(t_2) & t_2 \leq t \leq T, \end{cases}$$

$$x_2(t) = \begin{cases} x_2(0) & 0 \leq t \leq t_1, \\ x_2(t_1)\mathrm{e}^{t-t_1} & t_1 \leq t \leq t_2, \\ x_1(t_2)(t-t_2) + x_2(t_2) & t_2 \leq t \leq T, \end{cases}$$

$$\lambda_1(t) = \begin{cases} -\dfrac{1}{x_2(0)}\mathrm{e}^{t_1-t} & 0 \leq t \leq t_1, \\ \lambda_2(t) - \dfrac{x_2(t)}{(x_1(t_2))^2} & t_1 \leq t \leq t_2, \\ \times\left(\ln\left(\dfrac{x_2(t)}{x_2(T)}\right) - 1\right) - \dfrac{x_2(T)}{(x_1(t_2)^2} & t_2 \leq t \leq T, \end{cases}$$

$$\lambda_2(t) = \begin{cases} \dfrac{1}{x_2(0)}(t-t_1) - \dfrac{1}{x_2(t_1)} & 0 \le t \le t_1, \\ -\dfrac{1}{x_2(0)} e^{t_1 - t} & t_1 \le t \le t_2, \\ \dfrac{1}{x_1(t_2)} \ln\left(\dfrac{x_2(t)}{x_2(T)}\right) & t_2 \le t \le T, \end{cases}$$

where $t_2 = T - 2.79328213$, $x_1(t_1) = x_1(0)e^{t_1}$, $x_2(t_1) = x_2(0)$, $x_1(t_2) = x_2(t_2)(T - t_2 - 1)$, and $x_2(t_2) = x_2(t_1)e^{t_2 - t_1}$. Also, switch t_1 can be obtained by using a nonlinear solver for the following nonlinear equation:

$$\frac{x_2(0)}{x_1(0)} e^{-} t_1 = \frac{1}{T - 1 - t_1}.$$

9.4.2 *Discretization of the plant problem*

We discretize the following regularized problem:

$$\begin{aligned} \min \quad & J_\rho(u) = -\textstyle\int_0^T \ln(x_2(t))dt + \rho V(u) \\ \text{s.t.} \quad & \dot{x}_1(t) = u(t)x_1, \\ & \dot{x}_2(t) = (1 - u(t))x_1, \\ & x_1(0) = x_{1,0} > 0, \quad x_2(0) = x_{2,0} \ge 0, \\ & \qquad 0 \le u(t) \le 1, \end{aligned} \tag{9.95}$$

where $0 \le \rho < 1$ is the bounded variation penalty parameter and $V(u)$ measures the total variation of control u, as defined in Eq. (9.56). We also discretize the following adjoint equations:

$$\lambda_1'(t) = -\frac{\partial H}{\partial x_1} = (\lambda_2 - \lambda_1)u - \lambda_2, \tag{9.96}$$

$$\lambda_2'(t) = -\frac{\partial H}{\partial x_2} = \frac{1}{x_2}, \tag{9.97}$$

with $\lambda_1(T) = \lambda_2(T) = 0$ being the transversality conditions.

We assume the control u is constant over each mesh interval. We partition time interval $[0, T]$, by using $N+1$ equally spaced nodes, $0 = t_0 < t_1 < \cdots < t_N = T$. For all $k = 0, 1, \ldots, N$, we assume that $x_{1,k} = x_1(t_k)$ and $x_{2,k} = x_2(t_k)$. For the control, we denote $u_k = u(t)$ for all $t_k \leq t < t_{k+1}$ when $k = 0, \ldots N-2$ and $u_{N-1} = u(t)$ for all $t_{N-1} \leq t \leq t_N$. So we have $\boldsymbol{x}_1 \in \mathbb{R}^{N+1}$, $\boldsymbol{x}_2 \in \mathbb{R}^{N+1}$, while $u \in \mathbb{R}^N$. We use a left-rectangular integral approximation for objective functional J_ρ in problem (9.95). The discretization of problem (9.95) is then

$$\begin{aligned} \min \quad & J_\rho(\boldsymbol{u}) = \sum_{k=0}^{N-1} (-h \ln (x_{2,k})) + \rho \sum_{k=0}^{N-2} |u_{k+1} - u_k| \\ & x_{1,k+1} = x_{1,k} + h(u_k x_{1,k}) \quad \text{for all } k = 0, \ldots, N-1, \\ & x_{2,k+1} = x_{2,k} + h((1-u_k)x_{1,k}) \quad \text{for all } k = 0, \ldots, N-1, \\ & \qquad x_{1,0} > 0, \quad x_{2,0} \geq 0 \\ & 0 \leq u_k \leq 1 \quad \text{for all } k = 0, \ldots, N-1, \end{aligned}$$

where $h = \frac{T}{N}$ is the mesh size and the first component of $\boldsymbol{x}_1$ and $\boldsymbol{x}_2$ is set to being the initial condition associated with the state equations. Since PASA uses a gradient scheme for one of its phases, we need the cost functional to be differentiable. We perform a decomposition on each absolute value term in J_ρ to ensure that J_ρ is differentiable. We introduce two $N-1$ vectors $\boldsymbol{\zeta}$ and $\boldsymbol{\iota}$ whose entries are nonnegative. Each entry of $\boldsymbol{\zeta}$ and $\boldsymbol{\iota}$ is defined as

$$|u_{k+1} - u_k| = \zeta_k + \iota_k \quad \text{for all } k = 0, \ldots, N-2.$$

With this decomposition in mind, the discretized penalized problem is the following:

$$\begin{aligned} \min \quad & J_\rho(\boldsymbol{u}, \boldsymbol{\zeta}, \boldsymbol{\iota}) = \sum_{k=0}^{N-1} (-h \ln (x_{2,k})) + \rho \sum_{k=0}^{N-2} (\zeta_k + \iota_k) \\ & x_{1,k+1} = x_{1,k} + h(u_k x_{1,k}) \quad \text{for all } k = 0, \ldots, N-1, \\ & x_{2,k+1} = x_{2,k} + h(1-u_k)x_{1,k} \quad \text{for all } k = 0, \ldots, N-1, \\ & \qquad x_{1,0} > 0 \quad x_{2,0} \geq 0, \\ & 0 \leq u_k \leq \quad \text{for all } k = 0, \ldots, N-1, \\ & u_{k+1} - u_k = \zeta_k - \iota_k \quad \text{for all } k = 0, \ldots, N-2, \\ & \zeta_k \geq 0, \quad \iota_k \geq 0 \quad \text{for all } k = 0, \ldots, N-1. \end{aligned} \tag{9.98}$$

Note that for problem (9.98), we are minimizing the penalized objective functional with respect to three vectors, $\boldsymbol{u}, \boldsymbol{\zeta}$, and $\boldsymbol{\iota}$. The equality constraints associated with $\boldsymbol{\zeta}$ and $\boldsymbol{\iota}$ are linear constraints that PASA can interpret. The equality constraints associated with $\boldsymbol{\zeta}$ and $\boldsymbol{\iota}$ can be written like so:

$$\left[\boldsymbol{A} \middle| -\boldsymbol{I}_{N-1} \middle| \boldsymbol{I}_{N-1}\right] \begin{bmatrix} \boldsymbol{u} \\ \hline \boldsymbol{\zeta} \\ \hline \boldsymbol{\iota} \end{bmatrix} = \boldsymbol{0},$$

where $\boldsymbol{I}_{N-1}$ is the identity matrix with dimension $N-1$, $\boldsymbol{0}$ is the $N-1$ dimensional all zeros vector, and $\boldsymbol{A}$ is an $N-1 \times N$ sparse matrix defined on Eq. (9.7).

For finding the gradient of J_ρ in problem (9.98), we use Theorem 9.1. In order to compute the Lagrangian of problem (9.98), we rewrite the discretized state equations accordingly:

$$\begin{aligned} -x_{1,k+1} + x_{1,k} + h(u_k x_{1,k}) &= 0, \quad \text{for all } k = 0, \ldots, N, \\ -x_{2,k+1} + x_{2,k} + h(1-u_k)x_{1,k} &= 0, \quad \text{for all } k = 0, \ldots, N. \end{aligned}$$

The Lagrangian of problem (9.98) is then

$$\begin{aligned} \mathcal{L}(\boldsymbol{x}_1, \boldsymbol{x}_2, \boldsymbol{u}, \boldsymbol{\zeta}, \boldsymbol{\iota}) = &\sum_{k=0}^{N-1} (-h \ln(x_{2,k})) + \rho \sum_{k=0}^{N-2} (\zeta_k + \iota_k) \\ &+ \sum_{k=0}^{N-1} \lambda_{1,k}(-x_{1,k+1} + x_{1,k} + h(u_k x_{1,k})) \\ &+ \sum_{k=0}^{N-1} \lambda_{2,k}(-x_{2,k+1} + x_{2,k} + h((1-u_k)x_{1,k})), \end{aligned}$$

where $\boldsymbol{\lambda}_1, \boldsymbol{\lambda}_2 \in \mathbb{R}^{N-1}$ are the Lagrangian multiplier vectors. We take the partial derivative of $\mathcal{L}$ with respect to u_k and obtain

$$\frac{\partial \mathcal{L}}{\partial u_k} = h(x_{1,k}(\lambda_{1,k} - \lambda_{2,k})) \quad \text{for all } k = 0, \ldots, N-1.$$

By using Theorem 9.1, we obtain the following:

$$\nabla_{\boldsymbol{u},\boldsymbol{\zeta},\boldsymbol{\iota}} J_\rho = \nabla_{\boldsymbol{u},\boldsymbol{\zeta},\boldsymbol{\iota}} \mathcal{L} = \left[\begin{array}{c} h(x_{1,0}(\lambda_{1,0} - \lambda_{2,0})) \\ \vdots \\ h(x_{1,k}(\lambda_{1,k} - \lambda_{2,k}) \\ \vdots \\ h(x_{1,N-1})(\lambda_{1,N-1} - \lambda_{2,N-1}) \\ \hline \rho \\ \vdots \\ \rho \\ \hline \rho \\ \vdots \\ \rho \end{array}\right], \tag{9.99}$$

provided that condition (9.11) in Theorem 9.1 is satisfied. To satisfy (9.11), we take the partial derivative of the Lagrangian $\mathcal{L}$ with respect to $x_{1,k}$ and $x_{2,k}$ for all $k = 1, \ldots, N$, and note that we are not taking the partial derivative of $\mathcal{L}$ with respect to $x_{1,0}$ and $x_{2,0}$ because they are known values. Taking the partial derivative of $\mathcal{L}$ with respect to each state vector components yields the following expressions:

$$\frac{\partial \mathcal{L}}{\partial x_{1,k}} = h(\lambda_{1,k} u_k + \lambda_{2,k}(1 - u_k)) + \lambda_{1,k} - \lambda_{1,k-1} \quad \text{for } k = 1, \ldots, N-1 \tag{9.100}$$

$$\frac{\partial \mathcal{L}}{\partial x_{1,N}} = -\lambda_{1,N-1} \tag{9.101}$$

$$\frac{\partial \mathcal{L}}{\partial x_{2,k}} = -h\left(\frac{1}{x_{2,k}}\right) + \lambda_{2,k} - \lambda_{2,k-1} \quad \text{for } k = 1, \ldots, N-1 \tag{9.102}$$

$$\frac{\partial \mathcal{L}}{\partial x_{2,N}} = -\lambda_{2,N-1}. \tag{9.103}$$

For satisfying condition (9.11) from Theorem 9.1, we set the above equations equal to zeros and perform the following steps: solve for

$\lambda_{1,k-1}$ in Eq. (9.100); solve for $\lambda_{1,N-1}$ in Eq. (9.101); solve for $\lambda_{2,k-1}$ in Eq. (9.102); and solve for $\lambda_{2,N-1}$ in equation (9.103). Consequently, we generate a discretization for the adjoint equations (9.67) and (9.68) that also produces the transversality conditions (9.69) as follows:

$$\lambda_{1,k-1} = \lambda_{1,k} + h(\lambda_{1,k}u_k + \lambda_{2,k}(1-u_k)) \quad \text{for } k = 1,\ldots,N-1, \tag{9.104}$$

$$\lambda_{1,N-1} = 0, \tag{9.105}$$

$$\lambda_{2,k-1} = \lambda_{2,k} - h\left(\frac{1}{x_{2,k}}\right) \quad \text{for } k = 1,\ldots,N-1 \quad \text{and} \tag{9.106}$$

$$\lambda_{2,N-1} = 0. \tag{9.107}$$

9.4.3 *Numerical results of the plant problem*

We use PASA to numerically solve problem (9.65) with stopping tolerance set to being 10^{-10} and with our initial guess for the control being $u(t) = 0$ over the entire time interval $[0,T]$. We partition the time interval to where there are $N = 750$ mesh intervals with mesh size being $h = 0.00667$, and the discretization process is as described in Section 9.4.2 with $\rho = 0$. We are interested in solving the problem when the parameters are set so that a singular case occurs. King and Roughgarden[8] mention that from a biological standpoint, the initial weight of the reproductive part of a plant, $x_2(0)$, is always zero since germination involves vegetative growth only. Consequently, the only realistic situation where the exact solution to problem (9.65) contains a singular subarc is if the exact solution is of Case 2b with $x_{2,0} = 0$. However, we still would like to observe how PASA performs for all possible cases when the exact solution to problem (9.65) contains a singular subarc. Table 9.5 describes the parameter settings that we used for solving problem (9.65).

When parameters are set to the values shown in column three of Table 9.5, the exact solution to problem (9.65) will be of Case 2a where control u^* begins singular and switches to the purely reproductive case at $t_2 \approx 2.2067$. When parameters are set to the values shown in column four of Table 9.5, the exact solution to problem (9.65) is of Case 2b where control u^* begins purely reproductive,

Table 9.5. Numerical values of parameters used in computations for the plant problem.

Parameter	Description	Case 2a	Case 2b	Case 2c
T	Terminal time	5	5	5
$x_{1,0}$	Initial value for vegetative weight	4	1	1
$x_{2,0}$	Initial value for reproductive weight	1	10^{-4}	2

Table 9.6. Results from unpenalized solution to the plant problem for each case.

Case	$\|u^* - \hat{u}\|_{L^1}$	$\|u^* - \hat{u}\|_{L^\infty}$	t_1^*	$\hat{t}_1$	t_2^*	$\hat{t}_2$	$-J(u^*)$	$-J(\hat{u})$	Runtime
2a	0.017557	0.444444	NA	NA	2.2067	2.2	11.7874	11.7875	46.25
2b	0.017519	0.444444	0.2678	0.2667	2.2067	2.2	3.6009	3.6010	32.97
2c	0.013588	0.444444	1.5778	1.5733	2.2067	2.2	8.6130	8.6130	7.48

Note: The starred notation corresponds to the exact solution, while the hat notation corresponds to the approximated solution. Note $-J(u)$ is the left rectangular integral approximation of $\int_0^T \ln(x(t))dt$. We have that $\|u^* - \hat{u}\|_{L^\infty(0,T)} = 0.444444$ for each case. This is due to the approximated switch from singular to purely reproduction, $\hat{t_2}$, being off by one node. The Runtime column gives the time (in seconds) it takes for PASA to solve the problem.

switches to the singular case solution at $t_1 \approx 0.2678$, and switches back to being purely reproductive at $t_2 \approx 2.2067$. We want to emphasize that in Case 2b parameter settings, $x_{2,0} = 10^{-4}$ because we want this initial value to be a close approximation of zero without having any issues in computing the integral approximation for the objective functional $J(u) = \int_0^T (-\ln x_2(t))dt$. When parameters are set to the values shown in the last column of Table 9.5, the exact solution to problem (9.65) is of Case 2c where control u^* begins purely vegetative, switches to the singular case solution at $t_1 \approx 1.5778$, and switches back to being purely reproductive at $t_2 \approx 2.2067$.

For each case, we first observe the unregularized solutions that PASA obtained when solving for problem (9.65) to see if it is even necessary to penalize this problem. When numerically solving the unpenalized problem, we obtain solutions that are not oscillatory for all three cases. Figures and descriptions corresponding to the unregularized solutions are given in Appendix Section 9.A.1 and in Table 9.6. We find PASA's unpenalized solution to be a sufficient approximation to the true solution to problem (9.65) in Cases 2b

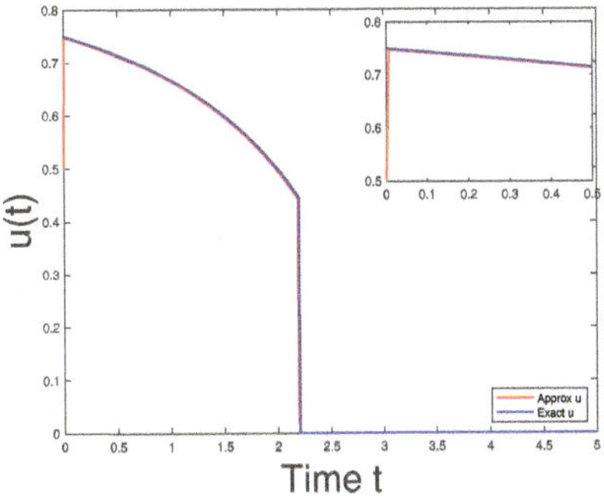

(a) u_ρ vs u^* for $\rho = 0$

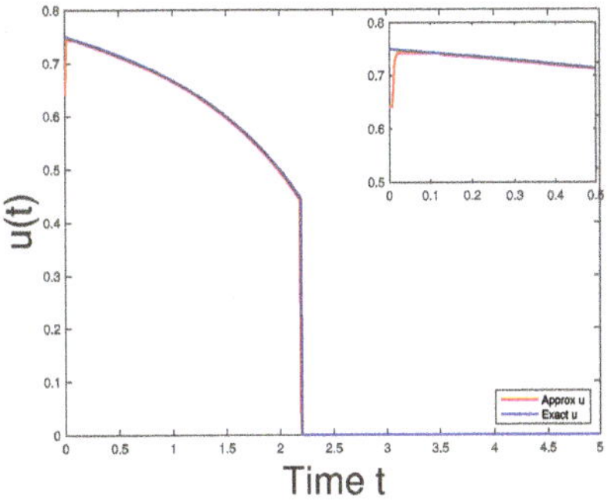

(b) u_ρ vs u^* for $\rho = 10^{-6}$

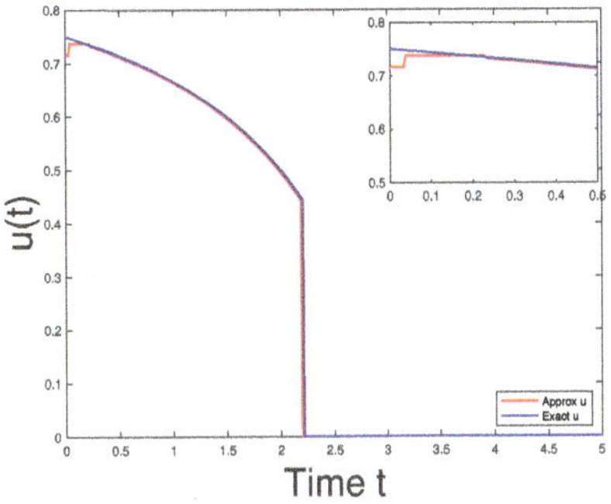

(c) u_ρ vs u^* for $\rho = 10^{-5}$

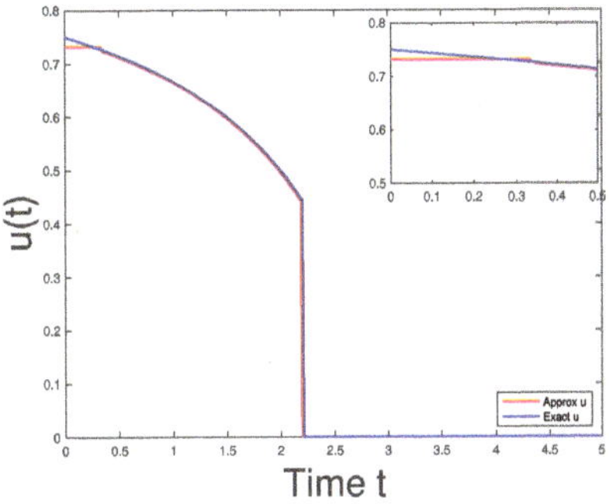

(d) u_ρ vs u^* for $\rho = 10^{-4}$

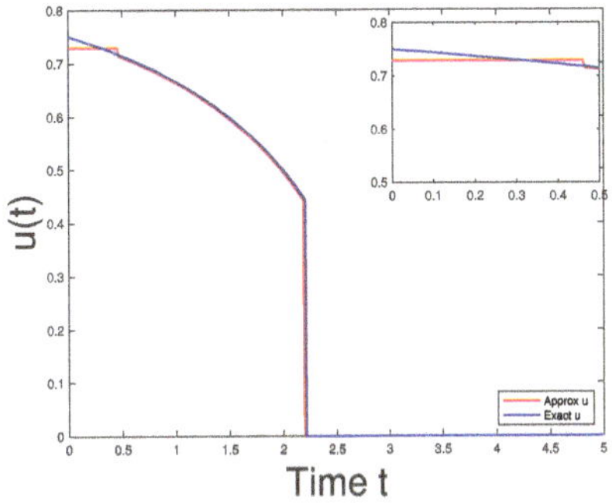

(e) u_ρ vs u^* for $\rho = 10^{-3}$

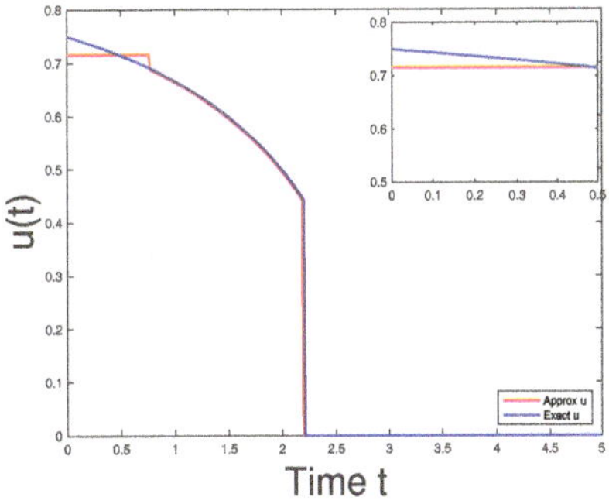

(f) u_ρ vs u^* for $\rho = 10^{-2}$

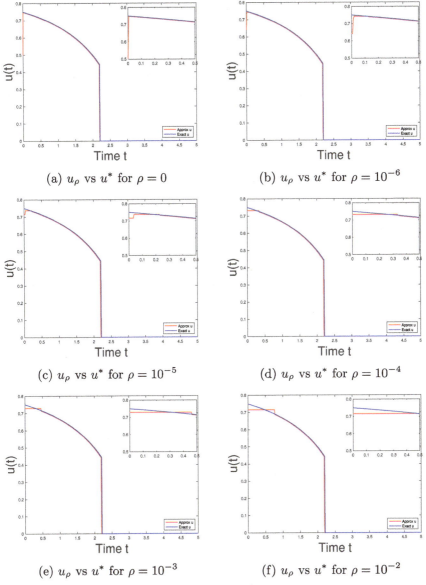

Fig. 9.7. Varying penalty to plant problem (Case 2a): Plots of regularized control u_ρ (red) vs optimal control u^* (blue) for tuning parameter $\rho \in \{0, 10^{-6}, 10^{-5}, 10^{-4}, 10^{-3}, 10^{-2}\}$. The top right corner of each figure is a zoomed-in plot of the same figure over the time interval $[0, 0.5]$.

and 2c. However, in Figure 9.7(a) the unpenalized approximation for Case 2a, $\hat{u}$, appears to have an unusual dip at $t = 0$. Based upon explicit solution (9.94) and the parameters setting given in Table 9.5, $u^*(0) = 0.75$, but for the approximated, unpenalized solution we have $\hat{u}(0) \approx 0.4980$. We use PASA to solve for Problem (9.65) for Case 2a with $N = 1,000$ and $tol = 10^{-12}$ to see if such changes would provide any improvements for the initial value. The unregularized solution that PASA obtained when $N = 1,000$ and $tol = 10^{-12}$ still possesses a dip at $t = 0$ where $u(0) \approx 0.4985$. The reason behind this initial jump is because Case 2a is a degenerate case. When looking at the conditions between becoming a solution of Case 2a, Case 2b, and Case 2c, one notices that they all share a condition that pertains to how the ratio of initial values relates with the fraction $\frac{1}{T-1}$. When having parameters set to where $\frac{x_{2,0}}{x_{1,0}} = \frac{1}{T-1}$, the discretization of problem (9.65) can cause any numerical solver to converge to a solution that does not begin singular.

Although there is no oscillations in Figure 9.7(a), we add a bounded variation penalty term to the objective functional of problem (9.65) to see if the approximated penalized control, u_ρ, does not dip as significantly as the unpenalized solution did at $t = 0$. As before, we partition time interval $[0, T]$ to where there are $N = 750$ mesh intervals and use the discretization method described in Section 9.4.2. We use PASA to solve the regularized problem (9.66) for varying values of the tuning parameter $\rho \in \{0, 10^{-9}, 10^{-8}, 10^{-7}, 10^{-6}, 10^{-5}, 10^{-4}, 10^{-3}, 10^{-2}, 10^{-1}\}$, with initial guess for our control being $u(t) = 0$ over the entire time interval. We have parameters set to being the third column in Table 9.5, with stopping tolerance set to being 10^{-10}. In Table 9.7, we record the L^1 and L^∞ norm errors between the exact solution u^* and the penalized solution that PASA obtained, u_ρ, the approximated value of $u_\rho(0)$, the approximated switching point t_2, and the runtime that was needed to solve for penalized problem (9.66). In column four of Table 9.7, we are finding $\|u^* - u_\rho\|_{L^\infty(0,2)}$ because otherwise the whole column would be 0.444444, which is what we observed in Table 9.6. Observe from Table 9.7, the penalty parameter starts to influence the behavior of u_ρ when $\rho \geq 10^{-6}$. However, it is worth noting that for penalty parameter values $\rho = 10^{-9}, 10^{-8}, 10^{-7}$, PASA obtained a solution comparable to the unpenalized solution at a faster rate. Based upon

Table 9.7. Varying penalty parameter for plant problem.

ρ	$\|u_\rho - u^*\|_{L^1(0,T)}$	$\|u_\rho - u^*\|_{L^\infty(0,2)}$	$u_\rho(0)$	Switch t_2	Runtime
0	0.01755676	0.25200195	0.4980	2.2	46.25
10^{-9}	0.01755676	0.25153629	0.4985	2.2	45.81
10^{-8}	0.01755672	0.24953915	0.5005	2.2	36.36
10^{-7}	0.01755623	0.23038155	0.5196	2.2	32.01
10^{-6}	**0.01755173**	0.10869044	0.6413	**2.2067**	27.54
10^{-5}	0.01760049	0.03367945	0.7163	2.2667	32.82
10^{-4}	0.01785072	**0.01809606**	**0.7319**	2.2667	27.60
10^{-3}	0.01928274	0.02181486	0.7282	2.3667	26.27
10^{-2}	0.02628408	0.03439004	0.7156	2.6067	23.42
10^{-1}	0.06203874	0.07234748	0.6777	2.3667	**9.96**

Note: Case 2a. $N = 750$, tol $= 10^{-10}$, and $h = 0.00667$.

Table 9.7, u_ρ was closest to u^* with respect to the L^1 norm when $\rho = 10^{-6}$, and u_ρ switched to being non-singular at the node that was closest to the true switching point. The penalized solution u_ρ that was obtained when $\rho = 10^{-4}$ had an initial value that was the closest to the true solution's initial value $u^*(0) = 0.75$; however, this improvement seems to be at the expense of increasing the L^1 norm error between u_ρ and u^*.

In Figures 9.7 and 9.8, we provide a chart of figures that pertain to u_ρ and a chart of figures that pertain to $u^* - u_\rho$ for $\rho = 0, 10^{-6}, 10^{-5}, 10^{-4}, 10^{-3}$, and 10^{-2}. Note that in Figures 9.7 and 9.8, the penalized solutions do not begin singular immediately; however, in comparison to the unpenalized solution, they give better approximations to $u^*(0)$. In Figure 9.7(d), the penalized solution associated with penalty parameter $\rho = 10^{-4}$ begins constant with constant value approximately being 0.7319 and switches to the singular case solution approximately at $t = 0.32$. Notice in Figures 9.7(e) and 9.7(f) that these penalized solutions also begin constant at values close to 0.7, but remain constant for a longer time period. Note also in the sixth column of Table 9.7 that for values $\rho \geq 10^{-5}$, u_ρ starts to overestimate the point at which the solution should switch from singular to non-singular. In Figure 9.7(b), we find that the unpenalized solution u_ρ associated with penalty parameter value $\rho = 10^{-6}$ did the best job in improving the approximated initial value without deviating too much from the exact solution. In Figure 9.9, we have the trajectories

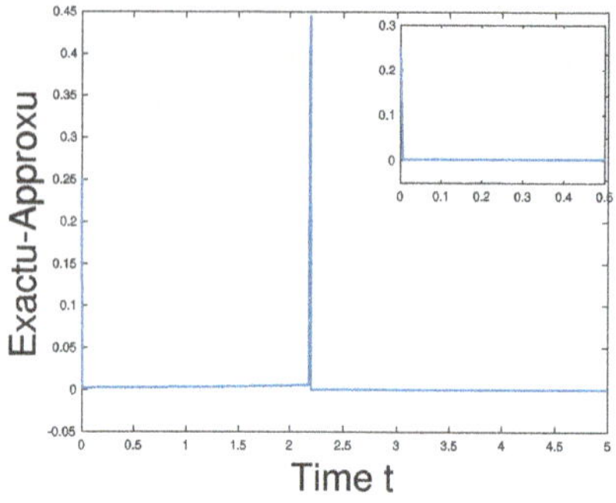

(a) $u^* - u_\rho$ for $\rho = 0$

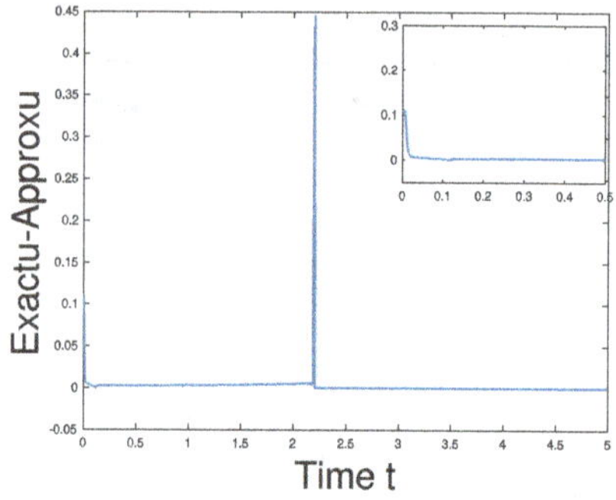

(b) $u^* - u_\rho$ for $\rho = 10^{-6}$

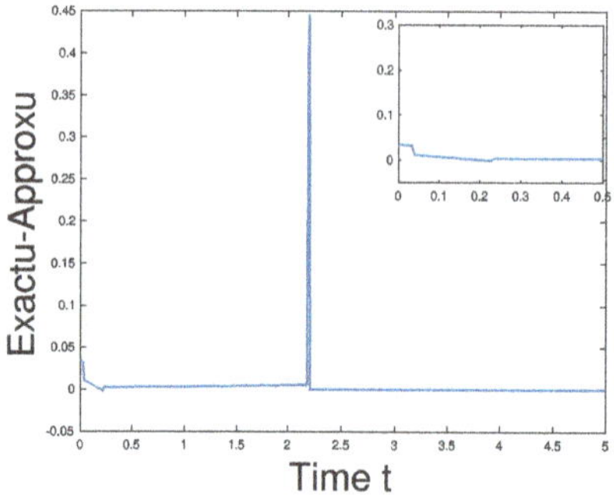

(c) $u^* - u_\rho$ for $\rho = 10^{-5}$

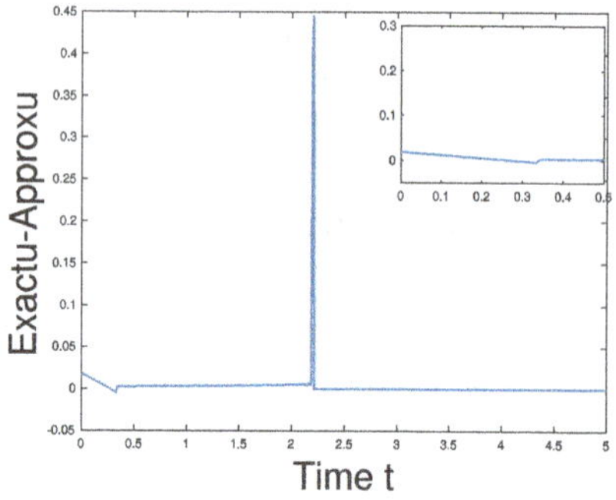

(d) $u^* - u_\rho$ for $\rho = 10^{-4}$

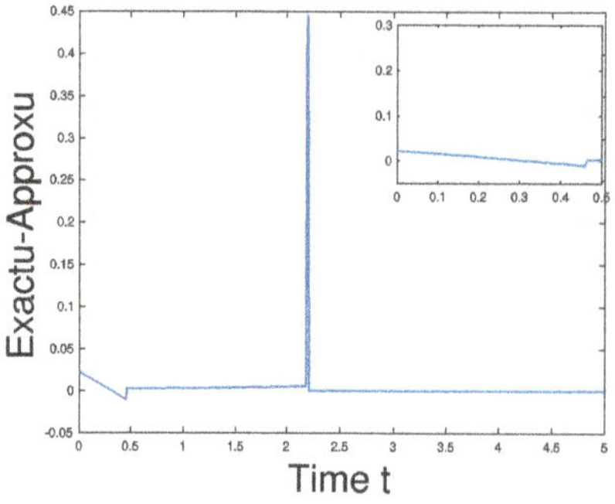

(e) $u^* - u_\rho$ for $\rho = 10^{-3}$

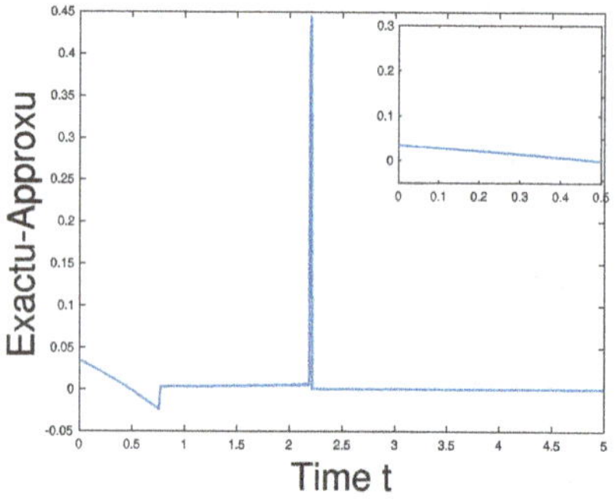

(f) $u^* - u_\rho$ for $\rho = 10^{-2}$

Fig. 9.8. Varying penalty to plant problem (Case 2a): Plots of $u^* - u_\rho$ (cyan) for tuning parameter $\rho \in \{0, 10^{-6}, 10^{-5}, 10^{-4}, 10^{-3}, 10^{-2}\}$. The top right corner of each figure is a zoomed-in plot of the same figure over the time interval $[0, 0.5]$.

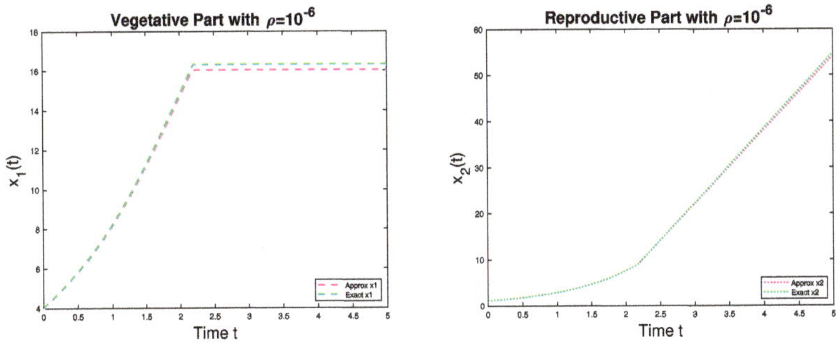

(a) Exact State x_1 (green) vs $x_{1,\rho}$ with $\rho = 10^{-6}$ (pink)

(b) Exact State x_2 (green) vs $x_{2,\rho}$ with $\rho = 10^{-6}$ (pink)

Fig. 9.9. Corresponding trajectories of x_1 and x_2 for penalized plant problem with $\rho = 10^{-6}$.

of x_1 and x_2 corresponding to u_ρ when $\rho = 10^{-6}$. The corresponding state solutions to the penalized control are comparable to the exact state solutions.

9.5 Example 3: The SIR Problem

Our next example is from Ledzewicz, Aghaee, and Schättler's article.[17] This example demonstrates how PASA can be applied to problems in the absence of a true solution. In Ref. 17, Ledzewicz *et al.* use an SIR model with demography to study the 2013–2016 outbreak of Ebola in West Africa. In an SIR model, the total population N is divided into the following three compartments: the susceptible class (S), the infectious class (I), and the recovered class (R). The dynamics in consideration are based on a modification of a system of equations that was first presented in Brauer and Castillo-Chavez's book[38]:

$$\dot{S} = \gamma N - \nu S - \beta \frac{IS}{N} + \rho R, \quad S(0) = S_0, \tag{9.108}$$

$$\dot{I} = \beta \frac{IS}{N} - (\nu + \mu) I - \alpha I, \quad I(0) = I_0, \tag{9.109}$$

$$\dot{R} = -\nu R - \rho R + \alpha I, \quad R(0) = R_0. \tag{9.110}$$

The Greek letters in the equations above represent constant parameters. In Eq. (9.108), the coefficient γ is the birth rate per unit time, and it is assumed that the population of newborns are immediately classified as being susceptible to the disease. Standard incidence is assumed in the model with transmission rate being β, and it is assumed that infected members recover from Ebola at rate α. Parameter ν is the natural death rate while parameter μ is the death rate due to infection. We emphasize here that for this model ρ is not the bounded variation tuning parameter, but instead ρ is the parameter that measures the rate at which a recovered individual becomes again susceptible to Ebola.

In Ref. 17, Ledzewicz, *et al.* construct an optimal control problem to gain understanding on how treatment and a theoretical vaccination of Ebola should be applied to best contribute to limiting the spread of the disease. For the state equations, they incorporate controls, u and v, into Eqs. (9.108)–(9.110), where u represents the vaccination rate and v represents the treatment rate. The modified dynamics are as follows:

$$\dot{S} = \gamma N - \nu S - \beta \frac{IS}{N} + \rho R - \kappa S u, \quad S(0) = S_0, \tag{9.111}$$

$$\dot{I} = \beta \frac{IS}{N} - (\nu + \mu) I - \alpha I - \eta I v, \quad I(0) = I_0, \tag{9.112}$$

$$\dot{R} = -\nu R - \rho R + \kappa S u + \alpha I + \eta I v, \quad R(0) = R_0, \tag{9.113}$$

where κ and η are denoted as being the efficacy of vaccination and treatment, respectively. For constructing of an objective functional, Ledzewicz *et al.*[17] have the following goal in mind: "*Given initial population sizes of all three classes, S_0, I_0, and R_0, find the best strategy in terms of the combined efforts of vaccination and treatment that minimizes the number of infectious persons while at the same time also taking into account the cost of vaccination and treatment.*" The objective functional in consideration for minimization is as follows:

$$J(u, v) = \int_0^T (aI(t) + bu(t) + cv(t))\mathrm{d}t \tag{9.114}$$

The objective functional J is intended to represent the weighted average of the number of infectious persons and costs of vaccination and treatment. The optimal control problem is as follows:

$$
\begin{aligned}
\min_{(u,v)\in\mathcal{A}} \quad & J(u,v) = \textstyle\int_0^T (aI(t) + bu(t) + cv(t))\mathrm{d}t \\
\text{s.t.} \quad & \dot{S} = \gamma N - \nu S - \beta\tfrac{IS}{N} + \rho R - \kappa S u, \\
& \dot{I} = \beta\tfrac{IS}{N} - (\nu+\mu)I - \alpha I - \eta I v, \\
& \dot{R} = -\nu R - \rho R + \kappa S u + \alpha I + \eta I v, \\
& S(0) = S_0, \quad I(0) = I_0, \quad R(0) = R_0,
\end{aligned} \tag{9.115}
$$

where $\mathcal{A}$ is the set of admissible controls, which is assumed as being the set of all Lebesgue measurable functions $u : [0,T] \to [0, u_{\max}]$ and $v : [0,T] \to [0, v_{\max}]$, where $u_{\max}$ is the maximum vaccination rate and $v_{\max}$ is the maximum treatment rate. The assumptions on $\mathcal{A}$ ensure existence of an optimal solution to problem (9.115), which follows from Fleming and Rishel's .[11]

For problem (9.115), Ledzewicz *et al.*[17] used Pontryagin's Maximum Principle[10] to find the first-order necessary conditions for optimality. By rewriting the state equations into a multi-input control-affine system of vector form and using Lie derivatives they were able to compute the switching functions for controls u and v as well as multiple derivatives of the switching function, allowing the Legendre–Clebsch Condition to be checked. Their methods of using Lie brackets are particularly useful in determining existence of a singular control when observing problems with many states and control variables, and we direct the reader to their article[17] and to Schättler and Ledzewicz's book [46] for more details on their procedures.

In Ref. 17, Ledzewicz *et al.* use the Tomlab algorithm PROPT,[47] a collocation solver, to obtain an approximate solution to problem (9.115) with the numerical values of parameters set to where a singular subarc is present in u^*, while v^* contains no singular subarc. In this section, we discuss how to discretize the penalized Ebola problem given in problem (9.116), where the penalty term involved is based on the total variation of the vaccination control u.[30] Additionally, in Appendix Section 9.A.2, we provide the MATLAB file, demoOC.m, to demonstrate how to use PASA to solve problem (9.116). We use PASA to solve both the unpenalized and penalized problems with parameters set to being what was used in Ledzewicz *et al.*[17] and remark about PASA's performance on this problem.

9.5.1 *Discretization of the SIR problem*

We discretize the following regularized problem:

$$\begin{aligned}
\min_{(u,v)\in\mathcal{A}} \quad & J_p(u,v) = \textstyle\int_0^T (aI(t) + bu(t) + cv(t))\mathrm{d}t + pV(u) \\
\text{s.t.} \quad & \dot{S} = \gamma N - \nu S - \beta \tfrac{IS}{N} + \rho R - \kappa S u, \\
& \dot{I} = \beta \tfrac{IS}{N} - (\nu + \mu) I - \alpha I - \eta I v, \\
& \dot{R} = -\nu R - \rho R + \kappa S u + \alpha I + \eta I v, \\
& S(0) = S_0, \quad I(0) = I_0, \quad R(0) = R_0,
\end{aligned} \tag{9.116}$$

where $0 \leq p < 1$ is the bounded variation tuning parameter and function V measures the total variation of the control which is defined in Eq. (9.56). Since we are numerically solving the problem with parameters set to where control v^* has no singular subarcs, it is likely that v need not be penalized. Additionally, we discretize the following adjoint equations:

$$\begin{aligned}
\dot{\lambda}_S = \lambda_S & \left(-\gamma + \nu + \beta \frac{I}{N} - \beta \frac{IS}{N^2} + \kappa u \right) \\
& + \lambda_I \left(\beta \frac{IS}{N^2} - \beta \frac{I}{N} \right) - \lambda_R \kappa u,
\end{aligned} \tag{9.117}$$

$$\begin{aligned}
\dot{\lambda}_I = -a + \lambda_S & \left(-\gamma + \beta \frac{S}{N} - \beta \frac{IS}{N^2} \right) \\
& + \lambda_I \left(-\beta \frac{S}{N} + \beta \frac{IS}{N^2} + (\nu + \mu + \alpha) + \eta v \right) - \lambda_R(\alpha + \eta v),
\end{aligned} \tag{9.118}$$

$$\dot{\lambda}_R = -\lambda_S \left(\gamma + \beta \frac{IS}{N^2} + \rho \right) + \lambda_I \left(\beta \frac{IS}{N^2} \right) + \lambda_R(\nu + \rho), \tag{9.119}$$

with transversality conditions being

$$\lambda_S(T) = 0, \quad \lambda_I(T) = 0, \quad \lambda_R(T) = 0. \tag{9.120}$$

First we assume that controls u and v are constant over each mesh interval. We partition time interval $[0, T]$, by using $n + 1$ equally spaced nodes, $0 = t_0 < t_1 < \cdots < t_n = T$. For all $k = 0, 1, \ldots, n$, we assume that $S_k = S(t_k)$, $I_k = I(t_k)$, and $R_k = R(t_k)$. For the controls we denote $u_k = u(t)$ and $v_k = v(t)$ for all $t_k \leq t < t_{k+1}$ when $k = 0, \ldots, n-2$. Additionally, we denote $u_{n-1} = u(t)$ and $v_{n-1} = v(t)$ for all $t_{n-1} \leq t \leq t_n$. We then have $\boldsymbol{S} \in \mathbb{R}^{n+1}, \boldsymbol{I} \in \mathbb{R}^{n+1}$, and $\boldsymbol{R} \in \mathbb{R}^{n+1}$, while $\boldsymbol{u} \in \mathbb{R}^n$ and $\boldsymbol{v} \in \mathbb{R}^n$. Furthermore, $\boldsymbol{N} \in \mathbb{R}^{n+1}$ where $N_k = N(t_k) = S_k + I_k + R_k$ for all $k = 0, 1, \ldots, n$. We use a left-rectangular integral approximation for approximating the objective functional $J_p(u, v)$ in problem (9.116), and we use forward Euler's method for discretizing the state equations. Since PASA uses a gradient scheme for one of its phases, we need the cost function to be differentiable. We suggest a decomposition of each absolute value term arising from the total variation term in J_ρ to ensure that J_ρ is differentiable. We introduce two $n-1$ vectors, $\boldsymbol{\zeta}$ and $\boldsymbol{\iota}$, whose entries are nonnegative. Each entry of $\boldsymbol{\zeta}$ and $\boldsymbol{\iota}$ is defined as

$$|u_{k+1} - u_k| = \zeta_k + \iota_k \quad \text{for all } k = 0, \ldots, n-2.$$

The discretization of problem (9.116) then becomes

$$\begin{aligned}
\min \quad J_p(\boldsymbol{u}, \boldsymbol{\zeta}, \boldsymbol{\iota}, \boldsymbol{v}) &= \sum_{k=0}^{n-1} h(aI_k + bu_k + cv_k) + p \sum_{k=0}^{n-1} (\zeta_k + \iota_k) \\
S_{k+1} &= S_k + h\left(\gamma N_k - \nu S_k - \beta \tfrac{I_k S_k}{N_k} + \rho R_k - \kappa S_k u_k\right) \\
&\quad \text{for } k = 0, \ldots, n-1, \\
I_{k+1} &= I_k + h\left(\beta \tfrac{I_k S_k}{N_k} - (\nu + \mu + \alpha) I_k - \eta I_k v_k\right) \\
&\quad \text{for } k = 0, \ldots, n-1, \\
R_{k+1} &= R_k + h(-\nu R_k - \rho R_k + \kappa S_k u_k + \alpha I_k + \eta I_k v_k) \\
0 \leq u_k &\leq u_{\max} \quad \text{for } k = 0, \ldots, n-1, \\
0 \leq v_k &\leq v_{\max} \quad \text{for } k = 0, \ldots, n-1, \\
u_{k+1} - u_k &= \zeta_k - \iota_k; \quad \text{for } k = 0, \ldots, n-2, \\
\zeta_k \geq 0, \iota_k &\geq 0 \quad \text{for } k = 0, \ldots, n-1,
\end{aligned} \tag{9.121}$$

where $h = \frac{T}{n}$ is the mesh size and the first components of S, I, and R are set to be the initial conditions associated with the state equations. Note that, for the above problem, we are minimizing the penalized objective function with respect to vectors $\boldsymbol{u}, \boldsymbol{\zeta}, \boldsymbol{\iota}$, and $\boldsymbol{v}$. The equality constraints associated with $\boldsymbol{\zeta}$ and ι are linear constraints that PASA can interpret. The equality constraints associated with $\boldsymbol{\zeta}$ and $\boldsymbol{\iota}$ are written as

$$\left[\boldsymbol{A}\middle|-\boldsymbol{I}_{n-1}\middle|\boldsymbol{I}_{n-1}\middle|\boldsymbol{O}_{n-1,n}\right]\begin{bmatrix}\boldsymbol{u}\\ \hline \boldsymbol{\zeta}\\ \hline \boldsymbol{\iota}\\ \hline \boldsymbol{v}\end{bmatrix} = \mathbf{0}, \tag{9.122}$$

where $\boldsymbol{I}_{n-1}$ is the identity matrix with dimension $n-1$, $\boldsymbol{O}_{n-1,n}$ is an $n-1 \times n$ all zeros matrix, $\mathbf{0}$ is the $n-1$ all zeros vector, and $\boldsymbol{A}$ is an $n-1 \times n$ dimensional sparse matrix give in Eq. (9.7).

For finding the gradient of J_p in problem (9.121), we use Theorem 9.1. The Lagrangian of problem (9.121) is as follows

$$\begin{aligned}\mathcal{L} = &\sum_{k=0}^{n-1} h(aI_k + bu_k + cv_k) + p\sum_{k=0}^{n-1}(\zeta_k - \iota_k)\\ &+ \sum_{k=0}^{n-1}\lambda_{S,k}\left(-S_{k+1} + S_k + h\left(\gamma N_k - \nu S_k\right.\right.\\ &\qquad\left.\left. - \beta\frac{I_k S_k}{N_k} + \rho R_k - \kappa S_k u_k\right)\right)\\ &+ \sum_{k=0}^{n-1}\lambda_{I,k}\left(-I_{k+1} + I_k + h\left(\beta\frac{I_k S_k}{N_k} - (\nu+\mu+\alpha)I_k - \eta I_k v_k\right)\right)\\ &+ \sum_{k=0}^{n-1}\lambda_{R,k}\left(-R_{k+1} + R_k + h\left(-\nu R_k - \rho R_k\right.\right.\\ &\qquad\left.\left. + \kappa S_k u_k + \alpha I_k + \eta I_k v_k\right)\right),\end{aligned} \tag{9.123}$$

where $\boldsymbol{\lambda}_S \in \mathbb{R}^{n-1}$, $\boldsymbol{\lambda}_I \in \mathbb{R}^{n-1}$, $\boldsymbol{\lambda}_R \in \mathbb{R}^{n-1}$ are the Lagrangian multiplier vectors. As suggested in Theorem 9.1, we use the gradient of the Lagrangian with respect to vectors $\boldsymbol{u}$, $\boldsymbol{\zeta}$, $\boldsymbol{\iota}$, and $\boldsymbol{v}$ to find $\boldsymbol{\nabla}_{\boldsymbol{u},\boldsymbol{\zeta},\boldsymbol{\iota},\boldsymbol{v}}\boldsymbol{J_p} \in \mathbb{R}^{2(n-1)+2(n-2)}$. This is necessary because based upon the state equations (9.111)–(9.113) of problem (9.116), S, I, and R can be viewed as functions of u and v. So when computing $\boldsymbol{\nabla}_{\boldsymbol{u},\boldsymbol{\zeta},\boldsymbol{\iota},\boldsymbol{v}}\boldsymbol{J_p} \in \mathbb{R}^{2(n-1)+2(n-2)}$, we should consider that $\boldsymbol{S}$, $\boldsymbol{I}$, and $\boldsymbol{R}$ depend on $\boldsymbol{u}$ and $\boldsymbol{v}$. We find the partial derivative of $\mathcal{L}$ with respect to u_k, ζ_k, ι_k, and v_k, and obtain the following:

$$\frac{\partial \mathcal{L}}{\partial u_k} = hb + h\kappa S_k(\lambda_{R,k} - \lambda_{S,k}) \quad \text{for all } k = 0, \ldots, n-1$$

$$\frac{\partial \mathcal{L}}{\partial \zeta_k} = p \quad \text{for all } k = 0, \ldots, n-2$$

$$\frac{\partial \mathcal{L}}{\partial \iota_k} = p \quad \text{for all } k = 0, \ldots, n-2$$

$$\frac{\partial \mathcal{L}}{\partial v_k} = hc + h\eta I_k(\lambda_{R,k} - \lambda_{I,k}) \quad \text{for all } k = 0, \ldots, n-1.$$

By Theorem 9.1, we compute $\boldsymbol{\nabla}_{\boldsymbol{u},\boldsymbol{\zeta},\boldsymbol{\iota},\boldsymbol{v}}\boldsymbol{J_p} \in \mathbb{R}^{2(n-1)+2(n-2)}$ as follows

$$\boldsymbol{\nabla}_{\boldsymbol{u},\boldsymbol{\zeta},\boldsymbol{\iota},\boldsymbol{v}}\boldsymbol{J_p} = \boldsymbol{\nabla}_{\boldsymbol{u},\boldsymbol{\zeta},\boldsymbol{\iota},\boldsymbol{v}}\boldsymbol{\mathcal{L}} = \left[\begin{array}{c} hb + h\kappa S_0(\lambda_{R,0} - \lambda_{S,0}) \\ \vdots \\ hb + h\kappa S_k(\lambda_{R,k} - \lambda_{S,k}) \\ \vdots \\ hb + h\kappa S_{n-1}(\lambda_{R,n-1} - \lambda_{S,n-1}) \\ \hline p \\ \vdots \\ p \\ \hline p \\ \vdots \\ p \\ \hline hc + h\eta I_k(\lambda_{R,0} - \lambda_{I,0}) \\ \vdots \\ hc + h\eta I_k(\lambda_{R,k} - \lambda_{I,k}) \\ \vdots \\ hc + h\eta I_{n-1}(\lambda_{R,n-1} - \lambda_{I,n-1}) \end{array}\right], \tag{9.124}$$

provided that the Lagrange multiplier vectors satisfy condition (9.11) in Theorem 9.1. To satisfy condition (9.11), we take the partial derivative of the Lagrangian $\mathcal{L}$ with respect to S_k, I_k, and R_k for all $k = 1, \ldots, n$. Note that we are not finding the partial derivative of $\mathcal{L}$ with respect to S_0, I_0, and R_0 since these entries are the known values. Taking the partial derivative of the Lagrangian, given in Eq. (9.123), with respect to S_k, I_k, and R_k yields the following:

$$\frac{\partial \mathcal{L}}{\partial S_k} = -\lambda_{S,k-1} + \lambda_{S,k} + h\lambda_{S,k}\left(\gamma - \nu - \beta\frac{I_k}{N_k} + \beta\frac{I_k S_k}{N_k^2} - \kappa u_k\right)$$

$$+ h\lambda_{I,k}\left(\beta\frac{I_k}{N_k} - \beta\frac{I_k S_k}{N_k^2}\right) + h\lambda_{R,k}(\kappa u_k), \tag{9.125}$$

$$\text{for } k = 1, \ldots, n-1 \quad \text{and}$$

$$\frac{\partial \mathcal{L}}{\partial S_n} = -\lambda_{S,n-1}, \tag{9.126}$$

$$\frac{\partial \mathcal{L}}{\partial I_k} = ha - \lambda_{I,k-1} + \lambda_{I,k}$$

$$+ h\lambda_{I,k}\left(\beta\frac{S_k}{N_k} - \beta\frac{I_k S_k}{N_k^2} - (\nu + \mu + \alpha) - \eta v_k\right)$$

$$+ h\lambda_{S,k}\left(\gamma - \beta\frac{S_k}{N_k} + \beta\frac{I_k S_k}{N_k^2}\right) + h\lambda_{R,k}(\alpha + \eta v_k), \tag{9.127}$$

$$\text{for } k = 1, \ldots, n-1 \quad \text{and}$$

$$\frac{\partial \mathcal{L}}{\partial I_n} = -\lambda_{I,n-1}, \tag{9.128}$$

$$\frac{\partial \mathcal{L}}{\partial R_k} = -\lambda_{R,k-1} + \lambda_{R,k} - h\lambda_{R,k}(\nu + \rho)$$

$$+ h\lambda_{S,k}\left(\gamma + \rho + \beta\frac{I_k S_k}{N_k^2}\right) - h\lambda_{I,k}\left(\beta\frac{I_k S_k}{N_k^2}\right), \tag{9.129}$$

$$\text{for } k = 1, \ldots, n-1 \text{ and}$$

$$\frac{\partial \mathcal{L}}{\partial R_n} = -\lambda_{R,n-1}. \tag{9.130}$$

For finding multiplier vectors $\boldsymbol{\lambda}_S$, $\boldsymbol{\lambda}_I$, $\boldsymbol{\lambda}_R$ that satisfy Theorem condition (9.11), we set the above equations (9.125)–(9.130) equal to zero and do the following: solve for $\lambda_{S,k-1}$ in Eq. (9.125), solve for

$\lambda_{S,n-1}$ in Eq. (9.126), solve for $\lambda_{I,k-1}$ in Eq. (9.127), solve for $\lambda_{I,n-1}$ in Eq. (9.128), solve for $\lambda_{R,k-1}$ in Eqs. (9.129), and solve for $\lambda_{R,n-1}$ in Eq. (9.130). After performing the following steps and rearranging terms, we generate a discretization of Eqs. (9.117)–(9.119) that satisfy the transversality conditions (9.120):

$$\begin{aligned}\lambda_{S,k-1} &= \lambda_{S,k} + h\lambda_{S,k}\left(\gamma - \nu - \beta\frac{I_k}{N_k} + \beta\frac{I_kS_k}{N_k^2} - \kappa u_k\right)\\ &\quad + h\lambda_{I,k}\left(\beta\frac{I_k}{N_k} - \beta\frac{I_kS_k}{N_k^2}\right) + h\lambda_{R,k}(\kappa u_k), \qquad (9.131)\\ &\quad \text{for } k = 1,\ldots,n-1, \text{ and}\end{aligned}$$

$$\lambda_{S,n-1} = 0, \qquad (9.132)$$

$$\begin{aligned}\lambda_{I,k-1} &= \lambda_{I,k} + ha + h\lambda_{S,k}\left(\gamma - \beta\frac{S_k}{N_k} + \beta\frac{I_kS_k}{N_k^2}\right)\\ &\quad + h\lambda_{I,k}\left(\beta\frac{S_k}{N_k} - \beta\frac{I_kS_k}{N_k^2}\right) + h\lambda_{R,k}(\alpha + \eta v_k), \qquad (9.133)\\ &\quad \text{for } k = 1,\ldots,n-1, \text{ and}\end{aligned}$$

$$\lambda_{I,n-1} = 0, \qquad (9.134)$$

$$\begin{aligned}\lambda_{R,k-1} &= \lambda_{R,k} + h\lambda_{S,k}\left(\gamma + \rho + \beta\frac{I_kS_k}{N_k^2}\right) - h\lambda_{I,k}\left(\beta\frac{I_kS_k}{N_k^2}\right)\\ &\quad - h\lambda_{R,k}(\nu + \rho), \quad \text{for } k = 1,\ldots,n-1, \text{ and} \qquad (9.135)\end{aligned}$$

$$\lambda_{R,n-1} = 0. \qquad (9.136)$$

9.5.2 *Numerical results of the SIR problem*

We first use PASA to numerically solve the unregularized problem given in (9.115) with stopping tolerance set to being 10^{-8}. Our initial guess for the vaccination and treatment control is $u(t) = 0$ and $v(t) = 0$ over the entire time interval, respectively. The numerical parameters, which are given in Table 9.8, are set to the values that are used in Ledzewicz *et al.*[17] because we want to see if our numerical results are comparable to theirs. Now, Ledzewicz *et al.*[17] are investigating optimal vaccine and treatment strategies for a theoretical

Table 9.8. Numerical values of parameters used in computations.

Parameter	Numerical value	Parameter description
γ	0.00683	Birth rate of the population
ν	0.00188	Natural death rate of the population
β	0.2426	Rate of infectiousness of the disease
μ	0.005	Disease induced death rate
α	0.00002	Rate at which disease is overcome
ρ	0.007	Resensitization rate
κ	0.3	Effectiveness of vaccination
η	0.1	Effectiveness of treatment
a	5	Weight of I used in cost functional
b	50	Weight for cost of vaccination
c	300	Weight for cost of treatment
T	50	Time horizon (weeks)
$u_{\max}$	1	Maximum vaccination rate
$v_{\max}$	1	Maximum treatment rate
S_0	1000	Initial condition for S
I_0	10	Initial condition for I
R_0	0	Initial condition for R

vaccine and treatment for Ebola, and one can infer from their parameter settings that they are not concerned with realistic limitations in widely developing and administering these theoretical controls. For example, based on these parameter values, a vaccination strategy being $u(t) = u_{\max} = 1$ implies that we are vaccinating approximately 97% of the susceptible population immediately at time t. When considering realistic limitations in developing and administering vaccines, one might want to consider setting $u_{\max}$ to a value that is less than one.

We partition the time interval to where there are $n = 750$ mesh intervals with mesh size being $h = 0.0667$, and use the discretization methods presented in Section 9.5.1 with penalty parameter set to being $p = 0$. The results corresponding to the approximated solution that PASA obtained for problem (9.115) are given in Figure 9.11. Observe in Figure 9.11(a) that the numerical solution associated with the vaccination control, $\hat{u}$, begins with a full dose segment (i.e., $\hat{u} = 1$), followed by an interval of many oscillations, and ends with the control being turned off. The many oscillations approximately start at time $t_{u,1} = 12.933$ and end at time $t_{u,2} = 36.533$. We looked

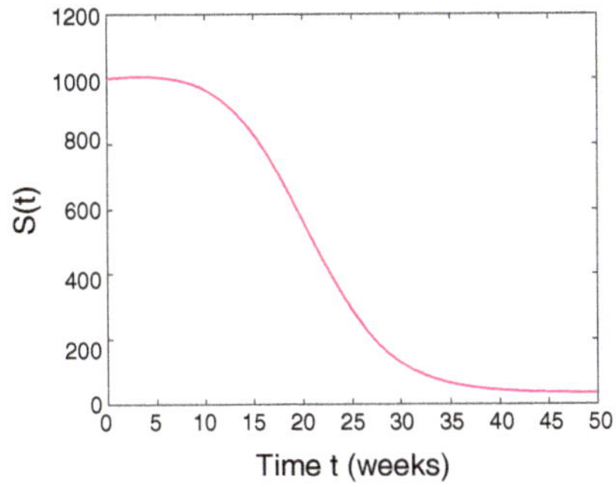

(a) Susceptible Class

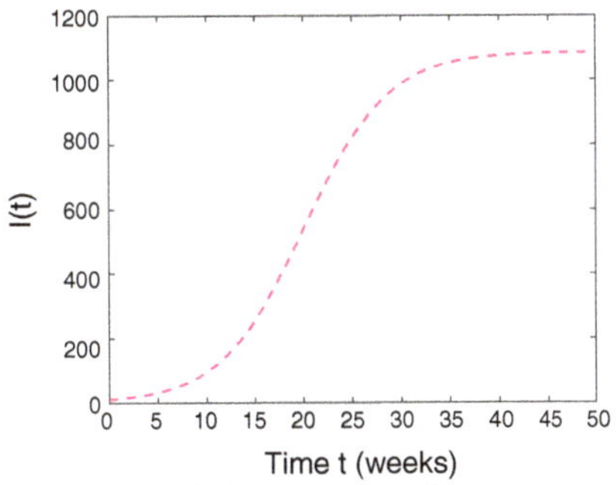

(b) Infectious Class

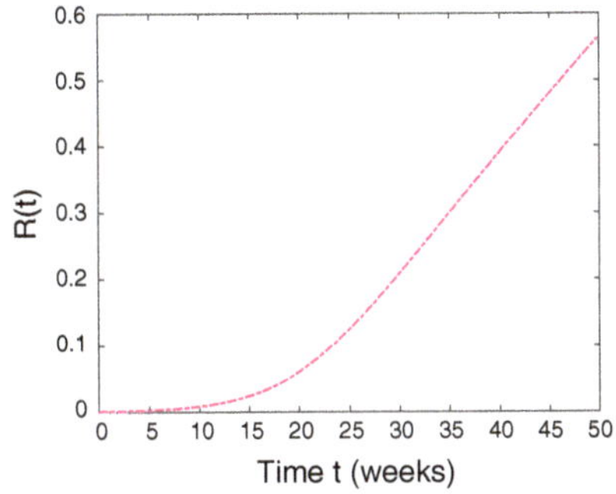

(c) Recovered Class

Fig. 9.10. Spread of disease without vaccination and treatment.

at the corresponding switching function to $\hat{u}$, which is $\Phi_u(t) = b + (\lambda_R(t) - \lambda_S(t))\kappa S(t)$ (see Ref. 17 for construction of switching function) to verify that the oscillating region corresponds to a singular subarc. It is necessary that the optimal control u^* is as follows:

$$u^*(t) = \begin{cases} 0 & \text{if } \Phi_u(t) > 0, \\ u_{\max} & \text{if } \Phi_u(t) < 0, \end{cases} \tag{9.137}$$

and $u^*(t)$ is singular if Φ_u is zero over an open interval of time. A plot of the switching function Φ_u is given in Figure 9.11(c).

Although we observe in Figures 9.11(c)–9.11(e) that the dynamics of S, I, and R corresponding to the approximated controls $\hat{u}$ and $\hat{v}$ give significantly favorable results in comparison to the dynamics associated with no application of vaccination or treatment, which is given in Figure 9.10, we recognize that the wild oscillations found in $\hat{u}$ makes it an unrealistic strategy to implement. In order to obtain a

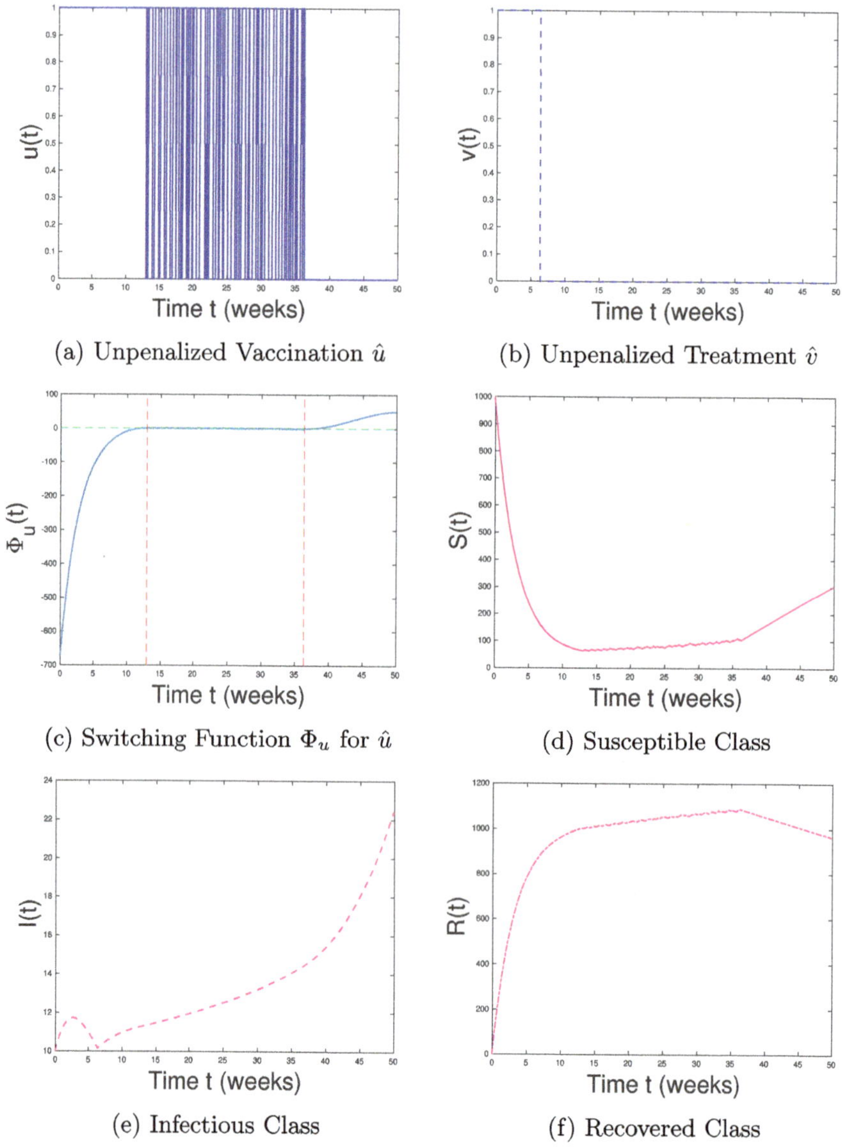

(a) Unpenalized Vaccination $\hat{u}$

(b) Unpenalized Treatment $\hat{v}$

(c) Switching Function Φ_u for $\hat{u}$

(d) Susceptible Class

(e) Infectious Class

(f) Recovered Class

Fig. 9.11. Spread of disease with unpenalized vaccination and treatment.

more realistic vaccination strategy, we penalize control u via a total variation penalty term.[30] We are not penalizing control v, since control v is a piecewise constant function that can be easily interpreted.

We use PASA to solve for penalized problem (9.116) with stopping tolerance set to 10^{-8} and with parameter value settings given in Table 9.8. Our initial guess for controls are $u(t) = 0$ and $v(t) = 0$ over the entire time interval. We partitioned $[0, T]$ to $n = 750$ mesh intervals of size $h = 0.0667$, and the discretization procedure is given in Section 9.5.1. Additionally in Section 9.A.2, we have a MATLAB file given that can be used for solving the problem. Note that this MATLAB file is associated with the most up to date version of PASA (SuiteOPT Version 2.0.0), but our experiments presented here are using an older version of PASA (SuiteOPT Version 1.0.0). We numerically solve for problem (9.116) for varying values of penalty parameter $p \in \{10^{-5}, 10^{-4}, 10^{-3}, 10^{-2}, 10^{-1}\}$. If the approximated control, u_p, corresponding to penalty parameter value p is no longer oscillating and if the sign of the approximated switching, Φ_{u_p}, aligns with Eq. (9.137), then we consider p as being a suitable penalty parameter value to use.

In Table 9.9, we record an approximation of the unpenalized cost functional, given in Eq. (9.114), being evaluated at the penalized solution u_p. The values corresponding to $J(u_p)$ are remarkably similar to the approximated objective value that corresponds to the unpenalized solution $\hat{u}$. We also record PASA's runtime performance (in

Table 9.9. Varying penalty results.

	p	$J(u_p)$	$t_{u,1}$	$t_{u,2}$	Runtime
$N = 750$	0	6572.955	12.933	36.533	16.77
$h = 0.0667$	10^{-5}	6572.956	13.067	36.400	17.07
tol $= 10^{-8}$	10^{-4}	6572.948	13.133	36.400	20.58
	10^{-3}	6573.018	12.933	36.533	19.12
	10^{-2}	6573.101	12.467	36.000	56.58
	10^{-1}	6575.429	12.333	37.067	10.47

Note: $J(u_p)$ is the approximated unpenalized cost functional value, $t_{u,1}$ and $t_{u,2}$ are the approximated switches corresponding to each solution. The Runtime column corresponds to the time (in seconds) it took to run PASA for each problem.

seconds) on solving problem (9.116), and we record the approximated switches of u_p, $t_{u,1}$, and $t_{u,2}$. When numerically solving problem (9.116), we observe oscillations in all cases except for when $p = 10^{-1}$. In Figure 9.12, we provide the plots of the penalized solution u_p and the plots of the corresponding switching function when the penalty parameter was set to being 10^{-3}, 10^{-2}, and 10^{-1}. Figures 9.12(a), 9.12(b), and 9.12(c) illustrate a trend of how increasing the tuning parameter p influences the penalized solution to where many oscillations no longer occur. We could increase the penalty parameter values larger than 10^{-1}; however, we did not do so because the approximated switching function associated u_p when $p = 10^{-1}$, given in Figure 9.12, gave indication that the structure of u_p aligns with Eq. (9.137). Meaning we have a non-oscillatory vaccination strategy that is not only more realistic to implement, but also satisfies the first-order necessary conditions of optimality.[10] In addition, we do not increase the tuning parameter to values larger than 10^{-1} due to the possibility of over-penalization. In terms of runtime performances, we have that PASA performed the fastest when the penalty parameter was set to being 10^{-1}.

We recognize in Figure 9.12 that the singular region associated with u_p when the bounded variation tuning parameter is set to $p = 10^{-1}$ is a rather crude approximation. However, if we want a smoother presentation of the singular region, we recommend either increasing the number of mesh intervals used for the discretization and/or tightening the stopping tolerance. In Figure 9.13, we provide a solution that PASA obtained in solving for penalized problem (9.116) with $p = 10^{-1}$ when time interval $[0, T]$ was partitioned to have 2000 mesh intervals with mesh size being $h = 0.025$. The approximated singular subarc that is given in Figure 9.13 is comparable to the approximation that Ledezewicz *et al.*[17] obtained.

Although these results are theoretical due to no vaccination for Ebola, optimal control theory is used to see how vaccination and treatment can influence the spread of the disease. We observe in Figure 9.10 that without treatment and vaccination, the susceptible class starts to decline due to many individuals becoming infectious, and few individuals recover from the disease. When incorporating the vaccination strategy and treatment strategy that is respectively,

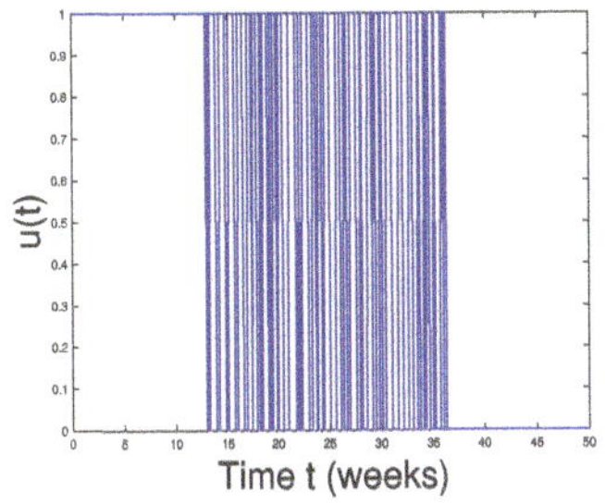

(a) u_p with $p = 0$

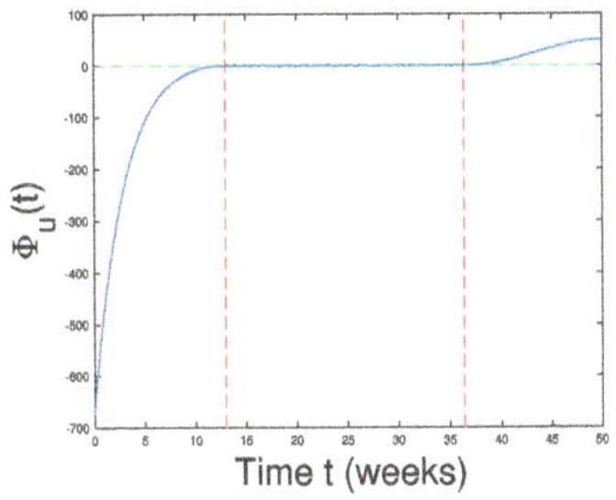

(b) Φ_{u_p} with $p = 0$

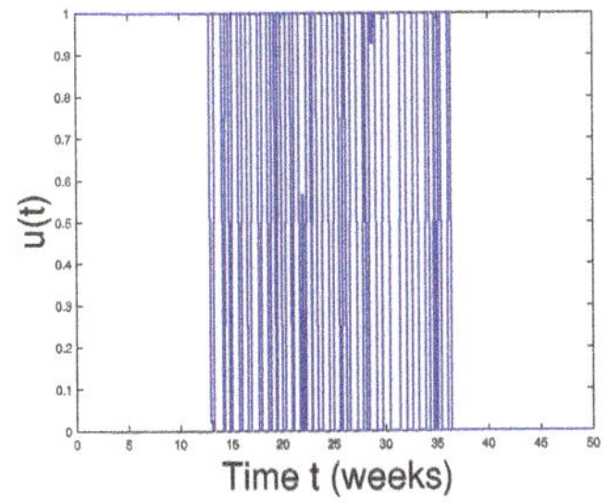

(c) u_p with $p = 10^{-3}$

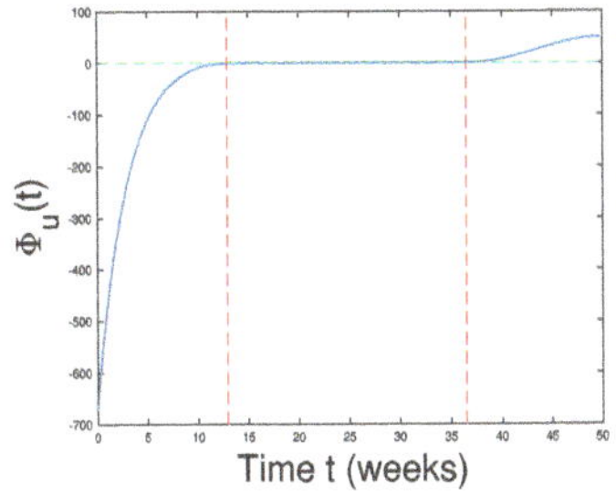

(d) Φ_{u_p} with $p = 10^{-3}$

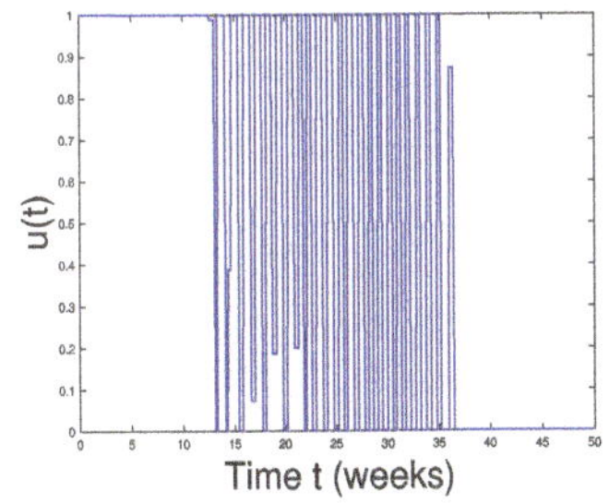

(e) u_p with $p = 10^{-2}$

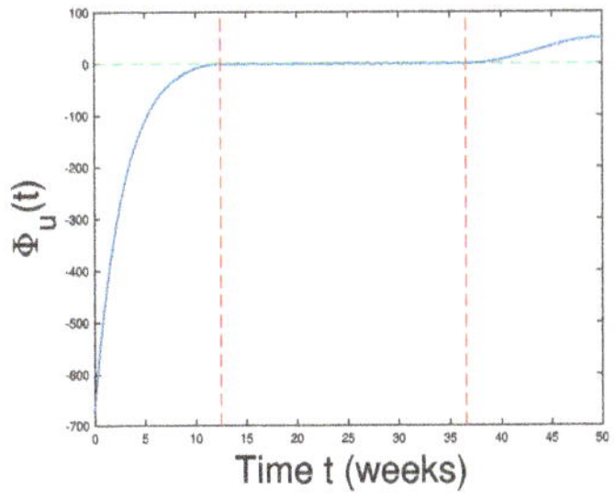

(f) Φ_{u_p} with $p = 10^{-2}$

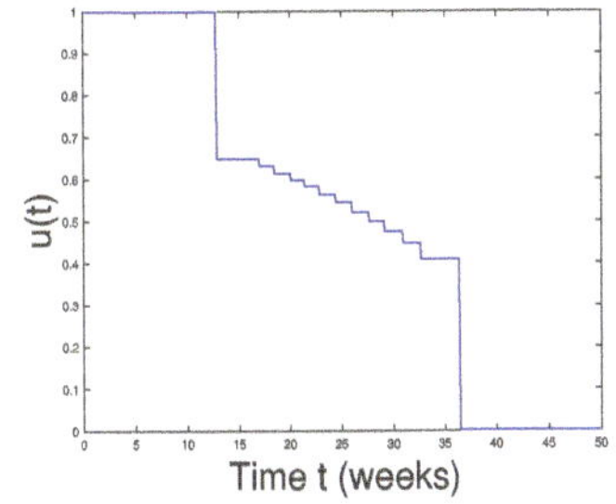

(g) u_p with $p = 10^{-1}$

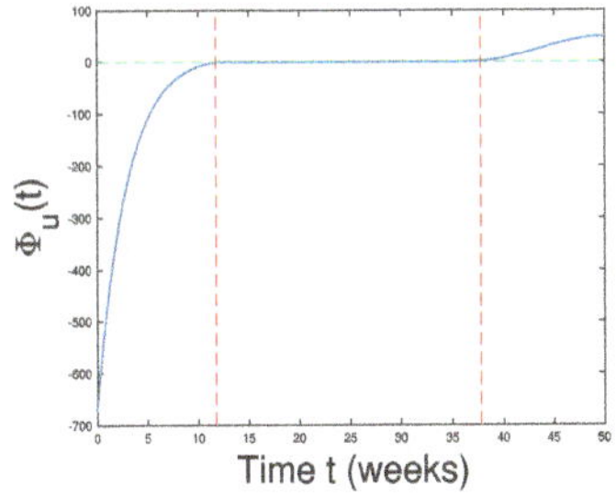

(h) Φ_{u_p} with $p = 10^{-1}$

Fig. 9.12. Plots of the regularized vaccination strategy u_p and its corresponding switching function ϕ_u for varying tuning parameter values for SIR Problem.

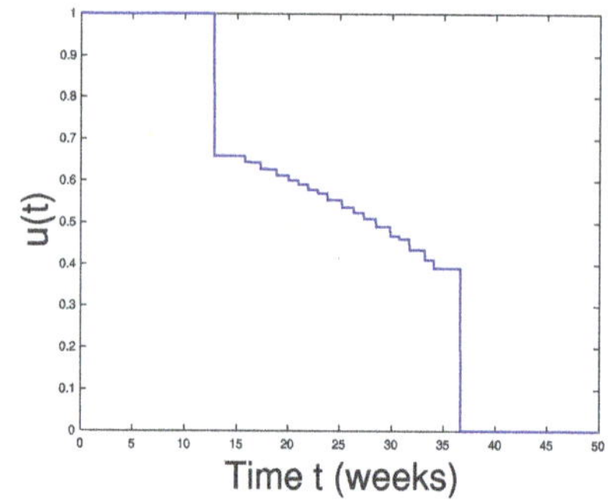

(a) Penalized Vaccination u_p

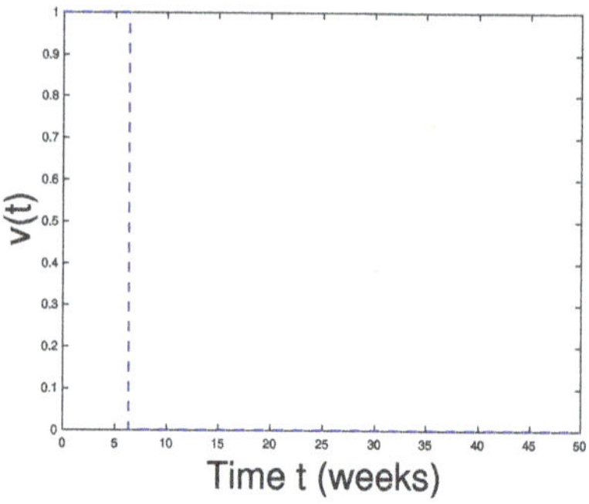

(b) Treatment v

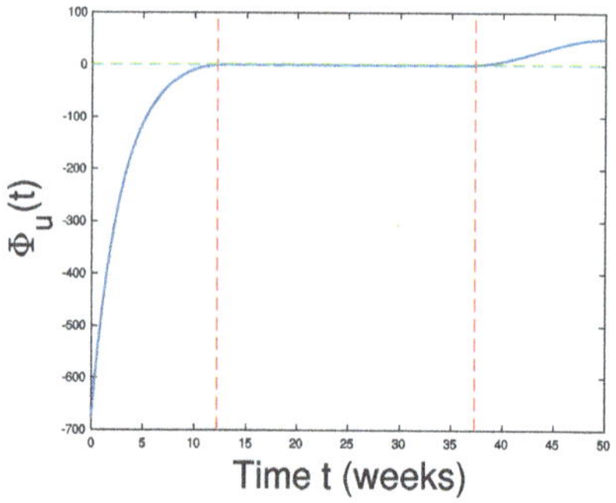

(c) Switching Function, Φ_{u_p} for u_p

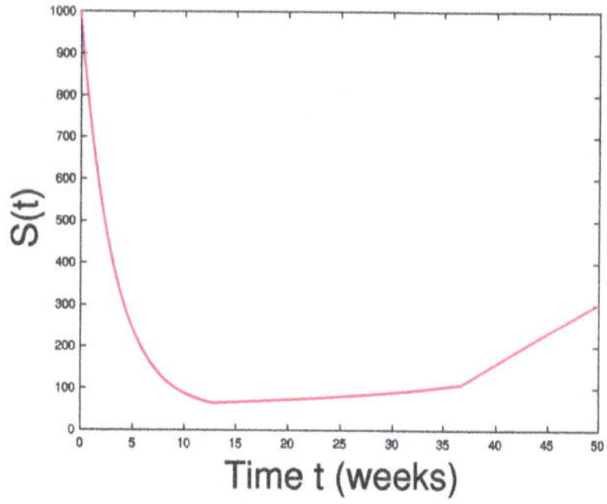

(d) Susceptible Class

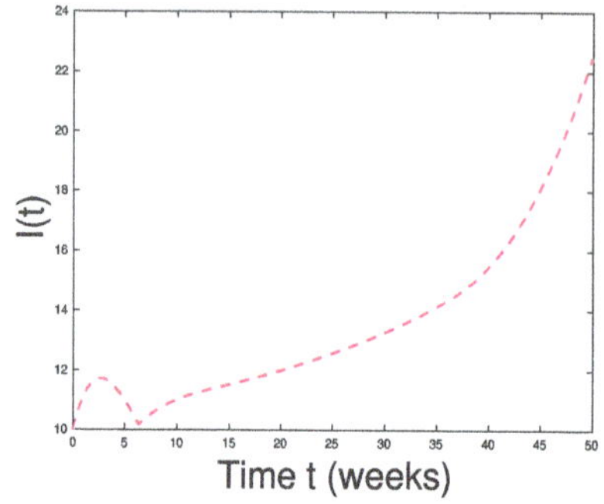

(e) Infectious Class

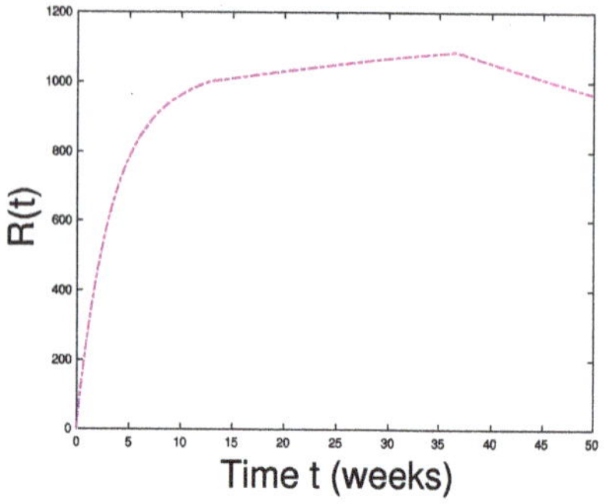

(f) Recovered Class

Fig. 9.13. Spread of the disease with regularized vaccination and unregularized treatment. Tuning parameter: $p = 10^{-1}$, mesh intervals: $n = 2,000$, and Error Tolerance: 10^{-8}.

given in Figures 9.13(a) and 9.13(b), we observe in Figures 9.13(a)–9.13(b) a large portion of the susceptible class transfer to the recovered class through vaccination. The positive influence that treatment has on the infectious class occurs approximately in the first 6 weeks when the treatment strategy is set to treat all infectious individuals. We believe that the weighted cost of treatment parameter value c is influences the optimal treatment strategy to switch to incorporating no treatment at $t_{v,1} \approx 6.4$ weeks. Reducing the weighted cost of treatment parameter would alter the optimal treatment strategy to extending the region(s) when $v^* = v_{\max}$.

9.6 Concluding Remarks

We demonstrate how Hager and Zhang's[31] Polyhedral Active Set Algorithm (PASA) can be implemented for numerically solving optimal control problems that are linear in the control. If problems of this form possess a singular subarc, we recommend regularizing such problems by using a penalty term based on the total variation of the control as suggested in Caponigro *et al.*.[30] Total variation regularization allows for PASA to yield approximations that do not oscillate wildly over the singular region. We provide a discretization method for a general optimal control problem that uses Euler's method for the state variables and uses the gradient of the Lagrangian of the discretized optimal control problem to approximate the gradient of the cost functional that is being used. Additionally, the use of the Lagrangian consequently aids in the discretization of the adjoint equations.

We then present in detail three optimal control problems that apply to biological models. For the first two problems, an explicit solution satisfying the first-order necessary conditions of optimality[10] is found. In the third problem, we can verify existence of a singular subarc for one of the control variables used. For all three examples, we use PASA to numerically solve each problem with parameters set to where the existence of a singular subarc is possible. For the example associated with the plant problem that was presented in King and Roughgarden's,[8] three cases are investigated where each case determines the beginning behavior of the optimal allocation strategy. PASA solves the unpenalized plant problem for all three cases, and in all three cases no oscillations are found. However, in the degenerate case, the case corresponding to the optimal control beginning singular, the unpenalized solution has an unusual dip at $t = 0$. For this case, we regularize the problem via total variation and obtain a penalized solution that serves to be a better approximation of the true solution. When using PASA to solve the unpenalized fishery problem[12,37] and to solve the unpenalized SIR problem,[17] we obtain numerical solutions that possessed oscillations along the singular region. We then apply a total variation regularization for varying values of the penalty parameter. We find that the oscillatory solution can be controlled by increasing the penalty parameter size.

In these examples, we find that observing the plots of the approximated penalized solution and of the corresponding switching function are effective heuristics for determining whether or not the penalty parameter size is over or under-penalizing the solution. For the fishery problem, we use additional information such as the approximated switching points and the L^1 norm difference between the true solution and the penalized solution to help determine what penalty parameter value is the most appropriate choice in comparison to all of the other penalty parameter values being tested. For the SIR problem, we find that the penalized solution that PASA obtains is comparable to the approximated solution that is presented in Ledzewicz *et al.*,[17] which was obtained via a collocation method called PROPT.[47]

Overall, the use of PASA and the use of total variation regularization make the process of numerically solving singular control problems more tractable. With the aid of PASA and total variation penalization, mathematical biologists can consider constructing an optimal control problem with a cost functional that possesses a more accurate representation of the cost in implementing the control, which aligns more with the principle of parsimony than when using an optimal control problem with quadratic dependence. Solving an optimal control problem with a more accurate representation of the cost functional ensures that the optimal control associated with the problem truly meets the criteria that are being considered for determining which control strategy is the best relative to the state dynamics being studied. Additionally, using an optimal control problem where the control variables appear linearly can potentially yield optimal control solutions with regions that are both easy to interpret and implement. Incorporating a more practical, uncomplicated, and accurate optimal control into a dynamical system that is intended to model some biological phenomenon can provide more compelling and elucidating results that are applicable to practice and worthwhile to study.

Acknowledgments

Support by the National Science Foundation (NSF) under Grants 1522629 and 1819002, and by the Office of Naval Research under Grants N00014-15-1-2048 and N00014-18-1-2100 is gratefully acknowledged. S. Atkins was supported in part by the University of Florida's Graduate Student Fellowship. S. Atkins also received student travel support for the CMPD5 conference from NSF CMPD5 Grant 1917506, The Center for Applied Mathematics (CAM), and University of Florida's CLAS. M. Martcheva was supported in part by NSF CMPD5 Grant 1917506 and from the University of Florida's Department of Mathematics. We also are grateful to Suzanne Lenhart for her helpful comments and suggestions of problems to explore.

Appendix 9

9.A.1 *Numerical Results Associated with the Unpenalized Plant Problem*

The unpenalized solutions that PASA obtained when parameters were set to be of Cases 2a, 2b and 2c are shown in Figures 9.A.14, 9.A.15, and 9.A.16, respectively; and numerical results associated with each case are provided in Table 9.6. As you can see in Figures 9.A.14(a), 9.A.15(a), and 9.A.16(a), the unpenalized control found for each case does not present any oscillatory artifacts. We see in Figures 9.A.15(b) and 9.A.16(b) that other than the discrepancies arising from the switching points, the unpenalized solutions for Case 2b and Case 2c only differ slightly from their respective exact solution along the singular region. The state solutions that correspond to the unpenalized solution for Cases 2a, 2b, and 2c, which are respectively shown in Figures 9.A.16(a)–9.A.16(d), Figures 9.A.15(c)–9.A.15(d) and Figures 9.A.16(c)–9.A.16(d), compare relatively well with the state solutions that correspond to u^*. In Figures 9.A.14(c), 9.A.15(c), and 9.A.16(c), we see that in each case the approximate solutions

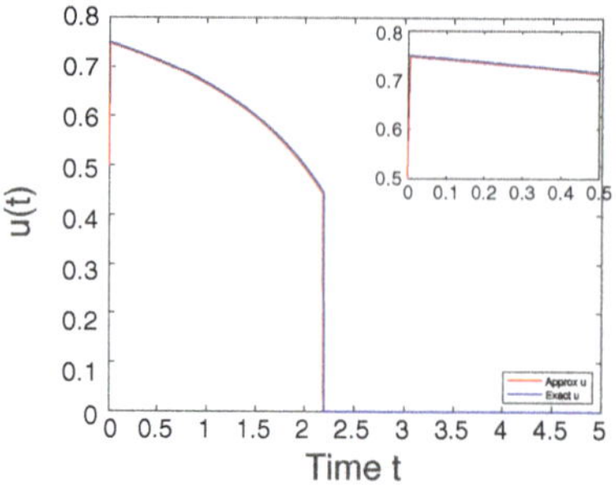

(a) Approximated Allocation Strategy for Unpenalized Problem. Inset shows zoomed in area of interest.

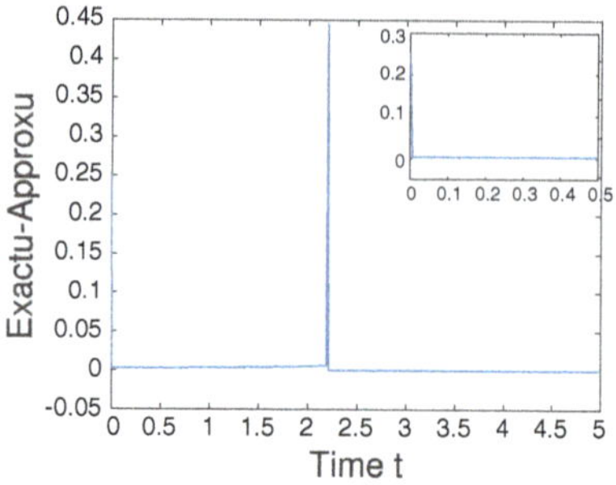

(b) $u^* - \hat{u}$ for Unpenalized Problem. Inset shows zoomed in area of interest.

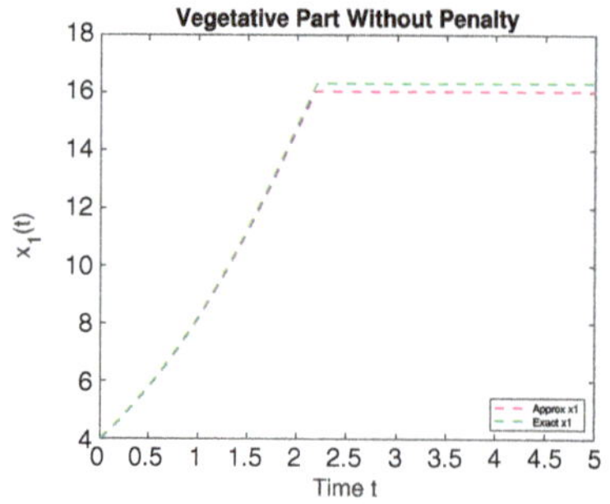

(c) Vegetative Part of Plant from Unpenalized $\hat{u}$

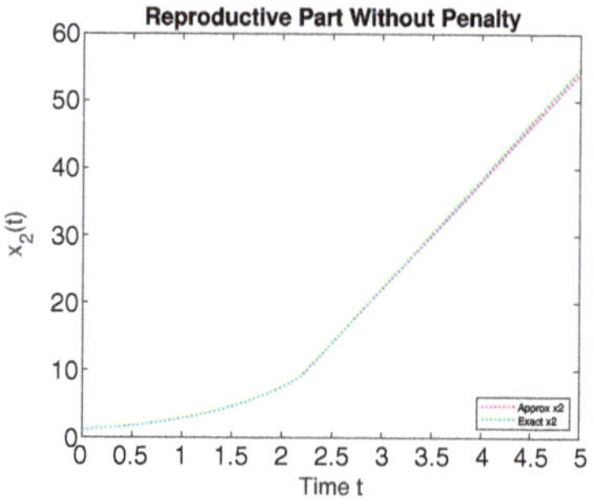

(d) Reproductive Part of Plant from Unpenalized $\hat{u}$

Fig. 9.A.14. Results from solving unpenalized plant problem (Case 2a).

associated with the vegetative parts of the plant do give slightly underestimated values along the interval $[1.5, T]$. This could be due to the small discrepancies in $\hat{u}$ along the singular region. Moreover, if we look at the analytic solution for x_1 on $[t_2, T]$ given in (9.90), we can see how errors between u^* and $\hat{u}$ and how switching prematurely to the non-singular case could contribute errors in x_1 along the non-singular region $[t_2, T]$. Now in Figures 9.A.14, 9.A.15, and 9.A.16 we see that solutions associated with the reproductive parts of plants give slightly underestimated values on the interval $[t_2, T]$. From the explicit solution of state variable x_2 given in Eq. (9.91), we see that state variable x_2 is a line on time interval $[t_2, T]$ with slope being $x_1(t_2)$. So for each case, the approximated solution for x_2 presents underestimated values along time interval $[t_2, T]$ because

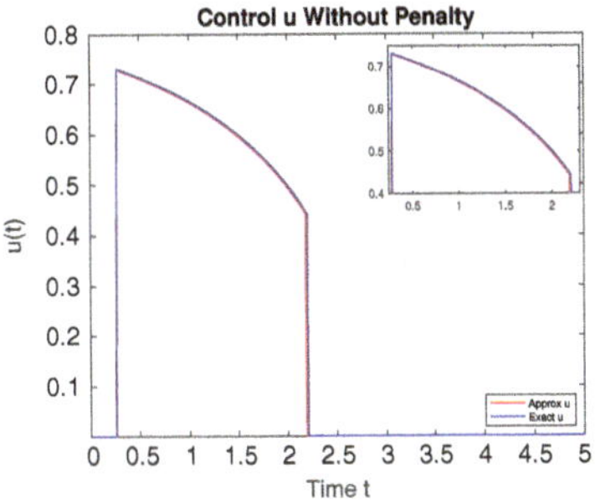

(a) Approximated Allocation Strategy for Unpenalized Problem. Inset shows zoomed in area of interest.

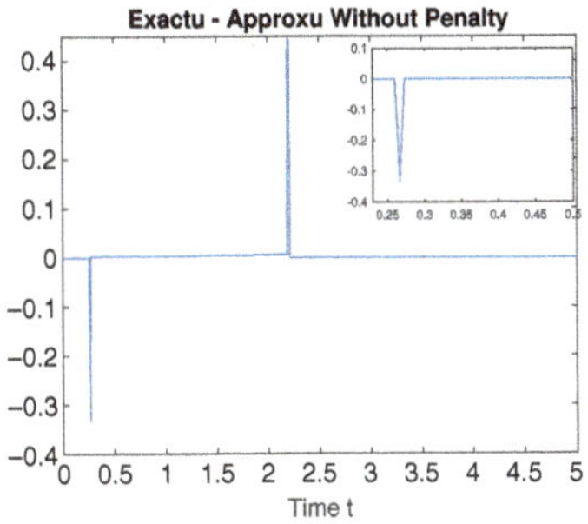

(b) $u^* - \hat{u}$ for Unpenalized Problem. Inset shows zoomed in area of interest.

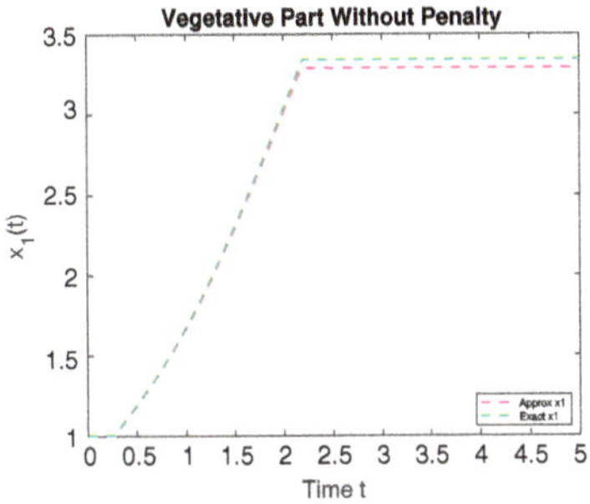

(c) Vegetative Part of Plant from Unpenalized $\hat{u}$

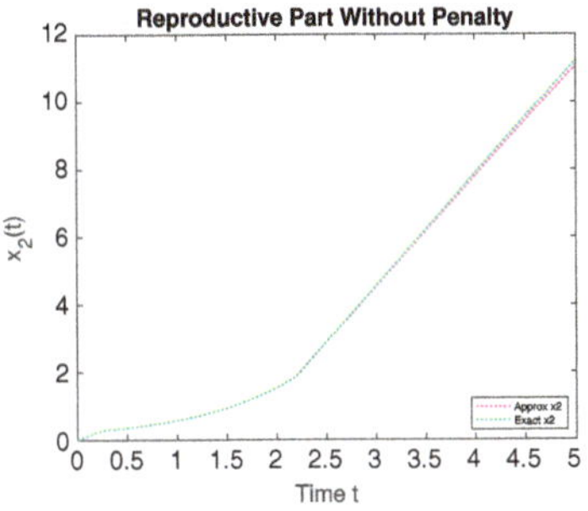

(d) Reproductive Part of Plant from Unpenalized $\hat{u}$

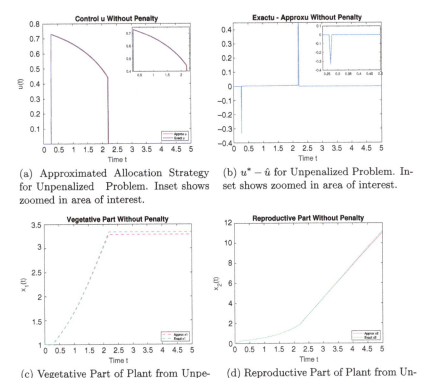

Fig. 9.A.15. Results from solving unpenalized plant problem (Case 2b).

the approximation of state variable x_1 yields underestimated values near t_2. Despite these differences, we find PASA's unpenalized solution to be a sufficient approximation to the true solution to problem (9.65) in Cases 2b and 2c.

9.A.2 *MATLAB code for the SIR problem*

The following is the MATLAB file called demoOC.m. It can be used in solving problem (9.116). This file is included when downloading the PASA package, and it is located in the following directory:

SuiteOPT/PASA/MATLAB. This demo file is associated with the most up-to-date version of PASA, which is SuiteOPT Version 2.0.0. To access PASA software which can be used on MATLAB for

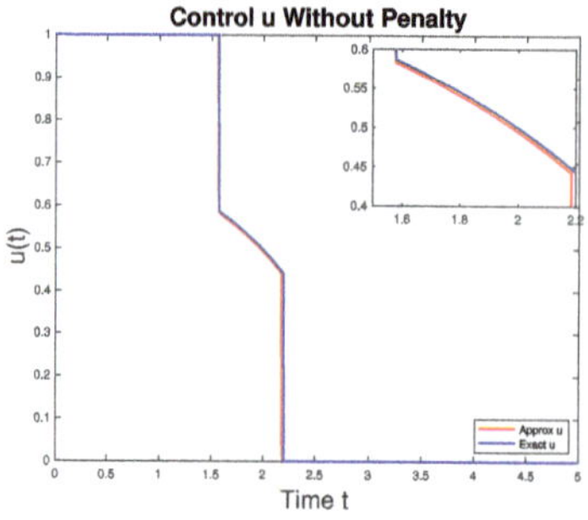

(a) Approximated Allocation Strategy for Unpenalized Problem. Inset shows zoomed in area of interest.

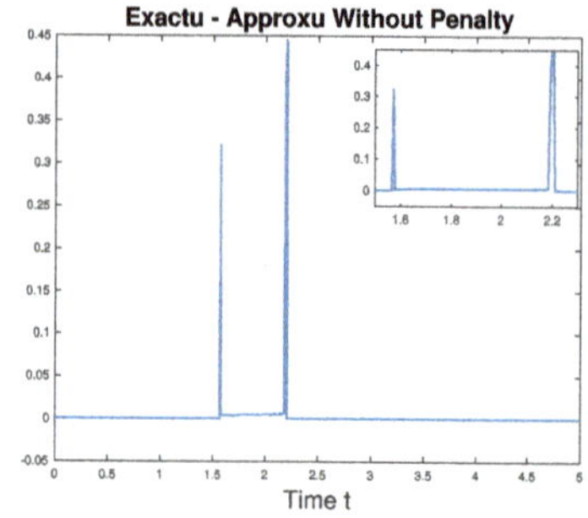

(b) $u^* - \hat{u}$ for Unpenalized Problem. Inset shows zoomed in area of interest.

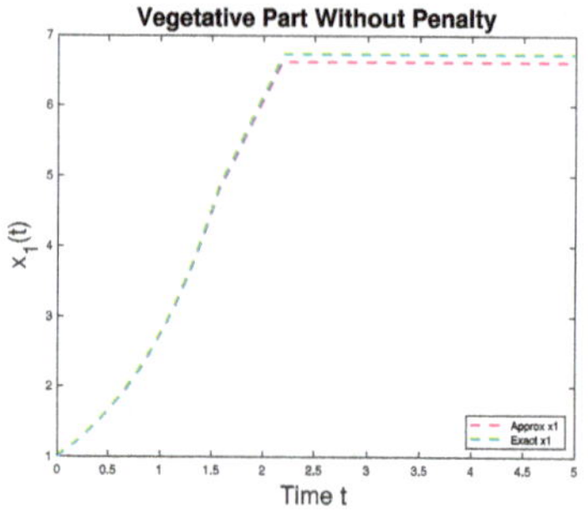

(c) Vegetative Part of Plant From Unpenalized $\hat{u}$

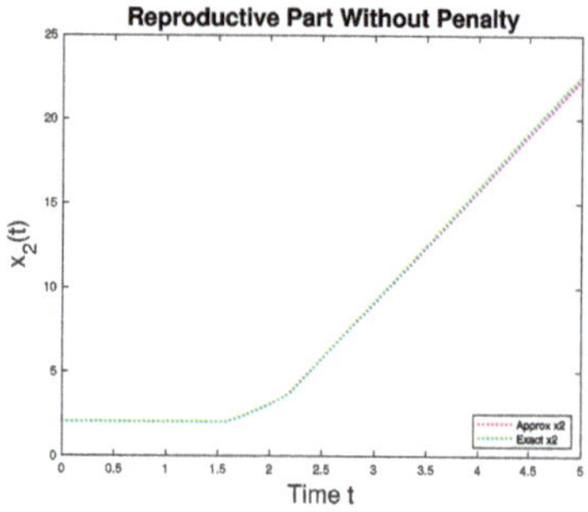

(d) Reproductive Part of Plant From Unpenalized $\hat{u}$

Fig. 9.A.16. Results from solving unpenalized plant problem (Case 2c).

Linux and Unix operating systems, download the SuiteOPT Version 2.0.0 software given on https://people.clas.ufl.edu/hager/software. For future reference, any updates to the software will be uploaded to this link, and to access older versions of the software, use the same link and then select "Software Archive." We recommend the reader to read MATLAB file readme.m which is located in the same directory as demoOC.m. The readme.m file goes into detail about the inputs that are used for running PASA. Note that we have state variables initialized as n-vectors rather than $n+1$-vectors because from the discretization of problem (9.116) given in Section 9.5.1, the $(n+1)$th mesh point of S, I, and R is not present in the following: the approximation of the penalized objective functional, the approximation of gradient of the objective functional, and the discretized adjoint equations. We only used the $(n+1)$th mesh points to generalize the transversality conditions for the adjoint equations. We still are discretizing $[0, T]$ with n mesh intervals, which is why the mesh size is set to being $h = T/n$.

```
% This test problem is an optimal control problem that arises in the
% treatment of the ebola virus (see *). An important feature of the problem is
% that it is singular. If the dynamics were discretized and the cost
% was optimized, then the solution oscillates wildly in the singular
% region, so the optimal control in that region would not be determined.
% The oscillations are removed using a penalty term based on the total
% variation of the optimal control. The dynamics are discretized using
% Euler's method with a constant control on each mesh interval. If u_i
% is the control on the i-th interval, then we add the constraint
%
% u_{i+1} - u_i = iota_i - zeta_i
%
% where zeta_i and iota_i >= 0. The penalty term that we add to the objective
% function is p * sum_{i = 1}^n iota_i + zeta_i, which is p times the total
% variation in the control. Besides the control u, which can be singular,
% there is another control v which is bang/bang. The optimization in this
% problem is over x = [u, zeta, iota, v] where u and v are both constrained
% to the interval [0, 1], while the only constraint on zeta and iota is the
% nonnegativity constraint. If there are n intervals in the mesh, then the
```

```
% dimension of x is 4*n - 2 since u and v both have n components while
% zeta and iota have n-1 components. For pasa, the constraints should be
% written in the form bl <= A*x <= bu and lo <= x <= hi. Thus a row of the A
% matrix corresponds to the constraint:
%
%           0 <= u_i - u_{i+1} + iota_i - zeta_i <= 0, 1 <= i <= n-1
%
% The state variable in the control problem is the triple (S, I, R)
% corresponding to the susceptible, infected, and recovered individuals.
% The control u is vaccination rate while v is referred to as treatment rate.
% The controls are linear in both the dynamics and the cost, which leads
% to the singularity of the solution.
%
%    * Optimal Control for a SIR Epidemiological Model with Time-varying
%    Populations by Urszula Ledzewicz, Mahya Aghaee, and Heinz Schaettler,
%    2016 IEEE Conference on Control Applications (CCA), pp. 1268-1273,
%    DOI: 10.1109/CCA.2016.7587981

function demoOC
    % ------- Initialize constant parameters ------- %
    global gamma beta nu rho kappa mu alpha eta T n a b c p
    clf ;
    gamma = 0.00683 ;  %birth rate of the population
    nu = 0.00188 ;     %natural death rate of the population
    beta = 0.2426 ;    %rate of infectiousness of the disease
    rho = 0.007 ;      %resensitization rate
    kappa = 0.3 ;      %effectiveness of vaccination
    mu = 0.005 ;       %disease induced death rate
```

```
    alpha = 0.00002 ;  %rate at which disease is overcome
    eta = 0.1 ;        %effectiveness of treatment
    T = 50 ;           %time horizon (weeks)
    n = 500 ;          % Dimension of u and v, the number of mesh intervals
    a = 5 ;            % Constant in cost function
    b = 50 ;           % Constant in cost function
    c = 300 ;          % Constant in cost function
    p = 1e-1 ;         % penalty parameter in cost function
    umax = 1 ;         % maximum vaccination rate
    vmax = 1;          % maximum treatment rate
    h = T/n ;          % Step size
    S = zeros (n, 1) ;
    I = zeros (n, 1) ;
    R = zeros (n, 1) ;

%% —— Store problem description in a structure which we call pasadata —— %
    % ———————— Setup sparse matrix A ———————— %
    A1 = spdiags([ones(n-1,1) -ones(n-1,1)], [0,1], n-1, n) ;
    A2 = speye(n-1) ;
    A = sparse (n-1, 4*n-2) ;
    % A1 corresponds to the u_i - u_{i+1} term in the contraint while
    % A2 is used for the zeta and iota terms in the constraint
    A(:, 1:3*n-2) = [A1 -A2 A2] ;

    % put A in the structure
    pasadata.A = A ;

    % If the constraint bl <= A*x <= bu is present, then pasa uses the
    % dimensions of A to determine the number of linear constraints
    % and the number of components in x. Thus if the constraint matrix lies
    % in the upper left nrow by ncol submatrix of a larger matrix Afull,
```

```
    % set pasadata.A = Afull(1:nrow, 1:ncol). In the example above, the
    % statement A = sparse (n-1, 4*n-2) specified the dimensions of A

    % ---------- store the bounds for A*x ---------- %
    pasadata.bl = zeros (n-1, 1) ;
    pasadata.bu = zeros (n-1, 1) ;

    % ---------- store the bounds for x ---------- %
    pasadata.lo = zeros(4*n-2,1) ;
    pasadata.hi = ...
            [umax*ones(n,1); inf*ones(n-1,1); inf*ones(n-1,1); vmax*ones(n,1)] ;

    % The codes to evaluate the cost function and its gradient appear below.
    % Store the name of the codes in the pasadata structure.
    pasadata.grad  = @grad ; % objective gradient
    pasadata.value = @cost ; % objective value

%% -------------------- User defined parameter values for pasa -------------------- %%
    % Type "pasa readme" for discussion of the parameters.
    % By default, there is no printing of statistics.
    pasadata.pasa.PrintStat = 1 ;    % print statistics for used routines
    pasadata.pasa.grad_tol = 1.e-8 ; % PASA stopping tolerance (1.e-6 default)

    % -------------------- Call pasa to determine optimal x -------------------- %
    [x, stats] = pasa (pasadata) ;

    % Since pasadata.pasa.PrintStat = 1, the statistics are displayed at
    % the end of the run. Since the stats structure was included as an
    % output, the corresponding numerical entries can be found in the
```

```
    % structures stats.pasa and stats.cg

    % ---------------------------------- Plot the states ---------------------------------- %
    u = x (1:n) ;            % extract the control u from the returned solution x
    v = x (3*n-1:4*n-2) ;  % extract the control v from the returned solution x
    [S, I, R] = state (u, v) ; % the state associated with the optimal controls

    t = linspace (0, T-h, n) ;
    subplot(3,2,1)
    plot (t, S,'linewidth',2) ;
    xlabel('Time')
    ylabel('S')
    subplot(3,2,2)
    plot (t, I, 'linewidth',2) ;
    xlabel('Time')
    ylabel('I')
    subplot(3,2,3)
    plot (t, R, 'linewidth',2) ;
    xlabel('Time')
    ylabel('R')

    % -------------------------- Plot the controls u and v . -------------------------- %
    subplot(3,2,5);
    plot (t, u,'linewidth',2) ;
    xlabel('Time')
    ylabel({'Control u','(Vaccination)'})
    subplot(3,2,6);
    plot (t, v, 'linewidth',2) ;
    xlabel('Time')
    ylabel({'Control v','(Treatment)'})

%% ---------------------------- User defined functions for pasa ---------------------------- %%
    % ----- Objective function ----- %
    function J = cost(x)
        h = T/n ;
        u = x(1:n) ;
```

```
        zeta = x(n+1:2*n-1) ;
        iota = x(2*n:3*n-2) ;
        v = x(3*n-1: 4*n-2) ;
        [S, I, R] = state (u, v) ;
        J = h*(a*sum(I) + b*sum(u) + c*sum(v)) ;
        J = J + p*sum(zeta + iota) ;
    end

    % —— Gradient of objective function —— %
    function g = grad(x)
        h = T/n ;
        u = x(1:n) ;
        v = x(3*n-1: 4*n-2) ;

        % Compute state and costate
        [S, I, R] = state (u, v);
        [lS, lI, lR] = costate (S, I, R, u, v) ;

        % Update gradient values
        for i=1:n
            Fu(i) = h*(b - lS(i)*kappa*S(i) + lR(i)*kappa*S(i)) ;
            Fv(i) = h*(c - lI(i)*eta*I(i) + lR(i)*eta*I(i)) ;
        end

        % Store gradient values in array g to return
        g = [Fu, p*ones(1,n-1), p*ones(1,n-1), Fv] ;
    end

    % —— State —— %
    function [S, I, R] = state (u, v)
        h = T/n;
        S(1) = 1000 ;
        I(1) = 10 ;
        R(1) = 0 ;

        for i = 1:n-1
            N = S(i) + I(i) + R(i) ;
            S(i + 1) = S(i) + h*(gamma*N - nu*S(i) - (beta*S(i)*I(i))/N ...
                + rho*R(i) - kappa*S(i)*u(i)) ;
```

```
            I(i + 1) = I(i) + h*(beta*S(i)*I(i)/N - (nu + mu + alpha)*I(i)...
                - eta*I(i)*v(i)) ;
            R(i + 1) = R(i) + h*( - nu*R(i) - rho*R(i) ...
                +  kappa*S(i)*u(i) + alpha*I(i) + eta*I(i)*v(i)) ;
        end
    end

    % ---- Costate ---- %
    function [lS, lI, lR] = costate (S, I, R, u, v)
        h = T/n ;
        lS(n) = 0 ;
        lI(n) = 0 ;
        lR(n) = 0 ;

        for i=n:-1:2
            N = S(i) + I(i) + R(i) ;
            lS(i-1) = lS(i)+ h*lS(i)*(gamma - nu - beta*(I(i)/N) ...
                        + beta*(S(i)*I(i)/(N^2)) - kappa*u(i))...
                        + h*lI(i)*(beta*(I(i)/N) - beta*(S(i)*I(i)/(N^2)))...
                        + h*lR(i)*(kappa*u(i)) ;
            lI(i-1) = lI(i)+ h*a +h*lS(i)*(gamma - (beta*S(i))/N ...
                        + (beta*S(i)*I(i))/N^2)...
                        + h*lI(i)*(( beta*S(i))/N - (beta*S(i)*I(i))/(N^2) ...
                        - (nu + mu + alpha) - eta*v(i))...
                        + h*lR(i)*(alpha + eta*v(i)) ;
            lR(i-1) = lR(i)+ h*lS(i)*(gamma + (beta*S(i)*I(i))/(N^2) + rho)...
                        + h* lI(i)*(- (beta*S(i)*I(i))/(N^2) ) ...
                        + h* lR(i)*(- nu - rho) ;
        end
    end
end
```

References

1. Joshi, H. R. (2002). Optimal control of an HIV immunology model, *Optim. Control Appl. Methods* **23**, pp. 199–213.
2. Kirschner, D., Lenhart, S., and Serbin, S. (1997). Optimal control of the chemotherapy of HIV, *J. Math. Biol.* **35**, pp. 775–792.
3. Jung, E., Lenhart, S., and Feng, Z. (2005). Optimal control of treatments in a two strain tuberculosis model, *Discrete Contin. Dyn. Syst.* **2**, pp. 473–482.
4. Ögren, P. and Martin, C. (1997). Vaccination strategies for epidemics in highly mobile populations, *Appl. Math. Comput.* **127**, pp. 315–322.
5. Neubert, M. (2003). Marine reserves and optimal harvesting, *Ecol. Lett.* **6**, pp. 843–849, doi: 10.1046/j.1461-0248.2003.00493.x.
6. Ding, W. and Lenhart, S. (2009). Optimal harvesting of a spatially explicit fishery model, *Natural Resource Modeling* **22**, pp. 173–211.
7. Gaff, H., Joshi, H. R., and Lenhart, S. (2007). Optimal harvesting during an invasion of a sublethal plant pathogen, *Environ. Develop. Econ.* **12**, 4, pp. 673–686.
8. King, D. and Roughgarden, J. (1982). Graded allocation between vegetative and reproductive growth for annual plants in growing seasons of random length, *Theoret. Popul. Biol.* **22**, 1, pp. 1–16, doi: 10.1016/0040-5809(82)90032-6.
9. Seierstad, A. and Sydæter, K. (1987). *Optimal Control Theory with Economic Applications*, Vol. 24 of Advanced Textbooks in Economics North-Holland Publishing Co., Amsterdam.
10. Pontryagin, L., Boltyanskii, V., Gamkrelidze, R., and Mishchenko, E. (1964). *The Mathematical Theory of Optimal Processes*, Macmillan, New York.
11. Fleming, W. H. and Rishel, R. (1975). *Deterministic and Stochastic Optimal Control*, Springer-Verlag, New York, NY.
12. Lenhart, S. and Workman, J. T. (2007). *Optimal Control Applied to Biological Models*, Chapman and Hall/CRC Press, Boca Raton, FL.
13. Miele, A. (1975). Recent advances in gradient algorithms for optimal control problems, *J. Optim. Theory Appl.* **17**, pp. 361–430.
14. Edge, E. and Powers, W. (1976). Function-space quasi-newton algorithms for optimal control problems with bounded controls and singular arcs, *J. Optim. Theory Appl.* **20**, 4, pp. 455–479.
15. Betts, J. T. (1998). Survey of numerical methods for trajectory optimization, *J. Guid. Control Dyn.* **21**, pp. 193–207.
16. Rao, A. (2010). A survey of numerical methods for optimal control, *Adv. Astronaut. Sci.* **135**, p. 32.

17. Ledzewicz, U., Aghaee, M., and Schättler, H. (2016). Optimal control for a SIR epidemiological model with time-varying populations, *IEEE Conference on Control Applications (CCA)*, pp. 1268–1273, doi: 10.1109/CCA.2016.7587981.
18. Ledzewicz, U. and Schättler, H. (2011). On optimal singular controls for a general SIR-model with vaccination and treatment, in *Proceedings of the 8th AIMS Conference*, pp. 981–990.
19. Shen, M., Xiao, Y., Rong, L., and Meyers, L. A. (2019). Conflict and accord of optimal treatment strategies for HIV infection within and between hosts, *Math. Biosci.* **309**, pp. 107–117.
20. Joshi, H. R., Herrera, G. E., Lenhart, S., and Neubert, M. (2009). Optimal dynamic harvest of a mobile renewable resource, *Nat. Resource Model.* **22**, 2, pp. 322–343.
21. McDanell, J. and Powers, W. (1971). Necessary conditions for joining optimal singular and nonsingular subarcs, *SIAM J. Control* **9**, pp. 161–173.
22. Robbins, H. (1967). A generalized Legendre-Clebsch condition for the singular cases of optimal control, *IBM J. Res. Dev.* **11**, 4, pp. 361–372.
23. Lewis, R. (1980). Definitions of order and junction conditions in singular optimal control problems, *SIAM J. Control Optim.* **18**, pp. 21–32.
24. Zelikin, M. I. and Borisov, V. (1994). *Theory of Chattering Control: with Applications to Astronautics, Robotics, Economics, and Engineering*, Birkhäuser, Boston, MA.
25. Foroozandeh, Z., De Pinho, M. D. R., and Shamsi, M. (2019). On numerical methods for singular optimal control problems: An application to an AUV problem, *Discrete Contin. Dyn. Syst. — B* **24**, 5, pp. 2219–2235, doi: 10.3934/dcdsb.2019092.
26. Van Wyk, E. J., Falugi, P., and Kerrigan, E. C. (2010). ICLOCS2: Imperial college London optimal control software, http://www.ee.ic.ac.uk/ICLOCS/. Accessed 15 September, 2020.
27. Patterson, M. A. and Rao, A. (2014). GPOPS-II: A MATLAB software for solving multiple-phase optimal control probles using hp-adaptive qaussian quadrature collocation methods and spares nonlinear programming, *ACM Trans. Math. Softw.* **41**, 1, pp. 1–37, doi: 10.1145/2558904.
28. Foroozandeh, Z., De Pinho, M. D. R., and Shamsi, M. (2017). A mixed binary non-linear programming approach for the numerical solution of a family of singular optimal control problems, *Int. J. Control* pp. 1–35, doi: 10.1080/00207179.2017.1399216.
29. Yang, X.-q., Teo, K. L., and Caccetta, L. (2001). *Optimization Methods and Applications*, Vol. 52 of Applied Optimization. Springer, New York, NY.

30. Caponigro, M., Ghezzi, R., Piccoli, B., and Trelat, E. (2018). Regularization of chattering phenomena via bounded variation controls, *IEEE Trans. Automatic Control* **63**, 7, pp. 2046–2060, doi: 10.1109/TAC.2018.2810540, arXiv:1303.5796.
31. Hager, W. W. and Zhang, H. (2016). An active set algorithm for nonlinear optimization with polyhedral constraints, *Sci. China Math.* **59**, 8, pp. 1525–1542, doi: 10.1007/s11425-016-0300-6, arXiv:1606.01992.
32. Bertsekas, D. P. (1976). On the Goldstein-Levitin-Polyak gradient projection method, *IEEE Trans. Automat. Control* **21**, pp. 174–184.
33. Calamai, P. and Moré, J. J. (1987). Projected gradient for linearly constrained problems, *Math Program* **39**, pp. 93–116.
34. Goldstein, A. A. (1964). Convex programming in Hilbert space, *Bull. Amer. Math. Soc.* **70**, pp. 709–710.
35. Levitin, E. S. and Polyak, B. T. (1966). Constrained minimization problems, *USSR Comput. Math. Math. Phys.* **6**, pp. 1–50.
36. McCormick, G. P. and Tapia, R. A. (1972). The gradient projection method under mild differentiability conditions, *SIAM J. Control* **10**, pp. 93–98.
37. Clark, C. W. (1990). *Mathematical Bioeconomics: The Optimal Management of Renewable Resources*, Wiley, New York.
38. Brauer, F. and Castillo-Chávez, C. (2001). *Mathematical Models in Population Biology and Epidemiology*, Springer, New York, NY.
39. Martcheva, M. (2015). *An Introduction to Mathematical Epidemiology*, Springer, New York, NY.
40. Conway, J. (1978). *Functions of One Complex Variable*, Springer-Verlag, New York.
41. Hager, W. W. (2000). Runge-Kutta methods in optimal control and the transformed adjoint system, *Numer. Math.* **87**, pp. 247–282.
42. Hager, W. W. and Rostamian, R. (1987). Optimal coatings, bang-bang controls, and gradient techniques, *Optim. Control Appl. Meth.* **8**, pp. 1–20.
43. Kelley, H., Kopp, R., and Moyer, H. (1967). Singular extremals, *Topics Optim.: Math. Sci. and Eng.* **31**, pp. 63–101.
44. Cohen, D. (1971). Maximizing final yield when growth is limited by time or by limiting resources, *J. Theor. Biol.* pp. 299–307, doi: 10.1016/0022-5193(71)90068-3.
45. Cohen, D. (1967). Optimizing reproduction in a randomly varying environment when a correlation may exist between the conditions at the time a choice has to be made and the subsequent outcome,

J. Theor. Biol., **16**, 1, pp. 1–14, doi: https://doi.org/10.1016/0022-5193(67)90050-1, http://www.sciencedirect.com/science/article/pii/0022519367900501.
46. Schättler, H. and Ledzewicz, U. (2012). *Geometric Optimal Control*, Springer, New York, NY.
47. Rutquist, P. and Edvall, M. (2010). *PROPT: Matlab Optimal Control Software*, Tomlab Optimization, Inc., Pullman, WA.

https://doi.org/10.1142/9789811263033_0010

Chapter 10

Phase Locking and Lyapunov Exponent Behavior in Brain Networks of Epileptic Patients*

Heather Berringer[†], Kris Vasudevan[¶], Paolo Federico[§], Michael Cavers[‡], and Elena Braverman[||,#]

[†]*University of Victoria*
[§]*Department of Clinical Neurosciences, Hotchkiss Brain Institute, Cummings School of Medicine, Calgary, Canada*
[‡]*University of Toronto at Scarborough*
[||]*University of Calgary*
[#]*maelena@ucalgary.ca*

Characterization of neuronal connectivity in human brains before, during, and after an epileptic seizure uses different signal theoretic and nonlinear methods. These methods analyze intracranial electroencephalogram (iEEG) data from multiple seizure types. Changes in brain neuronal network structure result in changes in phase synchronization. We have analyzed phase synchronization from iEEG data, which entailed applying the Hilbert transform to both unfiltered and frequency-filtered data, calculating metrics related to phase-locking studies, and image

*Heather Berringer and Elena Braverman presented partial contents of this chapter at the Fifth International Conference on Computational and Mathematical Population Dynamics held at Florida Atlantic University, Fort Lauderdale, Florida, USA from May 19 to 24, 2019.

¶Deceased on August 22, 2022.

processing. Furthermore, since the iEEG data are nonlinear representing a dynamical system, we have investigated the deterministic chaos with Lyapunov exponents. In this chapter, we report the results of our work and draw preliminary conclusions on the seizure behavior.

Keywords: Phase Locking Value, Lyapunov exponents, Epileptic seizures, Intracranial electroencephalogram, Hilbert Transform, Pre-ictal and Post-ictal periods

10.1 Introduction

Epilepsy is a common neurological disorder that is characterized by abnormal brain activity resulting in seizures.[1] Approximately 0.6% of the Canadian population has epilepsy, including those who take anticonvulsant medication or who have had a seizure within the past five years.[2] Epilepsy can affect anyone, regardless of age, gender, or ethnic background, and about half of all cases of epilepsy have no identifiable cause.[1]

Furthermore, specific types of epilepsy that occur frequently and are drug-resistant can cause serious or life-threatening complications. These include permanent brain damage, or even death, sometimes in the form of Sudden Unexpected Death in Epilepsy. Epilepsy is usually treated through medication, however, if this fails, then surgery may be performed. The goal of the surgery is to remove the part of the brain tissue, a lesion, causing the seizures.[2]

There are four main classifications of epilepsy: focal, generalized, focal and generalized, and unknown cause.[3] Focal seizures result from abnormal activity in one brain area, whereas generalized seizures involve all areas of the brain. Focal and generalized epilepsy is the type of epilepsy where people have both focal and generalized seizures. Unknown epilepsy is diagnosed when doctors are sure the patient has epilepsy, but are not sure if the seizures are generalized or focal. See Table 10.1 for the major types of seizures within the focal and generalized categories.

Some drug-resistant epilepsy patients undergo resection surgery. This provides neurosurgeons and epileptologists an opportunity to investigate the source and spreading mechanism of seizures. A conventional experimental tool they use is based on recording electroencephalograms (EEGs) intracranially with grid electrodes or depth electrodes or with a combination of both. The EEGs measure the

Table 10.1. A description of the major seizure types.[4,8]

Type	Name	Characteristics
Focal	Aware	The person knows what's happening during the seizure.
Focal	Impaired Awareness	Can cause confusion and make the person feel dazed.
Focal	Motor	Can cause involuntary movements.
Focal	Non-motor	Can cause changes in sensations, thoughts, or emotions.
Generalized	Absence (or non-motor)	Can cause rapid blinking or a few seconds of staring into space.
Generalized	Motor	Can cause the person to cry out, lose consciousness, fall, have muscle jerks or spasms. Includes tonic-clonic seizures.

fluctuating potentials in millivolts at different contact points of the electrodes as a result of many neurons participating in the conductance behaviour.

In the literature, two modeling approaches are pursued to understand the physical mechanism of the seizure onset, duration, and offset, of what is known as the ictal period. There is also considerable effort directed at the pre-ictal and the post-ictal periods. In the first approach, multi-neuronal models or neural mass models are considered. In the second approach, much of the effort is expended in extracting interpretable patterns embedded in the EEG data. This is usually done with defining signal-theoretic based, physically meaningful metrics that exploit the nonlinear and non-stationary behavior of the EEG data. It is the second approach we consider in this chapter. To this end, we processed and analyzed EEG data from two drug-resistant epilepsy patients both undergoing two styles of epileptic seizures, the focal to bilateral tonic-clonic seizure (FBTCS) and the focal impaired awareness seizure (FIAS).

In Section 10.2, we provide details of the acquisition geometry used by the neurosurgeons and epileptologists for the two patients.

In Section 10.3, we summarize the processing methods. In Section 10.4, we introduce two metrics, the phase-locking value and the Lyapunov exponent, and show how they are calculated for the data in question. In Section 10.5, we discuss the results of the metrics obtained for the FBTCS and the FIAS and the conclusions reached.

10.2 Acquisition Geometry

Epilepsy causes abnormalities in electrical activity in the brain, which is measured by the EEG device. In the drug-resistant cases in which surgery is required, the patient is monitored for a period of a few days with intracranial electroencephalogram (iEEG) recordings. iEEG electrodes are placed directly on the patient's brain, and due to their increased proximity to the neuronal populations and absence of the skull, result in clearer and less noisy data.

The recordings monitored the patients' epileptic events, from which we analyzed two seizure types: A FBTCS and a FIAS. The datasets studied are denoted by FBTCS1 and FIAS1 (Patient 1) and by FBTCS2 and FIAS2 (Patient 2). The FBTCS1 and FIAS1 datasets were acquired with 64 surface contact point in an 8-by-8 grid, and two sets of four-contact points depth electrodes, totalling to 72 contact points (see Figure 10.1).

The FBTCS2 and FIAS2 datasets had 108 contact points in total, with 12 depth electrodes in different brain regions ranging from 8–10 contact points each. See Table 10.2 for the regional placements and Figure 10.2 for the images of the electrodes.

10.3 Pre-Processing

Electrical signals from EEG recordings are not only nonlinear, but also exhibit unpredictable and complex behavior. Recently, an increase in understanding of the complex dynamics of low-dimensional nonlinear systems and chaos has led to the idea that EEG signals are generated by a nonlinear process.[5]

The FBTCS and FIAS data from patient 1 (FBTCS1 and FIAS1) were sampled at a sampling rate of 2,000 Hz, and the FBTCS and FIAS data for patient 2 (FBTCS2 and FIAS2), at 1,000 Hz

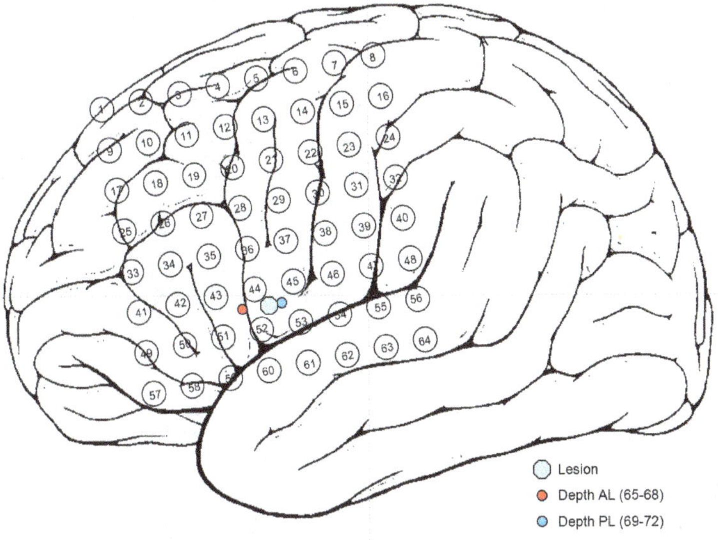

Fig. 10.1. The iEEG grid electrode configuration on the brain for the FBTCS and FIAS datasets 1, with the lesion region (light blue dot) being between the electrode regions 44 and 52. The red-filled and the blue-filled circles refer to two four-contact point depth electrodes.

Table 10.2. Brain region location of the electrodes for the FBTCS2 and FIAS2 datasets, with the seizure focus being the Left Hippocampus.

Channels	Brain region (Abbreviation)
1–10	Left anterior insula (LAI)
11–20	Left lateral orbitofrontal (LLOF)
21–28	Left mid cingulate (LMC)
29–38	Left amygdala (LAm)
39–46	Left anterior hippocampus (LAH)
47–54	Left posterior hippocampus (LPH)
55–64	Left mesial occipital (LMOc)
65–74	Left temporoparietal (LTP)
75–84	Right anterior insula (RAI)
85–92	Right amygdala (RAm)
93–100	Right anterior hippocampus (RAH)
101–108	Right posterior hippocampus (RPH)

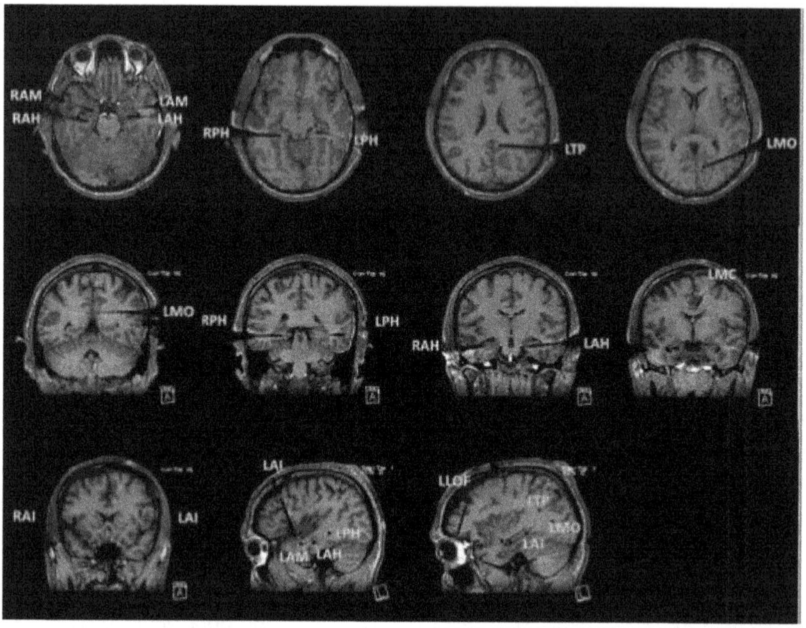

Fig. 10.2. The brain images showing the iEEG electrode configurations for the second data-set described in Table 10.2.

(see Figure 10.3). In the case of FIAS2 dataset, as shown in Figure 10.3(d) in red rectangular boxes, there are two time periods where the recorded signals are too noisy. We removed them and concatenated the three segments into one for our analysis. In addition to this, a small noisy glitch with the contact positions of the Left Anterior Insula region interfered with our analysis. So, we discarded the LAI region for any subsequent interpretation. We show in Figure 10.3(e) an example of the Left Anterior Hippocampus region corresponding to the contact positions 39 to 46 of the recorded iEEG with 1,070,000 samples. For PLV analysis, we use a subset with samples running from 300,001 to 900,000 samples.

For each dataset, we applied the same filters to clean the signals from noise and artifacts. However, here we applied a multi-windowed, iterative despiking algorithm to generate a despiked time-series for each electrode, based on the earlier work of Goring and Nikora[6] and Mori *et al.*[7] Next, we applied a high-pass filter, then finally a notch filter to remove the power artifacts at 60 Hz and multiplies of 60 Hz, using both the "eglab" software package and MATLAB signal theory

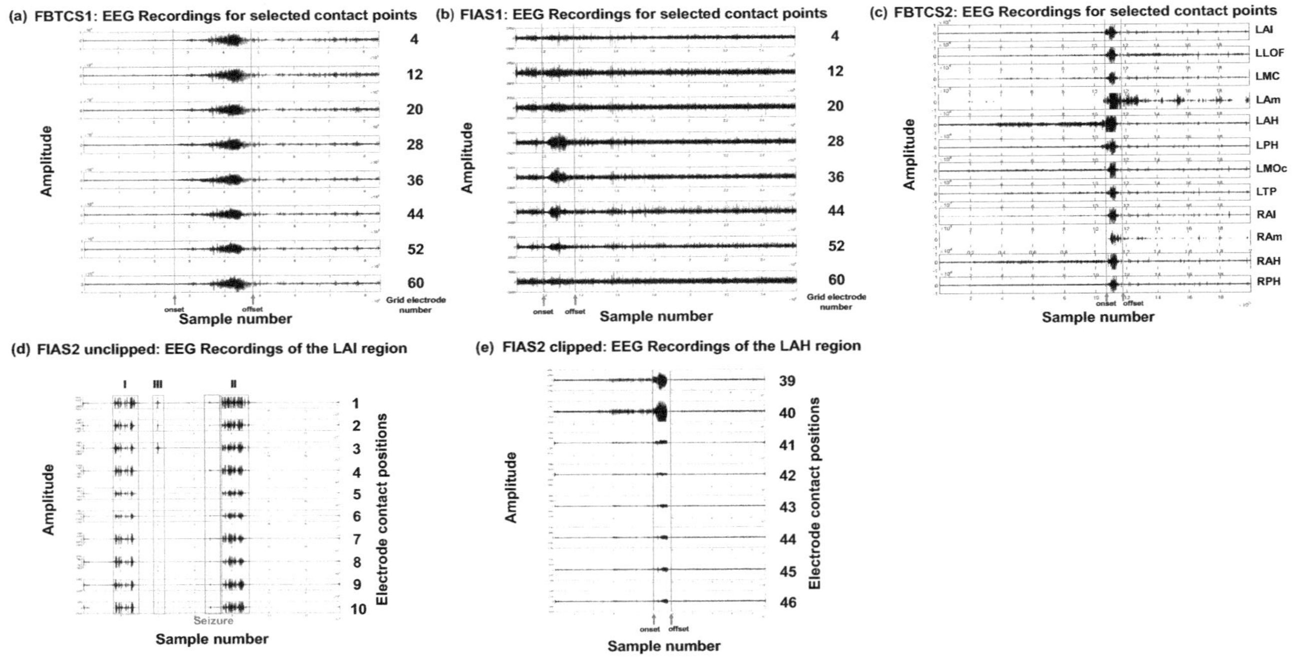

Fig. 10.3. Intracranial EEG recordings of the FBTCS and FIAS for two patients. (a) FBTCS1 for selected contact points, 4, 12, 20, 28, 36, 44, 52, and 60. (b) FIAS1 for selected contact points, 4, 12, 20, 28, 36, 44, 52, and 60. (c) FBTCS2 for the shallow contact points from the following 12 regions, LAI, LLOF, LMC, LAM, LAH, LPH, LMOc, LTP, RAI, RAM, RAH, and RPH. (d) FIAS2 from the contact points from the LAI region. The red rectangular windows correspond to noisy patches that interfered with our analysis. Patches I and II that appear at all contact positions are excluded from the entire data. Patch III appears only in the LAI region. We excluded the LAI region for any interpretation. We used the remaining data for our analysis. (e) FIAS2 for the contact points from the LAH region to illustrate the use of 1,070,000 samples. For PLV analysis, we show our results in Figures 10.5(d), 10.6(b), and 10.7(d) for a subset with samples from 300,001 to 900,000 samples.

package. Note that the signal from the 33rd contact point of the patient 1 data was too noisy, thus removed in the calculations or ignored during the analysis of results.

Since the FBTCS2 and FIAS2 datasets were from 12 depth electrodes, we processed the data using the difference modification (also known as bipolar modification) approach. This was done as follows: given contact points i to $i + k$ corresponding to sequential contact points in the same region from the same depth electrode, we computed k signals from the difference of each contact point neighbor. For example, given contact points 1 to 4 in the same region, $[x_1, x_2, x_3, x_4]$, we would get three new signals: $[x'_1 = x_2 - x_1, x'_2 = x_3 - x_2, x'_3 = x_4 - x_3]$. These three new signals would become our new difference modification dataset. This technique is used to avoid volume conduction and noise, in the sense that it isolates the unique components each signal is picking up, and avoids overlapping signal information. Using this difference modification, the datasets FBTCS2_diff and FIAS2_diff were created.

10.4 Metrics

The two primary metrics we used to follow the seizure evolution were the phase-locking value (PLV) and the Lyapunov exponent (LE). The PLV follows the spread of two signals' phase difference over time, in order to measure their phase synchronization. First developed by Lachaux *et al.*[8] and further adapted by Mormann *et al.*[9] for epilepsy studies, the PLV metric gauges the level of communication between two signals. Normalized PLV values approaching 1 indicate almost complete phase synchronization, meaning that the two signals are in close communication with each other. Low normalized PLV values approaching 0 imply that the two signals are not communicating with each other. We further discuss PLV in Section 10.4.1, along with its mathematical definition and calculations pertaining to our data in Section 10.4.1.2

Lyapunov exponents offer a method to numerically measure a level of chaoticity, associated with dynamical systems. By reconstructing a time series into p-dimensional trajectories, the divergence of nearby trajectories is measured and used to estimate the average Lyapunov exponents. If a single positive Lyapunov exponent occurs, the system is chaotic.[10] We discuss LE in Section 10.4.2 and its calculation pertaining to our data in Sections 10.4.2.2 and 10.4.2.3.

The main purpose of this chapter is to demonstrate that the two metrics would allow one to generate a complete picture of the seizure evolution and gain insight into the merits and downfalls of the methods.

10.4.1 *Phase-locking studies*

How does the brain flawlessly coordinate complex cognitive, thought, emotional, and perceptual processes? What causes an interruption of such coordination, resulting in neurological disorders such as epilepsy? Over the past few decades, researchers have diverged from the idea that distinct brain regions are only responsible for separate roles, and have begun to examine neuronal large-scale integration. Rather than considering anatomy and function to be fundamentally inseparable, it has emerged that collective interactions between brain areas support smooth integration of cognitive acts. This interregional communication system can be represented as a network, with neurons and neuronal populations dynamically interacting.[11]

In the context of the brain network, large-scale integration is defined to be the "synchronization of neural assemblies ... a process that spans multiple spatial and temporal scales in the nervous system."[12] Exact mechanisms involved are still being studied, and current methods to measure brain activity introduce interpretation issues.[11] Phase synchronization is believed to be crucial in large-scale integration, and is one of the most studied connectivity hypotheses, especially in common recording techniques such as EEG.[11]

In 1999, Lachaux *et al.*[8] released a ground-breaking paper that introduced a method for computing and studying phase synchrony between neuronal signals. Written in the frequency domain, a signal $x(t)$ can be separated into instantaneous amplitude and phase (see Section 10.4.1.1). They argue that phase "contains all the information about the temporal structure of neural activity,"[12] which then allows the phase to be the metric in the synchronization study.

The method for analyzing phase compares the distribution of phase difference between two signals over time, under the hypothesis that if two signals are in communication, their instantaneous phases will evolve together.[8,11] This difference is known as the phase-locking value. It is one of the most commonly used phase synchrony metrics, due to its simplicity of computation and retention of information.[13]

10.4.1.1 *Hilbert transform*

The calculation of the phase-locking value for two signals requires instantaneous phase extraction. The Hilbert transform, continuous wavelet transforms, short-time Fourier transform, and complex demodulation are examples of methods used to extract instantaneous phase.[14] Wavelet transforms and the Hilbert transform are most commonly used, due to their reliability and calculation speed.[11] In our PLV calculations, we have chosen to use the Hilbert transform, as it can deal with non-stationary and broadband signals.[13] This method of instantaneous phase and amplitude extraction does not depend on the choice of a specific frequency like other methods.[8]

Definition 10.1. The Hilbert transform of a real signal $x(t)$, denoted $H[x(t)]$, is

$$H[x(t)] = \tilde{x}(t) = \frac{1}{\pi} pv \int_{-\infty}^{\infty} \frac{x(\tau)}{t-\tau} \mathrm{d}\tau. \tag{10.1}$$

It is the convolution of $x(t)$ with $\frac{1}{\pi t}$.[15] Here, pv denotes the Cauchy principal value of the integral, to deal with the possibility of a singularity at $t = \tau$.[16] Specifically, the Cauchy principal value of (10.1) is

$$H[x(t)] = \frac{1}{\pi} \lim_{\epsilon \to 0+} \left(\int_{t-\frac{1}{\epsilon}}^{t-\epsilon} \frac{x(\tau)}{t-\tau} \mathrm{d}\tau + \int_{t+\epsilon}^{t+\frac{1}{\epsilon}} \frac{x(\tau)}{t-\tau} \mathrm{d}\tau \right). \tag{10.2}$$

A signal $x(t)$ and its Hilbert transform $\tilde{x}(t)$ together form an analytic signal, denoted $X(t)$,[16]

$$X(t) = x(t) + \mathrm{i}\tilde{x}(t), \tag{10.3}$$

which can be rewritten as

$$X(t) = A(t)\mathrm{e}^{\mathrm{i}\varphi(t)}. \tag{10.4}$$

Here, $A(t)$ is the instantaneous amplitude of the signal, and $\varphi(t)$ is the instantaneous phase,[15] interpreted as the angle between $x(t)$ and

$\tilde{x}(t)$. Rearranging (10.4), we obtain

$$A(t) = \sqrt{x^2(t) + \tilde{x}^2(t)} \quad \text{and} \quad \varphi(t) = \arctan\left(\frac{\tilde{x}(t)}{x(t)}\right). \tag{10.5}$$

In particular, this instantaneous phase extracted from the analytic signal will be used in our PLV calculation.

10.4.1.2 *PLV computations*

To deal with ever-prevalent noise from real signals, the PLV measures the spread of distribution of phase differences.[11] Given two analytic signals $x_i(t)$ and $x_j(t)$, their frequency domain representation is

$$x_i(t) = A_i(t)e^{i\varphi_i(t)}, \quad x_j(t) = A_j(t)e^{i\varphi_j(t)} \tag{10.6}$$

with $A_i(t), A_j(t)$ the signals' instantaneous amplitudes at time t and $\varphi_i(t), \varphi_j(t)$ their instantaneous phases at time t. Then, the phase difference of $x_i(t)$ and $x_j(t)$ would be

$$\Delta\varphi_{i,j}(t) = \varphi_j(t) - \varphi_i(t). \tag{10.7}$$

We then define the PLV mathematically as

$$\mathrm{PLV}_{i,j}(t) = \frac{1}{N}\left|\sum_{n=1}^{N} e^{i(\varphi_j(t,n)-\varphi_i(t,n))}\right| \tag{10.8}$$

which measures the phase difference over N trials with $\varphi_i(t,n)$, the instantaneous phase for signal $x_i(t)$ in trial n at time t. PLV values range between 0 and 1, with values approaching 1 indicating phase synchrony, and values approaching 0 indicating no phase dependency between the signals. With the data from the different seizures, we computed the phase-locking value for every pair of contact points throughout the entire dataset. For a window-by-window analysis of the seizure evolution, with each window corresponding to 2.5 (N = 5,000 points for FBTCS1 (Patient 1), N = 2,500 points for FBTCS2 (Patient 2)), the output was 168 matrices of size 72-by-72 for FBTCS1 (Patient 1) and 798 matrices of size 108-by-108 for FBTCS2 (Patient 2). Based on the contact point configuration for the FBTCS1 dataset, we further split the FBTCS1 data into depth and surface contact points. This resulted in two sets of 168 windows, one with matrices of size 64-by-64 (surface contact points), and one of size 8-by-8 (depth contact points). Note that these matrices are symmetric, as $\mathrm{PLV}_{i,j} = \mathrm{PLV}_{j,i}$.

We used "imagesc" with "colormap" in MATLAB to visualize the range of values by color. Blue corresponded to a low PLV value, yellow, a mid-range value, and red, a high PLV value. Since the phase-locking value of any signal to itself is 1, for each of the matrices, the diagonal was dark red indicating a PLV value of 1. Consider Figure 10.4.

The window-by-window approach gave crucial insight into the specifics of each contact point for every 2.5 s window of the time series. However, we modified the output in order to see the general pattern of the seizure periods in one image, in the form of a composite map. To do so, we took each window of surface contact point data, a 64-by-64 matrix for the FBTCS1 dataset, and computed a vertical sum for every column. That is, for a given column, we added every value belonging to that column, and then divided by the column length (in this case, 64). Therefore, each 64-by-64 matrix was converted to a row vector of length 64. To construct the composite map matrix, the transpose of the vertical sum vector became one column in the composite map matrix. In this way, we constructed a 64-by-168 matrix that captured the overall evolution of the seizure and different behavior during seizure periods. Note that we computed this composite map for window size $N = 5{,}000$, 10,000, and 20,000 samples for the FBTCS1 dataset. We used $N = 5{,}000$ to get the most detailed results corresponding to our sampling rate and to avoid a "smoothing effect" by taking too large of a window size, which results in some loss of information. As $N = 5{,}000$ corresponds to 2.5 s, we used 2.5 s as our window size for the other datasets and the remainder of the analysis.

The FBTCS2 data were processed exactly the same way as the filtered and notched FBTCS1 data, and the PLV computed identically as well. The only differences were the sampling rate and the number of channels of the new dataset. The sampling rate was 1,000 Hz, thus resulting in $N = 2{,}500$ points per window rather than 5,000. In addition, the new data had 108 channels of contact points. Consider Figure 10.4 to see the four windows of 108-by-108 PLV matrices at each of the stages of the seizure.

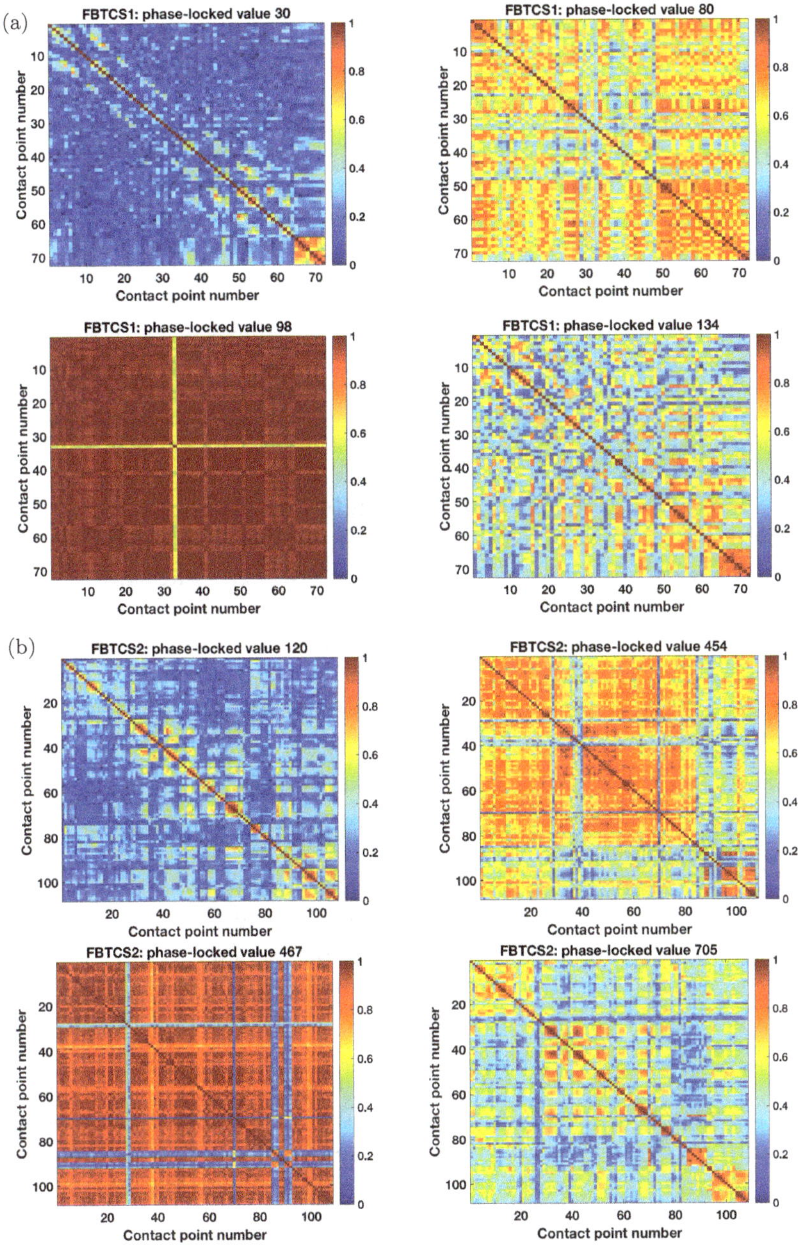

Fig. 10.4. (*Continued*)

Fig. 10.4. Phase-locking values for selected windows (or panels) of the data. (a) Phase-locking value 72-by-72 window matrices for FBTCS1. One entry of a matrix $PLV_{i,j}$, represents the phase-locking value between row i and column j. Top left (30th window) is from the pre-ictal period, top right (80th window) is from the ictal period, bottom left (98th window) is from the offset period, and bottom right (134th window) is from the post-ictal period. The yellow cross in the 98th window is from the noisy 33rd contact point, which was later removed from results. (b) Phase-locking value 108-by-108 window matrices and signal from one contact point. One entry of a matrix $PLV_{i,j}$ represents the phase-locking value between row i and column j. Top left (120th window) is from the pre-ictal period, top right (454th window) is from the ictal period, bottom left (467th window) is from the offset period, and bottom right (705th window) is from the post-ictal period.

10.4.2 *Lyapunov exponent*

In 1985, Wolf[10] presented an algorithm to estimate nonnegative Lyapunov exponents for experimental nonlinear time series. Babloyantz and Destexhe,[17] applied this algorithm[10] to EEG data drawn from a seizure. This is due to the discovery that low-dimensional nonlinear systems can exhibit a strong dependence on initial conditions, without any stochastic input.[18] EEG, and similarly iEEG, data are believed by some to reflect not only stochastic processes, but also the deterministic chaos of brain activity. In particular, Babloyantz and Destexhe,[17] Palus,[19] Iasemidis,[20] and Osowski[21] have shown the presence of chaotic behavior in EEG or iEEG data recorded during an epileptic seizure.[21]

The two characteristics of nonlinear systems that we are most interested in are the strong dependence on initial conditions, and the existence of attractors of fractal dimension (strange attractors).[20]

10.4.2.1 *Phase space*

To understand how these particular characteristics can be identified and quantified in the data, we first have to define the phase space portrait of a time series. Given a data segment signal $x(t)$, the phase space portrait is constructed by evaluating the signal at several points in time as follows:

$$X(t) = [x(t), x(t+\tau), \ldots, x(t+(p-1)\tau)], \tag{10.9}$$

where τ is the time lag, and p is the embedding dimension of the phase space.[20] That is, p is the smallest value such that the trajectory of the signal in the phase spaces does not self-intersect. Repeating this process by increasing t gives the phase space portrait. This portrait is useful as it can give a visual interpretation of the data over time, particularly when $p = 2$ or 3.[20] Note that each vector $X(t)$ represents an instantaneous state of the system in the phase space.[21] Following the evolution of $X(t)$ for increasing t gives a trajectory, or orbit, in the phase space.

An orbit in the phase space represents the temporal evolution of the system.[20] Orbital convergence toward a smaller region of a lower-dimensional phase space indicates the presence of an attractor and characterizes systems with energy loss. An attractor is a lower-dimensional area of the phase space that the future evolution of the signal is confined to.[20] An attractor that has self-similar structure is a fractal attractor, or a strange attractor. An attractor in which two orbits with similar initial conditions (represented by a set of points in the phase space) diverge at an exponential rate is a chaotic attractor.[20] Orbits belonging to a chaotic attractor are infinitesimal, whereas the attractor has finite size. Consequently, as time evolves, the orbits have to return into the attractor in what is known as the folding process.[20]

10.4.2.2 *Calculations*

We modeled our estimation of the Lyapunov exponent on Wolf's popular algorithm.[10] The procedure monitors the long-term evolution of two nearby orbits in the phase space. However, experimental data, in comparison to a linear system, only has one trajectory in the phase space. To circumvent this issue, the algorithm works in a reconstructed attractor, in which points with enough temporal separation can be considered to be from different trajectories.[10] Furthermore, a folding process will likely occur as the algorithm follows the long-term evolution of the orbits. We want to ensure the calculations are capturing the local stretch of the temporary orbital proximity, rather than the global fold of the attractor. Wolf's algorithm addresses this by replacing one of the trajectories right before folding occurs. The choice of replacement trajectory is a crucial and delicate step. The replacement is chosen to be as close as possible to the remaining trajectory and to have the same orientation as the replaced trajectory.[5]

First, we construct the phase space portraits of a pair of orbits. This is done by choosing an initial point, t_0, and using the method of delays based on the time lag τ to create a p-dimensional vector (with p the embedding dimension). This orbit will be considered the fiducial trajectory, which does not get replaced.[5] Next, we pick a second point that must lie on a different trajectory. It has to be at least one mean orbital period apart from the first orbit to be considered from a different trajectory.[10] One way of calculating the mean orbital period is to take the average time between folding.[5] Then, the algorithm follows the evolution of these trajectories. Right before folding is about to occur, a replacement point is chosen. Iasemidis *et al.*[18] suggest stricter conditions in the replacement trajectory selection, to compensate for the difference in noise levels between EEG and iEEG data. This entire process is repeated until the fiducial trajectory reaches the end of the data segment. We computed the Lyapunov exponent using Wolf's original algorithm. The choice of parameters that determine the orbits and the replacement trajectories is dependent on the given data. A detailed explanation of how we chose the parameters is given in Section 10.4.2.4.

10.4.2.3 *Explicit formula*

The Lyapunov exponent is calculated by

$$L = \frac{1}{N_a \Delta t} \sum_{i=1}^{N_a} \log_2 \frac{|X(t_i) - X(t_j)|}{|X(t_i + \Delta t) - X(t_j + \Delta t)|}, \qquad (10.10)$$

where N_a is the number of iterations necessary for convergence of L, Δt is the time evolution for the trajectory, $t_i = t_0 + (i-1)\Delta t$ for $i \in [1, N_a]$, $t_j = t_0 + (j-1)\Delta t$ for $i \in [1, N]$ (N is the length of the data segment and t_0, the initial point of the fiducial trajectory), $X(t_i)$, a vector of the fiducial trajectory, and $X(t_j)$, an appropriately chosen vector adjacent to $X(t_i)$ in the phase space.

10.4.2.4 *Choice of parameters*

The delicate aspect of calculating Lyapunov exponents, and the one that draws the most criticism, is choosing appropriate parameters for the data set. The overall goal is to pick values that open up the phase space orbits, and track appropriate adjacent points. Proper parameter optimization is still needed to ensure the accuracy of the results. The details of our initial parameter choices are given as follows. τ is the time delay in the delay reconstructed orbit. Our choice was based on Babloyantz and Destexhe's range for τ, $17\,dt \leq \tau \leq 25\,dt$ with dt the sampling rate of the data.[17] This range ensures that the delay reconstructed orbit has as simple an appearance as possible so that the algorithm can quantify orbital divergence as accurately as possible.[22] Note that the dependence of Lyapunov exponent on choice of τ is very weak.[22] We chose $\tau = 9$, which was consistent with Babloyantz and Destexhe.[17]

p is the embedding dimension of the reconstructed orbit. It should be chosen such that the dimension of the epileptic attractor in the ictal period is clearly defined,[20] in particular $p > 6$. However, it is difficult to know its exact value, especially over a long time period.[20] In practice, it is rare to have enough data to run a system requiring $p > 5$, and the phase space orbit fractal dimension calculations are notoriously unstable.[22] Furthermore, we found the Lyapunov exponent estimate was not significantly affected by choice of p, so after preliminary results with $p = 7$, we calculated L with $p = 5$ to help the algorithm run smoothly and in a timely manner.

Evolution time, *evolve*, is the number of steps for which each pair of vectors is followed through the phase space. According to Wolf, it should be small enough such that the orbital divergence is monitored a few times per orbit.[22] Based on our observation of 12 data points per mean orbital period, *evolve* was chosen to be 3, consistent with Wolf's suggestions. Due to the lack of datasets of infinitesimal length, we could not check orbital divergence at infinitesimal separations.[22] Thus, the Lyapunov exponent had a strong dependence on evolve.

N is the number of points per window, for which one Lyapunov exponent was calculated per window. It was chosen to be 20,000 for datasets sampled at 2,000 Hz and 10,000 for datasets sampled at 1,000 Hz, to obtain a meaningful portrait of the Lyapunov exponent behavior throughout the evolution of the seizure.

10.5 Results and Discussion

10.5.1 *Phase-locking values*

We computed the composite map using window-by-window matrices for four main datasets: the focal to bilateral tonic-clonic seizure

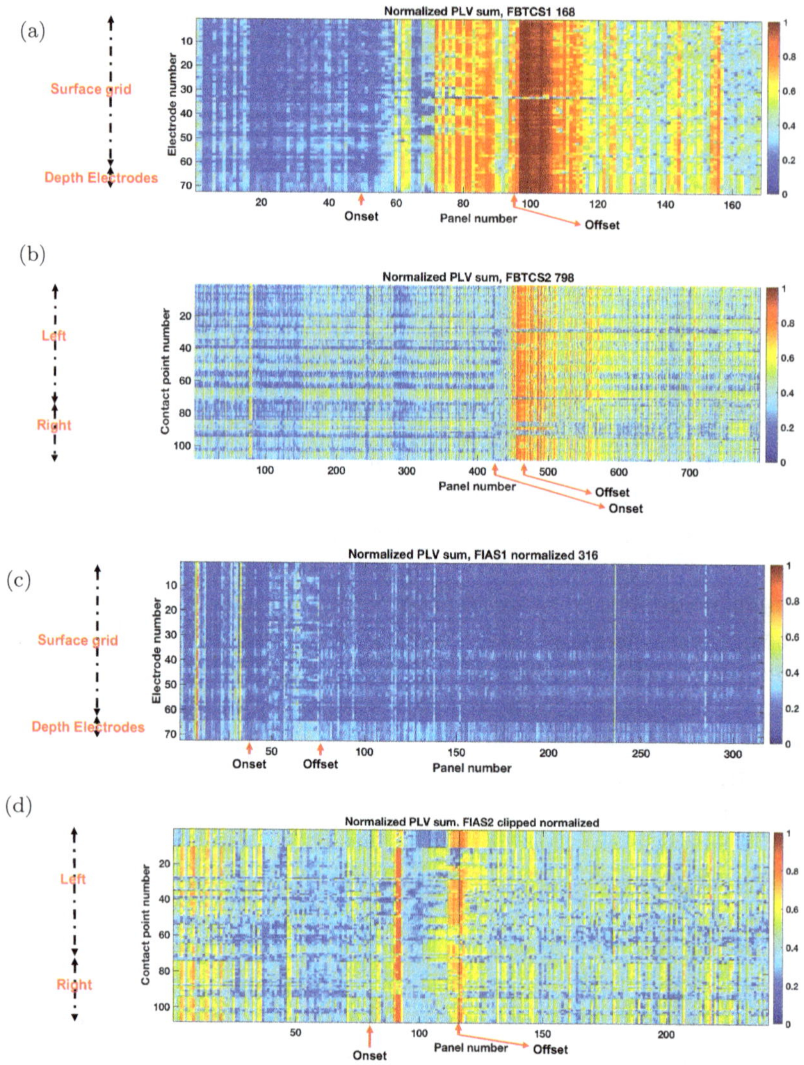

Fig. 10.5. PLV composite maps for the chosen FBTCS1, FBTCS2, FIAS1, and FIAS2 datasets. (a) FBTCS1 PLV composite map. This map includes both the surface grids (1 to 64) and the depth electrodes (65 to 72). The time locations in terms of the panel numbers for the seizure onset and offset are indicated as red arrows. Electrode 33 is quite noisy and should not be considered for any interpretation. (b) FBTCS2 PLV composite map. This map includes depth electrodes with contact points 1 to 74 on the left hemisphere and 75 to 108 on the right hemisphere of the brain, as shown in Figure 10.2. The time locations in terms of

Fig. 10.5. (*Continued*) the panel numbers for the seizure onset and offset are indicated as red arrows. (c) FIAS1 PLV composite map. This map includes both the surface grids (1 to 64) and the depth electrodes (65 to 72). The time locations in terms of the panel numbers for the seizure onset and offset are indicated as red arrows. Electrode 33 is quite noisy and should not be considered for any interpretation. The PLV maps are displayed on a 0 to 1 scale. The noisy panels from the patient's movements are not included. (d) FIAS2 PLV composite map. This map includes depth electrodes with contact points 1 to 74 on the left hemisphere and 75 to 108 on the right hemisphere of the brain, as shown in Figure 10.2. The time locations in terms of the panel numbers for the seizure onset and offset are indicated as red arrows. The PLV maps are displayed on a 0 to 1 scale. The noisy panels from the patient's movements are not included.

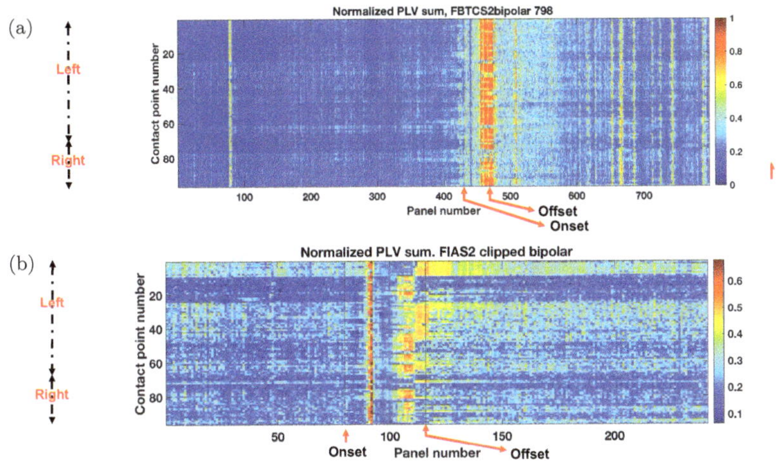

Fig. 10.6. PLV composite maps of the "bipolar" version of FBTCS2 and FIAS2 data sets, originally used in Figure 10.5. (a) FBTCS2 with the seizure onset and offset zones. (b) FIAS2 with the seizure onset and offset zones.

(FBTCS1 and FBTCS2) data, and the focal impaired awareness seizure (FIAS1 and FIAS2) data (see Figure 10.5).

We also calculated composite maps (see Figure 10.6) for a differencing modification applied to FBTCS2 and FIAS2 data.

For both datasets, FBTCS1 and FBTCS2, we also computed the composite map of the filtered data, according to frequency bands defined in Table 10.3 (see Figure 10.7).

The FBTCS are usually the most violent type of seizure, characterized by abrupt loss of consciousness, body stiffening and shaking,

Table 10.3. Standard frequency bands in computational neuroscience.

Range (Hz)	Frequency band
0.1–1	Slow
1–3	Delta
4–8	Theta
8–12	Alpha
12–30	Beta
25–100	Gamma
80–200	Ripple 1
200–400	Ripple 2

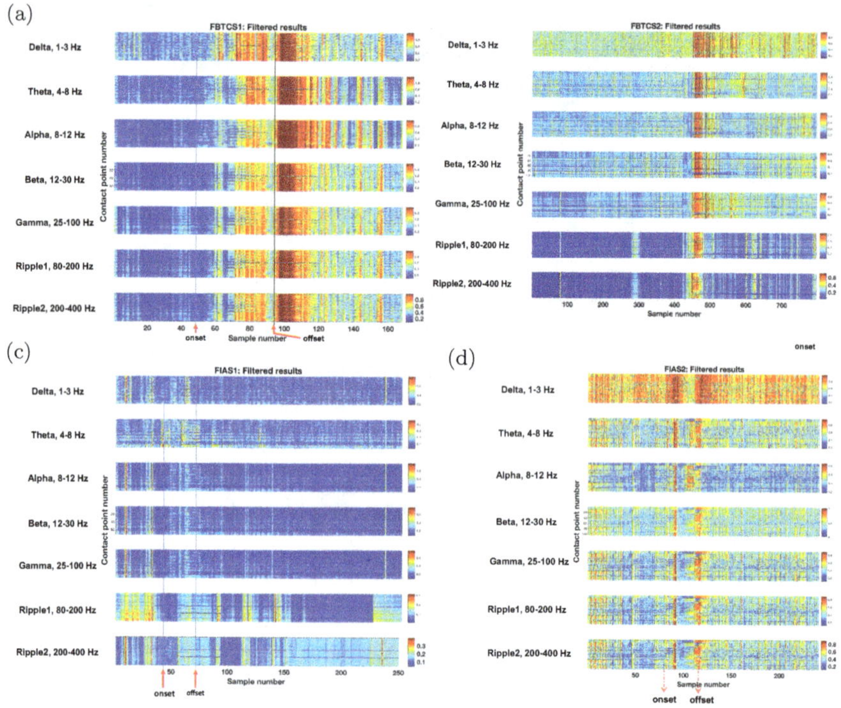

Fig. 10.7. PLV composite maps of the filtered data of the chosen FBTCS1, FBTCS2, FIAS1, and FIAS2 datasets. The filter bands used are summarized in Table 10.3. The seizure onset and offset zones are indicated in the figures. (a) FBTCS1. (b) FBTCS2. (c) FIAS1. (d) FIAS2.

tongue biting, and loss of bladder control. It had distinct behavior at each seizure period, so the results were split up accordingly. Here, we use FBTCS1 as an example (see Figure 10.5(a)).

Pre-ictal: During the pre-ictal period, the PLV values were low: ranging between approximately 0.1 and 0.4. This indicated that the neuronal populations had low phase synchronization, and thus were ready to receive and interpret new signals, consistent with normal brain activity.

Ictal: After the seizure onset, there were approximately 13 s before the PLV behavior changed from pre-ictal values, which corresponded to roughly half of the seizure duration. After 13 s, flashes of approximately 0.6 phase-locking occurred, then raised to 0.7 after about 10 more seconds. Next, the values increased to about 0.8–1 for almost every contact point, indicating a complete phase-locking of all the neuronal populations. This lasted until an abrupt de-synchronization at the offset, with PLV values dropping to 0.5–0.6. There was a (horizontal) periodic block of contact points with lower phase-locking values: these corresponded to the contacts further away from the lesion on the brain.

The violent "flashing" behavior of the synchronization mirrored the violent and sporadic movements of the tonic-clonic event.

Offset period: About 10 s after the offset of the seizure, there was a period of widespread high phase-locking values nearing 1 lasting for about 14 s. This was the period in which the patient was regaining consciousness, but was dazed from the epileptic event. The completeness of the synchronization reflected the utter spread and severity of the seizure, as it dominated the entire brain. At this level of phase-locking, the neuronal populations were at their least independent and lacked the ability to rapidly process complex cortical processes such as emotions, thoughts, and the coordination of actions.

Post-ictal: After the offset period, the PLV values dropped to approximately 0.4–0.6 during the post-ictal period. These values were higher than the pre-ictal period, indicating that the neuronal populations were still moderately phase-locked and recovering from the epileptic event. Further analysis of the data recordings well-beyond the time of the seizure indicated PLV values returning to the pre-ictal low values.

We filtered the time series from the FBTCS1 into seven separate frequencies (see Figure 10.7). These frequencies were chosen to replicate the commonly separated frequency bands in computational neuroscience (see Table 10.3).

Aside from the delta band, which exhibits phase-locking values higher than the other bands by approximately 0.1, the frequency bands are all very similar. In particular, the gamma, ripple 1, and ripple 2 bands are essentially identical.

Differences begin to emerge just before the seizure offset, specifically at the theta and alpha bands. They show slightly increased PLV values immediately before the offset in comparison to the other bands. Furthermore, they exhibit more uniform and higher valued synchronization during the offset period, and more pronounced post-ictal behavior. These results were based on the FBTCS1 filtered and notched data.

Consistent with the FBTCS1 frequency band results, all of these bands exhibited almost identical behavior throughout the duration of the seizure. The delta band had overall slightly higher PLV values, attributable to the sampling rate at this low frequency. As with the FBTCS1 data, the theta and alpha bands exhibited longer and higher PLV values during the ictal period. See Figure 10.7.

The FBTCS2 data were processed exactly the same as the filtered and notched FBTCS1 data, and the PLV computed identically as well. The only differences were the sampling rate and the number of channels of the new dataset. The sampling rate was 1,000 Hz, thus resulting in $N = 2{,}500$ points per window rather than 5,000. In addition, the new data had 108 channels of contact points. Consider Figure 10.4 to see the four windows of 108-by-108 PLV matrices at each of the stages of the seizure.

The FBTCS2 PLV composite map (Figure 10.5) exhibited similar trends to the FBTCS1 dataset. In particular, it had lower pre-ictal PLV synchronous values, ranging between 0.2 and 0.5. Some anomalous behavior begins around the onset, however, when we analyzed the window-by-window PLV matrices, we noticed similar behavior to that of the corresponding FBTCS1 ictal period behavior. Namely, it exhibited a general increase in synchrony levels as well as pulsing synchronous behavior. Near the end of the ictal period, there was almost complete synchronization, corresponding to the seizure offset.

From then on, strange signal behavior occurs then carries through the remainder of the dataset (see contact points 28, 70, and 92, for example). The cause of this behavior is undetermined, but could likely be noise artifacts. In the post-ictal period, there was mid-range synchrony, with values reaching slightly higher than pre-ictal values, accompanied by flashes of high levels of synchrony (around 0.7).

The focal impaired awareness seizure is less violent, and more local than the focal to bilateral tonic-clonic seizure. Its primary symptoms include confusion and an overall dazed feeling. Hence, we observed much more subtle phase-locking behavior with the FIAS datasets, FIAS1 and FIAS2. We observed both FIAS1 and FIAS2 flashes of higher phase-locking values in the pre-ictal period, in coincidence with the patient's movements.

Other than the flashes in the pre-ictal period, the general pre-ictal period exhibited very low (0.1–0.25) PLV values, which increased to approximately 0.4 about halfway through the ictal period. Different from the FBTCS datasets (FBTCS1 and FBTCS2), there was no intense phase-locking observed at the offset period. Instead, the neurons, which did not reach the same levels of synchronization during the seizure, were able to respond quickly and return to the normal levels of phase synchronization in the post-ictal period. Again, there was a periodic decoupling occurring at contact points furthest away from the seizure lesion, even more pronounced in these datasets (see Figure 10.5). We stress the subtleties of the phase-locking values for this seizure type, and note the importance of knowledge of the seizure onset and offset times in order to properly analyze the PLV results.

Also, similar to FBTCS2_diff dataset, we computed their PLV composite maps using the FIAS2_diff datasets. In general, we see the same behavioral trends at each of the ictal periods when compared to the original composite map, but with overall lower and more uniform PLV values. See Figures 10.6(a) and 10.6(b).

The PLV results show distinct behavior for the different seizure types. In the FBTCS1, FBTCS2, FIAS1, and FIAS2 datasets, the pre-ictal period exhibited overall low PLV values. However, both FIAS datasets exhibited sensitivity to noise artifacts in the pre-ictal periods. While both FBTCS1 and FIAS1 datasets showed an increase in PLV values halfway through the ictal period, the FIAS1 dataset showed much more subtle and localized phase synchronization. In the post-ictal period, the FIAS phase-synchronization levels returned to the low values observed pre-ictally. On the other hand, the ictal and offset period for the FBTCS data both reached significantly higher PLV values, across the entire brain. This suggests that the level of synchronization during the ictal period affects how quickly the neuronal behavior returns to normal after the seizure offset. Both FBTCS1 and FIAS1 datasets also exhibited periodic lower PLV values in the surface contact points further from the lesion. We saw consistent trends in PLV behavior at the different seizure stages for the difference modification composite maps. This invariance in overall synchronous behavior is encouraging, and supports our observations about the tonic-clonic and impaired awareness seizures, respectively.

The frequency band filtering of both FBTCS1 and FBTCS2 datasets highlighted the theta and alpha bands, which had slightly longer high phase-locking periods and higher PLV values overall.

10.5.1.1 *Issues and robustness*

One of the most common drawbacks of the PLV calculation, especially when applied to inherently noisy EEG data, is its sensitivity to volume conduction. Volume conduction is the instantaneous projection of one signal onto another, so that the signal is recorded as originating from more than one source.[11] This produces zero-lag synchronization: the recording of a signal by one contact point overlaps with the recording of that same signal by another contact point, creating apparent instantaneous synchronization. This issue is particularly prevalent in EEG recordings due to their lack of spatial resolution.[12] However, measuring signals directly from the surface of the brain, like with intracranial EEG (iEEG) data, generally avoids severe volume conduction.

Due to the definition of PLV, as the number of contact points or sample sizes increase, the PLV becomes increasingly computationally consuming. Bruña *et al.*[11] used exponential properties to modify the PLV expression (10.8) and reduce the PLV computational cost by up to a factor of 100. Bruña *et al.*[11] also adapted the PLV formula to compute a corrected counterpart of the imaginary part of PLV calculation to deal with volume conduction and zero-lag synchronization. For our computations, we did not implement the method of Bruña *et al.*,[11] since the datasets are not large in size.

Another issue that arises is that the distribution of phase is non-uniform, and can create spurious results.[14] These biases are introduced by the transformation that extracts the phase, as the transform method divides an analytic signal by its envelope.

Crucial to our data analysis is the robustness of the methods used. To ensure that our results originated from our data and were invariant to the addition of random noise, we executed two statistical tests. We added Gaussian distributed noise with 0 mean and standard deviations 0.1, 0.05, 0.01, and 0.005 to our data. In all cases, throughout the entire seizure duration, there was no significant difference between the results with data and without. Thus, our results were invariant to random noise.

We also created a null model, by "shuffling" our data around and applying the metrics. That is, for each window, the rows and columns of the matrices were permuted randomly, resulting in a scrambled matrix of data. Then, we computed the PLV matrices with this shuffled data, and observed an output that reflected the scrambling: PLV values remained the same, but the patterns scattered. Thus, the patterns of our results could be safely assumed to fall from the data itself.

Celka studied the impact of Gaussian, Uniform, and Laplace noise on PLV. He found that significant random noise can decohere the signals, and that phase unwrapping can lead to difficulties in extraction of the instantaneous phase.[13]

10.5.2 *Lyapunov exponent*

We computed the Lyapunov exponent for all of the datasets, FBTCS1, FBTCS2, FIAS1, and FIAS2. The parameters used for estimation are kept the same for all four cases, although they are not optimized for individual datasets. By construction, this metric is computed for each individual time series. In our case, this corresponds to Lyapunov exponent estimates for each contact point throughout the seizure. To track the contact point behavior, we examined the Lyapunov exponents of all of the contact points, as well as those closest to the lesion (see Table 10.4 and Figure 10.8).

Table 10.4. Parameters used for Alan Wolf's MATLAB code.

Parameter	Value(s) used
τ	9
ndim	5
ires	10
maxbox	8,000
dt	0.0005×12
evolve	3
dismin	$0.02 \times$ range(Data)
dismax	$0.15 \times$ range(Data)
thmax	30

Of the four examples studied, the FBTCS1 case stands out in that the pre-ictal period is annotated by higher LE values than the ictal period is (see Figure 10.8(a)). There is certain evidence for a chaotic attractor suggesting a deterministic character of brain activity. As Babloyantz and Destexhe[17] had observed, there is a clear transition of going from a high LE in the pre-ictal period to a low LE in the ictal period. It is interesting to see the post-ictal period containing intermittent low-dimensional chaos. These observations are noted for the brain zones in the vicinity of the lesions. In the case of FBTCS2, the fluctuating behaviour of LEs in the pre-ictal and ictal periods for the chosen contact points makes the interpretation somewhat difficult. (See Figures 10.8(b), 10.8(c), and 10.8(d)). The mean plots of the results over the contact points considered reveal a general trend with higher LE values in the pre-ictal period and lower LE values in the ictal period as enclosed in these figures.

In the case of FIAS1 and FIAS2, the pre-ictal zone is not analyzed since it is contaminated by the physical movement of the two patients. However, the ictal zones denote the lowering of the LE value corresponding to a deterministic order (see Figures 10.8(e), 10.8(f)). During the post-ictal periods, the mean plots suggest an excursion of the brain regions into a mixture of deterministic order states and chaotic states. In the case of FIAS2, in the pre-ictal period before the seizure there are time periods with low LE values. Further investigations are necessary.

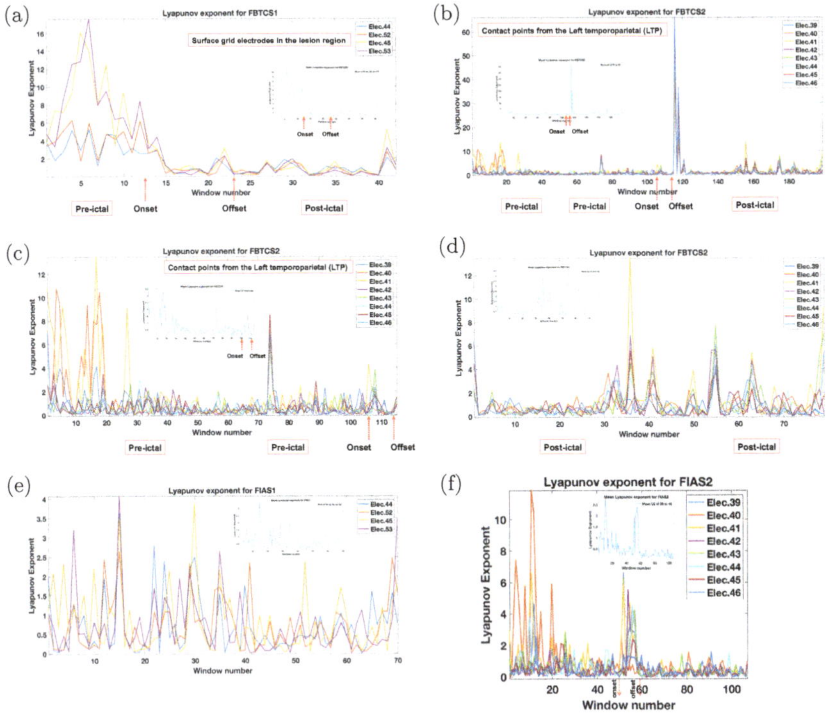

Fig. 10.8. Lyapunov exponent (LE) profiles of selected electrodes or contact points of the chosen FBTCS1, FBTCS2, FIAS1, and FIAS2 datasets. Each panel in the profiles for FBTCS1 and FIAS1 is derived from 20,000 samples. In the case of FBTCS2 and FIAS2, the number of samples per panel is 10,000. All LE profile computations use the same parameter set as that listed in Table 10.4. (a) FBTCS1. The seizure onset and offset zones and the pre-ictal and the post-ictal zones are clearly indicated. For a meaningful comparison of results, we show four electrodes (44, 52, 45, and 53) in the lesion region of the brain for this patient. (b) FBTCS2. For analysis purposes, we use the processed data from the contact points 39 to 46. The seizure onset and offset zones and the pre-ictal and the post-ictal zones are clearly indicated. (c) FBTCS2. Same as (b) except that only the results preceding the anomalous peak period are shown. (d) FBTCS2. Same as (b) except that only the results following the anomalous peak period are shown. (e) FIAS1. The seizure onset and offset zones and the pre-ictal and the post-ictal zones are clearly indicated. For a meaningful comparison of results, we show four electrodes (44, 52, 45, and 53) in the lesion region of the brain for this patient. (f) FIAS2. For analysis purposes, we use the processed data from the contact points 39 to 46 corresponding to the left hemisphere. The seizure onset and offset zones and the pre-ictal and the post-ictal zones are clearly indicated.

Conclusion

Focal to bilateral tonic-clonic seizure and focal impaired awareness seizure of two drug-resistant patients awaiting resection surgery are distinctly two different types of seizures, FBTCS being characterized by distinct pre-ictal, ictal, and post-ictal periods and FIAS being marked by an anomalous amplitude signature during the ictal period. Although data were collected intracranially in both cases, the acquisition parameters, as defined by the respective neuro-surgical teams, were distinctly different. Yet, the two metrics, phase-locking value and Lyapunov exponent, we keyed on here provide an excellent avenue to understand the two types of seizures. As an initial guide to analysis of the data, results of the video-monitoring of the patients were used. Video-monitoring during the recording session suggests to the epileptologists the onset and offset times of any seizure. The onset and offset time bracket the ictal period. In the FBTCS case, the seizure begins locally but spreads to both hemispheres of the brain.

Results of our analysis of the data confirm the usefulness of the two metrics, and lead to the following conclusions:

(a) The pre-ictal period of the FBTCS is annotated by low phase-locking values in the composite map. During the ictal period, there was an increase in the phase-locking values, coupled with an oscillatory behavior of the PLVs. Interestingly, at the offset, the phase-locking values maintain a value of 1 for a short time corresponding to full synchrony. It is important to note that this behavior is independent of the acquisition geometry.

(b) In contrast to the FBTCS, the FIAS observed in both patients lead to a characteristic PLV signature during the ictal period. In both instances of the FIAS, there were high PLV anomalies which were attributed to the movement of the patients and other noise artifacts. Without the video recording and the observer's notes on the physical movement of the patients, analysis of the PLV results would be difficult. Furthermore, since the video recording show higher amplitude iEEG oscillations on some channels on the left-side of the hemisphere, as noted with FIAS2, amplitude effects will have to be brought into the PLV calculations.

(c) Although extensive optimization of the parameters for the estimated Lyapunov exponent has not been carried out, present results restricted to FBTCS1 and FBTCS2 indicate that high Lyapunov exponent values define the pre-ictal period, low exponent values are characteristics of the ictal period, and intermediate values between the high and the low are the earmarks of the post-ictal period. Similar observations were made by Babloyantz and Destexhe[17] in their study on the low-dimensional chaos in an instance of epilepsy. Since some of the pre-ictal Lyapunov exponent profiles also contain a low value, a question arises of whether or not this could be used as a predictor. Additional work is in progress on this.

Acknowledgments

We would like to thank the Department of Mathematics and Statistics and the Hotchkiss Brain Institute of the Cummings School of Medicine for access to and use of the data in this work. Heather Berringer expresses her gratitude to Natural Sciences and Engineering Research Council (NSERC USRA) for a Summer Research Fellowship, which made this work possible. Elena Braverman (NSERC grant ID: RGPIN-2020-03934) and Paolo Federico (CIHR grant ID: CIHR MOP-136839) thank the respective funding agencies for financial support and travel funding. We sincerely acknowledge the data support and useful discussions provided by Daniel Pittman and Joseph Peedicail of the Alberta Health Services.

References

1. Mayo Clinic Staff, Epilepsy (2018). Mayo Clinic. https://www.mayoclinic.org/diseases-conditions/epilepsy/symptoms-causes/syc-20350093. Accessed 19 December 2019.
2. Epilepsy Canada (2016). Epilepsy facts. http://www.epilepsy.ca/epilepsy-facts.html. Accessed 19 December 2019.
3. Lava N. (2017). Types of seizures and their symptoms. WebMD. https://www.webmd.com/epilepsy/types-of-seizures-their-symptoms#1. Accessed 19 December 2019.
4. Centers for Disease Control and Prevention (2018). Epilepsy: Types of seizures. https://www.cdc.gov/epilepsy/about/types-of-seizures.htm. Accessed 19 December 2019.

5. Marshall W. (2008). Predicting epileptic seizures using nonlinear dynamics. MS thesis in Statistics, University of Waterloo, Waterloo, Ontario, Canada.
6. Goring G. D. and Nikora I. V. (2002). Despiking acoustic Doppler velocimeter data, *J. Hydraul. Eng.* **128**, 1, pp. 117–126, doi:10.1061/(ASCE)0733-9429(2002).
7. Mori N., Suzuki T., and Kakuno S. (2007). Noise of acoustic Doppler velocimeter data in bubbly flows, *J. Eng. Mech.* **133**, 1, pp. 122–125, doi:10.1061/(ASCE)0733-9399.
8. Lachaux P. J., Rodriguez E., Martinero J., and Varela J. F. (1999). Measuring phase synchrony in brain signals, *Human Brain Mapp.* **8**, pp. 194–208.
9. Mormann F., Lehnertz K., David P., and Elger E. C. (2000). Mean phase coherence as a measure for phase synchronization and its application to the EEG of epilepsy patients, *Phys. D Nonlinear Phenom.* **144**, 3–4, pp. 358–369.
10. Wolf A., Swift J., Swinney H., and Vastano J. (1985). Determining Lyapunov exponents from a time series, *Physica D* **16**, pp. 285–317.
11. Bruña R., Maestú F., and Pereda Em. (2018). Phase-locking value revisited: Teaching new tricks to an old dog, *J. Neural. Eng.* **15**, pp. 1–11.
12. Varela F., Lachaux P. J., Rodriguez E., and Martinerie J. (2001). The Brainweb: Phase synchronization and large-scale integration, *Nat. Rev.* **2**, pp. 229–239.
13. Celka P. (2007). Statistical analysis of the phase-locking value, *IEEE Signal. Proc. Lett.* **14**, pp. 577–580.
14. Kovach K. C. (2017). A biased look at phase locking: Brief critical review and proposed remedy, *IEEE Trans. Signal. Proc.* **65**, pp. 4468–4480.
15. King F. (2009). *Hilbert Transforms.* Encyclopaedia of Mathematics and Its Applications, Cambridge University Press, Cambridge, England.
16. Feldman M. (2011). *Hilbert Transform Applications in Mechanical Vibration*, John & Wiley Sons Inc., Hoboken, NJ, USA
17. Babloyantz A. and Destexhe A. (1986). Low dimensional chaos in an instance of epilepsy. *Proc. Natl. Acad. Sci. USA* **83**, pp. 3513–3517.
18. Iasemidis D. L. and Sackellares C. J. (1991). The evolution with time of the spatial distribution of the largest Lyapunov exponent on the human epileptic cortex, in D. Duke and W. Pritchard (eds.), *Measuring Chaos in the Human Brain. World Scientific*, Singapore, pp. 49–82.
19. Palus M. (1995). Testing for nonlinearity using redundancies: quantitative and qualitative aspects. *Physica D.* **80**, pp. 186–205.

20. Iasemidis D. L., Sackellares C. J., Zaveri P. H., and Williams J. W. (1990). Phase space topography and the Lyapunov exponent of electrocortigrams in partial seizures. *Brain Topogr.* **2**, pp. 187–201.
21. Osowski S., Świderski B., Cichocki A., and Rysz A. (2007). Epileptic seizure characterization by Lyapunov exponent of EEG signal. *COMPEL* **26**, 5, pp. 1276–1287.
22. Wolf A. (2013). Lyapunews: Instructions for using the algorithm, presented in Physica 16D. https://www.researchgate.net/publication/259459893_Lyapunews_-_Instructions_for_using_the_algorithm_presented_in_Physica_16D.

https://doi.org/10.1142/9789811263033_bmatter

Index

N

O

P

Q

R

S